ANNALS OF THE NEW YORK ACADEMY OF SCIENCES

Volume 264

CARRIERS AND CHANNELS IN BIOLOGICAL SYSTEMS

Edited by Adil E. Shamoo

The New York Academy of Sciences
New York, New York
1975

Library of Congress Cataloging in Publication Data

Main entry under title:

Carriers and channels in biological systems.

(Annals of the New York Academy of Sciences ; v. 264)
"Papers . . . of an international conference . . . held by the New York Academy of Sciences on February 24-27, 1975."
1. Biological transport--Congresses. 2. Membranes (Biology)--Congresses. 3. Ion-permeable membranes--Congresses. 4. Bilayer lipid membranes--Congresses.
I. Shamoo, Adil E. II. New York Academy of Sciences.
III. Series: New York Academy of Sciences. Annals ;
v. 264. [DNLM: 1. Biological transport--Congresses.
2. Cell membrane permeability--Congresses. W1 AN626YL
v. 264/QH601 C316]
Q11.N5 vol. 264 [QH509] 508'.ls [574.8'75] 75-44053
ISBN 0-89072-018-5

BiC/MP
Printed in the United States of America
ISBN 0–89072–018–5

ANNALS OF THE NEW YORK ACADEMY OF SCIENCES

Volume 264
December 30, 1975

CARRIERS AND CHANNELS IN BIOLOGICAL SYSTEMS*

Editor and Conference Chairman
ADIL E. SHAMOO

Advisory Committee
CLAY ARMSTRONG, GEORGE EISENMAN, ERIC HEINZ, AND DANIEL TOSTESON

CONTENTS

* This series of papers is the result of an international conference entitled Carriers and Channels in Biological Systems, held by The New York Academy of Sciences on February 24–27, 1975.

Part IX. Permeation Mechanism across Biological Membranes: III

Part X. The Lipid Bilayer Membrane as an Assay for Ion-Conductance-Inducing Materials Isolated from Biological Membranes

Financial assistance was received from:

- BURROUGHS WELLCOME COMPANY
- HOFFMANN-LA ROCHE INC.
- MERCK SHARP & DOHME RESEARCH LABORATORIES
- NATIONAL INSTITUTE OF GENERAL MEDICAL SCIENCES
- SMITH KLINE & FRENCH LABORATORIES
- UPJOHN INTERNATIONAL, INC.

INTRODUCTORY REMARKS

The field of membrane transport and its related areas, specifically carriers and channels in membranes, is exciting and is at the brink of major discoveries. As a matter of fact, some breakthroughs may have already occurred. We hope that this conference will facilitate such breakthroughs and will generate new knowledge that, mixed with the enthusiasm and excitement of the researcher, will lead to greater understanding of membrane-related processes.

The purpose of this conference is to bring together the various facets and approaches to this subject. In more practical terms, we hope to bring about a better grasp and appreciation of the various methods of understanding membrane transport on the molecular level. Gains can be made if a more multidisciplinary approach is undertaken and lines of communication are strengthened.

I would like to thank The New York Academy of Sciences, and its President, Dr. Feigelson, for sponsoring the meeting; Dr. Herbert J. Kayden, Chairman of the Conference Committee, for his advice throughout the course of development of the program; Ms. Ellen Marks, Conference Director, for her help in the overall planning and executing of the minute details of the Conference; members of the Advisory Committee, Drs. Tosteson, Armstrong, Heinz, and Eisenman, for their help in planning the list of speakers and the scientific format of the Conference; Dr. Robert Blumenthal for his help, which has extended from the inception of the program, and for chairing the panel discussion during the Conference; Dr. Paul Mueller, our afterdinner speaker, for a hilarious and cheerful evening; Bill Boland, Executive Editor, and Justine Cullinan, Associate Editor, for their assistance in handling the editorial details of this book; and Ms. Marianne Torti, my secretary, for her devotion and patience in typing and handling the correspondence during the preparation for the Conference. Finally, I extend many thanks to our speakers and attendants for providing us with a successful conference.

Adil E. Shamoo
Conference Chairman

PART I. MECHANISM OF ACTION OF CARRIERS AND CHANNELS IN MEMBRANES

FUNCTIONS OF THE LIPID IN BILAYER ION PERMEABILITY

D. A. Haydon

Physiological Laboratory
University of Cambridge
Cambridge, England

INTRODUCTION

The extensive studies over the past few years of the mechanisms of ion transport across artificial lipid membranes have revealed a great deal both about the ionophores themselves and about the properties which the lipid membranes must have in order that the ionophores should function efficiently. For a carrier molecule, for example, it is of little value that it may have a high ion selectivity if it will not combine with the membrane or, if it does combine, it will not move or bind ions in the membrane. For pore-forming molecules, binding to different membranes may be very similar, but the efficiency of the transport process may depend critically on the state of ionization of the lipid or on the membrane thickness. In this introductory paper it is proposed to review the various properties of lipid membranes which we now know to have special relevance to ion transport by polypeptides and other small macromolecules, and to illustrate the discussion with detailed descriptions of selected topics. It is thus hoped to emphasize the considerable range of factors which contribute to the success or otherwise of attempts to incorporate biological membrane components in artificial lipid bilayers.

Since most of the findings described in this paper are for "black" lipid membranes, i.e., those formed using a hydrocarbon solvent, it is perhaps pertinent to summarize at the outset current knowledge of the interrelationship between these membranes and those of the lamellar phase of lecithin and the planar solventless membranes studied by Montal and Mueller.[1,2] As seen in TABLE 1, the thickness, as deduced from x-ray data, of the hydrocarbon region of a leaflet of the fully hydrated lamellar phase is not significantly different from the thickness deduced from the specific capacitance of a planar

TABLE 1

THICKNESSES OF ARTIFICIAL LIPID BILAYERS FORMED FROM EGG LECITHIN

System	Specific Capacitance (μF/cm^2)	Hydrocarbon Region Thickness (Å)
Lecithin-hexadecane (black film)*	0.603	31 ($\epsilon = 2.14$)
Lecithin (no solvent)†	~0.76	~25 ($\epsilon = 2.14$)
Lecithin (liquid crystal)‡	–	~26 (x-ray)

* From Fettiplace *et al.*[3]
† From Fettiplace[4] and Fettiplace *et al.*[5]
‡ From Lecuyer *et al.*,[6] Reiss-Husson,[7] and Small.[8]

solventless membrane, assuming a dielectric constant of 2.14. Although the data of Fettiplace[4, 5] cannot be claimed as definitive, the close agreement of the two thicknesses helps greatly to establish the validity of using a bulk dielectric constant for aliphatic hydrocarbon in the interpretation of the capacitance data. The somewhat smaller capacitance and larger thickness of the lecithin-hexadecane black film can be attributed to the retention of some solvent, although just how much is difficult to say. It may be that a small amount of solvent has a large effect on the capacitance (and hence on the thickness) or it may be that the area per molecule given previously[3] is about 10% too low.

The above check on the validity of membrane thicknesses calculated from specific capacitances places on a firmer basis the earlier findings that black films formed from shorter chain alkanes such as decane are indeed of greater thickness than the solventless structures, and that this is because they retain considerable amounts of solvent.[3, 9, 10]

The various roles that the lipid may play in bilayer ion permeability may be conveniently considered under the broad headings of (1) combination of the ionophore with the membrane and (2) functioning of the ionophore in the membrane.

Under (1), the potential relevance of studies such as those of Gulik-Krzywicki *et al.*[11] concerning the combination of proteins with phospholipid liposomes is obvious, although the importance of the sign of the charged groups, which was brought out in this work, has not so far seemed of comparable importance for the known ion transporting molecules. It may turn out to be generally unimportant owing, perhaps, to the necessarily lipophilic character of an ionophore and also to the screening of any ionic groups in the relatively high electrolyte concentrations often employed. The nature of the nonpolar part of the membrane is, by contrast, known to have a considerable influence on the uptake of certain ionophores. The necessity for cholesterol in the case of the polyenes (e.g., nystatin and amphotericin B)[12, 13] and the adverse effects of cholesterol on the uptake of valinomycin[14] are well documented.

The state of the hydrocarbon region of the bilayer, i.e., liquid or solid, might also be expected to influence the binding of ionophores. The experiments of Krasne *et al.*[15] are of special interest in this respect. Cooling the bilayer to the point where it appeared solid rather than liquid eliminated conduction by nonactin but not by gramicidin A. Obviously the solidification of the membrane did not cause the desorption of the gramicidin, but whether the nonactin conduction ceased because it could no longer diffuse across the hydrocarbon region or whether it ceased because it was simply not taken up by the membrane is not clear.[16] The result for gramicidin prompts the question of whether or not the polypeptide induces conduction when placed in the presence of an already solidified membrane. If it does not, it has to be concluded that the freezing of the membrane may trap pore-forming molecules, at least for long periods, in a functioning but nonequilibrium state.

The binding of an ionophore by a membrane is clearly a necessary, but not sufficient condition for ion transport. Conditions in the membrane may well prevent the ionophore from functioning as efficiently or normally as in a biological system. Thus, under heading (2), factors such as membrane fluidity, surface potential, membrane thickness and specific lipid-polypeptide interactions are known to be important. The significance of membrane fluidity for the functioning of carriers is obvious; the surface potential affects the uptake of ions by the ionophores in the membrane and this in turn determines selectivity between ions of different valence and sign; membrane thickness can greatly influence the conduction efficiency of an assembly of pore-forming molecules; and specific interactions may completely change the response of a pore-forming molecule to applied potential. The effects of membrane fluidity have been well described in a number of places.[15, 17] Ways in which the other factors may operate are illustrated in the remainder of this paper.

Surface Potential

The relationship of the ion permeability of a lipid membrane to the change in the electrostatic potential across its surface has been discussed by several authors.[13, 16–25] In earlier work, only the diffuse and Stern layer potentials were considered, but it is now recognized that the potential change arising from the oriented dipoles of the polar head groups may be of similar or even greater importance.[16, 24, 25] This is clear for carrier transport from the fact that either the ion or the ion-carrier complex must pass through all of the potential changes due to the presence of the polar groups (Figure 1). For the classical pore where, as far as the permeating ions are concerned, the lipid polar groups are considered to be shunted, the position is less obvious.

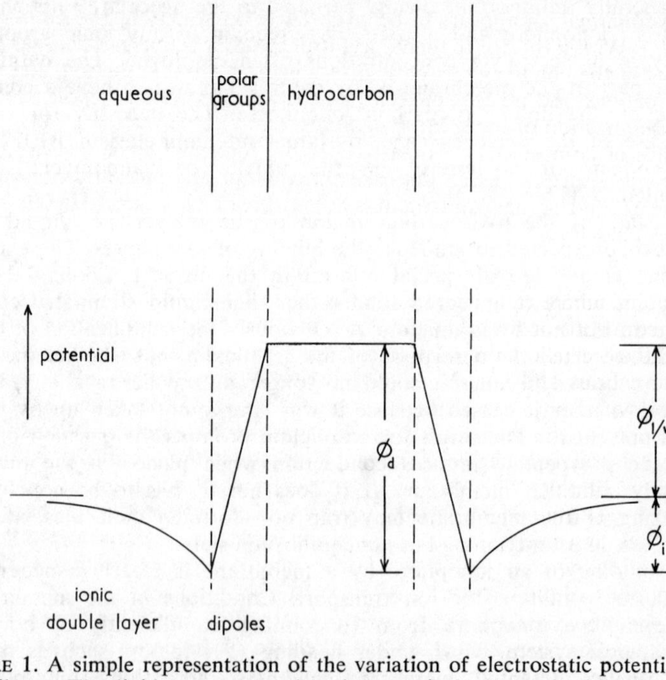

Figure 1. A simple representation of the variation of electrostatic potential across a lipid bilayer. The lipids are assumed to be negatively charged while the oriented dipoles of the polar groups are taken to be positive inwards.

The Debye-Hückel reciprocal length parameter is likely to be sufficiently large that diffuse layer potentials will affect the ion concentrations over the entrance to a pore. However, the range of influence of the lipid polar group dipoles is much less, and the permeating ions could be unaffected by their presence.

While the importance of the surface potential (or, at least, the diffuse layer part of it) for the functioning of biological membranes has long been appreciated, attempts at quantitative tests of the hypothesis have been made only recently, and these so far only with artificial membranes. If consideration is restricted to carrier systems, elementary theory suggests that for constant electrolyte concentration, carrier concentration in the membrane and mobility of ion-carrier complex, the membrane conductance G(0) in the limit of zero applied potential is given by[16]

$$G(0) \propto \exp\left[\frac{-zF\phi_{1/w}}{RT}\right] \qquad (1)$$

where z is the valence of the ion and $\phi_{1/w}$ is the total potential difference between the aqueous phase and the interior of the membrane, as indicated in FIGURE 1. $\phi_{1/w}$ may be split into the ionic double layer (ϕ_i) and dipole (ϕ_d) contributions. Thus

$$\phi_{1/w} = \phi_i + \phi_d \qquad (2)$$

It has been amply demonstrated that changes in G(0) for a given ion-carrier system may be accurately accounted for in terms of the changes in $\phi_{1/w}$ measured for monolayers of the appropriate lipids at air or hydrocarbon/water interfaces.[24] Moreover, in certain systems, when ϕ_d is assumed constant and specific adsorption of ions is minimal, ϕ_i appears to be given fairly accurately by the Gouy-Chapman theory.[21, 24] Within the above theoretical framework it is thus possible to understand variations in the conductance for a given ion, and the selectivity of the bilayer as between cations and anions, in terms of the average surface dipole moment and degree of ionization of the membrane lipids.

Surface or compensation potentials, $\Delta V_{1/w}$, for some lipids and lipid mixtures are shown in TABLE 2. These values are not equal to $\phi_{1/w}$ but are the changes which occur in $\phi_{1/w}$ when the lipid monolayer is spread on a clean aqueous solution surface.[25] Therefore

$$\Delta V_{1/w} = \phi_{1/w} - \phi_{a/w} \qquad (3)$$

where $\phi_{a/w}$ is numerically equal to the so-called χ-potential of water (or of the appropriate aqueous solution). Differences in $\Delta V_{1/w}$, all other factors being equal, should therefore reflect the relative intrinsic permeabilities of different lipid membranes to cations and anions, i.e., the permeabilities that would be found if the cation and the anion were identical except for the sign of their charge. That this is accurately so for the permeability of egg phosphatidyl choline and monoolein bilayers to K^+-nonactin has been shown. Thus, the difference in $\Delta V_{1/w}$ for the two lipids is ~120 mV, whereas the difference in the conductance corresponds to ~115 mV.[25] (It should be noted, however, that the introduction of cholesterol destroys this agreement, presumably owing either to its tendency to reduce the solubility of the nonactin in the membrane or to its effect on the fluidity, and hence on the mobility of the ion-carrier complex.)

For the phosphatidyl serine and phosphatidyl inositol given in TABLE 2, the ionic double layer potentials are likely to be in the region of −125 mV. The dipole potentials for all the phospholipids, and the cholesterol, shown in TABLE 2 are therefore very similar and, apart from the relatively small

TABLE 2

SURFACE POTENTIALS OF LIPID MONOLAYERS AT AREAS PER MOLECULE SIMILAR TO THOSE IN THE BILAYER

Lipid	Surface Potential $\Delta V_{1/w}$ (mV)*
Phosphatidyl choline (egg)	441†
Phosphatidyl ethanolamine (egg)	420‡
Phosphatidyl serine	320‡
Phosphatidyl inositol	300‡
Cholesterol	390§
Phosphatidyl choline + cholesterol (~3:1)	420†
Glyceryl monooleate	321†
Monooleyl glyceryl ether	411†

* The sign of the potential is for the air (or hydrocarbon) relative to the aqueous phase.

† Substrate 0.1 M NaCl, pH ~ 6.

‡ From Papahadjopoulos.[45] Substrate 0.145 M NaCl, pH 6.

§ From Adam *et al.*[46] Substrate 0.02 M HCl.

ionic double layer potentials, the intrinsic relative permeabilities of the corresponding bilayers should also be very similar. The absolute magnitude of these relative permeabilities to cations and anions may be estimated. The direct approach is simply to measure the membrane conductance in the presence of equal concentrations of permeating, oppositely charged, but otherwise identical ions. This has been attempted using lecithin-cholesterol membranes with tetraphenyl boride and tetraphenylphosphonium ions.[27] The specific conductance for the negative boride was some 10^4-fold larger than for the positive tetraphenylphosphonium ion, from which $\phi_{1/w}$ may be calculated to be ~115 mV.

An alternative method is based on a knowledge of $\phi_{a/w}$, or the χ-potential of water. The most recent and probably best estimate of this is 40 ± 100 mV.[28] Bearing in mind that a different sign convention is used for present purposes, and taking $\Delta V_{1/w}$ as 420 mV, $\phi_{1/w}$ for lecithin: cholesterol (~3:1 mole ratio) may be calculated from Equation 3 to be between 280 and 480 mV. This result reinforces the idea that the lipid bilayer has a strong intrinsic selectivity for anions over cations. The quantitative discrepancy between the two estimates of $\phi_{1/w}$ given above is not surprising, but also is not readily explicable. Despite first appearances, there may be significant differences between tetraphenyl boride and tetraphenylphosphonium ions or their interactions with the bilayer which invalidate their use for present purposes. On the other hand, estimates of the χ-potential for water have, in the past, been so widely varied that complete confidence in the most recent result may not be fully justified.

In addition to affecting the conductance and the selectivity, the dipole potential of either the natural lipids, or a change in this quantity produced by the adsorption of dipolar molecules, may act in much the same way as an ionic double layer potential in changing the polarization or depolarization required to stimulate conduction in a voltage-dependent system such as a nerve axon. This point is illustrated in FIGURE 2. Provided the effect is asymmetrical, both an increased negative ionic double layer potential, or an increased negative dipole potential tend to increase the field in the membrane and, although the resting potential is not changed, a larger than normal depolarization could be needed to produce an action potential.

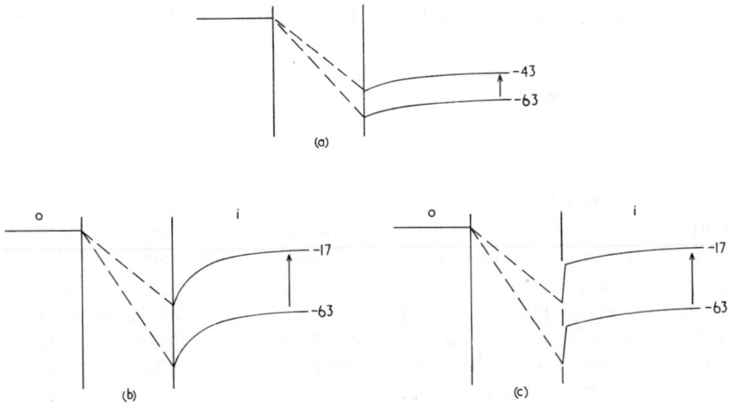

FIGURE 2. An illustration of the way in which surface potentials might influence the stimulation of an action potential in a nerve axon. (a) A normal axon having a resting potential of —63 mV which is stimulated by depolarization to —43 mV. A small negative surface charge density is assumed to be present on the inner surface of the membrane. (b) The resting potential is the same but ion adsorption on the inner surface is assumed to increase the negative charge. In order to restore the field in the membrane to that which produced stimulation in the normal axon, a larger depolarization is now required. (c) This is analogous to (b) except that the field in the membrane is assumed to be enhanced in the resting state by the adsorption of dipolar molecules on the inside.

MEMBRANE THICKNESS

The thickness of artificial membranes increases with the chain length of the lipid[4, 29–31] and, in the case of black films, tends to vary inversely with the chain length of the solvent.[3, 9] Together, these two effects make it possible for artificial bilayers formed from C_{16} to C_{22} lipids to have hydrocarbon thicknesses ranging from approximately 26 to 53 Å. In such membranes the ion transferring characteristics of polypeptides and macrocyclic molecules are known to vary considerably. Thus, in phosphatidyl choline + decane membranes the conductance per mole of the carrier molecule valinomycin in the system decreases with increasing membrane thickness. For fatty acid chain lengths from C_{14} to C_{22} (corresponding to a thickness increase of about 10 Å) the conductance has been found to decrease by about an order of magnitude.[30]

When the membrane conductance arises from pore formation by gramicidin, the effect of thickness appears to be rather larger. Accurate data supporting this statement are not easy to obtain, as there is no reliable means of making a succession of membranes of different thickness such that they each contain the same amount of the polypeptide[16] and, furthermore, there is as yet no way of monitoring the concentration of gramicidin in the membranes once they are formed. Results for a limited range of thickness may, however, be deduced from a comparison of the current-voltage curves for compressible and incompressible membranes. Two such membranes are glyceryl monooleate + decane and glyceryl monooleate + hexadecane. The current-voltage curves for these membranes containing gramicidin, in 0.5 M NaCl, have been shown, normalized at low applied potential, by Haydon and Hladky.[16] For the incompressible membrane, the current-voltage curve is nearly linear and, in this respect, resembles the corresponding curve for the single channels.[32] The curve for the compressible membrane is strongly nonlinear, the conductance increasing with applied potential, in contrast to the single channel current-voltage relationship, which is ohmic and identical to that for the incompressible membrane. It is inferred that the rise in the conductance in the compressible membrane reflects chiefly the formation of additional channels from the inactive gramicidin known to be present.[16] Since the variation of the membrane thickness with the applied potential has been determined,[9] it is possible to deduce how, for a membrane containing a constant amount of gramicidin, the conductance varies with thickness. This is shown in FIGURE 3. As can be seen, the rate of change of the conductance ratio corresponds to an increase of nearly two orders of magnitude for a thickness decrease of 10 Å, and, for the range 47 to 31 Å, which can readily occur in black films, to over three orders of magnitude. Over the latter range, the rate constant for closure of the channels has been measured. From this data and the extrapolated conductance ratio, the variation of the channel opening rate constant may be inferred (TABLE 3). The equilibrium constant (which is proportional to the conductance ratio) is thus seen to increase with decreasing membrane thickness partly because the closure rate constant decreases but mainly because the opening rate constant increases. (The proportionality constant, which is identical for the equilibrium and opening rate constants, arises because, although Bamberg and Läuger[33] have determined them for one thickness of a lecithin-decane system, no absolute measurements of these quantities have yet been made in glyceryl monooleate membranes.) The long extrapolation indicated in FIGURE 3 as necessary to obtain the data of TABLE 3 leaves some doubt

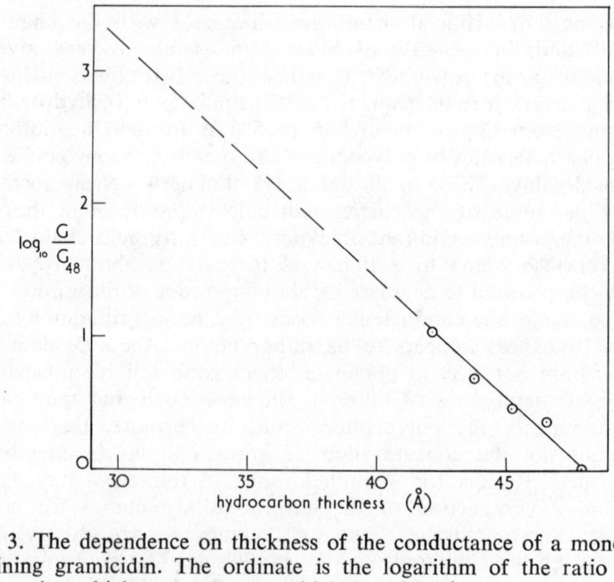

FIGURE 3. The dependence on thickness of the conductance of a monoolein black film containing gramicidin. The ordinate is the logarithm of the ratio of the conductance at a given thickness to that at a thickness of 48 Å.

TABLE 3
GRAMICIDIN CONDUCTANCE PARAMETERS AS A FUNCTION OF MEMBRANE THICKNESS FOR MONOOLEIN BLACK FILMS

Hydrocarbon Solvent	Hydrocarbon Region Thickness (Å)	Constant × k* ($mol^{-1}sec^{-1}$)	k† (sec^{-1})	Constant × K‡ (mol^{-1})
n-decane	47	2.5	2.5	1
n-tetradecane	40	12.8	0.77	16.6
n-hexadecane	31	335	0.45	737

* k = channel opening rate constant.
† k = channel closing rate constant.
‡ *K* = channel opening rate constant/channel closing rate constant.

as to whether or not the thickness effect is as large as has been concluded from the analysis. For what it is worth, the magnitude of the effect is confirmed very roughly by the substantially smaller stoichiometric concentrations of gramicidin in the bulk phases needed to produce similar levels of conductance in glyceryl monooleate + hexadecane as compared with glyceryl monooleate + decane membranes.[47]

Membranes containing alamethicin, which is also well established as a pore-forming molecule, might also be expected to show a strong dependence of conductance on thickness. As with gramicidin, however, there are technical difficulties in demonstrating the effect directly. A 1,000-fold change in conductance in a membrane containing alamethicin may arise from either a twofold change in the concentration of the adsorbed polypeptide or from a change of some 27 mV in the applied potential. Although experimental conditions should be controllable to within these limits, the margin for error is small and convincing data would not be readily obtained. Certainly, no clear evidence has yet been presented.

The reason for the relative instability of gramicidin channels in the thicker membranes is not known for certain. However, one obvious explanation, that energy is required to deform or dimple the lipid so as to expose the entrances to the pores, happens to account semiquantitatively for the observations. If it is accepted that the gramicidin channel is approximately 28 Å in length,[25, 34–37] it is clear that when the hydrocarbon region of the membrane is 48 Å thick, dimples 10 Å deep on each side are required to expose the pore (FIGURE 4). If the mean diameter of the dimple is 7 Å and if the effective tension in the surface of the dimple is 5 dyne/cm, the free energy of their formation is 2.2×10^{-13} erg. This is sufficient to reduce the channel opening rate constant by some 200-fold. Together with the observed increase in the closure rate constant of about 5×, this accounts for the change in the equilibrium constant.

LIPID COMPOSITION AND VOLTAGE-DEPENDENT PERMEABILITY

The means by which the lipid may influence the voltage dependence of permeability through its ionic double layer and dipole potentials has been

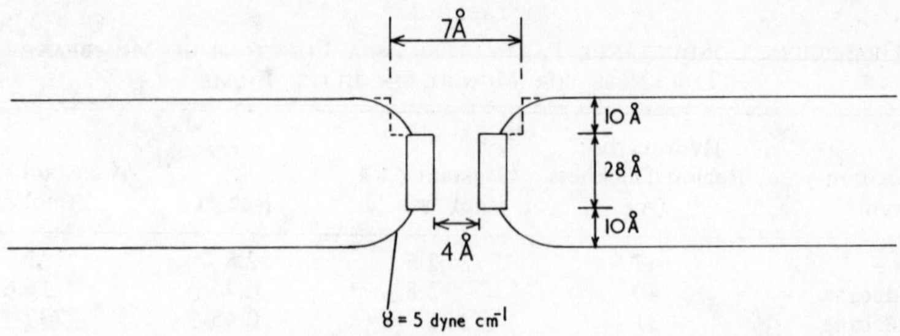

FIGURE 4. An illustration of the dimpling that a lipid membrane may have to undergo in order that gramicidin pores may form. For the influence of this phenomenon on the channel opening rate constant, see text.

mentioned briefly in an earlier section. Specific interactions between lipid and polypeptide may also affect the voltage dependence of permeability. In the experiments described by Bean[38] on EIM it is probable that all these effects played some part. The behavior of alamethicin too is sensitive, in some respects, to the nature of the membrane lipid. This system, while still complex, has the merit that it is somewhat simpler than those involving EIM, and the *et al.*,[39] Gordon and Haydon[40, 41] in which specific interactions seem important.

When alamethicin is introduced to one side of some lipid bilayers, a conductance increase may be obtained by applying either positive or negative potentials (though of different magnitude) to the polypeptide side of the membrane (FIGURE 5). Eisenberg *et al.* reported, however, that with certain phosphatidyl ethanolamine membranes no response was obtained when the alamethicin side was negative. This finding has been subsequently confirmed by Dr. Gordon in our laboratory.

Two possible explanations immediately offer themselves. The first is that the alamethicin is able to diffuse across some membranes, but not those formed from phosphatidyl ethanolamine, and that it is the material which has done so that responds when the compartment to which the polypeptide was originally added is made negative. In this mechanism, therefore, the alamethicin is supposed to respond to one sign only of the membrane field and, in phosphatidyl ethanolamine membranes, the absence of polypeptide on one side restricts the stimulation of permeability to one direction of the field. Assuming this explanation to be correct, the nature of the lipid determines whether or not the polypeptide may penetrate to the far side of the membrane. The second explanation reveals the lipid in another role but, before this can be advanced, it is necessary to examine the evidence concerning the interaction of the alamethicin with the applied field.

There is general agreement that conduction is stimulated when the alamethicin side of the membrane becomes more positive, and it has been proposed by several authors that this is because the polypeptide binds inorganic cations. Thus Mueller and Rudin,[42] Eisenberg *et al.*[39] and others have found that the membrane conductance is proportional to a high power of the inorganic electrolyte concentration. While this is consistent with the binding of ions, it has been shown by Gordon and Haydon[40] that the effect appears to originate from the dependence of the adsorption of the alamethicin at the membrane surface on approximately the square root of the electrolyte concentration,

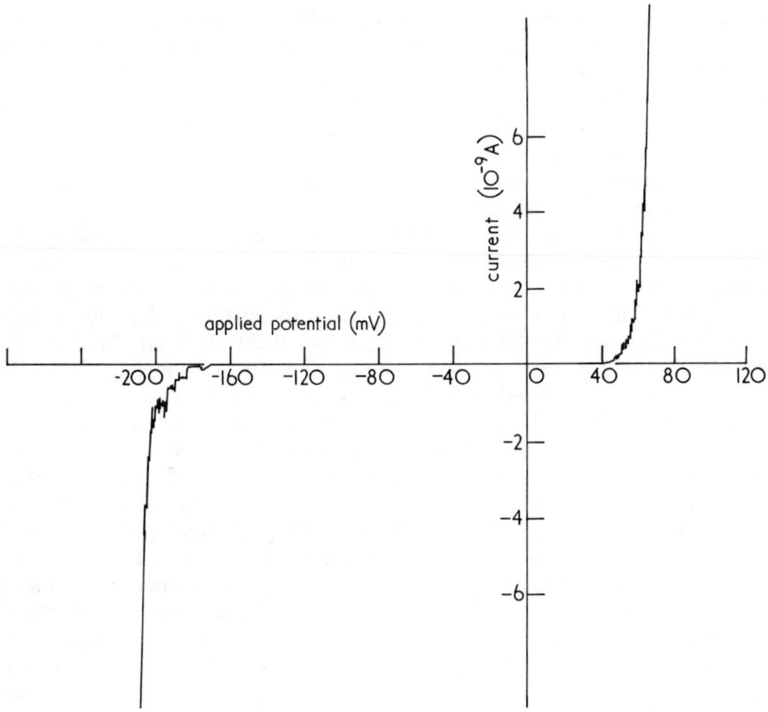

FIGURE 5. The current-voltage relationship for a membrane formed from monoolein + cholesterol in hexadecane, in 2 M KCl. Alamethicin (10^{-6} M) was added to only one of the two aqueous compartments. The sign of the potential is given for the alamethicin side of the membrane. (T = 20° C).

and does not necessarily imply any ion binding. Pressman[43] found that alamethicin extracted alkali metal ions into nonpolar media, and concluded that this resulted from a binding of the ions by the polypeptide. Unless a nonspecific binding by the glutamyl carboxyl is invoked an alternative explanation for this data is less easy to see. It is significant, however, that no clearly defined complexes with alkali metal ions (such as are known for the actins, valinomycin, and other ionophores) have yet been reported.

More direct evidence against ion binding comes from the observation that the membrane conductances show the same potential dependence for a number of quite different electrolytes including sodium, potassium, calcium,* tetramethylammonium and tris chlorides, and do so over a range of electrolyte concentrations.[40] The implication of these observations is that, if ion binding occurs, it must, under all these varied circumstances, remain constant and at a high level. The improbability that this would occur, however, seems such that the ion binding hypothesis must be discounted.

In the absence of substantial evidence that the alamethicin may carry a net positive charge, it would appear that the polypeptide or, at least, its

* It has not been found possible to reproduce the effects of di- and trivalent cations on the slope of the current-voltage curve,[48] as reported by Cherry *et al.*[44]

aggregates, must respond to the applied field through the possession of a dipole moment. If the whole membrane potential were to fall across a dipole of length 5 Å, the current-voltage data require that the moment per alamethicin molecule should be about 17 Debye units. For a polypeptide of the size of alamethicin such a dipole moment is, in general, quite feasible, and surface potential studies of alamethicin monolayers at air/water interfaces[49] show that in these conditions the molecule has a moment, normal to the surface, of the required order of magnitude.

If it is accepted that the alamethicin molecule, or its aggregate, reacts to applied potentials because it is dipolar rather than because it carries a net charge, several features of the membrane conductance are more readily accounted for. Thus, it is no longer necessary to postulate that the molecules must diffuse through the bilayer and establish themselves on the far side in order that conduction may occur when the direction of the field is reversed. Instead, the molecules need merely to orient themselves in the membrane in the opposite sense. If this occurred, however, it is unlikely that the threshold potential for the onset of conduction (FIGURE 5) would be the same for the two field directions since the interactions of the alamethicin molecule with the membrane would almost certainly favor one orientation rather than the other. Moreover, the difference between the threshold potentials for the two field directions should depend on the nature of the lipid. Thus, the finding of Eisenberg *et al.*[39] that for bacterial phosphatidyl ethanolamine membranes no conduction occurs when the polypeptide side of the membrane is negative, may be regarded as simply an extreme instance in which the interaction of the alamethicin with the lipid is so much stronger in one orientation than the other that it is not significantly affected by any field that the membrane will tolerate. The difference in the interaction free energies for the two orientations needed to account for this observation does not have to be greater than about 10 kcal for the whole conducting aggregate, and hence is relatively trivial in comparison to the hydrogen and lipophilic bonding capacity for the eight or nine molecule assembly. A mathematical analysis of the conduction mechanism for a dipole model of alamethicin has been given by Gordon and Haydon.[40]

A further consequence of the dipole mechanism, and the different interaction free energies of the conducting aggregate in its two orientations with respect to the lipid, is that the properties of the aggregate might vary according to which direction of the field produced the conduction. For membranes of glycerol monooleate + cholesterol this appears to be so (FIGURE 6). Although the levels of conduction exhibited by a single complex are the same for the two field directions, the probability of their occurrence is quite different and, in fact, more levels are detectable when the alamethicin side of the membrane is negative than when it is positive. These observations appear to constitute strong evidence that the conduction which occurs when the polypeptide side of the membrane is negative does not originate from alamethicin which has penetrated the bilayer and is responding as it would have done on the side from whence it came.

Although not directly connected with the potential-dependence of alamethicin conductance, it is also of interest that the rate constants for transition between the levels of a single current burst are very sensitive to the nature of the membrane lipid. An extreme example is a comparison of results for phosphatidyl ethanolamine on the one hand and glyceryl monooleate + choles-

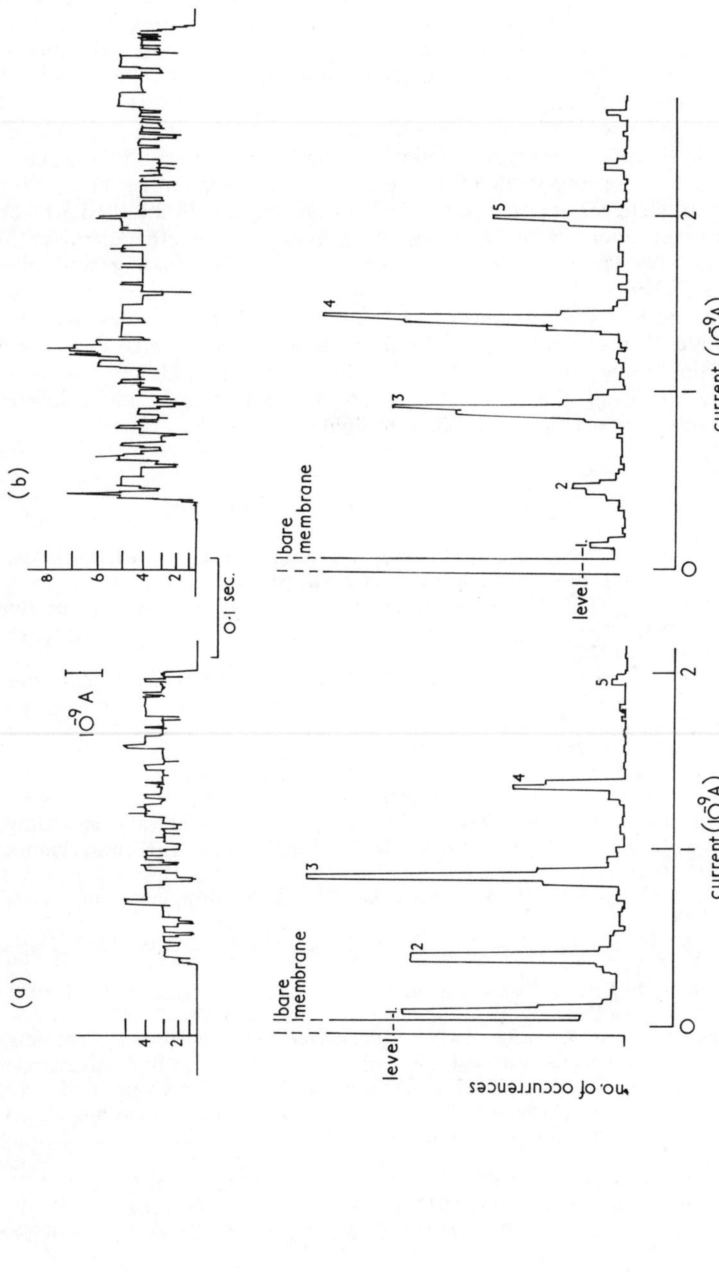

FIGURE 6. Single current bursts, and the corresponding current level distributions (from more extended records) for alamethicin in monoolein-cholesterol-hexadecane membranes in 2 M KC1. 3×10^{-8} M alamethicin was present on one side only of the membrane. (a) Alamethicin side of the membrane positive; (b) alamethicin side negative.

terol on the other. In the glyceride membranes the transitions, both upward and downward, occur about 100 times more frequently than in the phospholipid membranes. However, both rate constants are affected simlarly by the change of lipid so that the overall level probability is unchanged.[41] Although, in the example given, the membrane thicknesses are not the same, it has been shown that it is through its chemical constitution, rather than through its effect on membrane thickness that the lipid chiefly acts.[41] As with the switching on of the conductance, therefore, the specific interaction of the lipid with the polypeptide seems to be important.

The main purpose of the foregoing discussion of conduction by alamethicin is to illustrate the point that, for a potential-dependent permeability, the nature of the lipid may seriously affect the extent of the membrane polarization or depolarization necessary to open the channels. Indeed, if the dipole mechanism is correct, there is at least one lipid, phosphatidyl ethanolamine, for which inaccessibly large depolarizations are required. This finding, and those discussed in earlier sections, combine to emphasize the importance of working with appropriate lipids in biological membrane reconstitution studies. Not only may the natural transport proteins fail to function in foreign lipids but, more confusingly, they may function quite abnormally. In addition, membrane components, which in the natural state are not involved in ion transport, may become so in association with different lipids.

References

1. Montal, M. & P. Mueller. 1972. Proc. Nat. Acad. Sci. U.S.A. **69:** 3561–3566.
2. Montal, M. 1973. Biochim. Biophys. Acta **298:** 750–754.
3. Fettiplace, R., D. M. Andrews & D. A. Haydon. 1971. J. Membrane Biol. **5:** 277–296.
4. Fettiplace, R. 1974. Ph.D. dissertation.
5. Fettiplace, R., L. G. M. Gordon, S. B. Hladky, J. Requena, H. P. Zingsheim & D. A. Haydon. 1975. Techniques in the formation and examination of "black" lipid bilayer membranes. *In* Methods in Membrane Biology. E. D. Korn, Ed. Vol. **4:** Plenum Press. New York, N.Y.
6. Lecuyer, H. & D. G. Dervichian. 1969. J. Mol. Biol. **45:** 39–57.
7. Reiss-Husson, F. 1967. J. Mol. Biol. **25:** 363–382.
8. Small, D. M. 1967. J. Lipid Res. **8:** 551–557.
9. Andrews, D. M., E. D. Manev & D. A. Haydon. 1970. Spec. Discuss. Faraday Soc. **1:** 46–56.
10. Hanai, T., D. A. Haydon & J. Taylor. 1964. Proc. Roy. Soc., Ser. A **281:** 377–391.
11. Gulik-Krzywicki, T., E. Schtecter, V. Luzzati & M. Faure. 1969. Nature **223:** 1116–1121.
12. Norman, A. W., R. A. Demel, B. de Kruyff, W. S. M. Geurts van Kessell & L. L. M. van Deenen. 1972. Biochim. Biophys. Acta **290:** 1–14.
13. Finkelstein, A. & R. Holz. 1973. Aqueous pores created in thin lipid membranes by the polyene antibiotics nystatin and amphotericin B. *In* Membranes. G. Eisenman, Ed. Vol. **2:** 377–408. Marcel Dekker, Inc. New York. N.Y.
14. Hunter, M. J. 1974. The use of lipid bilayers as cell membrane models: An experimental test using the ionophore, valinomycin. *In* Drugs and Transport Processes. B. A. Callingham, Ed.: 227–240. Macmillan. London. England.
15. Krasne, S., G. Eisenman & G. Szabo. 1971. Science **174:** 412–415.
16. Haydon, D. A. & S. B. Hladky. 1972. Q. Rev. Biophys. **5:** 187–282.
17. Szabo, G., G. Eisenman, R. Laprade, S. M. Ciani & S. Krasne. 1973. Experi-

mentally observed effects of carriers on the electrical properties of bilayer membranes-equilibrium domain. *In* Membranes. G. Eisenman, Ed. Vol. **2:** 179–328. Marcel Dekker, Inc. New York, N.Y.
18. Chandler, W. K., A. L. Hodgkin & H. Meves. 1965. J. Physiol. **180:** 821–836.
19. Gilbert, D. L. & G. Ehrenstein. 1969. Biophys. J. **9:** 447–463.
20. Neumcke, B. 1970. Biophysik **6:** 231–240.
21. McLaughlin, S. G. A., G. Szabo, G. Eisenman & S. M. Ciani. 1970. Proc. Nat. Acad. Sci. U.S.A. **67:** 1268–1275.
22. McLaughlin, S. G. A., G. Szabo & G. Eisenman. 1971. J. Gen. Physiol. **58:** 667–687.
23. Muller, R. U. & A. Finkelstein. 1972. J. Gen. Physiol. **60:** 285–306.
24. Haydon, D. A. & V. B. Myers. 1973. Biochim. Biophys. Acta **307:** 429–443.
25. Hladky, S. B. & D. A. Haydon. 1973. Biochim. Biophys. Acta **318:** 464–468.
26. Aveyard, R. & D. A. Haydon. 1973. An Introduction to the Principles of Surface Chemistry. Cambridge University Press. Cambridge, England.
27. LeBlanc, O. H. Jr. 1970. $TPMA_3$. Abstracts of Biophysical Society. 14th Annual Meeting. Baltimore, Maryland. The Biophysical Society.
28. Nedermeijer-Denessen, H. J. M. & C. L. De Ligny. 1974. J. Electroanal. Chem. **57:** 265–266.
29. Taylor, J. & D. A. Haydon. 1966. Discuss. Faraday Soc. **42:** 51–59.
30. Stark, G., R. Benz, G. W. Pohl & K. Janko. 1972. Biochim. Biophys. Acta **266:** 603–612.
31. Requena, J. & D. A. Haydon. Proc. Roy. Soc. London, Ser. A. In press.
32. Hladky, S. B. & D. A. Haydon. 1972. Biochim. Biophys. Acta **274:** 294–312.
33. Bamberg, E. & P. Lauger. 1973. J. Membrane Biol. **11:** 177–194.
34. Urry, D. W., M. C. Goodall, J. D. Glickson & D. F. Mayers. 1971. Proc. Nat. Acad. Sci. U.S.A. **68:** 1907–1911.
35. Veatch, W. R., E. T. Fossel & E. R. Blout. 1974. Biochemistry **13:** 5249–5256.
36. Veatch, W. R. & E. R. Blout. 1974. Biochemistry **13:** 5257–5264.
37. Fossel, E. T., W. R. Veatch, Y. A. Ovchinnikov & E. R. Blout. 1974. Biochemistry **13:** 5264–5275.
38. Bean, R. C. 1973. Protein-mediated mechanisms of variable ion conductance in thin lipid membranes. *In* Membranes. G. Eisenman, Ed. Vol. **2:** 409–477. Marcel Dekker, Inc. New York. N.Y.
39. Eisenberg, M., J. E. Hall & C. A. Mead. 1973. J. Membrane Biol. **14:** 143–176.
40. Gordon, L. G. M. & D. A. Haydon. 1975. Phil. Trans. Roy. Soc. London, Ser. B. **270:** 433–447.
41. Gordon, L. G. M. & D. A. Haydon. To be published.
42. Mueller, R. & D. O. Rudin. 1968. Nature **217:** 398–404.
43. Pressman, B. C. 1968. Fed. Proc. **27:** 1283–1288.
44. Cherry, R. J., D. Chapman & D. E. Graham. 1972. J. Membrane Biol. **7:** 325–344.
45. Papahadjopoulos, D. 1968. Biochim. Biophys. Acta **163:** 240–254.
46. Adam, N. K., F. A. Askew & J. F. Danielli. 1935. Biochem. J. **29:** 1786–1801.
47. Hladky, S. B. Private communication.
48. Gordon, L. G. M. Private communication.
49. Gordon, L. G. M. & D. A. Haydon. Unpublished results.

Discussion

Dr. Tosteson: Do you have any evidence on the ion-selectivity of the alamethicin channels opened by the opposite signs of potential?

DR. HAYDON: The alamethicin channel is remarkably unselective, apart from surface charge effects, and I don't think that the opening from opposite sides makes any difference to that state of affairs. Both are nonselective. The conductances of the channels themselves are exactly the same; it's only the probabilities of the different levels that seem changed.

DR. S. MCLAUGHLIN (*State University of New York, Stony Brook, L.I., N.Y.*): Dr. Haydon, what is the origin of the dipole potential that seems to occur somewhere at the surface of glycerol monooleate and phospholipids?

DR. HAYDON: We thought of trying to pin this down at one stage, but it is incredibly difficult to devise experiments to do this. Since such a wide variety of lipids seem to give a similar sort of dipole potential and these lipids include, for example, cholesterol, which has a fairly simple polar group, I can only think that it must arise from the carbon-oxygen bonds between the hydrocarbon chains and the inside boundary of the head group. This bond is the one thing that all the lipids seem to have in common, and they all seem to have the same sign of potential.

DR. S. CHAN (*California Institute of Technology, Pasadena, Calif.*): We have studied the aggregation properties of alamethicin in solution as a function of ionic strength and we have found that the aggregation does go up as you increase the ionic strength of the solution, which is consistent with your idea. Secondly, by analyzing this aggregation data, we were able to show that the equilibrium concentration of the decamer varies as the ninth power of the alamethicin concentration under the experimental conditions of Hall and Eisenberg.

DR. HAYDON: That's very comforting.

RESOLUTION AND RECONSTITUTION OF ION-TRANSPORT SYSTEMS*

E. Racker, A. F. Knowles, and E. Eytan

Section of Biochemistry, Molecular and Cell Biology
Cornell University
Ithaca, New York 14850

Introduction

During the past five years, we have resolved and reconstituted ten membrane systems that involve ion translocation (Table 1). The first five are

Table 1

Reconstitutions

1.	Proton pump of mitochondria (ATP-driven)
2.	Oxidative phosphorylation site III
3.	Oxidative phosphorylation site I
4.	Oxidative phosphorylation site II and III
5.	Respiratory control (site III)
6.	Proton pump of *Halobacter halobium*
7.	Ca^{++} pump of sarcoplasmic reticulum
8.	Na^{+},K^{+} pump of electric eel
9.	ADP/ATP transporter
10.	Pi transporter

directly concerned with mitochondrial oxidative phosphorylation. The ADP/ATP and the Pi transporters are required indirectly because these ions have to be translocated into the matrix space where oxidative phosphorylation takes place. The proton pump of *H. halobium* participates in ATP generation in intact bacteria.[1] Only the Ca^{++} pump of the sarcoplasmic reticulum and the Na^{+} pump of the plasma membrane are involved primarily in ion translocation and not in ATP generation.

We would like to explain how we have become involved in the reconstitution of these systems. The first five systems and the proton pump of *H. halobium* were reconstituted in the hope of shedding light on the mechanism of oxidative phosphorylation. We reconstituted the last two systems mainly to enable us to study systems that require adenine nucleotides or Pi inside of vesicles which are impermeable to ions, as shown in Figure 1. On top we show how the adenine nucleotide transporter could be used in the reconstitution of a plasma membrane pump with the physiological orientation, pumping Na^{+} out and K^{+} in. Below we show how we could reconstitute oxidative phosphorylation in the orientation of intact mitochondria. The motivation

* This investigation was supported by Grant No. CA-08964 from the National Cancer Institute and Grant No. BC-156 from the American Cancer Society.

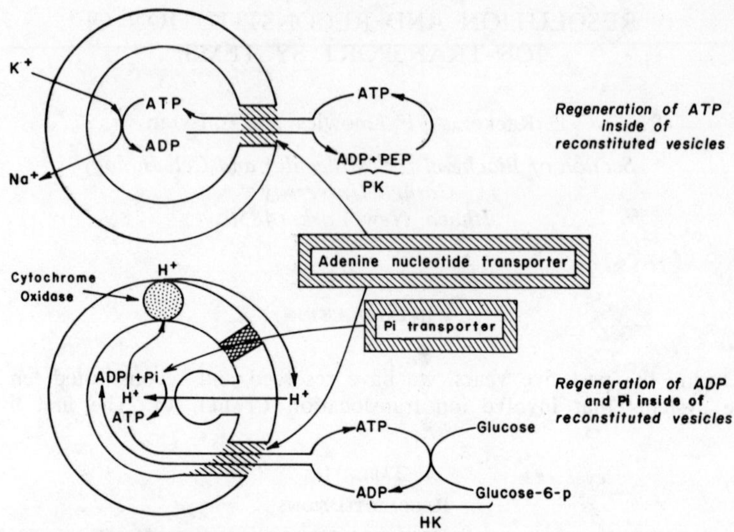

FIGURE 1. The role of ADP and Pi transporters in reconstitutions.

for our studies on the Ca^{++} and Na^{+} pump will become apparent after the discussion of the mechanism of oxidative phosphorylation.

RECONSTITUTION OF OXIDATIVE PHOSPHORYLATION

According to Mitchell,[2] the function of the respiratory chain is the translocation of protons resulting in the formation of a proton motive force consisting of a membrane potential and a pH gradient. The proton motive force is used by the oligomycin-sensitive ATPase, which functions as a proton pump in reverse to form ATP from ADP and Pi. The experiments described below are strong evidence in favor of the chemiosmotic hypothesis of Mitchell.

Reconstitution of the Mitochondrial Proton Pump

We added phospholipids in the presence of 2% cholate to purified preparations of the oligomycin-sensitive ATPase and then removed the detergent and ammonium sulfate by dialysis. The vesicles thus formed catalyze, in the presence of valinomycin, an ATP-driven and uncoupler-sensitive translocation of protons[3] as shown in FIGURE 2. They also catalyze uncoupler-sensitive ^{32}Pi-ATP exchange and ATP-dependent ANS†-fluorescence enhancement. The mitochondrial ATP-driven proton pump is complex. As shown in TABLE 2, it contains, in addition to soluble F_1, several other coupling factors, two phospholipids

† Abbreviations: ANS = 1-anilino-8-naphthalene sulfonate, DCCD = *N,N'*-dicyclohexylcarbodiimide, PMS = *N*-methyl phenazonium methosulfate, PE = phosphatidylethanolamine, PC = phosphatidylcholine, HP = hydrophobic protein, and E–P = phosphoenzyme.

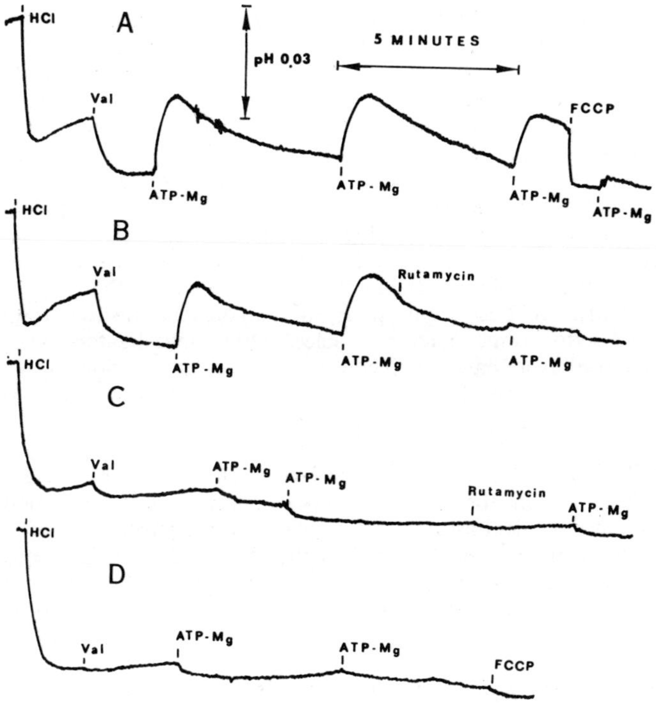

FIGURE 2. The reconstitution of the mitochondrial ATP-driven proton pump. Experimental conditions were as described by Kagawa *et al.*[3]

TABLE 2

COMPONENTS OF OXIDATIVE PHOSPHORYLATION IN RECONSTITUTIONS

Components		Assays
	Mitochondrial Proton Pump	
1. Coupling factors: F_1, F_2, OSCP (F_5), F_6		ATP-driven proton translocation
2. Hydrophobic protein fraction		^{32}Pi-ATP exchange
3. Phospholipids: Phosphatidylethanolamine [4] Phosphatidylcholine [1]		ATP-induced ANS fluorescence enhancement
	Respiratory Chain	
1. Cytochrome oxidase	→	Ascorbate-PMS → O_2
2. Complex I	→	DPNH → Q_1
3. Succinate-cytochrome *c* reductase, cytochrome *c*-cytochrome oxidase	→	Succinate → O_2

and a hydrophobic protein fraction isolated from mitochondria. The coupling factors are available in highly purified or homogeneous form. We have made good progress with the purification of the hydrophobic protein and have obtained a preparation of the oligomycin-sensitive ATPase with high specific activity and good yield. However, the best preparation of the hydrophobic protein (about eight-fold purified) contains, in addition to the major protein band with a molecular weight of about 28,000, several other protein bands.[4] In the absence of a homogeneous preparation, it is difficult to conclude which of the bands are associated with biological activity.

We have advocated the use of a biological activity for assay during purification of the hydrophobic protein. We measure either the oligomycin- (or DCCD)-sensitive ATPase activity or the oligomycin-sensitive respiration of reconstituted cytochrome oxidase vesicles.[5] Other investigators who have not used a biological assay have obtained much cleaner preparations of hydrophobic proteins than we have. For example, Cattell *et al.*[6] and Stekhoven *et al.*[7] have prepared a homogeneous proteolipid with a molecular weight of about 10,000 to 14,000 that was isolated with the aid of an assay measuring [^{14}C]DCCD binding. No reports on the biological activity of these hydrophobic proteins described in the literature have appeared, and we found none of them active in biological assays. We are pursuing the purification of biologically active hydrophobic proteins from mitochondria and have just embarked on purification of a similar fraction from chloroplasts in the hope that a more stable and homogeneous preparation can be obtained from this source.

Reconstitution of Oxidation Segments with Respiratory Control

Cytochrome oxidase preparations isolated after solubilization with cholate[8, 9] are suitable for reconstitutions. Preparations that have been exposed to deoxycholate have been uniformly unsuitable for reconstitutions. It is a pity that so many interesting physical-chemical studies have been performed with such biologically impotent preparations. Cytochrome oxidase incorporated into mixtures of phospholipids that contain phosphatidylcholine or phosphatidylethanolamine plus cardiolipin[5] exhibit the phenomenon of respiratory control. Ragan and Hinkle[10] have reconstituted site I vesicles with complex I of Hatefi *et al.*[11] and Hinkle and Leung[12] have reconstituted site II vesicles. These vesicles show properties very similar to cytochrome oxidase vesicles, one being that they do not respire maximally unless both nigericin and valinomycin are added. Since nigericin collapses the proton gradient by a K^+/H^+ exchange and K^+-valinomycin collapses the membrane potential, these experiments support Mitchell's formulation of the mechanism of respiratory control.

Reconstitution of Oxidative Phosphorylation

Oxidative phosphorylation is catalyzed by vesicles reconstituted with respiratory enzymes together with the oligomycin-sensitive ATPase complex.[13] The reconstitution of site III oxidative phosphorylation with cytochrome oxidase, cytochrome *c*, and the ATPase complex is shown in TABLE 3. As in the case of the ^{32}Pi-ATP exchange (Experiment 1), oxidative phosphorylation requires both phosphatidylcholine and phosphatidylethanolamine. If the optimal

TABLE 3

RECONSTITUTION OF SITE III OXIDATIVE PHOSPHORYLATION*

Components during Reconstitution	^{32}Pi-ATP Exchange (nmol AT^{32}P/min/mg protein)
Experiment 1	
Soy phosphatidylethanolamine (PE)	8
Soy phosphatidylcholine (PC)	0.3
PE + PC (4:1)	28
PE + PC (2.5:2.5)	22
PE + PC (2.5:2.5) + cardiolipin (0.15)	50
	Oxidative Phosphorylation (P:O)
Experiment 2	
PE + PC (4:1)	0.38
PE + PC (2.5:2.5)	0.11
PE + PC (2.5:2.5) + cardiolipin (0.15)	0.33
Experiment 3	
Hydrophobic protein 0.04 mg/μmol P-lipid	0.32
0.08 mg/μmol P-lipid	0.52
0.16 mg/μmol P-lipid	0.28
0.20 mg/μmol P-lipid	0.14

* Experimental conditions for these experiments were described in publications by Kagawa *et al.*[3] and Racker and Kandrach.[13] The values in parenthesis give the ratio of phospholipids (5 μmol total).

ratio is not used, marked stimulation by cardiolipin is observed (Experiment 2). We shall come back later to the problem of phospholipid specificity and to the role of phosphatidylethanolamine. As shown in Experiment 3, the optimal amount of hydrophobic protein has to be established in each system since inhibition is observed when too much was added. This inhibition is readily explained by the uncoupling effects of hydrophobic proteins described earlier.[5] With complex I, we have reconstituted the first site of oxidative phosphorylation[14] and recently site II by incorporating succinate-cytochrome *c* reductase[15] together with cytochrome oxidase, cytochrome *c* and the oligomycin-sensitive ATPase into liposomes.[16] These vesicles catalyzed oxidative phosphorylation with succinate as substrate. The P:O ratio with succinate was about twice as high as the P:O ratio with ascorbate-PMS as substrate, indicating a functional reconstitution of site II.

RECONSTITUTION OF LIGHT-DRIVEN ATP FORMATION WITH BACTERIORHODOPSIN

It was shown by Oesterhelt and Stoeckenius[17] that in *H. halobium,* light activation of rhodopsin results in an efflux of protons. Danon and Stoeckenius[1] observed light-dependent increases in the ATP level in *H. halobium.* We have incorporated bacteriorhodopsin into liposomes and shown a light-dependent proton uptake by the vesicles.[18] By a fortunate coincidence, the direction of

TABLE 4
RECONSTITUTION OF LIGHT-ACTIVATED PROTON PUMP CATALYZING ATP FORMATION*

Additions	ATP Production (nmol/20 min/mg bacteriorhodopsin)
Complete system in the light	594
Complete system in the light + 1799	0
Complete system in the light + rutamycin	170
Complete system in the dark	23

* Experimental conditions were as described by Racker and Stoeckenius.[18]

proton movement observed was the same as in submitochondrial particles (from outside to inside), and we could therefore incorporate into the liposomes the oligomycin-sensitive ATPase complex of mitochondria together with bacteriorhodopsin. The vesicles catalyzed light-driven and rutamycin- and uncoupler-sensitive ATP generation as shown in TABLE 4. An important control is shown in the last line. Virtually no [^{32}P]-labeled ATP was detected in the dark, showing that the ^{32}Pi-ATP exchange reaction, catalyzed by the oligomycin-sensitive ATPase complex, was effectively suppressed by the presence of a large excess of hexokinase and glucose that prevented a steady-state level of ATP.

RECONSTITUTION OF MITOCHONDRIAL ADP AND PI TRANSPORTERS

ADP Transporter

We have reconstituted the ADP/ATP transporter of mitochondria.[19] During the past year, we have fractionated the transporter after solubilization with cholate and observed a marked stimulation of the ADP/ATP exchange by various salts (e.g., KCl).[20] Since this stimulation was markedly diminished by valinomycin and not observed in the case of the ATP/ATP exchange, we suspect that the ADP/ATP exchange is electrogenic in the reconstituted system.

Pi Transporter

Reconstitution of the phosphate transporter has recently been achieved[21] in our laboratory with fractions from a cholate extract of bovine heart mitochondria. We are now pursuing the purification and characterization of the active component.

RECONSTITUTION OF THE Ca^{++} AND Na^{+}, K^{+} PUMPS

We have reconstituted both the Ca^{++} pump of sarcoplasmic reticulum[22] and the Na^{+}, K^{+} pump of the plasma membrane. In the latter case, we used the Na^+, K^+-ATPase from electric eel and incorporated the protein by exposure of the phospholipid-protein mixture to sonic oscillation,[23] into vesicles without use of a detergent. Goldin and Tong[24] and Hilden *et al.*[25] have reconstituted Na^+ pump liposomes by the cholate dialysis procedure. The usefulness of reconstitution is illustrated by the interesting observation made by Goldin and Tong[24] that vesicles reconstituted with the Na^+, K^+-ATPase isolated from dog kidney, catalyze transport of Na^+ but not of K^+. They present evidence that a Na^+/Cl^- symport mechanism is operative that obviously has important physiological implications with respect to the operation of the pump in the kidney. In contrast, vesicles reconstituted in our laboratory with the eel enzyme, appear to operate by a Na^+/K^+ antiport mechanism. Optimal activity was obtained when a 4:1 ratio of phosphatidylethanolamine to phosphatidylcholine was used in reconstitution. This ratio has also been optimal in all of our other reconstitutions thus far tested.

Ample indirect evidence has been available for years implicating the Ca^{++}-ATPase and Na^+, K^+-ATPase in the mechanism of ion translocation. But the question remained whether the pump required a single protein or whether other "coupling factors" were operative, as in the case of the mitochondrial proton pump. To determine this it was very important to perform experiments with the purest ATPase preparation available. We have recently obtained experimental evidence that there is a second component present in Ca^{++}-ATPase of sarcoplasmic reticulum[26] which participates in Ca^{++} translocation.[27] We have suspected this for some time because of the observation, illustrated in TABLE 5, that there is a discrepancy between the specific ATPase activity and the capacity to catalyze Ca^{++} translocation in different fractions of the preparation. It can be seen that fraction R_{3c}, which has the highest specific ATPase activity, is considerably less active in transport than fraction R_{3e}, which has a lower specific ATPase activity, but an efficiency of pumping similar to that of sarcoplasmic reticulum vesicles. Fraction R_{3a}, which has an ATPase activity comparable to that of fraction R_{3a}, has only 5% to 10% of the transport activity. As shown in TABLE 6, the transport activity of fraction R_{3a} can

TABLE 5

COMPARISON OF ATPASE ACTIVITY AND Ca^{++} TRANSLOCATION*

Fraction	ATPase (μmol/min/mg protein)	Reconstituted Vesicles: Ca^{++} Translocation (nmol/min/mg protein)	Reconstituted Vesicles: ATPase (nmol/min/mg protein)	Ca^{++}/ATP Ratio
R_{3a}	5.6	50	160	0.31
R_{3b}	8.0	242	630	0.38
R_{3c}	11.0	365	930	0.39
R_{3d}	8.6	646	708	0.91
R_{3e}	5.0	826	490	1.7

* The fractions of Ca^{++}-ATPase were obtained essentially as described by MacLennan[26] using smaller addition of ammonium acetate to collect more fractions. Reconstitution was performed by the cholate dialysis procedure with 0.2 M K oxalate as described previously.[22] The ATPase activity of the isolated fractions was assayed at 37°, while that of the reconstituted vesicles was measured at room temperature as was the ion transport activity.

TABLE 6

STIMULATION OF Ca^{++} TRANSPORT BY HEAT-STABLE FACTOR*

	Ca^{++} Translocation	ATPase	
	(nmol/min/mg protein)		Ca^{++}/ATP Ratio
R_{3a}	50	160	0.31
R_{3a} + 3 μg-heated supernatant fraction	144	180	0.80
R_{3a} + 6 μg-heated supernatant fraction	180	130	1.40

* The R_{3e} ATPase fraction was heated for 5 minutes at 70° and then centrifuged.

be increased severalfold by including during reconstitution a heated extract of fraction R_{3e}, which by itself is completely inactive. Since the ATPase activity of the reconstituted vesicles is not altered considerably by the heated fraction, there is a marked increase in the efficiency of the pump operation. Although we cannot make a definitive statement about the identity of the activating component, we believe that we are dealing with the proteolipid described by MacLennan *et al.*[28] The reason for this statement is that (a) the major component in the heated extract of R_{3a} migrated in acrylamide gels similarly to the proteolipid, and (b) a proteolipid fraction prepared freshly according to MacLennan *et al.*[28] also stimulated transport of fraction R_{3a}. However, in contrast to the heated R_{3e} extract, the purified proteolipid was unstable on storage.

WHAT CAN WE LEARN FROM RECONSTITUTION EXPERIMENTS?

One of the most important informations about a pump is the minimum number of components required for its operation. The example just quoted illustrates how reconstitution experiments can help in answering this question. Another example is the elucidation of the symport mechanism in the Na^+ pump of dog kidney mentioned earlier. In addition there are two other problems that can be experimentally approached in reconstituted systems.

Role of Asymmetry

The importance of the role of asymmetry is best illustrated by the failures we encountered when we first attempted to reconstitute the third site of oxidative phosphorylation according to the chemiosmotic mechanism and the respiratory assembly of submitochondrial particles. As illustrated in FIGURE 3, for the reconstitution of liposomes with the orientation of submitochondrial particles, cytochrome oxidase should be oriented with cytochrome *a* on the inside and cytochrome a_3 on the outside. Cytochrome *c*, which reacts with cytochrome *a*, should also be inside. Ascorbate reduces lipophilic PMS, which enters the vesicles, reduces cytochrome *c* and releases protons. The electrons are shuttled back across the membrane from cytochrome *c* to cytochrome *a* and via the "copper wire" to cytochrome a_3 on the outside, where they interact with oxygen and two protons to form water. The proton motive

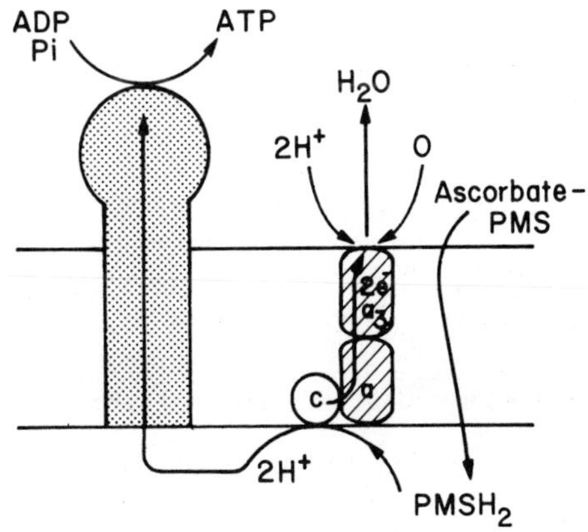

FIGURE 3. Mechanism of site III oxidative phosphorylation.

force thus formed is used by the proton pump of the oligomycin-sensitive ATPase to generate ATP from ADP and Pi on the outside of the vesicles. But what happens when there are cytochrome oxidase molecules present that are oriented in the opposite direction and cytochrome *c* is available on the outside? Obviously, the proton motive force will collapse and no ATP formation can occur. This is exactly what happened in the early experiments when we did not remove all traces of cytochrome *c* and the majority of cytochrome oxidase molecules were assembled in the mitochondrial orientation. There was no chance of forming a proton gradient. Since we had no way of imposing a unidirectional orientation of cytochrome oxidase, we chose to achieve asymmetry by removing all traces of cytochrome *c* from the outside by efficient washing and by adding to the closed vesicles polylysine, an effective competitor of cytochrome *c*. Then we observed oxidative phosphorylation. The uncoupling effect of cytochrome *c*, which we observe only with reconstituted vesicles and not in isolated submitochondrial particles, is much more readily explained in terms of the chemiosmotic than of the chemical hypothesis.

Not in all instances do we need to impose such artificial asymmetry because some of the substrates are ionic and do not get across the membrane. For example, the reconstitution of complex I into liposomes is random[14]; about one-half is oriented as in mitochondria, the other half is oriented as in submitochondrial particles. Fortunately, the vesicles are impermeable to DPNH so that only the flavins on the outside of the membrane can be reduced. Similarly, we do not need to worry about complications caused by scrambling when the assay depends on the utilization of impermeant reagents such as ATP, as in the case of the Ca^{++} and Na^{+} pumps.

Phospholipid Requirements

As shown in TABLE 7, we have reconstituted several pumps with acetyl phosphatidylethanolamine. It can be seen that the proton pump of bacteriorhodopsin operates in vesicles that contain the acetylated phospholipid as the only lipid constituent. In contrast, the reconstituted Ca^{++} pump does not operate with this phospholipid unless an alkyl chain containing a NH_2 group is incor-

TABLE 7

RECONSTITUTION OF PUMPS WITH ACETYL PHOSPHATIDYLETHANOLAMINE*

Reconstituted System	PE	Acetyl PE	Acetyl PE + Stearylamine
	(nmol [or atoms]/min/mg protein)		
Bacteriorhodopsin pump	2,360	2,400	–
Ca^{++} pump	114	0	109
^{32}Pi-ATP exchange (+PC)	133	0	112

* Reconstitutions were performed as described by Knowles *et al.*[29]

porated during reconstitution of the vesicles. Stearylamine and oleylamine are the most effective, but a careful titration must be made since they inhibit when added in excess.[29] Similar results were obtained with the reconstituted mitochondrial proton pump when we measured ^{32}Pi-ATP exchange, except that phosphatidylcholine was required in addition to acetylphosphatidylethanolamine and stearylamine. It therefore appears that NH_2 groups are required for the mitochondrial proton pump and the Ca^{++} pump, but not for the bacteriorhodopsin proton pump. Since the latter depends on the formation of closed vesicles the problem does not appear to lie in the formation of compartments. This conclusion is supported by the observation that the ATPase activity is also affected by the presence of NH_2 groups. We observed pronounced charge influences on ATPase activity. The effects of stearylamine could be counterbalanced by equivalent amounts of a negatively charged hydrophobic anion such as dicetylphosphate.[29]

We would like to mention here an experimental discrepancy between our work and that of Warren *et al.*[30] We could repeat their interesting work on the replacement of the intrinsic phospholipids of Ca^{++}-ATPase preparations by synthetic dioleoyl phosphatidylcholine and observed respectable ATPase activity. However, we failed to obtain preparations that were reconstitutively competent with respect to Ca^{++} pumping when a phosphatidylcholine was the only lipid.

We have recently succeeded in stripping the phospholipids from the Ca^{++}-ATPase without loss of potential reconstitutive activity. The preparation was passed through a Sephadex column in the presence of cholate. Analysis for phosphorus indicated that over 98% of the phospholipids were removed. As shown in TABLE 8, there was reconstitution of Ca^{++} translocation with this preparation when phosphatidylethanolamine was present. Phosphatidylcholine alone was consistently ineffective, but markedly stimulated transport activity when included during reconstitution.

Channels or Carriers

We should like to cite two examples that illustrate how reconstitution experiments can help in deciding whether we are dealing with a channel or a carrier type of translocation. Bacteriorhodopsin can be reconstituted into synthetic phospholipids with known transition temperatures.[31] An analysis of the proton pump suggests that it operates by a channel mechanism. The advantage of this system of analysis is that it can be performed in the presence of control transporters that are known to function either as channels or as carriers. When the proton pump was tested above the transition temperature of the phospholipid in the presence of nigericin, which is a mobile carrier catalyzing the exchange of H^+ against K^+, the proton gradient was collapsed. Below the transition temperature (e.g., at 4°) the bacteriorhodopsin pump was as active in the presence of nigericin as in its absence. In contrast, in the presence of a channel-former such as gramicidin there was no gradient formed even at 4°.

We have carried out similar experiments with the Ca^{++} pump, but the system is more complex because of the need for phosphatidylethanolamine mentioned earlier. We need more and better synthetic phospholipids with known transition temperature to arrive at a definite conclusion. However, we performed another type of experiment with the reconstituted Ca^{++} pump which convinced us that at least one of the proposed carrier mechanisms was incorrect.

In this experiment, vesicles were reconstituted by the cholate dialysate procedure and phosphorylated with γ-labeled [^{32}P]ATP. A control containing the same ingredient but without cholate was carried through the same dialysis and phosphorylation procedure. The same amount of phosphoenzyme was formed in the reconstituted and nonreconstituted enzyme. This experiment suggests that in the reconstitution, all ATPase molecules are oriented in the membrane in the same direction, which is also true in sarcoplasmic reticulum.[32] Moreover, we showed that there was no translocation of the ^{32}P to the inside of the vesicles, indicating that the ATPase does not rotate and discharge the phosphate group as has been suggested. Obviously, this experiment does not rule out several other modes of carrier-type mechanism but it has eliminated one in particular. We mention this example to illustrate another advantage of reconstitution experiments. By varying the phospholipid composition during reconstitution, we can make liposomes that are either permeable to Pi or oxalate, like sarcoplasmic reticulum, or impermeable, as the liposomes used in the last experiment we described. Although objections can be raised that we have departed too far from physiological conditions, it seems unlikely to us that the Ca^{++} pump could be operative in the impermeable liposomes by an entirely different mechanism.

TABLE 8

RECONSTITUTION OF DELIPIDATED CA^{++}-ATPASE*

	Ca^{++} Translocation (nmol/min/mg protein)
Crude soybean phospholipids	216
Mitochondrial phosphatidylcholine (PC)	0
Egg phosphatidylcholine	0
Mitochondrial phosphatidylethanolamine	183
+ mitochondrial PC (1:1)	337
+ egg PC (1:1)	448

* The total amount of phospholipids in all experiments was 12 μmol with 150 μg of Ca^{++}-ATPase. Reconstitution was performed by the cholate dialysis procedure with 0.2 M K oxalate.[5]

Mechanism of Action

We would like to conclude this presentation by describing some recent experiments that have brought us back full cycle to work on the mechanism of enzyme action. We found this a welcome relief after 20 years of work with "dirty" membranes.

Masuda and deMeis[33] showed that sarcoplasmic reticulum vesicles are phosphorylated when incubated with ^{32}Pi. We have confirmed this work and performed similar experiments with the purified enzyme.[32] We then added to the phosphoenzyme simultaneously, Ca^{++} and ADP, and found that ATP was formed stoichiometrically with phosphoenzyme. As shown in Table 9, Ca^{++} and ADP were required for the reaction. Ca^{++} was needed at rather high concentrations, in contrast to the low concentrations required for ATP hydrolysis. In Table 10, we show that the ATP formed is hydrolyzed by boiling in 1 N HCl and the reaction takes place in the presence of hexokinase, which rapidly transfers the γ phosphate group of ATP to glucose. This not only serves as a further identification of ATP but also rules out any contribution to ^{32}[P]ATP formation by an exchange reaction. The fact that the reaction stops when stoichiometry to phosphoenzyme is approached also speaks against an exchange reaction. The ATP that is formed in this reaction is not firmly bound to the membrane. After removal of the enzyme by centrifugation, it remains in the supernatant.

These experiments raise some serious thermodynamic questions. Where is the energy for ATP formation derived from? Previous experiments on the formation of phosphoenzyme from Pi with Na^+, K^+-ATPase by Lindenmayer *et al.*,[34] Albers *et al.*,[35] and Post *et al.*[36] have been interpreted in terms of a low-energy product. Although the phosphoryl group appears to be attached to the same aspartyl group that is phosphorylated by ATP, the two forms of the enzyme appear to be quite different. The phosphoenzyme formed from

Table 9

ATP Formation in the Second Cycle and its Dependency on Ca^{++} and ADP*

	First Incubation	Second Incubation	
	Phosphoenzyme Formed (nmol/mg)	Addition	ATP Formed (nmol/mg)
First cycle	1.71	Trichloroacetic acid	0
		$CaCl_2$ (6.7 mM)	0
		ADP (5 mM)	0
		$CaCl_2$ + ADP	0.93
Second cycle	2.53	Trichloracetic acid	0
		$CaCl_2$ (6.7 mM)	0
		ADP (5 mM)	0
		$CaCl_2$ + ADP	1.76

* Experimental conditions were as described by Knowles and Racker.[32]

TABLE 10

FORMATION OF ATP BY Ca^{++}-ATPASE AND ITS UTILIZATION BY HEXOKINASE*

Treatment of Product	Organic ^{32}P Formed – Hexokinase (nmoles/mg protein)	Organic ^{32}P Formed + Hexokinase (nmoles/mg protein)
Experiment 1		
None	0.54†	0.40
Boiled 7 minutes in 1 N HCl	<0.05	0.40
Experiment 2		
None	0.71†	0.92
Boiled 7 minutes in 1 N HCl	0.05	0.89

* Experiments were performed as described in Table 9.[31]

† Phosphoenzyme production determined in separate samples was 0.94 nmol/mg protein in Experiment 1 and 1.0 in Experiment 2.

ATP (E_1—P) could be discharged by ADP, while E_2—P formed from Pi did not form ATP under comparable conditions.[36] Post has calculated[37] from equilibria data the apparent free energy of hydrolysis of E_2—P and concluded that the intermediate is "low energy." On the other hand, Taniguchi and Post,[38] studying "microsomal" preparation of kidney Na^+, K^+-ATPase, have obtained evidence for ATP formation from E_2—P on addition of large amounts of Na^+. Assuming that the ATP formed in these experiments is also free and that the microsomes do not catalyze an exchange reaction, we can conclude from their experiments as well as from ours that net ATP can actually be formed from ADP and Pi stoichiometrically with phosphoenzyme. Where does the energy for this reaction come from?

Our first approach to this problem was to explore the possibility that the enzyme itself contained some preformed energy source, e.g., firmly bound ATP or a thiolester within the protein structure. We could not detect any firmly bound ATP. The possibility of a preformed high-energy group within the enzyme was ruled out by reisolation of the enzyme after it had gone through a cycle of E—P and ATP formation. The outcome of these experiments was clear-cut: ATP formation was as large in a second cycle as it was in the first (TABLE 9).

We therefore propose, as the only remaining alternative, that the energy for ATP formation is derived from the binding energy of Ca^{++} to the protein. A schematic outline of a tentative mechanism involving specific conformational changes of the protein is shown in FIGURE 4. Obviously this scheme should not be taken too literally. The major point it illustrates is that Mg^{++} induces a conformation of the protein that keeps the phosphoryl group (formed perhaps via a thiolester) in a hydrophobic region. Ca^{++} induces a conformation that exposes the phosphoryl group to water and thereby renders it "high-energy" and accessible to ADP. The resulting Ca^{++} enzyme must be converted back to the original conformation (by reprecipitation) before another cycle of events can be initiated.

This is the working hypothesis which enables us to design future experiments. Our first aim is to find a hydrophobic reagent that will react with the active site without denaturation of the enzyme. We hope that a classical

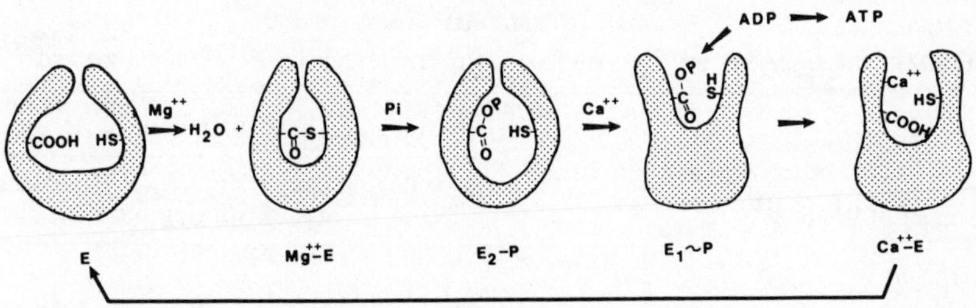

FIGURE 4. Mechanism of ATP formation from ADP and Pi by Ca^{++} ATPase.

chemical and physical-chemical approach to the active site of the enzyme will reveal a great deal about the mechanism of ion translocation.

SUMMARY

1. Oxidative phosphorylation was reconstituted with a mitochondrial proton pump (oligomycin-sensitive ATPase) and segments of the oxidation chain (cytochrome oxidase or DPNH-Q_1 reductase). A proton pump of bacteriorhodopsin substituted for the respiratory chain components, giving rise to light-induced ATP formation.

2. Since oxidative phosphorylation has thus become a special case of the problem of ion translocation in general, we have investigated and reconstituted other pumps. The reconstituted Ca^{++} pump of sarcoplasmic reticulum consists of two factors, the Ca^{++}-dependent ATPase and a heat-stable coupling factor.

3. Other information obtained from reconstitution experiments include the role of asymmetry in organized membranes and the specificity of protein-phospholipid interaction.

4. Purified preparations of Ca^{++}-ATPase catalyze the formation of ATP from Pi and ADP in a stepwise reaction stoichiometric with the enzyme and dependent on Ca^{++}.

REFERENCES

1. DANON, A. & W. STOECKENIUS. 1974. Proc. Nat. Acad. Sci. U.S.A. **71:** 1234–1238.
2. MITCHELL, P. 1966. Biol. Rev. Cambridge Phil. Soc. **41:** 445.
3. KAGAWA, Y., A. KANDRACH & E. RACKER. 1973. J. Biol. Chem. **248:** 676–684.
4. CHIEN, T. F. 1974. Fed. Proc. **33:** 1291.
5. RACKER, E. 1972. J. Mem. Biol. **10:** 221–235.
6. CATTELL, K. J., I. G. KNIGHT, C. R. LINDOP & R. B. BEECHY. 1970. Biochem. J. **125:** 169.
7. STEKOVEN, F. S., R. F. WAITKUS & H. T. B. VAN MOERKERK. 1972. Biochemistry **11:** 1144–1150.
8. YONETANI, T. 1967. *In* Methods in Enzymology, Vol. 10. R. W. Estabrook and M. E. Pullman, Eds.: 332–335. Academic Press. New York, N.Y.
9. KOBOYAMA, M., F. C. YONG, & T. E. KING, 1972. J. Biol. Chem. **247:** 6375–6388.

10. Ragan, C. I. & P. C. Hinkle. 1975. J. Biol. Chem. In press.
11. Hatefi, Y., A. G. Haavik, L. R. Fowler, & D. E. Griffiths. 1962. J. Biol. Chem. **237:** 2661.
12. Hinkle, P. C. & K. H. Leung. 1974. *In* Membrane Proteins in Transport and Phosphorylation. M. E. Klingenberg, E. Quagliariello and N. Siliprandi, Eds.: 73–78. North-Holland Publishing Co. Amsterdam, The Netherlands.
13. Racker, E. & A. Kandrach. 1973. J. Biol. Chem. **248:** 5841–5847.
14. Ragan, C. I. & E. Racker. 1973. J. Biol. Chem. **248:** 6876–6884.
15. Yu, C. A., L. Yu & T. E. King. 1974. J. Biol. Chem. **249:** 4905–4910.
16. Racker, E., T. F. Chien & A. Kandrach. FEBS Lett. In press.
17. Oesterhelt, D. & W. Stoeckenius. 1973. Proc. Nat. Acad. Sci. U.S.A. **70:** 2853–2857.
18. Racker, E. & W. Stoeckenius. 1974. J. Biol. Chem. **249:** 662–663.
19. Shertzer, H. G. & E. Racker. 1974. J. Biol. Chem. **249:** 1320–1321.
20. Shertzer, H. & E. Racker. To be submitted.
21. Banerjee, R. Unpublished observations.
22. Racker, E. 1972. J. Biol. Chem. **247:** 8198–8220.
23. Racker, E. 1973. Biochem. Biophys. Res. Commun. **55:** 224–230.
24. Goldin, S. M. & S. W. Tong. 1974. J. Biol. Chem. **249:** 5907–5915.
25. Hilden, S., H. M. Rhee & L. E. Hokin. 1974. J. Biol. Chem. **249:** 7432–7440.
26. MacLennan, D. H. 1970. J. Biol. Chem. **245:** 4508–4518.
27. Racker, E. & E. Eytan. J. Biol. Chem. In press.
28. MacLennan, D. H., C. C. Yip, G. H. Iles, & P. Seaman. 1972 Cold Spring Harbor Symp. Quant. Biol. **37:** 469–478.
29. Knowles, A. F., A. Kandrach, E. Racker & H. G. Khorana. 1975. J. Biol. Chem.
30. Warren, G. B., D. A. Toon, N. J. M. Birdsall, A. G. Lee & J. C. Metcalf. 1974. Prov. Nat. Acad. Sci. U.S.A. **71:** 622–626.
31. Racker, E. & P. C. Hinkle. 1974. J. Mem. Biol. **17:** 181–188.
32. Knowles, A. F. & E. Racker. 1975. J. Biol. Chem. **250:** 3538–3544.
33. Masuda, H. & L. de Meis. 1973. Biochemistry. **12:** 4581–4585.
34. Lindenmayer, G. E., A. H. Laughter & A. Schwartz. 1968. Arch. Biochem. Biophys. **127:** 187–192.
35. Albers, R. W., G. J. Koval & G. J. Siegel. 1968. Mol. Pharmacol. **4:** 324–336.
36. Post, R. L., S. Kume & F. N. Rogers. 1973. *In* Mechanisms in Bioenergetics. G. F. Azzone, *et al.* Eds.: 203–218. Academic Press. New York, N.Y.
37. Post, R. L., Personal communication.
38. Taniguchi, K. & R. L. Post. 1974. Fed. Proc. **33:** 1289.

Discussion

Dr. Pressman: Since this is a mixed audience that may have very different backgrounds I'd like to address myself to those sections of Dr. Racker's presentation dealing with his advocacy of the Mitchell hypothesis. The proton movements that he has chronicled are an interesting, necessary component of the Mitchell hypothesis, but they are not sufficient to represent a striking defense in and of themselves. It's difficult to write redox and hydrolytic reactions of biology without invoking the appearance and disappearance of charges and protons. There are several serious deficiencies in the Mitchell hypothesis that many of us still believe are very important. First, the energy requirements forming ATP under conditions that mitochondria can form it are much too large to be accounted for by the relatively small gradients of hydrogen ions and charges that the Mitchell mechanisms seem to be able to generate. The pH gradient is usually no larger than about one pH unit. Recently a whole series of measurements of mitochondrial membrane potentials made by a number of techniques, and we lay claim to one, have established the fact that the mitochondrial membrane potentials seem to be relatively modest, certainly no larger than about 60 millivolts. According to the Mitchell mechanism there should be two charge separations in the mitochondrial membrane per equivalent of ATP; we have measured six, and that's too much of a good thing.

DR. RACKER: Witt claims four protons per ATP. The more the merrier.

DR. PRESSMAN: Not with a mechanism that calls for two; six is as bad as zero if the mechanism calls for two. At any rate a large number of people still have serious reservations about the Mitchell hypothesis.

DR. RACKER: Thayer and Hinkle (1973. J. Biol. Chem. **248:** 5395; 1973. Fed. Proc. **32:** 669) have published two impressive papers on proton translocation in submitochondrial particles. Their values approach 2 for the H^+/ATP ratio. These measurements are more meaningful than those in mitochondria since they are not complicated by other reactions, such as calcium translocation, which must have influenced your measurements.

In chloroplast the ΔpH is much greater than 1, close to 3 (1975. Portis *et al.* Biochem. Biophys. Res. Commun. **64:** 877). Remember that Mitchell's formulation of 1961 includes both the ΔpH and the membrane potential. In recent papers Thayer and Hinkle (1975. J. Biol. Chem. **250:** 5330, 5336) have shown that ATP synthesis can be driven by a combination of artificially imposed ΔpH and membrane potential.

DR. PRESSMAN: If you've read our critiques, you will realize that we've never ignored the fact that the Mitchell hypothesis calls for a combination of energy contributions by both the proton and membrane potentials. This used to be Mitchell's point whenever he was cornered about the thermodynamic inadequacy of the gradients.

DR. RACKER: He wasn't cornered by anyone in 1961, and he included both ΔpH and the membrane potential in the original formulation.

DR. PRESSMAN: Whenever he's been cornered by the inadequacy of the observed gradients to account for the tremendous potential against which ATP can be made, (we have obtained a figure of 16 kilocalories in mitochondria), he has always claimed that he never committed himself as to where the energy comes from. He said that if one was going to talk about the proton gradient, then he wanted the energy in the membrane potential, but as soon as one talked about the membrane potential, he wanted it in the proton gradient. That's the point I want to make. I'm intrigued by the fact, that you've taken up the same position, Dr. Racker, but whatever you're addressing yourself to, the answer is in the other component of the Mitchell hypothesis.

DR. RACKER: Dr. Pressman, will you explain the experimental fact that cytochrome oxidase vesicles reconstituted with only one protein and two phospholipids require for maximal respiratory rate both valinomycin to collapse the membrane potential and nigericin to collapse the ΔpH.

DR. PRESSMAN: If you write the reactions out on paper you will see that protons and charges appear and disappear. The point is that the whole system goes together and forms protons and charges, but you knew this when you wrote the reaction without having to incorporate it in a vesicle. It was an experiment that couldn't fail; it does not necessarily prove that this is the mechanism by which ATP is formed. You are of course aware of the alternate hypothesis that there are proton movements in parallel with. . . .

DR. RACKER: Cytochrome oxidase is a single protein well characterized; it contains no proton pump.

DR. PRESSMAN: There's an old principle of chemistry known as mass action, namely, when the products pile up and you have no way of getting rid of them, they tend to slow down the reaction. This is all the explanation you need for that experiment.

DR. SHAMOO: Dr. Racker, have you ruled out the possibility that this proteolipid added to the reconstitution of calcium uptake, which carried the enhancement you see in Ca^{++} uptake, may be due to making the vesicles nonleaky? Or is there some magic factor such as an energy transduction coupling mechanism?,

DR. RACKER: We have not ruled out such a possibility, but it seems unlikely since we have a Ca^{++} trap, namely, oxalate, inside the vesicles.

DR. GREEN: I would like to put in perspective the studies in reconstitution. First, let me pay tribute to Dr. Racker's monumental work in putting together the systems that carry out coupled reactions like oxidative phosphorylation or active transport. It's very important to point out, however, that the fact of reconstitution or whether the coupling was carried out by the Mitchell hypothesis wouldn't make any difference as far as the mystery of the mechanism before the reconstitution.

We applaud Dr. Racker's great enthusiasm for the Mitchell model, but it is true that there are a great many people, including myself, who have very great reservations about it. I would like to put the problem in a very sharp experimental form. If Mitchell's initial model is correct, you should not be able to get coupling in a nonmembranous system. However, there are reports from our own laboratory and from other laboratories that this coupling is possible without a membrane.

DR. RACKER: I fully agree with you, Dr. Green. Mitchell's hypothesis must be abandoned if you can get oxidative phosphorylation in a system without membranes. We cannot.

THE KINETIC AND EQUILIBRIUM COMPONENTS OF SELECTIVE IONIC PERMEABILITY MEDIATED BY NACTIN- AND VALINOMYCIN-TYPE CARRIERS HAVING SYSTEMATICALLY VARIED DEGREES OF METHYLATION*

George Eisenman, Sally Krasne, and Sergio Ciani

Department of Physiology
UCLA Center for the Health Sciences
Los Angeles, California 90024

Introduction

In less than a decade since neutral macrocyclic antibiotics were first recognized to act as highly selective carriers of ions across natural and reconstituted membranes[1-6] our knowledge of their intimate molecular processes has developed explosively,[7-10] in no small part because of the relative simplicity of the underlying chemical reactions and physical phenomena involved. There is now convincing evidence, as was thought quite early,[3, 11, 12] that these lipophilic molecules do indeed carry ions across membranes merely by loading and unloading them at the membrane-solution interfaces, shuttling them back and forth across the membranes by diffusion. We have, of course, learned that only in certain limiting cases are the processes totally comprehensible in terms of the equilibrium chemistry of the formation of 1:1[13-15] or higher (e.g., 2:1, 3:1) complexes.[16, 17] More generally, the rates of unloading of certain carriers in certain lipids are sufficiently similar to the rate of translocation across the membrane that kinetic factors also come into play,[7, 9, 10, 18-22] with the kinetic effects being strongly influenced by the dipole potential and charge of the lipid.[9, 22] Additionally, the interpretation of recent data has led to a refinement of the model allowing for the shape[23] of the diffusion barrier and for some dependence of the loading and unloading reactions on the applied voltage.[24]

Although replacement of the previous, simplest picture of carrier action by one with such richness of detail makes intuitive comprehension increasingly difficult, except by specialists, we feel that the extended model (to be used here) more satisfactorily reflects reality in that it is required for an accurate (and physically reasonable) fit of all the data. We believe that the accuracy and reproducibility of the measurements obtainable with these simple molecules justifies not only our theoretical and experimental labor in using the extended model, but also the effort demanded of the reader! We hope the present paper successfully demonstrates this for the nactin- and valinomycin-type carriers, and we would like to point out at the outset that, despite the tedious characterization needed to verify and measure the additional parameters of the extended model, at the end it will still be possible to grasp intuitively all the essential conclusions.

* Supported by National Science Foundation Grant GB 30835, United States Public Health Service Grant NS 09931, and intramural support from the UCLA Campus Computing Network.

We will present here results from an ongoing study in our laboratory, part of which has been[9, 22] and part of which will be[25] presented elsewhere in more detail, directed at an analysis of the molecular factors underlying ion-selective permeation of membranes. We will restrict ourselves to the effects on ionic selectivity (as seen in membrane potentials) of systematically varying the degree of methylation for two series of typical cation carriers: the naturally occurring macrotetralide actins (nonactin, monactin, dinactin, trinactin, and tetranactin—each of which has one more methyl group than the preceding) and the synthetic[26] valinomycin homologs (lactate-valinomycin, valinomycin, hydroxyisovalerate-valinomycin—each of which differs from the preceding by six methyl groups). Concerning the ion selectivity of channels, we note only that the characterization of loading and unloading of carriers can serve at least as a conceptual prototype for the mechanism by which ions are trapped and released in permeating via channels and that the equations governing selective ion permeation through channels and carriers are similar (see Reference 27) under certain conditions.

Selectivity in Membrane Potential

The carrier-mediated selectivity of bilayers among monovalent cations can be conveniently characterized by the observed (zero-current) membrane potential, V_0, using the familiar expression

$$V_0 = \frac{RT}{F} \ln \frac{a_i' + \frac{P_j}{P_i} a_j'}{a_i'' + \frac{P_j}{P_i} a_j''} \qquad (1)$$

P_j/P_i, whose precise meaning depends on the system studied and the model proposed to describe it, defines the relative selectivity between ionic species, J^+ and I^+, whose activities in solution (′) and (″) on the two sides of the membrane are indicated by a_i', a_j'', a_i', a_j''. Equation 1 is identical to the familiar Goldman-Hodgkin-Katz equation[28, 29] for anion-impermeant systems. It has been widely used to study cell membranes[27, 30, 31] and ion-exchanger membranes[32] and has been shown under certain conditions to be applicable to gramicidin channels as well (cf. Equations 64 and 65 of Reference 33). It was first deduced to hold for carrier-mediated permeation in bilayers by Ciani *et al.*[12, 13] under rather general conditions except for the restriction that the complexation reactions were assumed to be rapid compared to the translocation of the complex, this being defined as the "equilibrium domain."[8] In this situation Equation 1 is particularly meaningful since the permeability ratios are expected, and observed, to be true constants and their interpretation in terms of the model shows them to be simply related to the equilibrium chemistry of ion complexation.[14, 15] We subsequently found that when the rates of complexation become significantly slow, Equation 1 is still a useful way to describe the potential even though the permeability ratios become dependent upon kinetic parameters and are functions of the voltage.[9, 22] In this "kinetic domain" the "apparent" permeability ratios defined as $\lim_{V \to 0} (P_j/P_i)$ are expected (and observed) to depend also upon the nature of the lipid, as we shall demonstrate in considerable detail below.

The "Equilibrium Domain"

For those lipids and carrier molecules that satisfy the condition of being in the "equilibrium domain," the permeability ratio of Equation 1 is a true constant, which we now designate as $(P_j/P_i)_{eq}$ given by

$$(P_j/P_i)_{eq} = \text{Const} \tag{2}$$

The value of this constant is, moreover, essentially identical to the ratio of equilibrium constants for ion extraction into the bilayer $(\bar{K}_j/\bar{K}_i)$ or for salt extraction (K_j/K_i) into appropriate model solvents for "isosteric" complexes (whose mobilities and partition coefficients are the same regardless of the ions complexed; see Reference 14 for details). Typical observations for systems in the equilibrium domain are given in FIGURE 1, which presents experimental data for nonactin, monactin, dinac-

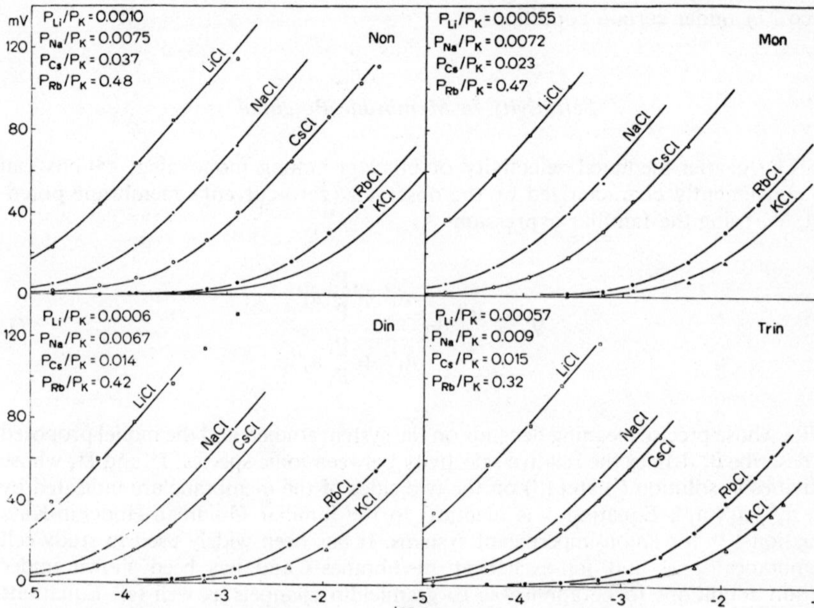

FIGURE 1. Membrane potentials in alkali chloride-potassium nitrate mixtures in the presence of nonactin, monactin, dinactin, or trinactin. Bilayer membranes were formed from asolectin in 10^{-2}M alkali metal chloride solution at 10^{-7}M concentration of the indicated macrotetralide actin (non-, mon-, din-, trin-). The curves are drawn according to Equations 1 and 2 for the indicated (constant) values of the permeability ratios. (From Szabo *et al.*[15] By permission of the *Journal of Membrane Biology*.)

tin, and trinactin in asolectin bilayers.[15] Note the agreement between the observed potentials, plotted as points, with the theoretical curves drawn according to Equation 1 with the indicated (constant) values of the permeability ratios. This agreement verifies that, in this lipid and for these carriers, the permability ratios are indeed constants, independent not only of the widely varied solution conditions, but moreover of variations in transmembrane potential in excess of 100 mV.

Certain trends in the selectivity with increasing methylation can be seen already in FIGURE 1. In particular, the relative preference of Cs^+ to Na^+ decreases substantially in going from nonactin to trinactin and the relative K^+ to Rb^+ preference increases somewhat. The trends in selectivity parallel the observed changes with methylation in selectivity measured at equilibrium in salt extraction experiments and correspond to changes (cf. FIGURE 13 of Reference 14) in the aforementioned (in relation to Equation 2) ratio K_j/K_i. These effects of methylation on selectivity led us to undertake the present studies, which will not only confirm these trends in different lipids, but also demonstrate an inversion in the selectivity between Na and Cs in the case of tetranactin.

The Kinetic Domain

In contrast to the above simple situation, studies by Lauger, Stark, and Benz *et al.*[18–21] on different phospholipids as well as those by Ciani *et al.*[9] and Laprade *et al.*[22] on glycerol dioleate (and monooleate) bilayers with the more strongly complexing carriers indicate that it is possible to encounter situations (called the "kinetic domain") for which the rates of loading and unloading of the complexes at the interface become comparable to the rates of transport of the complexes across the membrane interior. Ciani *et al.*[9] have deduced the expression, defined by Equation 1, for the permeability ratios under these circumstances and have shown that these ratios are neither expected nor observed to be constants but rather depend upon the transmembrane potential. Experimentally, this feature is illustrated in FIGURE 2, which refers to trinactin and glycerol diooleate bilayers instead of the asolectin.† Note that the data points do not fall on the dashed curves (which have the Nernstian limiting slopes expected from Equation 1). Rather, they fall on the solid curves whose slopes can be either sub-Nernst or supra-Nernst (as seen in the inset). Theoretically, this behavior is consistent with the expectations for the "kinetic domain"[9, 22] and its main features can also be accounted for, at least qualitatively, by the simple Eyring single-barrier model introduced by Läuger and Stark.[7, 18–21] Although we will shortly replace such a simple model with an extended version that requires a more cumbersome formalism, it is helpful to recall its consequences for a more direct illustration of some basic points. In terms of the Eyring single-barrier model and for the "kinetic domain," the permeability ratio in Equation 1 is given by an "apparent permeability ratio," $(P_j/P_i)_{app}$

$$\left(\frac{P_j}{P_i}\right)_{app} = \left(\frac{P_j}{P_i}\right)_{eq} \left[\frac{1 + 2w_i \cosh \dfrac{\phi}{2}}{1 + 2w_j \cosh \dfrac{\phi}{2}}\right] \tag{3}$$

where $\phi = FV_0/RT$ and w_i and w_j are defined as the ratios of rate constants for the translocation of the complex across the bilayer to those for unloading the complex at the membrane-solution interface ($w_i = k_{ms}/k_D$ in Läuger's[7] terminology). The "apparent permeability ratio" is seen to differ from the equilibrium permeability

† The interior of such GDO decane bilayers due to dipole effects appears to be about 160 mV more negative than those made of PE decane.[34, 35]

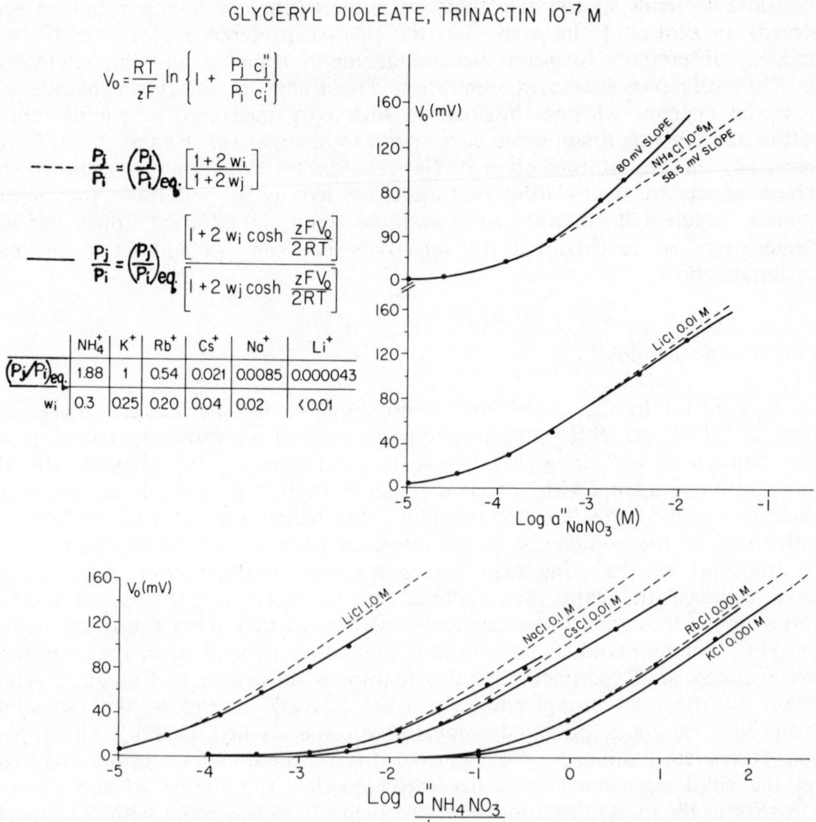

FIGURE 2. Membrane potentials of GDO bilayers in ion mixtures in the presence of trinactin. The filled circles represent experimental points. The solid curves are drawn according to Equations 1 and 3 with the values of the parameters $(P_j/P_i)_{eq}$ and w_j given in the figure. The dashed curves, drawn according to Equations 1 and 4 and giving limiting slopes of 58.5 mV/decade, correspond to the expected behavior if the "apparent permeability ratios," $(P_j/P_i)_{V\to 0}$, are assumed constant and equal to those measured at vanishingly small voltages (see the equations in the figure). (From Ciani *et al.*[9] By permission of the publisher.)

ratio by the bracketed term, whose value depends on the kinetic parameters w_i and w_j and also upon the transmembrane potential V_0.

In the limit $(V \to 0)$ of small transmembrane potentials Equation 3 reduces to the simpler form:

$$\left(\frac{P_j}{P_i}\right)_{V\to 0} = \left(\frac{P_j}{P_i}\right)_{eq} \left[\frac{1+2w_i}{1+2w_j}\right] \quad (4)$$

The apparent "low-voltage permeability ratio" $(P_j/P_i)_{V\to 0}$ is practically independent of voltage but still differs from the equilibrium permeability ratio $(P_j/P_i)_{eq}$ by the bracketed term, which depends upon the kinetic parameters w_i and w_j. From this it is clear that, although the low-voltage permeability ratio, $(P_j/P_i)_{V\to 0}$, measured in this limit does indeed become a constant, it is no longer identical to the "intrinsic" or equilibrium permeability ratio measured for a membrane in the "equilibrium domain," differing from it by a kinetic contribution, which in general is expected (and observed) to depend on the nature of the lipid.

A pictorial representation of the Eyring single-barrier model is presented in FIGURE 3 and is described in detail in the legend. In FIGURE 3 it is seen that the kinetic parameter, w_i, corresponds to the difference in height (as given by RT ln k_{ms}/k_D) of the central barrier relative to the interfacial barrier.

In terms of the pictorial model in FIGURE 3, the "equilibrium domain" (where

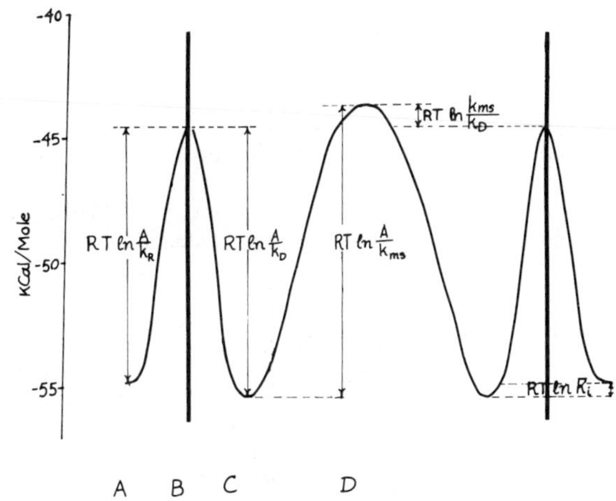

FIGURE 3. A schematic representation of the single-barrier Eyring model of the energy barriers surmounted by an ion and a carrier in the course of ion transport across a bilayer membrane. The membrane-solution interface is indicated by the heavy lines. *Left to right:* the levels of energy at which a K^+ ion finds itself in proceeding from an hydrated state in water (A), through its transition state (B) into its equilibrium position in the complex inside the carrier at the interface (C), as well as at the top of the hill (D) to be surmounted by the complex in crossing the interior. In this figure we have used the terminology of Läuger *et al.*[7, 18] so that k_R is the rate constant for the reaction to form the complex between an ion in solution and a carrier in the membrane, k_D, is the rate constant for dissociating the ion from the carrier at the interface, and k_{ms} is the rate constant for translocation of the complex across the membrane interior. K_i in our terminology[8, 9, 13] is the equilibrium constant of the interfacial reaction and is given by the ratio k_R/k_D. A is a constant, the frequency factor in the Eyring rate theory,[31] whose value is approximately 6.21×10^{12} sec^{-1}.

$w_i \ll 1$) is merely that situation in which for any experimentally reasonable applied voltage the central barrier is enough higher than the interfacial barrier that it still remains the rate-limiting barrier on the entire system (the effect of an applied voltage is simply to tilt the barriers, with 100 mV corresponding to 2.3 kcal on the ordinate). For "isosteric complexes" the height of the barrier to translocation (D) above the bottom of the well (C) is the same for all ionic species‡; so that the significant parameter in determining $(P_j/P_i)_{eq}$ is the depth of the well (C) vs. the energy level (A) of the ion in solution (RT ln $\bar{K}_i$). This is why selectivity in the "equilibrium

‡ Differences in mobilities (translocation rates) as for "nonisosteric" complexes would yield different heights (corresponding to different values of RT ln k_{ms}).

domain" represents the ratio of $\bar{K}_i$ to $\bar{K}_j$. (For a very lucid exposition of the roles of equilibrium and kinetic factors [as seen in wells and peaks] in ionic permeability, see Hille.[27]) The "kinetic domain" corresponds to the situation ($w_i = k_{ms}/k_D \gg 0$) when the interfacial barriers are higher than the central peak, either in the absence of an applied voltage or when they become so in its presence. The degree of "saturation" of the conductance-voltage characteristic or deviations from Nernstian behavior of V_0 reflect this voltage-dependence, as described in Equation 3. We will confine our analysis of kinetic effects in this paper to these steady-state properties, although further kinetic details are obtainable from relaxation studies which depend upon the individual rate constants (i.e., k_R, k_D, k_{ms}), as examined elsewhere.[18-22]

The Extended Model

The extended model that will be used to interpret our data is schematized in Figure 4. It extends the simple Eyring-type model of Figure 3 by approxi-

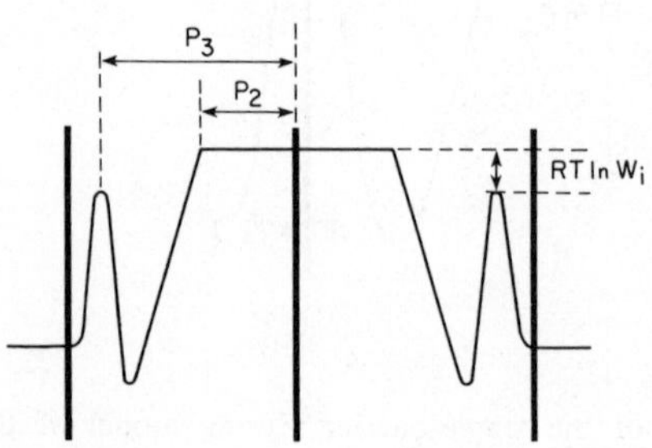

Figure 4. Schematic diagram of the "extended" model, identifying the various parameters discussed in the text. The heavy vertical lines denote the membrane-solution interfaces and the plane halfway across the membrane. The ordinate is free energy, as in Figure 3; but note that RT ln w_i is not strictly equal (as in the Eyring single-barrier model) to the indicated difference between the barrier heights but is related to it by an additive constant, which for isosteric complexes is the same for all ions, as will be discussed elsewhere.[25]

mating the diffusion barrier by a trapezoid of "half-width" P_2 and by including a voltage-dependence of the interfacial reactions defined by the locus of the "plane" of the reaction, P_3 (P_2 and P_3 are measured along the voltage gradients). These two refinements are suggested, respectively, by the work of Hall *et al.*[23] and of Hladky.[24] Indeed, Hladky has already used this model[24] to analyze the conductance-voltage behavior for NH_4^+ and K^+ in nonactin and trinactin in GDO bilayers; and we have elsewhere derived its expectations for the membrane potential in ionic mixtures.[25] This extended model leads to the following Equation 5 in place of the previous Equation 3:§

$$\left(\frac{P_j}{P_i}\right)_{app} = \left(\frac{P_j}{P_i}\right)_{eq} \frac{1 + 2w_i\phi\left(\dfrac{\cosh P_3\phi}{\sinh P_2\phi}\right)}{1 + 2w_j\phi\left(\dfrac{\cosh P_3\phi}{\sinh P_2\phi}\right)} \tag{5}$$

§ Note that for simplicity of presentation we have deliberately left off the indices i and j from the P_2 and P_3 parameters because one of our principal experimental findings will be that P_2 and P_3 may be considered the same for all ion species. This, of course, simplifies things considerably, but the detailed justification of this goes beyond our scope and will be found elsewhere.[25]

In the limit of low transmembrane potentials, Equation 5 simplifies to:

$$\left(\frac{P_j}{P_i}\right)_{V\to 0} = \left(\frac{P_j}{P_i}\right)_{eq} \frac{1 + \dfrac{2w_i}{P_2}}{1 + \dfrac{2w_j}{P_2}} \tag{6}$$

which corresponds to Equation 4 and can be seen to modify the kinetic parameter in the "low-voltage permeability ratio" by the incorporation of the barrier width (P_2) into it. Note, of course, that the parameter (P_3) for the voltage-dependence of the reaction does not appear in this low-voltage limit. Thus, only one additional parameter, P_2 (the barrier shape) appears in the "low-voltage permeability ratio" in the extended model.

For the conductance-voltage behavior of membranes interposed between solutions sufficiently dilute for the kinetics to be independent of ion concentrations, the extended model yields the following theoretical expression for the ratio of conductance at a given voltage, G, to the limiting (ohmic) conductance observed at low voltage, G_o¶

$$\frac{G}{G_0} = \frac{2(P_2 + 2w_i) \sinh \dfrac{\phi}{2}}{\sinh P_2\phi + 2w_i\phi \cosh P_3\phi} \tag{7}$$

in place of the familiar expression for the simpler Läuger-Stark *et al.*[7, 18] model

$$\frac{G}{G_0} = \frac{2(1 + 2w_i) \sinh \dfrac{\phi}{2}}{\left(1 + 2w_i \cosh \dfrac{\phi}{2}\right)} \tag{8}$$

Note that for the "equilibrium domain," where $w_i \ll 1$, Equation 7 reduces to the simple limiting expression

$$\frac{G}{G_0} = \frac{2P_2 \sinh \dfrac{\phi}{2}}{\sinh P_2\phi} \tag{9}$$

which depends only upon the barrier width (P_2). Also note that for very strong kinetic effects ($w_i > P_2$), the behavior of Equation 7 depends only on P_3 and w_i, being quite insensitive to P_2. The existence of these two limiting relationships is quite helpful experimentally in producing initial guesses as to the values of P_2 and P_3 since one can use ions expected to be either in the equilibrium domain (e.g., Na^+) or strongly in the kinetic domain (e.g., K^+), respectively, to evaluate these separately.

Selectivity In Zero-Current Membrane Conductance

We note, in passing, that selectivity can also be defined in terms of the membrane conductances measured at zero-current in solutions containing a

¶ This is the limiting expression for low salt concentrations, to which the experimental data of this paper are confined; the full expression with the explicit dependence on c_i will be given elsewhere (see the appendix by Ciani of the article by Krasne and Eisenman[25]).

single permeant cation. In the "equilibrium domain" the ratio of conductances produced by two different ions at comparable concentrations is always expected (and observed) to be identical to the permeability ratio.[8, 13, 15] In the "kinetic domain" it is also expected (and observed) that conductance ratios, $G_L^0(J)/G_L^0(I)$, in the "linear" region of low ion concentrations where conductance is proportional to the ion concentrations, will be equal to the low-voltage permeability ratios of Equation 5, so that

$$\frac{G_L^0(J)}{G_L^0(I)} = \left(\frac{P_j}{P_i}\right)_{V\to 0} = \left(\frac{P_j}{P_i}\right)_{eq}\left[\frac{1 + 2w_i/P_2}{1 + 2w_j/P_2}\right] \quad \textbf{(10)}$$

Thus, the same kinetic contribution appears in the apparent conductance ratios as in the permeability ratios, as is discussed in more detail elsewhere.[25]

The Effects of Methylation on the Permeability Ratios and Conductance-Voltage Behavior of Typical Carriers

Nonactin Trinactin, and Tetranactin

In order to illustrate the manner in which kinetic and equilibrium factors influence ion selectivity, it is useful to describe in some detail the behavior of the nonactin-type molecule for which we have found the most severe kinetic limitations—namely, tetranactin, the most highly methylated of the nonactin homologs. We shall then compare the results of the analysis for tetranactin with those of the same type of analysis for the less methylated species, trinactin and nonactin, which have one and four less methyl groups, respectively, than tetranactin. A detailed report on this analysis as well as on the general membrane permeability effects of tetranactin will appear elsewhere[25]; and the reader is referred to that report for the details of the experimental methods, data analysis, and the sensitivity of the results to the set of assumptions made.

The experiments (i.e., zero-current potential and conductance-voltage measurements in two types of lipid bilayers, PE decane and GDO decane) reported here were the simplest ones we could design that would enable us to characterize accurately the equilibrium and kinetic contributions to the selectivity as defined by the permeability ratio measured by membrane potential.

Tetranactin

The Membrane Potential at Zero Current. Figure 5 illustrates the zero-current membrane potential, V_0, observed in mixed ion solutions for PE decane bilayers (*top*) and GDO decane bilayers (*bottom*), respectively. The experiments were performed in our usual way in two ion mixtures[15] by adding the nitrate of one cation (NH_4NO_3) to one side of the membrane which was formed in the chloride of each of the cations at the concentrations indicated on the figure. || The experimental data are plotted as points (their reproducibil-

|| Not only do we find this to be a more convenient way of measuring permeability ratios than the often used[4, 5, 36] "bi-ionic" situation, but also it has the advantage of permitting a characterization of the conductance-voltage behavior in the single salt before adding the second cation. The procedure, furthermore, yields more information then the bi-ionic potential since it also gives values of the slope, which enables one at a glance to diagnose whether kinetic effects are present or absent.

TABLE 1

PERMEABILITY RATIOS FOR TETRANACTIN IN PE BILAYERS

	Li	Na	K	Rb	Cs	Tl	NH_4
$(P_i/P_{NH_4})_{eq}$	5.6×10^{-6}	4.7×10^{-3}	0.48	0.09	2.2×10^{-3}	0.88	1.0

ity is better than the size of the symbols); and the curves represent the theoretical expectations of Equations 1 and 2 for the "equilibrium domain" in PE and of Equations 1 and 5 for the "kinetic domain" in GDO.

For PE, the data at the top can be seen to agree well with the solid curves drawn according to the expectations of Equation 1 with constant permeability ratios (whose values are listed in TABLE 1). This strongly suggests that the permeability ratios so measured reflect solely equilibrium components and have no important kinetic contribution. As such, these permeability ratios should be independent of the lipid; and we can compare these data with those seen in asolectin bilayers (FIGURE 1) for the less methylated homologs. The observed behavior for tetranactin in PE is seen to be nicely consistent with the trends illustrated for the lower homologs for asolectin in FIGURE 1. In particular, the trend of decreasing Cs^+-Na^+ selectivity with increasing methylation for nonactin through trinactin culminates in tetranactin with an actual reversal of Na^+-Cs^+ selectivity. (The data of FIGURE 5 [*top*] indicate that Na^+ is more permeable than Cs^+.) This effect of methylation on equilibrium selectivity will be verified and discussed further when we come to FIGURE 10, where equilibrium-permeability ratios for nonactin, trinactin, and tetranactin will be compared for PE membranes as well as for GDO membranes.

In contrast to this simplicity, the situation for GDO bilayers appears to be much more complex, as can be seen in the lower portion of the figure. First, the observed membrane potentials (points) do not fall on the dashed curves expected if P_j/P_i were independent of voltage. Rather, they show substantial departures from the limiting Nernst slope expectations of these dashed curves. This is presumptive evidence for a kinetic effect. That this is indeed so is verified by the agreement with the solid curves drawn for "apparent permeability ratios" having the voltage-dependence given by Equation 5, for the values of the kinetic parameters w_i given in the second row of TABLE 2. Second, the apparent low-voltage permeability ratios $(P_j/P_i)_{V\to 0}$ in GDO bilayers (defined by the fit of the data to the dashed curves), which are listed in the first row of TABLE 2, can be seen to differ from those for PE (as can also be seen

TABLE 2

PERMEABILITY RATIOS FOR TETRANACTIN IN GDO BILAYERS

	Li	Na	K	Rb	Cs	Tl	NH_4
$(P_i/P_{NH_4})_{V\to 0}$	2.9×10^{-5}	3.1×10^{-2}	0.05	0.52	2.1×10^{-2}	2.9	1.0
w_i	0	0.012	0.54	0.08	0	0.24	1.65
$(P_i/P_{NH_4})_{eq}$	2.7×10^{-6}	3.1×10^{-3}	0.37	0.07	2.0×10^{-3}	0.66	1.0

NOTE: $P_2 = 0.342$; $P_3 = 0.45$.

in FIGURE 5 from the different spacings of the curves for the various ions in PE vs. GDO). This is most easily seen by comparing the low-voltage permeability ratios for GDO in row 1 of TABLE 2 with those for PE in TABLE 1 (which are independent of voltage). Further discussion of these differences will be delayed until we come to FIGURE 11, where they will be related to the data for trinactin and nonactin as well, but it should be noted that such a lipid-dependence of apparent permeability ratio is to be expected if the GDO data reflect additional kinetic components not seen in PE. That this is indeed

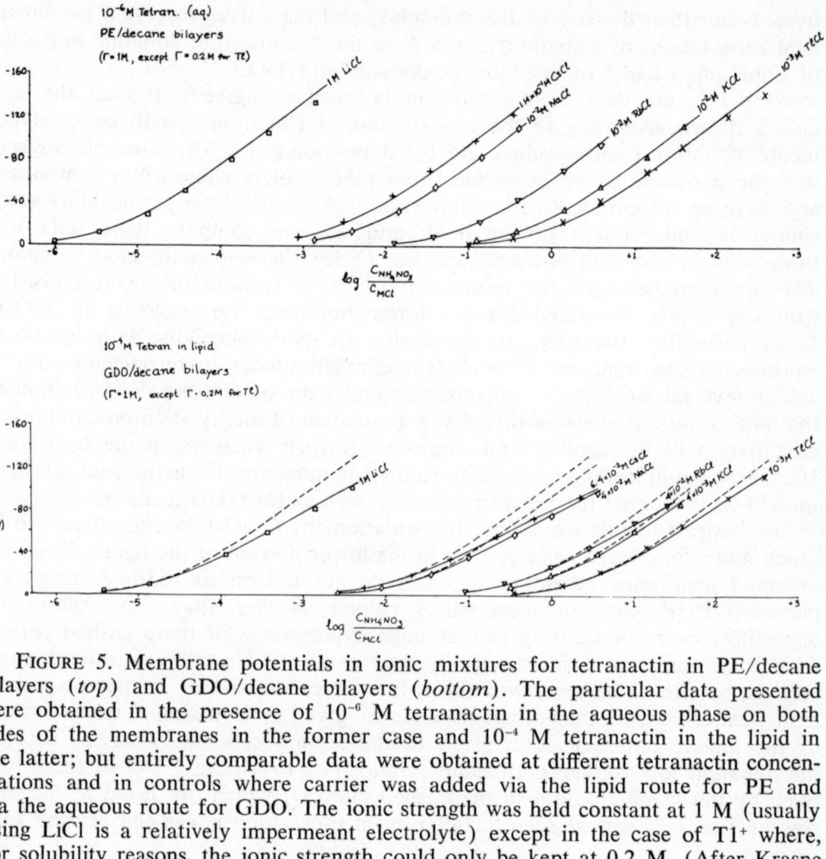

FIGURE 5. Membrane potentials in ionic mixtures for tetranactin in PE/decane bilayers (*top*) and GDO/decane bilayers (*bottom*). The particular data presented were obtained in the presence of 10^{-6} M tetranactin in the aqueous phase on both sides of the membranes in the former case and 10^{-4} M tetranactin in the lipid in the latter; but entirely comparable data were obtained at different tetranactin concentrations and in controls where carrier was added via the lipid route for PE and via the aqueous route for GDO. The ionic strength was held constant at 1 M (usually using LiCl is a relatively impermeant electrolyte) except in the case of Tl^+ where, for solubility reasons, the ionic strength could only be kept at 0.2 M. (After Krasne *et al.*[25])

the case can be seen by the close similarity between the directly observed $(P_j/P_i)_{eq}$ values for PE in TABLE 1 and the calculated (using Equation 6) values of $(P_i/P_j)_{eq}$ for GDO listed in the third row of TABLE 2. This result verifies the expectation that the equilibrium component of selectivity should depend solely on the carrier molecule and be independent of the lipid.

The Conductance-Voltage Behavior. From measurements of the current-voltage characteristic we have characterized the membrane conductance as a function of voltage.[25] Following Läuger and Stark[18] this characterization is conveniently presented by plotting G/G_0 (the ratio of conductance at a given voltage to the limiting [ohmic] conductance measured at low voltage) as a function of voltage. Such plots for tetranactin and the indicated cations are presented in FIGURE 6 for PE and GDO bilayers on the left and right, respectively. All measurements were made at sufficiently low ionic concentration so that the conductance-voltage behavior was independent of ion concentration. In this way, we avoid studying here the concentration-dependent kinetic contribution and restrict ourselves only to that part of the kinetic contribution that is independent of ion concentration.[9, 18, 22] The points are the experimentally observed values; while the curves have been drawn according to the expectations of the theoretical Equations 7 and 9.

It is immediately apparent that the tetranactin-mediated conductance-voltage behavior differs strikingly between the two different lipid compositions. In PE membranes the conductance is an increasing function of voltage for all ions, differing very little from ion to ion. Indeed, it is possible to encompass all of the data by the "equilibrium domain" conductance behavior (i.e., for $w_i = 0$) for all ions given in Equation 9 with values of the trapezoidal barrier half-width, P_2, ranging between 0.36 and 0.40 for all ions.** This, of course, is essentially what one would expect for the "equilibrium domain" for isosteric complexes. In contrast, in GDO membranes the conductance-voltage curve differs markedly from ion to ion, ranging from the "hyperbolic" ("equilibrium domain") limiting curve for Li and Cs (which, incidentally, is virtually the same as that seen in PE, P_2 being .34 for those ions in GDO) to the curves for Rb, Tl, K, and NH_4, which show increasing kinetic limitations corresponding to the increasing values of the kinetic parameter w_i given in TABLE 2.

From the close agreement between the observed data points and the theoretical curves, it is clear that the model satisfactorily describes the experimental data, which is not surprising in view of the number of parameters available. More importantly, the systematic trends seen in the corresponding data of FIGURES 7 and 8 for nonactin and trinactin (to be described in the next section) and the internal consistency with the V_0 data indicate that the model is doing much more for us than merely curve-fitting.

Therefore, before analyzing further the kinetic and equilibrium components of the permeability selectivity of tetranactin, we will first summarize the corresponding data for nonactin and trinactin, and then analyze the series nonactin, trinactin, tetranactin together. In this way certain important similarities between these molecules will become apparent (e.g., they are characterized by the same shape of trapezoidal barrier, P_2; and the voltage-dependent "reaction plane" parameter, P_3, for their complexation is the same). Also, the systematic changes in w_i which occur with increasing methylation will become apparent and will be able to be analyzed separately from the effects of methylation on the equilibrium selectivity, $(P_j/P_i)_{eq}$.

Nonactin and Trinactin

The conductance-voltage behaviors for nonactin and trinactin in GDO/decane bilayers are summarized in FIGURE 7, which also contains for

** The trend among the ions is actually best accounted for by a single trapezoidal barrier (P_2 = .36) and very small, but systematic kinetic effects, as described elsewhere.[25]

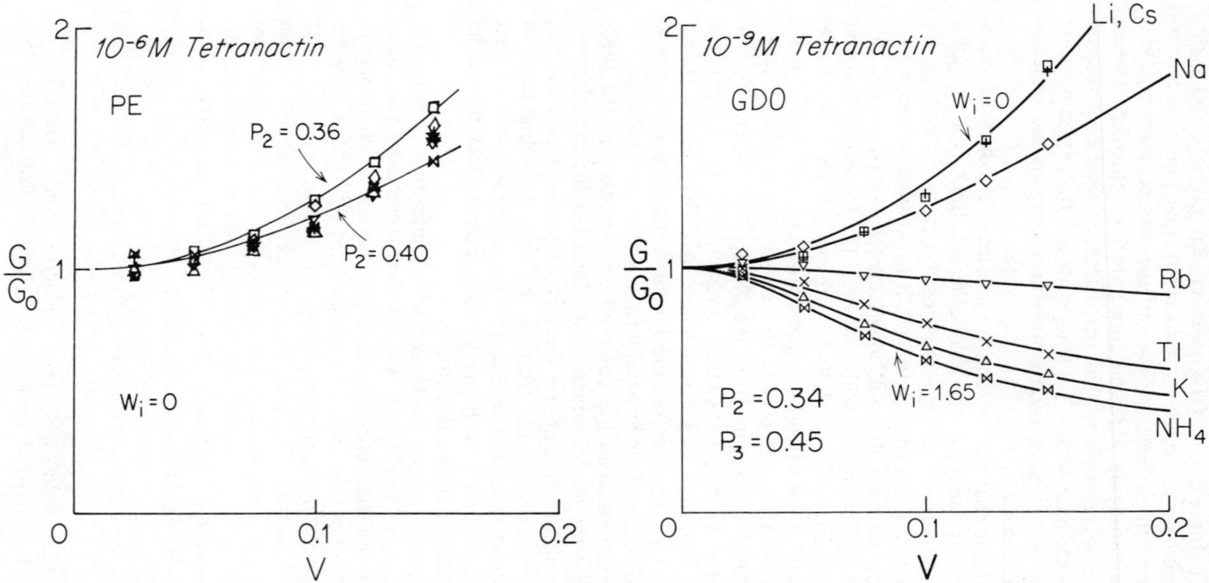

FIGURE 6. Conductance-voltage behavior of tetranactin in PE/decane bilayers (*left*) and GDO/decane bilayers (*right*). The current-voltage data were obtained with a four-electrode system using applied saw-tooth voltages over a wide range of frequencies for which the current-voltage curves were demonstrated to be independent of the sweep frequency (and for which concentration polarization was negligible). Sweep rates of 30–60 mV/sec were most commonly used. The experimental data are plotted as points for the indicated cations Li (+), Cs (□), Na (◇), Rb (▽), T1 (×), K (△), NH_4 (⋈). (The same symbols will be used throughout.) The curves were drawn according to Equation 7 for GDO and according to the "equilibrium domain" limit given by Equation 9 for PE. For the PE data note that the entire range of data can be fitted by a parameter P_2, whose extreme value ranges from 0.36 (*upper curve*) to 0.40 (*lower curve*). Actually, these data are best described by a single P_2 parameter ($P_2 = 0.36$) and very small values of w_i (as given elsewhere[23]), which are entirely negligible from the point of view of selectivity corrections. For GDO bilayers the data are best-fitted, assuming a single value for P_2 (0.34) and a single value for P_3 (0.45) and values of w_i that range from the equilibrium domain for Li and Cs ($w_i = 0$) through the various values of w_i given in the second row of TABLE 2 and corresponding to the kinetic domain for the remaining ions. (After Krasne *et al.*[25])

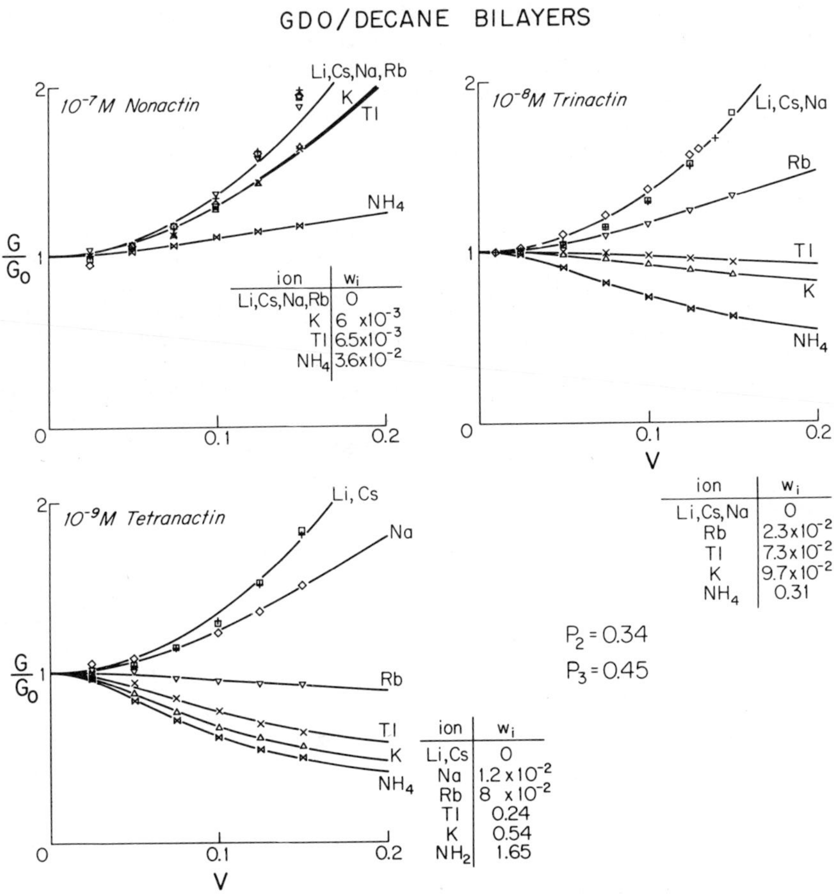

FIGURE 7. The conductance-voltage behavior of nonactin,[25] trinactin,[22] and tetranactin[25] in GDO/decane bilayers. The data here are obtained and plotted in the same manner as was done in FIGURE 6 (*right*) for the indicated carrier concentrations. Note that all curves have been fitted with the same value of the parameters P_2 (0.34) and P_3 (0.45) and with the values of w_i summarized in the insets and in TABLE 3 (where they are given, however, as values w_i/P_2). The symbols are the same as in FIGURE 6.

comparison the data for tetranactin. As in FIGURE 6, the points are experimentally measured, while the curves have been drawn according to the expectations of Equation 5 using *the same values* for barrier shape ($P_2 = 0.34$) and reaction plane ($P_3 = 0.45$) for all three carrier molecules, and with the indicated values of the kinetic parameter w_i, which varies from ion to ion and carrier to carrier in a way that we hope is becoming apparent. That the barrier shape (P_2) and the reaction plane (P_3) are independent of the degree of methylation for these molecules is physically reasonable since this degree of methylation should not change greatly the external dimensions or flexability of the carrier molecules (see References 37, 38 and 45). Clearly, methylation produces a pronounced effect on the kinetic parameter, w_i, for with increasing methylation more ionic species depart, and more severely, from the "equilibrium domain" curve (corresponding to $w_i = 0$), which is uppermost in each subfigure (values for w_i are given for tetranactin in TABLE 2, and of w_i/P_2 for all the nactins in TABLE 3. The values of log w_i will be plotted in FIGURE 9). Thus, for nonactin, four cations (Li, Cs, Na, and Rb) fall on the same hyperbolic "equilibrium domain" curve; and for this carrier the departures from this limiting behavior are not great, even for K^+ and Tl^+, and even the largest kinetic effect, seen for NH_4^+, corresponds to a quite small (from the point of view of selectivity) value of $w_i = .036$. As we progressively add methyl groups in going to tetranactin, however, the kinetic effects become increasingly pronounced, only Li^+ and Cs^+ remaining on the "equilibrium domain" curve for tetranactin, where even Na^+ shows a clear departure (corresponding to a small, but detectable value of $w_i = .012$); while for NH_4^+ a strongly "saturating" behavior is observed,

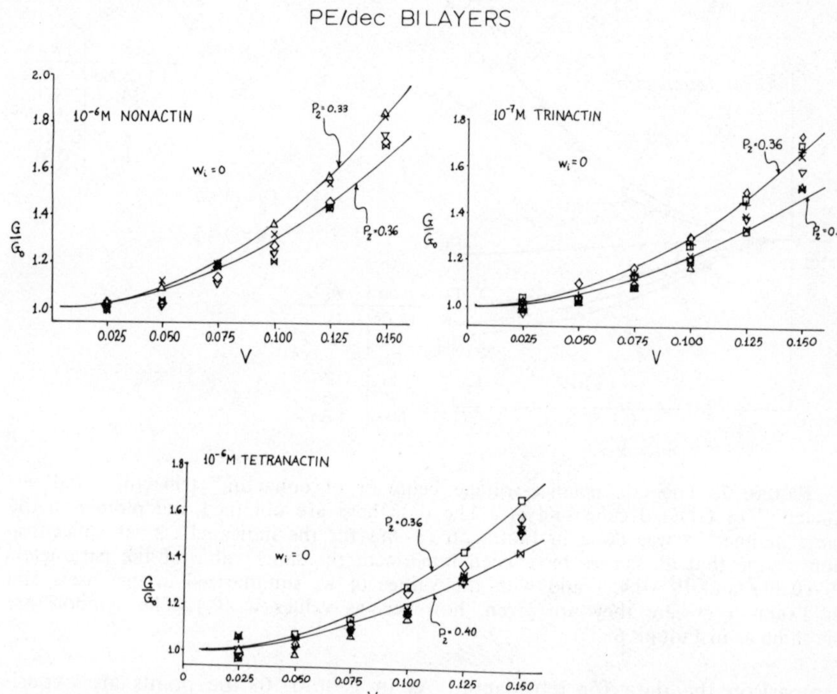

FIGURE 8. The conductance-voltage behavior of nonactin, trinactin, and tetranactin in PE/decane bilayers. These data were obtained and are plotted in the same manner as those of FIGURE 6 (*left*). In all cases the data points can be described by curves corresponding to the equilibrium domain of Equation 9 with the values of the parameter, P_2, ranging from the limits indicated on the figure. The symbols are the same as in FIGURE 6. However, we believe the truest representation corresponds to a value of $P_2 = 0.36$ with the departures from this in the cases of trinactin and tetranactin being due to very small kinetic effects increasing systematically for the most strongly complexed ions, as is discussed elsewhere.[25]

TABLE 3

VALUES OF KINETIC PARAMETER, w_i/P_2

	Non-	Trin-	Tetra-	Lac-Val	HyIv-Val
NH_4	0.106	0.912	4.814	(0)	(0.40)
Tl	0.019	0.215	0.709	(0)	(0.77)
K	0.018	0.282	1.584	(0)	(10.9)
Rb	0	0.066	0.234	(0.04)	(21.4)
Na	0	0	0.034	(0)	(0)
Cs	0	0	0	(0)	(2.3)
Li	0	0	0	(0)	–

NOTE: For Non, trin, Tetra: $P_2 = 0.342$, $P_3 = 0.45$; for Lac-Val, HyIv-Val: $P_2 = 0.365$, $P_3 = 0.46$.

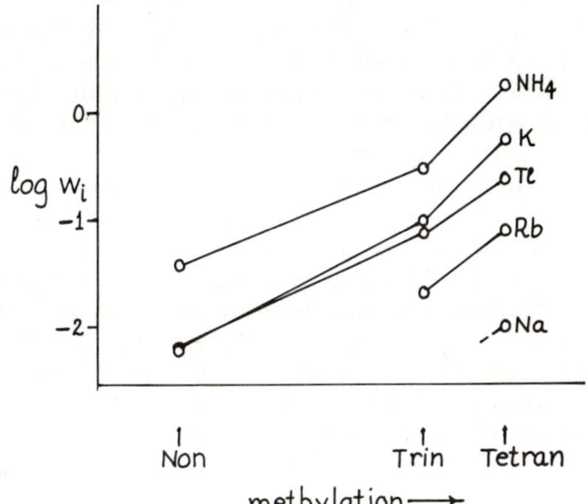

FIGURE 9. Effect of methylation on the magnitude of the parameter w_i.

w_i having attained a value of 1.65. That these trends are very systematic is already apparent, but can be seen even more clearly in FIGURE 9 and TABLE 3. Before looking at these more closely, we will first describe the conductance-voltage characteristics for PE bilayers.

These are summarized in FIGURE 8 and are beautifully dull, as required by theory if PE is in the equilibrium domain for all these carriers, as we are proposing. Thus, we find it is possible to encompass all of the observed conductance-voltage behavior in PE with a single "equilibrium domain" type "hyperbolic" curve corresponding to a barrier half-width, P_2, not significantly different from .34.

COMPARISON OF THE SELECTIVITIES OF NONACTIN, TRINACTIN, AND TETRANACTIN AND THEIR "KINETIC" AND "EQUILIBRIUM" COMPONENTS

The Kinetic Component

As far as kinetics are concerned, TABLE 3 summarizes the measured values of the kinetic parameter w_i/P_2, and FIGURE 9 plots log w_i as a function of the increasing methylation of the molecules nonactin, trinactin, and tetranactin. Clearly, w_i increases systematically with the degree of methylation. Since $w_i = k_{ms}/k_D$, its value reflects the ratio of the rate constant of translocation of the complex (k_{ms}) to that for unloading the ion at the member-solution interface (k_D). Intuitively, one expects k_{ms} to be nearly constant (or if anything to decrease) with increasing methylation; therefore, we can immediately infer that k_D is inversely proportional to w_i. The observed trends for w_i seen in FIGURE 9 may also therefore be taken as approximately illustrating the inverse trends in k_D; and we may conclude that as far as kinetics are concerned, the principal effect of methylation is to decrease the rate of unloading of the carrier.

The Equilibrium Component

With regard to equilibrium, FIGURE 10 characterizes the effects of methylation on the equilibrium selectivity $(P_j/P_i)_{eq}$ for both PE bilayers (*left*) and GDO bilayers (*right*). Comparing these figures it should be apparent that the equilibrium component of selectivity is independent of the lipid, as demanded by the model, since the pattern of equilibrium permeability ratios can be seen to be virtually identical for these two lipids, despite the large differences in the kinetic components. Furthermore, the changes in equilibrium selectivity pattern produced by increasing methylation, previously inferred for asolectin and PE bilayers, are verified by the values of $(P_j/P_i)_{eq}$ in GDO bilayers. Most noteworthy of these is the crossover of the Na^+ and Cs^+ selectivity curves, mentioned at the outset of this paper. This verifies that increasing methylation leads to changes in selectivity sequence from Eisenman's IV (K $>$ Rb $>$ Cs $>$ Na $>$ Li) to Eisenman's V (K $>$ Rb $>$ Na $>$ Cs $>$ Li), which is expected from simple "field strength" theory[39, 40] if increasing methylation increases the "field strength" of the ligands, which is certainly in the proper direction for either the expected Kirkwood-Westheimer field effect or for the Taft induction effect for a CH_3 group, although perhaps surprisingly large.

The trends in equilibrium and kinetic selectivity are, furthermore, concordant with each other since, with the sole exception of Tl^+ (for reasons to be discussed elsewhere,[25] where the selectivity of the "transition state" will be analyzed), the cations whose rate constant of unloading are the lowest are also the ones most preferred at equilibrium. Such behavior is analogous to the situation described by Diebler *et al.*[41] and Grell *et al.*[42] for the complexation in simple solvents. Indeed, the fact that a significantly detectable value of w_i is seen for Na^+ in tetranactin whereas for Cs^+ w_i is zero is also consistent with the observed preference of Na^+ over Cs^+ for the equilibrium selectivity.

The Apparent Selectivity

Concerning the "apparent selectivity," which reflects both equilibrium and kinetic components, the $(P_i/P_j)_{V\to 0}$ data of the central portion of FIGURE 10 for GDO bilayers indicate the apparent complexity (in particular the NH_4^+ crossovers of Tl^+ and K^+) that results from a combination of kinetic and equilibrium effects, which, taken individually, are systematic and simple. It is worth reiterating that the complexities of the "apparent" selectivity behavior

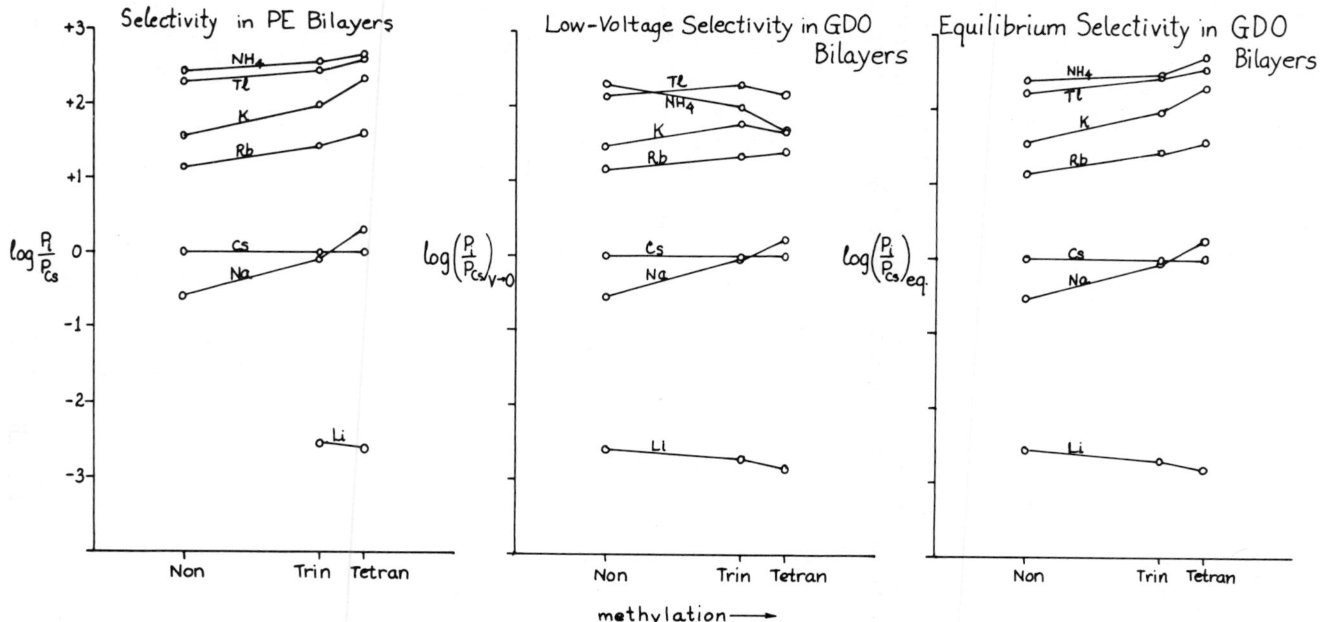

FIGURE 10. The effect of methylation on the "apparent" and "equilibrium" selectivity for the nactins in PE bilayers (*left*) and GDO bilayers (*middle and right*). The abscissa in each figure ranks the indicated nactins in their direction of increasing methylation. In the left figure, the ordinate plots the logarithm of the indicated permeability ratios (P_j/P_i) observed in PE bilayers, which under the present assumptions corresponds to the equilibrium ratio $(P_j/P_i)_{eq}$; the middle figures plot the low-voltage permeability ratio $(P_j/P_i)_{V\to 0}$ observed in GDO bilayers; and the right figure plots the equilibrium selectivity calculated for GDO bilayers using Equation 5 and the values of w_i given in TABLE 3 as well as the values $P_2 = 0.34$ and $P_3 = 0.45$ (as determined from the conductance-voltage data).

seen in the "raw" permeability ratios for GDO bilayers (*center*), as well as the apparent differences between these and the selectivities seen in PE bilayers (*left*), are completely reconciled by extracting the equilibrium component of the permeability ratios (*right*) using the kinetic parameters independently deduced from the conductance-voltage experiments. This verifies our initial supposition that the equilibrium component of selectivity should be independent of the type of lipid. (This can also be seen from the satisfactorily close agreement between the values of $(P_j/P_i)_{eq}$ for GDO vs. PE bilayers [cf. row 3 of TABLE 2 and TABLE 1, respectively]). Since the differences in "apparent selectivity" observed in GDO bilayers vs. PE bilayers have been shown here to result from measurable kinetic factors, it seems likely that similar kinetic effects underlie certain discrepant sequences that have been reported for the antibiotic action of the macrotetralide actins.[43]

VALINOMYCIN AND ITS LESS HIGHLY METHYLATED (LAC-VAL) AND MORE HIGHLY METHYLATED (HYV-VAL) HOMOLOG

A very similar situation for the effects of methylation on kinetic and equilibrium components of carrier-mediated ionic permeability is encountered in a series of homologs of valinomycin having different degrees of methylation. Although our characterization of these molecules has not yet been completed in the detail or with the precision attained for the nactins, the provisional results obtained so far demonstrate that even more dramatic effects of methylation are seen in the valinomycin-type carriers. This is illustrated in FIGURES 11, 12, and 13.

On the left side of FIGURE 11 are plotted the "apparent permeabilities" in the limit of low voltage $(P_j/P_i)_{V\to 0}$ observed in GDO bilayers for lac-val (cyclo[D-Val-L-Lac-LVal-D-Lac]$_3$) and hylv-val (cyclo[D-Val-L-HyIv-LVal-D-HyIv]$_3$) and in GMO bilayers†† for valinomycin (cyclo[D-Val-L-Lac-LVal-D-HyIv]$_3$). The less methylated homolog, lac-val, and the more methylated hyIv-val differ from valinomycin in having six fewer and six more methyl groups, respectively; and the permeability ratios in FIGURE 11 are accordingly plotted as a function of the increasing methylation of the carrier molecule. A substantial change in "apparent" selectivity sequence can be seen to occur as a function of increasing the methylation of the carrier since the sequence of low-voltage permeability ratios, $(P_j/P_i)_{V\to 0}$, shifts from the Eisenman sequence III (Rb > K > Cs > Na > Li) characteristic of lac-val to the "lower field strength" Eisenman sequence I (Cs > Rb > K > Na > Li) in hyIv-val (large changes are also seen in the permeabilities of Tl^+ and NH_4^+ relative to K^+).

Although both I and III are "permitted" sequences in the simplest "field strength" model,[39, 40] the direction of the change with increasing methylation disturbed us since it implied that for the valinomycin carriers the effective "field strength" of the site was decreasing with increasing methylation. This is opposite in direction to that observed for methylation of the nactins and also opposite to what would be expected for Kirkwood-Westheimer or Taft

†† Control experiments indicate that the values of $(P_j/P_i)_{V\to 0}$ from valinomycin in GDO bilayers are sufficiently close to those in GMO that (for completeness and because no trends are obscured thereby) we include the GMO data for Val in this figure.

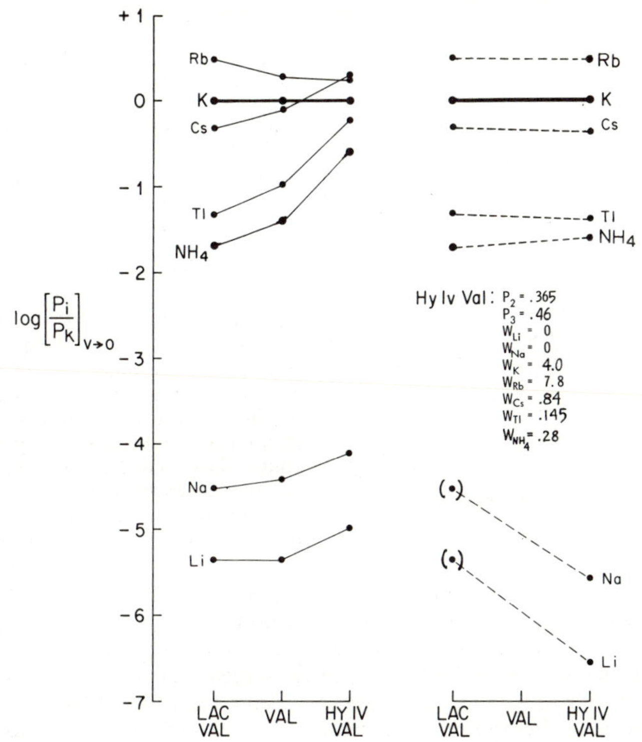

FIGURE 11. (*Left*) "apparent" permeability ratios $(P_j/P_i)_{V\rightarrow 0}$ are plotted for valinomycin and its two homologs, lac-val and HyIv-Val, as a function of increasing degree of methylation. The permeability ratios were measured in GDO/decane bilayers in the case of lac-val and HyIv-Val and in GMO (glyceryl monooleate)/decane bilayers in the case of valinomycin. The concentrations of the carrier were: lac-val 10^{-6} M, val 10^{-7} M, and HyIv-Val 3.6×10^{-8} M except for the measurements of Na and Li for which the HyIv-Val concentration was 3.6×10^{-7} M. (*Right*) calculated "equilibrium" permeability ratios $(P_j/P_i)_{eq}$ are plotted for lac-val and HyIv-Val. The kinetic contributions are totally negligible for lac-val in this lipid so that the "equilibrium" permeability ratios for lac-val are the same as the "apparent" permeability ratios. For HyIv-Val the kinetic parameters tabulated in the figure were estimated from the conductance-voltage data of FIGURE 12 and TABLE 3. Note that the apparent decrease of permeability ratios for Na^+ and Li^+ relative to K^+ seen on the right side probably corresponds to the fact that the lac-val measurements reflect trace contaminants of K^+ in the Na and Li salts, and the correction for kinetics in HyIv-Val allows one to project the permeability ratio below the K^+-contaminant level. For this reason the data points for Na^+ and Li^+ in lac-val have been parenthesized, and it is of some interest that the kinetic limitations on K^+ permeation actually enable one to see further into the "mud" level for HyIv-Val with Li^+ and Na^+.

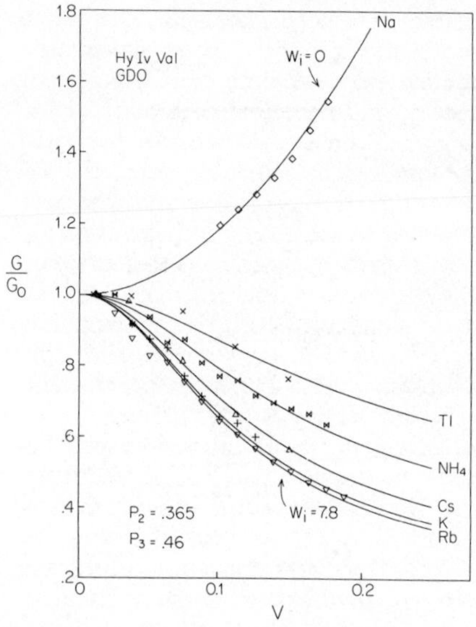

FIGURE 12. Conductance-voltage behavior for HyIv-valinomycin in GDO bilayers. The experimentally observed data are plotted as points; while the curves have been drawn according to Equation 7 with the values of the parameters, w_i, indicated in FIGURE 11 and in TABLE 3. Note that we have excluded the lowest voltage data points in the case of Na^+ in fitting the curve, owing to a 10% contribution to the resting conductance produced by K^+ contamination in the solution. The effect of the K^+ contaminant becomes increasingly negligible with increasing voltage since its conductance curve is strongly saturating so that the effect of a K^+ contaminant becomes less and less important as the voltages increase. This is the justification for our reliance on the higher voltage points in fitting the data to the curve.

effects. It could be accounted for by the large kinetic effects of methylation seen with the valinomycin homologs, as we shall now show.

The extended model satisfactorily describes the data in the valinomycins and yields the values of kinetic parameters measured from the conductance-voltage behaviors presented in FIGURES 12 and 13. From these, we have been able to calculate the approximate values of the equilibrium permeability ratios $(P_j/P_i)_{eq}$, which are plotted at the right side of FIGURE 11. Although we do not yet know these values as precisely‡‡ as for the nactins, nevertheless

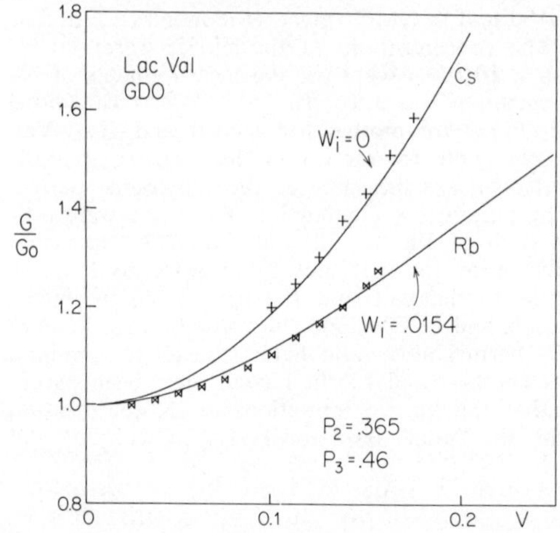

FIGURE 13. Conductance-voltage behavior of Lac-Val in glycerol dioleate bilayers.

‡‡ These values are approximate because in the case of extreme kinetic limitations, such as those observed for hylv-val, the inferred values of w_i are extremely sensitive to the precise value used for P_3.

it is quite clear that the "anomalous" behavior has been removed. The provisional magnitudes (as indicated by parentheses) of the kinetic parameters for lac-val and hylv-val are given in the last two columns of TABLE 3 for comparison with those for nonactin, trinactin, and tetranactin. Again, as has been noted for the nactins, and presumably for the same reasons, the most preferred ions from the point of view of equilibrium selectivity are also the ones exhibiting the greatest kinetic limitations,§§ thus implying that the rate of dissociation of the complex is inversely related to the strength of binding.

That this interpretation is indeed correct is verified, first, by the observation that the membrane potential (V_0) behavior in two ion mixtures for Lac-Val in GDO bilayers shows no sign of kinetic contributions (the permeability ratios being constant, as in FIGURE 1 or FIGURE 5 [*top*]). In contrast, the membrane potential behavior for Hylv-Val shows voltage-dependent permeability ratios, as in FIGURE 5 (*bottom*), implying strong kinetic effects.

These observations are strikingly verified by the conductance-voltage behavior illustrated in FIGURES 12 and 13 for Hylv-Val and Lac-Val, respectively, in GDO bilayers. The interpretation of these figures follows the same reasoning used for FIGURES 7, 8, and 9 for the nactins, and we may therefore be considerably more succinct here. FIGURE 13 confirms that, as far as membrane potentials are concerned, Lac-Val is effectively in the "equilibrium domain" in GDO bilayers for all ions, since even for the most kinetically limited ion, Rb^+, the kinetic parameter w_i alters the "apparent" from the "true" permeability ratio by less than 5%. For Lac-Val we may therefore take the measured permeability ratios to be identical to the equilibrium ratios $(P_j/P_i)_{eq}$ and we have done so in plotting the right side of FIGURE 11.

FIGURE 12, in contrast, indicates that in GDO bilayers Hylv-Val exhibits striking kinetic effects, w_i ranging from values as large as 7.8 for Rb^+ to as small as 0 for Na^+. Using the values of these kinetic parameters (TABLE 3), we have calculated (using Equation 6) the equilibrium component of the permeability ratio for Hylv-Val, and plotted these "corrected" values on the right of FIGURE 11, thereby removing the "anomalous" behavior.

Additionally, the data of FIGURES 12 and 13 indicate that the parameters P_2 and P_3 are independent of the cation and are the same for Lac-Val and Hylv-Val. Interestingly, they are also quite comparable to those characteristic of the nactins. Thus, for these valinomycins the provisional values for the half-width of the trapezoidal barrier ($P_2 = .365$) is quite close to that ($P_2 = .34$) seen for the macrotetralide actins, whereas the value for the "reaction plane" parameter for the valinomycin homologs ($P_3 = 0.46$) is also quite close to that ($P_3 = 0.45$) found for the nactins.

SUMMARY AND CONCLUSIONS

1. "Equilibrium" and "kinetic" parameters in the permeability ratios measured from membrane potentials are identified and defined. For carrier-mediated permeation, the equilibrium component of the permeability ratio,

§§ For the valinomycins, the anomaly, noted for Tl^+ with the nactins, is absent. Doubtless, this is related to the presence of "supra-Ia" Tl^+ selectivity[44] for the nactins and its absence in the valinomycin, as will be discussed elsewhere.[25]

$(P_j/P_i)_{eq}$, is expected to be intrinsic to the carrier molecule and independent of the lipid if isostericity can be assumed. The main factor in the kinetic component for carrier-mediated permeation, and the only one in the Eyring single-barrier model, is w_i, the ratio of the rate constant for translocation of the complex to that for unloading the ion at the membrane-solution interface.

2. A more general expression than heretofore available for the membrane potential has been deduced by extending the Eyring and Nernst-Planck treatments for carrier-mediated permeation of bilayers to include a trapezoidal barrier shape and a voltage-dependence of the complexation reaction, suggested by the work of Hall *et al.* and of Hladky. The voltage-dependence is rationalized by the convention of a reaction plane located somewhat internally to the membrane-solution interface. The results of a derivation by Ciani of the membrane-potential behavior expected for this model, given elsewhere, are presented here. In this "extended" formulation, the membrane potential is still conveniently described by an expression formally identical to the Goldman-Hodgkin-Katz equation. However, the permeability ratios, besides being functions of the voltage and of the same "kinetic" and "equilibrium" parameters as in the Eyring single-barrier model, are now also dependent upon the barrier width (P_2) and locus of the reaction plane (P_3). In the useful limit of small transmembrane potentials, the expression for the permeability ratio further simplifies to depend only on the ratio w_i/P_2.

3. Since the conductance-voltage characteristic in the case of sufficiently low permeant ion concentrations, previously inferred by Hladky for this extended model, depends upon the same parameters (w_i, P_2 and P_3), its experimental characterization for single salts of Li, Na, K, Rb, Cs, NH_4, and Tl and for nonactin, trinactin, tetranactin, valinomycin, lactate-valinomycin, and hydroxyisovalerate-valinomycin has been carried out in PE and GDO membranes to check the consistency of the model as well as to help resolve the "apparent" permeability ratio into its "equilibrium" and "kinetic" components by providing more accurate estimates of P_2 and P_3 than can be made from membrane potential data alone. For the nactin carriers (non-, trin- and tetra-) and for all of the ions (Li, Na, K, Rb, Cs, NH_4, Tl) carried across both GDO and PE bilayers, the same barrier shape is observed. This corresponds to a trapezoidal barrier across which 68% of the applied potential drop occurs (and which may be identified therefore loosely as being of "half-width," $P_2 = 0.34$). More limited data on lac-val and hylv-val indicate that a similar barrier shape ($P_2 = 0.36$) is also observed for these carriers.

4. The conductance-voltage behavior observed in GDO bilayers also indicates for all the nactin carriers and ions studied that a single value of the parameter ($P_3 = 0.45$) can be used to account for the voltage-dependence of the rate constants of the reactions. This can be rationalized by the assumption that the same fraction (about 10%) of the voltage drops between the interfaces and the plane of the reaction. More limited data on lac-val and hylv-val indicate a very similar locus for the "reaction plane" ($P_3 = 0.46$).

5. For all of these molecules the observed behavior in PE membranes is consistent with their being essentially in the "equilibrium domain." In all cases the observed conductance-voltage characteristics were "hyperbolic" in this lipid, and the membrane-potential behaviors in two-ion mixtures were characterized by constant (voltage-independent) permeability ratios. As a consequence no kinetic parameters could be measured for this lipid, but the "barrier shape" parameter ($P_2 = 0.36$) was found to be quite similar to that for GDO bilayers.

6. The findings (paragraphs 4 and 5) that single values for the reaction plane (P_3) and barrier shape (P_2) parameters were characteristic for all of the carrier molecules and carried ions are not only pleasingly concordant with physical intuition, but also confer simplicity to the interpretation of the data. Thus, all the kinetic differences from ion to ion and from carrier to carrier can be attributed to differences in the single parameter, w_i, which is the same as in the more primitive Eyring single-barrier model and is simply the ratio of rate constants for translocation to unloading of the complex.

7. The trends with increasing methylation in both the nactin and valinomycin series of carrier molecules are characterized experimentally and interpreted in the light of the extended model. Two effects of methylation are apparent: one on kinetic factors and the other on equilibrium factors. The overall behavior of the system, which can sometimes appear to be quite complex, is resolvable in terms of systematic and relatively simple effects of methylation on each of these terms separately.

8. The following detailed effects of methylation are deduced. First, with regard to kinetics, there is a systematic increase in the parameter, w_i, with increasing methylation in both series of carriers. The trends are consistent with the notion that the effect of increasing methylation is principally to decrease the rate of dissociation of the complex. The effects of these kinetic changes are reflected in the differences in "apparent permeability" ratios seen between PE and GDO bilayers, for which typical "equilibrium domain" and "kinetic domain" behavior, respectively, are observed in their membrane-potential and conductance-voltage behavior.

Second, with regard to equilibrium, there are seen in the nactins systematic changes in equilibrium selectivity with increasing methylation, which result in a change from Eisenman's selectivity sequence IV ($K > Rb > Cs > Na > Li$ to the "higher field strength" sequence V ($K > Rb > Na > Cs > Li$) between trinactin and tetranactin. This is as predicted by the "field strength" model, assuming the methyl group increases the dipole moment of the ligand oxygens either by a Kirkwood-Westheimer field effect or by a Taft induction effect, and is not easily accounted for on steric considerations. For the valinomycin series, the present analysis of the data is not yet at a level of sufficient accuracy to determine whether any trends in equilibrium selectivity exist, but an apparent discrepancy in the direction of the effects of methylation has been reconciled as due to kinetic effects.

Third, for both the valinomycin and nactin series of carriers (with the sole exception of Tl^+ in the nactins) the most preferred ions—equilibrium-wise—are also the ones most rate-limited kinetically, as would be expected if the rate constant for unloading the complex were inversely related to the strength of complexation with the carrier.

Acknowledgments

We thank Drs. W. Simon and K. Ando for their gift of tetranactin and Dr. Hans Bickel and Miss Barbara Stearns for the gifts of nonactin and trinactin. We are also indebted to Drs. V. I. Ivanov and Yu. Ovchinnikov for their gift of the valinomycin analogs, hydroxyisovalerate-valinomycin, and lactate-valinomycin; and we thank Dr. A. A. Lev for reading the manuscript and for his perceptive comments. Lastly, analysis of the various parameters of the model was accomplished using the facilities of the U.C.L.A. Campus Computing Network, to which we communicated via APL*PLUS interactive computing systems with considerable help from Mr. Chris Clausen.

REFERENCES

1. BANGHAM, A. D., M. M. STANDISH, J. C. WATKINS & G. WEISSMANN. 1967. Proc. Symp. Biophys. Physiol. Biol. Transport. Frascati, 1965; Protoplasma **63:** 183–187.
2. CHAPPELL, J. B. & A. R. CROFTS. 1966. *In* Regul. Metab. Processes Mitochondria, Proc. Symp., Bari, 1965. : 293–315.
3. STEFANAC, Z. & W. SIMON. 1966. Chimia **20:** 436.
4. MUELLER, P. & D. O. RUDIN. 1967. Biochem. Biophys. Res. Commun. **26:** 398–404.
5. LEV, A. A. & E. P. BUZHINSKY. 1967. Tsitologiya **9:** 102–106.
6. LARDY, H. A., S. N. GRAVEN & S. ESTRADA-O. 1967. Fed. Proc. **26:** 1355–1360.
7. LÄUGER, P. 1972. Carrier-mediated ion transport. Science **178:** 24–30.
8. EISENMAN, G., G. SZABO, S. CIANI, S. G. A. MCLAUGHLIN & S. KRASNE. 1973. Ion binding and ion transport produced by lipid-soluble molecules. Prog. Surf. Membr. Sci. **6:** 139–241.
9. CIANI S. M., G. EISENMAN, R. LAPRADE & G. SZABO. 1973. Theoretical analysis of carrier-mediated electrical properties of bilayer membranes. *In* Membranes—A Series of Advances. G. Eisenman, Ed. Vol. **2:** 61–177. Marcel Dekker. New York, N.Y.
10. HAYDON D. A. & S. B. HLADKY. 1972. Ion transport across thin lipid membranes: A critical discussion of mechanisms in selective systems. Quart. Rev. Biophys. **5:** 187–282.
11. PRESSMAN, B. C., E. J. HARRIS, W. S. JAGGER, & J. H. JOHNSON. 1967. Proc. Nat. Acad. Sci. U.S.A. **58:** 1949–1956.
12. EISENMAN, G., S. M. CIANI & G. SZABO. 1968. Fed. Proc. Fed. Amer. Soc. Exp. Biol. **27**(6): 1289–1304.
13. CIANI, S. M., G. EISENMAN, & G. SZABO. 1969. J. Membr. Biol. **1:** 1–36.
14. EISENMAN, G., S. M. CIANI & G. SZABO. 1969. J. Membr. Biol. **1:** 294–345.
15. SZABO, G., G. EISENMAN & S. CIANI. 1969. J. Membr. Biol. **3:** 346–382.
16. MCLAUGHLIN, S. G. A., G. SZABO, S. M. CIANI & G. EISENMAN. 1972. J. Membr. Biol. **9:** 3–36.
17. IVANOV, V. T. Ann. N.Y. Acad. Sci. This monograph.
18. LÄUGER, P. & G. STARK. 1970. Biochim. Biophys. Acta **211:** 458–466.
19. STARK, G., B. KETTERER, R. BENZ & P. LÄUGER. 1971. Biophys. J. **11:** 981–994.
20. STARK, G. & R. BENZ. 1971. J. Membr. Biol. **5:** 133–153.
21. BENZ, R., G. STARK, K. JANKO & P. LÄUGER. 1973. Valinomycin-mediated ion transport through neutral lipid membranes: Influence of hydrocarbon chain length and temperature. J. Membr. Biol. **14:** 339–364.
22. LAPRADE, R., S. M. CIANI, G. EISENMAN & G. SZABO. 1975. The kinetics of carrier-mediated ion permeation in lipid bilayers and its theoretical interpretation. *In* Membranes—A Series of Advances. G. Eisenman, Ed. Vol. **3:** 127–214. Marcel Dekker, New York, N.Y.
23. HALL, J. E., C. A. MEAD & G. SZABO. 1973. A barrier model for current flow in lipid bilayer membranes. J. Membr. Biol. **11:** 75–97.
24. HLADKY, S. B. 1974. The energy barriers to ion transport by nonactin across thin lipid membranes. Biochim. Biophys. Acta **352:** 71–85.
25. KRASNE, S. & G. EISENMAN. 1975. The influence on ion selectivity of structural variations of the ionophore and of the lipid : I. Tetranactin and the methyla-

tion of nonactin-type carriers (with an appendix by S. CIANI). J. Membr. Biol. Submitted for publication.
26. OVCHINNIKOV, YU. A. 1971. Rational synthetic routes to depsipeptides from the standpoint of their structure-function relations. *In* XXIIIrd International Congress of Pure and Applied Chemistry. **2:** 121–171.
27. HILLE, B. 1975. Ionic selectivity of Na and K channels of nerve membranes. *In* Membranes—A Series of Advances. G. Eisenman, Ed. Vol. **3:** 255–323. Marcel Dekker, New York, N.Y.
28. GOLDMAN, D. E. 1943. Potential, impedance, and rectification in membranes. J. Gen. Physiol. **27:** 37–60.
29. HODGKIN, A. L. & B. KATZ. 1949. The effect of sodium ions on the electrical activity of the giant axon of the squid. J. Physiol. (London) **108:** 37–77.
30. WOODBURY J. W., 1971. Eyring rate theory model of the current-voltage relationships of ion channels in excitable membranes. *In* Chemical Dynamics: Papers in Honour of Henry Eyring. J. O. Hirschfelder, Ed.: 601–617. John Wiley & Sons. New York, N.Y.
31. EYRING, H., R. LUMRY & J. W. WOODBURY. 1949. Rec. Chem. Prog. **10:** 100.
32. EISENMAN, G. 1965. Some elementary factors involved in specific ion permeation. *In* Proceedings of the XXIIIrd International Congress of Physiological Sciences. Tokyo, Japan. **87:** 489–506.
33. LÄUGER, P. 1973. Ion transport through pores: A rate-theory analysis. Biochim. Biophys. Acta **311:** 423–441.
34. SZABO, G. & G. EISENMAN. 1973. Biophys. Soc. Abstr. **13:** 175a.
35. HLADKY, S. B. & D. A. HAYDON. 1973. Membrane conductance and surface potential. Biochim. Biophys. Acta **318:** 464–468.
36. ANDREOLI, T. E., M. TIEFFENBERG & D. C. TOSTESON. 1967. J. Gen. Physiol. **50:** 2527–2545.
37. IITAKA, Y., T. SAKAMAKI & Y. NAWATA. 1972. Chem. Lett.: 1225.
38. KILBOURN, B. T., J. D. DUNITZ, L. A. R. PIODA & W. SIMON. 1967. J. Mol. Biol. **30:** 559–563.
39. EISENMAN, G. 1961. On the elementary atomic origin of equilibrium ionic specificity. *In* Symposium on Membrane Transport and Metabolism. A. Kleinzeller & A. Kotyk, Eds.: 163–179. Academic Press. New York, N.Y.
40. EISENMAN, G. 1969. Theory of membrane electrode potentials. *In* Ion-Selective Electrodes. R. A. Durst, Ed.: 1–56. National Bureau of Standard. Special Publication No. 314.
41. DIEBLER, H., M. EIGEN, G. ILGENFRITZ, G. MAAS & R. WINKLER. 1969. Pure Appl. Chem. **20:** 93–115.
42. GRELL, E., T. FUNCK & F. EGGERS. 1975. *In* Membranes—A Series of Advances. G. Eisenman, Ed. Vol. **3:** 1–126. Marcel Dekker. New York, N.Y.
43. GRAVEN, S. N., H. A. LARDY, D. JOHNSON & A. RUTTER. 1966. Biochemistry **5:** 1729–1735.
44. EISENMAN, G. & S. KRASNE. 1974. *In* MTP International Review of Science, Biochemistry Series. C. F. Fox, Ed. Vol. 2,: 1–33. Butterworths. London, England.
45. FELDBERG, S. W. & G. KISSEL. 1975. Charge pulse studies of transport phenomena in bilayer membranes. J. Membr. Biol. **20:** 269–300.

DISCUSSION

DR. A. MAURO (*Rockefeller University, New York, N.Y.*): Dr. Eisenman, you are dealing with essentially a homogeneous solution since the potential energy or energy barriers are only a function of X and not a function of the surface coordinates.

DR. EISENMAN: That is absolutely true.

DR. MAURO: For a nonhomogeneous system you would have to carry out this elegant analysis for a subsystem two and a subsystem three by the Eyring transition state theory.

DR. EISENMAN: Certainly.

DR. HAYNES (*University of Miami, Coral Gables, Fla.*): You have shown that when you increase the complexation affinity of your ionophore, or if you increase the rate of transfer across the membrane, the interfacial reaction eventually becomes rate-limiting.

DR. EISENMAN: That is right.

DR. HAYNES: Do you really think you can describe the distributions of the ionophore across the plane of the membrane in terms of some smooth type of curve?

DR. EISENMAN: No. It's purely a formalism.

DR. HAYNES: Well then, could you just as easily deal with a three- or four-stage model?

DR. EISENMAN: Absolutely; you could represent the trapezoidal barrier presumably by a double-peaked one. The point is that the trapezoid is a simplification that is not too far off from what you'd expect from the image charge. It's a realistic picture.

DR. HAYNES: Again with continuous models.

DR. EISENMAN: Yes.

DR. HAYNES: When one actually builds real stick models of these things, the valinomycin or nonactin looks quite large compared to the phospholipids through which they are supposed to be diffusing.

DR. EISENMAN: I agree that it's a simplified model, but it is the most complex of the simplified models that has been used so far in embodying a Nernst-Planck regime and a barrier shape. You could have done it with a four-barrier Eyring model as an equivalent description. Then you would interpret the data slightly differently, but you would still put those two barriers the same distance apart.

PART II. THE ROLE OF ION-TRANSPORT MEDIATORS IN MEMBRANES

ROLE OF IONOPHORES IN ENERGY COUPLING

D. E. Green

Institute for Enzyme Research
University of Wisconsin
Madison, Wisconsin 53706

The theory that underlies the paired moving charges (PMC) model of energy coupling in biological systems[1-5] provides a totally new perspective for the role of ionophores. The model predicts a role for ionophores not only in cation transport, but also in catalysis, energy coupling, and control of coupling function. I intend to develop the model to the point that the basis for postulating these multiple uses of ionophores becomes evident and then to provide the experimental evidence that supports this expanded role of ionophores.

The PMC Model

Energy coupling involves interaction between two transmembrane reaction centers—an exergonic center in which a driving chemical reaction takes place, and an endergonic center in which a driven chemical reaction takes place.[1, 3] The chemical reaction in both centers proceeds vectorially so that the reaction is initiated on one side of the membrane and terminates at the other side. By chemical reaction is meant not only the classical process by which two reactants form a new chemical species, but also a process such as the one in which a cation is separated from its anion on one side of the membrane and reunited with another anion on the other side of the membrane. The simplest form of energy coupling (symport coupling) depends upon the interaction of a negatively charged species in the driving chemical reaction with a positively charged species in the driven chemical reaction.[3, 6, 7] The characteristic feature of the substrates for the coupled driving and driven chemical reactions is their ease of polarizability.[3] Consider the coupling of electron transfer in an electron transfer complex (the exergonic center) to the ionophore-mediated transport of K^+ in the endergonic center (FIGURE 1). The separation of electron from proton is paired to the ionophore-mediated separa-

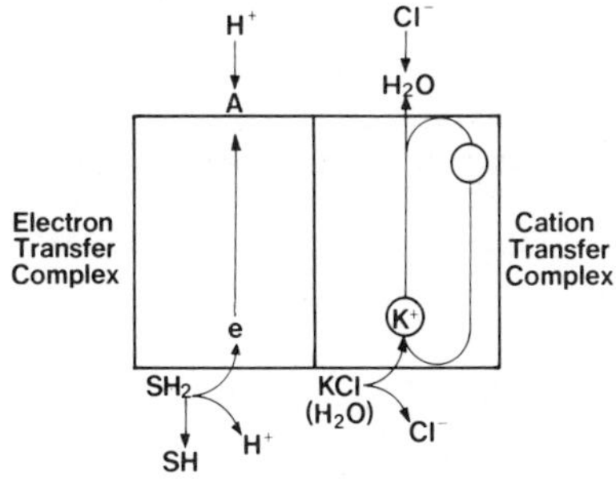

FIGURE 1. Coupling of electron transfer in an electron transfer complex (the exergonic center) to the ionophore-mediated transport of K^+ in the endergonic center. SH_2, hydrogen donor for the electron transfer reaction; A, terminal electron acceptor for the electron transfer reaction; O, Ionophore.

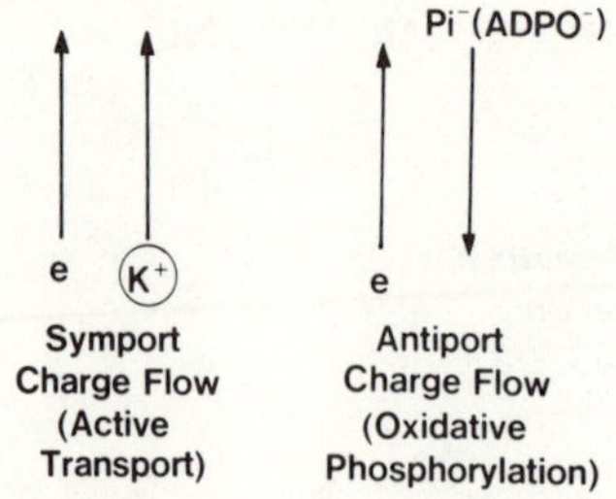

FIGURE 2. Symport versus antiport energy coupling.

tion of cation from anion. When paired, charge separation in the two centers is essentially isoenergetic; however, the energetic barrier virtually excludes unpaired charge separation. This means not only that the membrane-dependent separation of electron from proton is inextricably tied in to the separation of K^+ from its anion, but also that the movement of the electron away from its proton mediated by the electron transfer chain is inextricably tied in to the ionophore-mediated movement of K^+ away from its cation. We shall refer to this essentiality of pairing, both in charge separation and coupled charge movement, as the principle of respiratory control.[3]

In coupled processes such as oxidative phosphorylation, the movement of the negatively charged electron in the exergonic center is coupled to the successive movement of two negatively charged species (Pi^- and $ADPO^-$) in the endergonic center. Kemeny has enunciated the principle that all energy coupling has to be symport.[6, 7] How can this principle be reconciled with antiport coupling as in oxidative phosphorylation (FIGURE 2)? Antiport coupling is possible by virtue of the intervention of a linkage system of two interconnected, circulating positive charges[3, 4] (FIGURE 3). In this fashion, antiport coupling reduces to two symport charge flows proceeding in opposite directions.

In oxidative phosphorylation, electron flow is coupled to the synthesis of ATP by union of ADP and Pi (FIGURE 4). The separation of electron and proton is paired successively with the separation of Pi^- from its proton, and then of ADP^- from its proton. The energetically downhill movement of the electron in the electron transfer chain drives the energetically uphill movement of Pi^- in a phosphoryl transfer chain. At the end of its trajectory, the electron reacts with the final acceptor (A) with uptake of a proton; simultaneously, Pi^- reacts with ROH (bound adenosinemonophosphate)[8] to form ROP with release of a hydroxyl group. The second electron drives the movement of ADP^-. Once again, the reaction of the electron with the final acceptor is coupled to the proton-requiring reaction of ADP^- with ROP (forming ATP and ROH).

At the time of preparation of this article, a new dimension of antiport coupling has been recognized. The direct coupling of two negatively charged species (the electron and Pi^- or the electron and ADP^-) has to be directionally

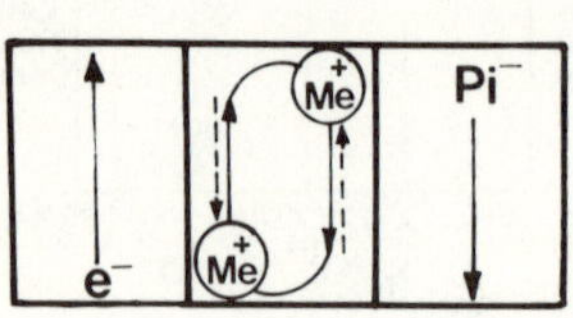

FIGURE 3. The linkage system for mediating antiport energy coupling. The linkage system consists of two circulating positive charges (contained within ionophores) whose movements are geared together so that the movements are always equal and in opposite directions.

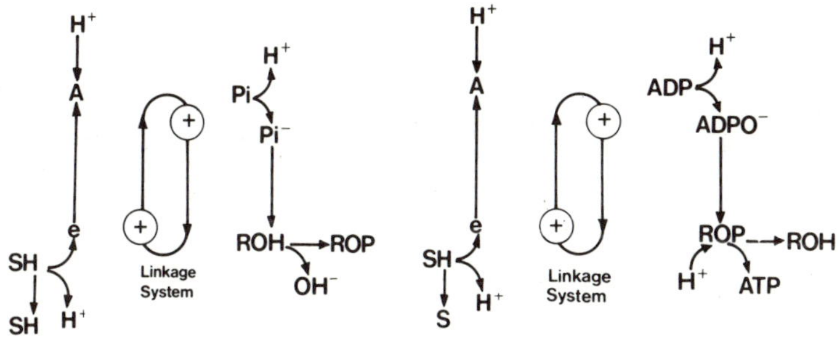

FIGURE 4. The coupling of electron flow to the synthesis of ATP by union of ADP and Pi. ROH represents bound adenosinemonophosphoric acid.

as depicted in FIGURE 2. The intervention of a linkage system to transform antiport into symport charge flow then becomes mandatory according to the principle of Kemeny. But if the moving charge is not Pi^-, but rather $[Mg^{2+} \cdots P^-]$ and similarly not ADP^-, but rather $[Mg^{2+} \cdots ADP^-]$, then symport flow of the electron with either $[Mg^{2+} \cdots P^-]$ or $[Mg^{2+} \cdots ADP^-]$ can take place without the intervention of a linkage system. The linkage of Mg^{++} to Pi^- or ADP^- in effect substitutes for a linkage system. This simplified version of antiport coupling by which the negatively charged species is integrated within a positively charged complex ion should be borne in mind whenever reference is made in the text to the older version of antiport coupling.

Charge-Separating Devices

There are two charge-separating devices in energy-coupling systems—the electron transfer chain, which separates the electron from proton; and ionophores, which separate cation from anion (FIGURE 1) or anion from proton (FIGURE 4) or anion from cation, as in the case of anionic ionophores such as iodonium compounds, which are anion carriers.[9] Wherever energy coupling takes place in systems devoid of an electron transfer chain, we may assume that the charge-separating device is always an ionophore.

An important distinction has to be made between ionophores concerned with the transmembrane transport of cations (FIGURE 1) and ionophores concerned with the activation of anions as in oxidative phosphorylation (FIGURE 4). Both transport ionophores and catalytic ionophores are molecules that can wrap around a metal ion and screen the charge of the metal to the point that the ionophore-metal ion adduct is soluble in nonpolar media. Moreover, in both types of ionophores, the metal ion can form coordination links with polar groups in the interior. Thus, in these two basic respects, there is no distinction between the two types of ionophores. Catalytic ionophores function only when linked to a specific protein; and the activation of anions, such as phosphate, is mediated both by the metal and by the protein. This activation is in fact a special case of catalysis. The activation of anions appears to involve the generation of a composite ion involving both the metal ion and the anion. Both are contained within the ionophore interior. The ionophore-metal ion-anion adduct may be transferred from one protein to another in

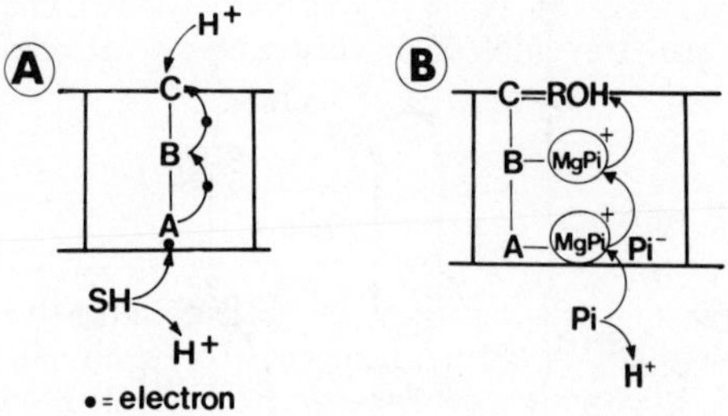

FIGURE 5. The electron transfer chain versus the ionophore-dependent phosphoryl transfer chain.

a transfer chain just like the electron in the electron transfer chain (FIGURE 5) or the activated anion may be transferred from the ionophore to an acceptor system as in the terminal stage of oxidative phosphorylation.

Kinases versus Energy-Coupling Systems

There is a large group of enzymes known as kinases[10] that catalyse ATP-energized synthetic reactions such as phosphorylation of sugars, acylation of coenzyme A, and peptide bond formation. We may think of kinase reactions as coupled reactions in which the hydrolysis of ATP is coupled to some endergonic synthetic reaction. These reactions are also governed by the pairing principle and the principle of respiratory control. As I shall discuss later, the charge-separating device in kinases appears to be an ionophore. The only distinction that exists between energy coupling in kinases and energy coupling in organelles such as mitochondria is the separability of the two reaction centers. In kinases, these two centers are part of the same protein; in organelles, the two centers are usually separable (the ATPase of the sarcoplasmic reticulum in fact conforms more closely to the kinase pattern than to the mitochondrial pattern in respect to the separability of the two centers[11]). Perhaps more fundamental than this question of the separability of the two centers is the fact that the separated charges move considerable distances in energy-coupling organelles (the width of a membrane), but do not move to any comparable degree in kinases. The point to be emphasized, however, is that the basic principles of energy-coupling paired-charge separation, respiratory control, symport coupling, and two reaction centers are identical for kinases and for organelles; and moreover, the same tactical devices for charge separation are found in both types of energy-coupling systems.

Predictions of the PMC Model with Respect to the Role of Ionophores in Energy Coupling

The PMC model not only assigns a central role for ionophores in all categories of energy coupling, but asserts that all charge separations other than those mediated by electron transfer chains must be ionophore-mediated. The model thus predicts a multiplicity of ionophores in energy-coupling organelles, both for ion transport and catalysis; the presence of ionophores in kinase systems; the requirement of ionophores for antiport coupling; and finally, the high probability that the control of energy coupling is exercised via the control of ionophores, since the essence of energy coupling involves charge separation and the ionophores are the molecular instruments of charge separation.

Multiplicity of Ionophores in Mitochondria

One can make a rough estimate of the total number of ionophoric species to be found in mitochondria, assuming a different species for each unique function.[4] At least 20 different species could be expected to be present. Blondin has isolated ten chromatographically pure species from the inner mitochondrial membrane[12] and Løvaas and Blondin have isolated six additional chromatographically pure species from the soluble matrix protein.[13] This count of 16 different ionophoric species does not include any of the phospholipid ionophores or free unsubstituted fatty acids with ionophoretic capability. The present indications are that the number of isolatable species will probably be larger than the estimated number of different species. This would mean that there are more ionophore-mediated functions than we are presently aware of.

Blondin, in this monograph,[12] considers in detail the procedures that he has developed for the separation, isolation, and characterization of the different ionophoric species in mitochondria and also the physical criteria by which the uniqueness of each chromatographically pure fraction has been established. Sufficient to say that the decisive test of purity is whether a unique structural assignment can be made. This has been accomplished in a sufficient number of cases to give us confidence that there is a 1:1 correspondence between the number of chromatographically resolvable peaks and the number of ionophoric species.

The functional criterion of an ionophore is the capacity to move a metal ion between two aqueous phases separated by a nonpolar phase. This capacity can readily be evaluated in a Pressman cell[14] (FIGURE 6). For routine evaluation of ionophoric capability, we use the induction of mitochondrial swelling by monovalent or divalent metal salts.[12, 15] The rate of swelling in this assay system is proportional to the amount of ionophore introduced. The Pressman cell and the swelling technique both test the ability of a test species to move metal ions between two aqueous phases separated by a nonpolar liquid barrier. There are other functional criteria for ionophoric capability, such as enhanced fluorescence of aminonaphtylsulfonic acid[16] and enhanced partition of a metal ion into an organic phase.[17] All these criteria have been routinely used in the documentation of the different isolated species as ionophores.

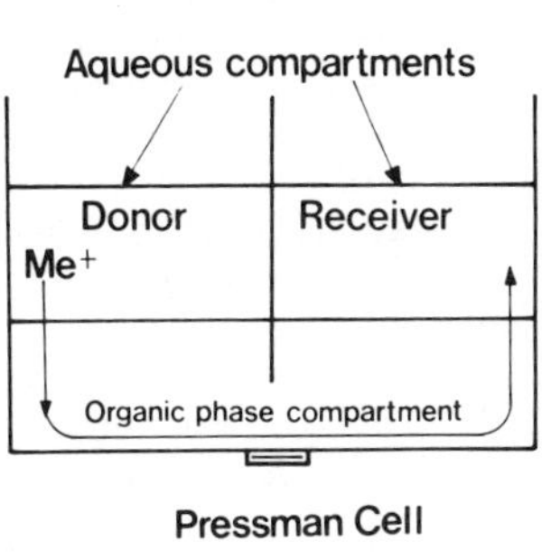

FIGURE 6. The Pressman cell for evaluating electrophoretic competence.

TABLE 1

OCTADECADIENOIC ACID FAMILY OF MITOCHONDRIAL IONOPHORES

	Δ		Δ
9-HOD	10,12	9-KOD	10,12
9-MHOD	10,12	9-MKOD	10,12
13-HOD*	9,11	13-KOD	9,11

NOTE: HOD = hydroxyoctadecadienoic acid; KOD = ketooctadecadienoic acid; MHOD = methyl ester of HOD; MKOD = methyl ester of KOD.
* Na^+/H^+ exchange activity.

The butanol:acetic acid:water mixture (4:1:5) which Blondin has successfully used for the extraction of divalent metal ionophores from mitochondria[18, 19] points up a critical facet of mitochondrial ionophores. These are not monodispersed in the lipid phase. The present indications are that these ionophores are contained within or linked to proteins. The isolation of the intrinsic ionophores requires the use of reagents such as butanol and acetic acid, as well as trypsin, which can detach the ionophores from their protein containers. Ionophores such as the K^+/Na^+ ionophore are even more recalcitrant with respect to extractability, and more drastic reagents have to be employed for their isolation.[18] It would take us too far afield to consider the different modalities of protein-ionophore containment. These modalities have to be appropriate for both the transport and catalytic functions of the intrinsic ionophores.

Many of the intrinsic ionophores of mitochondria are members of the same structural family. The six members of one such structural family, the family of C_{18} dienoic acids, are listed in TABLE 1. One of this family has been characterized as the Na^+-specific nigericin-type ionophore which is responsible for Na^+/proton exchange in mitochondria.[20] Note the variable features in this family: the position of the hydroxy group in the 9 or 13 position, the replacement of a hydroxy by a keto group, and the replacement of a free carboxylic group by a methyl ester. When the double bond characteristics of the different isomers have been defined, we may find even more variability in respect to this feature, i.e., whether *cis-cis, cis-trans,* or *trans-trans.* The C_{18} dienoic acid family of ionophores is localized in the intrinsic proteins of the inner membrane.[18, 19] Another comparable, though different family of substituted fatty acid ionophores has been extracted from the soluble proteins of the matrix, space.[13] There is suggestive evidence that this second group of ionophores with five different species are members of a C_{20} dienoic acid family. By virtue of their localization in the soluble fraction, we may infer that this set of ionophores is of the category of catalytic ionophores.

Not all the ionophores isolated by the butanol:acetic acid:water extraction procedure are substituted long-chain dienoic acids or methyl esters. Blondin finds indications of ionophoric species that are clearly not fatty acid derivatives.[48] The possibility of additional structural families of ionophores is therefore still open.

The amounts of the different isolated ionophoric species are usually some multiple[1–5] of the amount of cytochrome c_1. In other words, these species are present in an amount equivalent to at least one molecule per electron transfer chain.[18]

Phospholipids as Ionophores and Lipid Cytochrome c

In the tabulation of ionophoric species isolatable from mitochondria, phospholipids were excluded from consideration. In fact, the isolation procedures were designed to eliminate phospholipids at the very start. There were tactical reasons for setting aside the phospholipids in the quest for new ionophoric species. But there is no denying the fact that all common phospholipids of natural origin without exception have substantial ionophoretic capability.[22] Tyson *et al.*[22] have studied in some detail the ionophoretic competence of cardiolipin in a Pressman-type cell. Its activity with Ca^{++} compares favorably with that of X537A, one of the most active of all divalent metal ionophores[23] (FIGURE 7). But cardiolipin, just as X537A, is equally active with both monovalent and divalent metal ions. The so-called high-affinity Ca^{++} ionophore of the mitochondrion is highly sensitive to both ruthenium red and lanthanum.[24, 25] Cardiolipin shares this sensitivity (FIGURE 8). At a 1:1 molar ratio of cardiolipin and either of the two reagents, the movement of Ca^{+} is completely, or very nearly completely suppressed. However, the high-affinity Ca^{++} ionophore of the mitochondrion is specific for Ca^{++} and shows no activity with either monovalent ions or Mg^{++}. Thus, clearly on the basis of specificity alone, cardiolipin cannot be identified with the high-affinity Ca^{++} ionophore. The amount of cardiolipin in mitochondria is relatively high. About 16% of the phospholipid phosphorus is accountable for by cardiolipin.[26] Clearly, not all the cardiolipin in mitochondria can function as Ca^{++} ionophore.

Phospholipids are the structural components of the lipid bilayer phases of biological membranes.[27] By virtue of this commitment to the bilayer structure, which would reduce to negligible proportions the flip-flop or tumbling maneuver[28] by which phospholipids could act as ionophores in the lipid phase, it becomes difficult, if not impossible, to invoke any ionophoretic role for phospholipids. Is there any escape from this dilemma?

Cytochrome *c* is known to form complexes with acidic phospholipids,

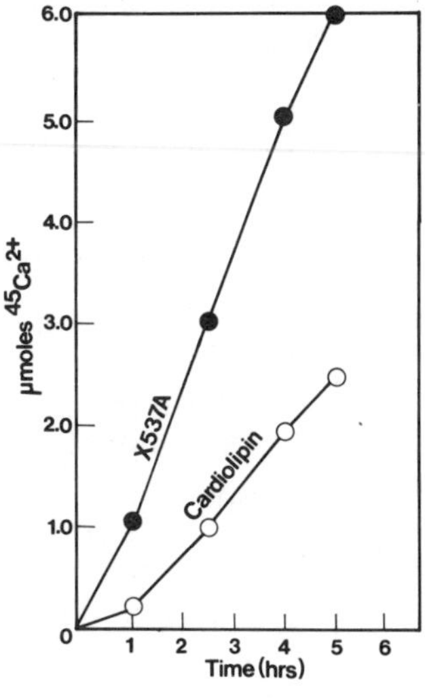

FIGURE 7. Ionophoretic capability of cardiolipin versus X537A with Ca^{++} as the divalent metal ion. The donor compartment of the Pressman cell (2 ml of water) contained 10 mM $^{45}CaCl_2$ and 25 mM trimethylammonium tricine (pH 8.3). The receiver compartment (2 ml) contained 25 mM trimethylammonium citrate (pH 5.4). The organic phase (6 ml of chloroform) was 0.5 mM in ionophore. Aliquots (20 μl) were taken from the receiver compartment and analyzed for $^{45}Ca^{2+}$ by liquid scintillation spectroscopy. The chloroform phase was presaturated with methanol and water according to the procedure of Bligh and Dyer. A single phase comprising $CHCl_3/CH_3OH/H_2O$ in the ratio 1:2:0.8 is phase-separated by addition of chloroform and water (1:1). The ionophore is dissolved in the organic phase after removal of the upper layer.

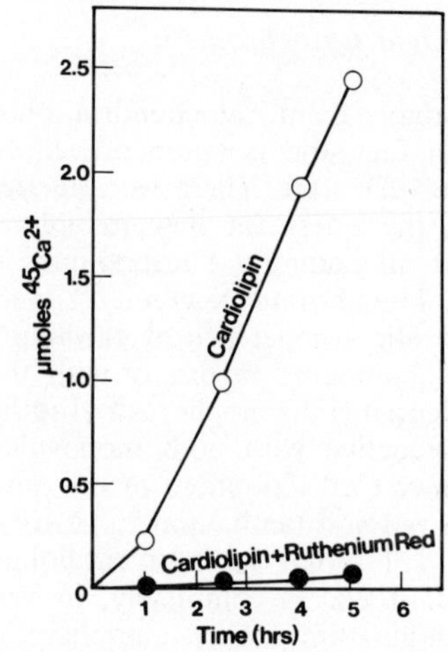

FIGURE 8. Sensitivity of cardiolipin to ruthenium red in the transport of Ca^{++} in the Pressman cell. The details of the composition of the three compartments are as described in the legend for FIGURE 7. Ruthenium red (1.5 mM) was added to the donor compartment. The molar ratio of ruthenium red to cardiolipin was 1:1.

such as cardiolipin, which are soluble in nonpolar solvents such as heptane.[29, 30] Moreover, such complexes, referred to as lipid-cytochrome *c*, are known to be present in mitochondria.[29] For example, extraction of mitochondria with 90% acetone leads to the extraction of the bulk of cytochrome *c*.[31] The phospholipids of lipid-cytochrome *c*, unlike free cardiolipin, would not be committed to a structural role in the bilayer phase. In fact, lipid-cytochrome *c* would automatically be a mobile species in the membrane by virtue of its structure (FIGURE 9). Tyson *et al.*[22] have shown that lipid-cytochrome *c* formed by interaction of cytochrome *c* with either soybean phospholipid or mitochondrial phospholipid has Ca^{++} ionophoretic ability greater than that of either of these two phospholipid mixtures when tested in a Pressman cell (FIGURE 10). Some 10 years ago, Penniston *et al.* discovered that cytochrome *c* can restore the capacity for energized translocation of Ca^{++} in aged and swollen mitochondria, and this restoration took place even with ATP as the energy source.[32] Clearly, then, under appropriate conditions, lipid-cytochrome *c* can function as an electrogenic Ca^{++} ionophore in mitochondria.

Lipid-cytochrome *c* represents a new genre of ionophore, namely a polyionophore. Each molecular unit consists of a protein with an outer shell of zwitterionic and acidic phospholipids. Since there have to be at least eight acidic phospholipid molecules per molecule of cytochrome *c*, the molecular unit of lipid-cytochrome *c* has the character of a polyionophore.

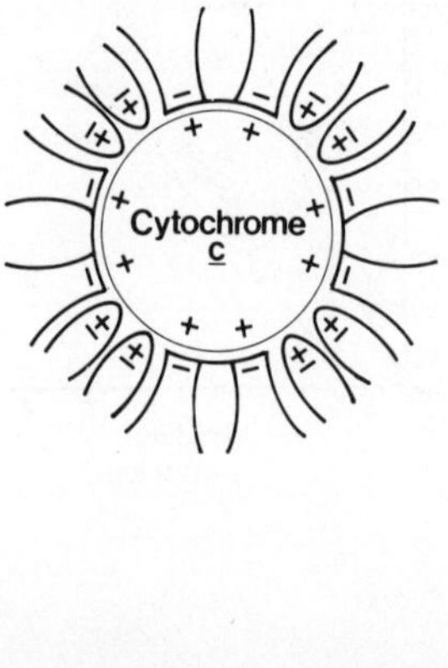

FIGURE 9. Diagrammatic representation of the structure of lipid-cytochrome *c* (cardiolipin + lecithin, linked to cytochrome *c*).

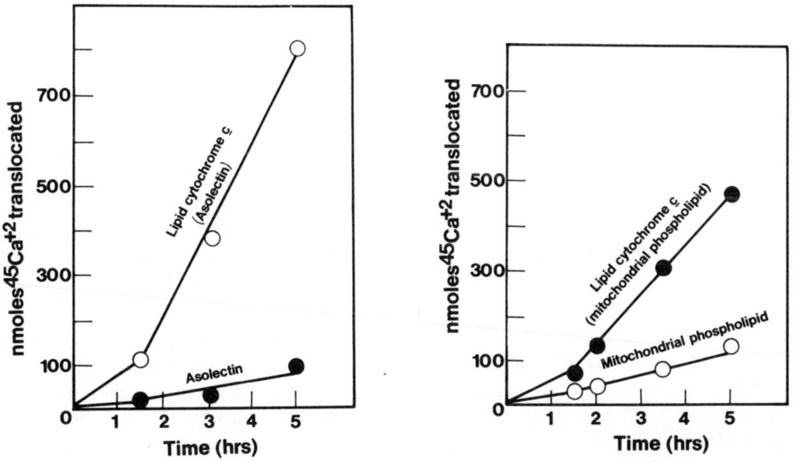

FIGURE 10. Ionophoretic capability of lipid cytochrome *c* versus lipid with Ca^{++} as the divalent metal ion in the Pressman cell: (*left*) Asolectin; (*right*) mitochondrial phospholipid. At *left* the concentration of asolectin in the chloroform layer was 1.8 mM and that of reduced cytochrome *c*, 60 μM. At *right* the concentration of mitochondrial phospholipid in the chloroform layer was 2.4 mM and that of reduced cytochrome *c*, 60 μM. The other details are as in the legend for FIGURE 7.

One additional parameter of lipid-cytochrome *c* deserves mention. The possibility that phospholipids can potentiate the partition not only of divalent cations, but also of anions such as inorganic phosphate, has been experimentally verified.[33] The data of TABLE 2 establish that lipid-cytochrome *c* formed by interaction of reduced cytochrome *c* with Asolectin can significantly affect

TABLE 2

EXTRACTION OF $^{32}P_I$ INTO HEPTANE FROM 30% ETHANOL BY LIPID CYTOCHROME *c* (Fe^{++})

Additions to Aqueous Phase (30% ethanol) before Extraction with Heptane	Cpm/100 μl Heptane: Asolectin	Cpm/100 μl Heptane: Mitochondrial Phospholipid
Phosphate buffer	47	33
Phosphate + phospholipid	88	53
Phosphate + phospholipid + cytochrome *c*	985	240

NOTE: Two ml of 30% ethanol were mixed with 4 ml of heptane, vortexed, and the phases separated and clarified by sedimentation. The concentrations in the aqueous phase prior to extraction were: trimethylammonium ^{32}P phosphate (3.7 mM); reduced cytochrome *c* (0.1 mM); asolectin (3 mM); and mitochondrial phospholipid (3.6 mM). The pH of the phosphate buffer was 7.3. Cytochrome *c* was reduced with ascorbate in six-fold excess. There was essentially quantitative transfer of both phospholipid and cytochrome *c* into the heptane phase. The recovery of P_i in the heptane phase with asolectin cytochrome *c* was 20% of the theory for one molecule of P_i per molecule of cytochrome *c*.

the partition of inorganic phosphate from water into heptane. Whether this partition is metal-dependent remains to be determined.

There is yet another way by which acidic phospholipids might participate in mitochondrial ion transport. A significant moiety of the total complement of mitochondrial phospholipid is not contained within the bilayer lipid phase. This moiety is intrinsic to the lipid contained within transmembrane proteins such as the Criddle protein of molecular weight 29,000[34, 35] and also within polyprotein complexes[36] as the F_1 of Racker.[37] This phospholipid contained within protein may play an entirely different role than does bilayer phospholipid. It is not excluded, therefore, that phospholipids in catalytic amounts associated with specific proteins may function as transport or catalytic ionophores.

The role of bilayer phospholipid in the control of mitochondrial coupling function involves a completely new dimension of ionophores. The inner membrane can exist in two configurational states[38–40]—a state appropriate for oxidative phosphorylation (the aggregated configuration) and a state in which oxidative phosphorylation is excluded (the orthodox configuration). In the aggregated configuration, the inner membrane is maximally expanded; in the orthodox configuration, the inner membrane is maximally contracted (FIGURE 11). This profound geometric change in the cristae is probably referable to a corresponding geometric change in the phospholipids of the inner membrane induced by a change in the concentration of divalent ions in the matrix space. Lecithin accounts for about 40% of the total phospholipid of the inner membrane.[21] Model studies have shown that lecithin can transport Pi between two aqueous phases separated by an organic phase.[41] This ionophoric transport of Pi shows an absolute requirement for divalent metal which can be satisfied by Mg^{++} or Mn^{++}, but not by Ca^{++}. In fact, Ca^{++} is a highly effective inhibitor of this transport. Arsenate and uncoupler can substitute for Pi and can be transported by lecithin in lieu of Pi. Moreover, Hg^{++}, like Ca^{++}, completely inhibits the lecithin-mediated transport of Pi. From these model studies, it can be deduced that lecithin can exist in two states within the bilayer continuum of the inner membrane—in what we may describe as the Mg^{++} and Ca^{++} states, respectively. In the Mg^{++} state, the divalent metal and Pi are encapsulated within the interior of the phospholipid (this interiorization of the ions would be mandatory for Mg^{++}-mediated transport of Pi); in the Ca^{++} state, the divalent metal and Pi are external to the phospholipid (this externalization of the ions would explain the inhibition of Pi transport by Ca^{++}). Lecithin would undoubtedly be in a much more expanded state when combined with Mg^{++} and Pi than when

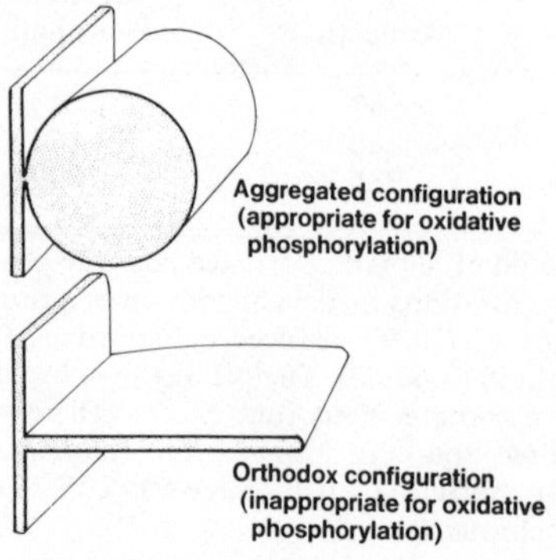

FIGURE 11. Two configurational states of the mitochondrion.

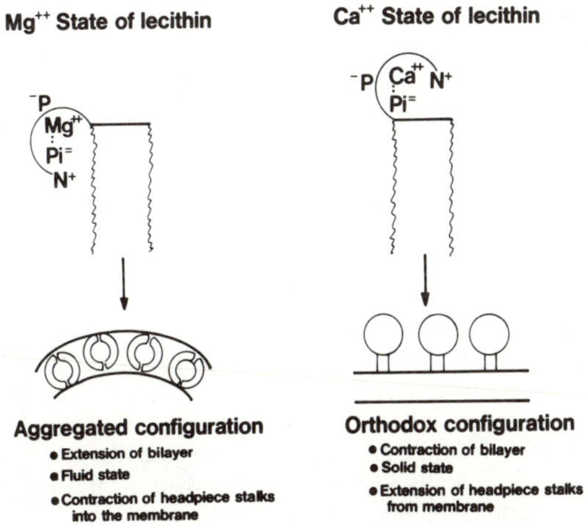

FIGURE 12.

combined with Ca^{++} and Pi (FIGURE 12). This Mg^{++}-induced expansion of lecithin and the Ca^{++}-induced contraction of lecithin could readily account for the aggregated-orthodox configurational transition. Thus, the ionophoric capability of lecithin can be expressed in a novel way, even though lecithin is structurally immobilized within the bilayer continuum. Moreover, reagents like Hg^{++}, arsenate, uncoupler, etc., can modulate the geometric state of the membrane via their interaction with bilayer lecithin. Vanderkooi[45] has deduced from conformational analysis of phosphatides involving mapping and minimization of the intramolecular energy that zwitterionic phospholipids can exist in an extended conformation in which the polar groups are extended away from the fatty chains, and a contracted conformation in which the polar groups nestle closely to the fatty chains. We are identifying the extended conformation of the lecithin with the Ca^{++} state and the contracted conformation with the Mg^{++} state.

Ionophores in Other Organelles and in Kinases

By application of the same extraction and isolative procedures and the same functional criteria for ionophoretic capability to chloroplasts,[42] sarcoplasmic reticulum,[43] and bacterial membranes,[44] unambiguous evidence has been obtained for multiple nonphospholipid ionophores in each of these membrane systems, both neutral and acidic. This program has only recently been initiated and it is still too early to indicate the structural characteristics of the different ionophoric species, except to say that their chromatographic properties closely resemble those of the species isolated from mitochondria. Although our initial efforts were directed towards the isolation of ionophores which were not phospholipids, we recognized, belatedly, that phospholipids were playing a key role as the prosthetic ionophores of a variety of enzymes, including dehydrogenases, kinases, ATPase and ATP synthetase. Three notable examples of this association can be cited. Blondin[48] found that the electrogenic K^+/Na^+

ionophore of the mitochondrial inner membrane could be identified with a derivative of phosphatidyl inositol. The active phospholipid was not extractable by organic solvents from the protein in which it was contained until the protein was succinylated—a token of the tightness of binding of the phospholipid to the protein. The β-hydroxybutyrate dehydrogenase of the mitochondrial inner membrane can be resolved into an apodehydrogenase which is inactive in catalyzing the oxidation of β-hydroxybutyrate by NAD^+ unless supplemented specifically with lecithin.[51–53] This reconstitution is inhibited by Cd^{++} and by 2,4-dinitrophenol.[51] Maina and Green[41] have shown that lecithin can mediate the transport of NAD^+ between two aqueous phases separated by an organic phase in the presence of Zn^{++}—the very metal found in highly purified preparations of the apodehydrogenase. Moreover, this transport of NAD^+ is inhibited by Cd^{++} and by 2,4-dinitrophenol. On the basis of the results of this model study, it is in order to classify β-hydroxybutyrate dehydrogenase as a metallo ionophoroprotein with lecithin as the ionophore and Zn^{++} as the tightly associated prosthetic metal. Finally, Blondin and Løvaas[46] have demonstrated that yeast hexokinase contains tightly bound phospholipid which is extractable into organic solvents only after the protein has been succinylated, as well as a nonphospholipid acidic ionophore. The molar ratio of phospholipid:protein corresponds to one molecule of phospholipid per molecule of hexokinase (molecular weight of 100,000). The combination of the prosthetic phospholipid and an acidic nonphospholipid ionophore mediates the partition of Ca^{++} from an aqueous to an organic phase and also the Mg^{++}-dependent transport of Pi between two aqueous phases separated by an organic phase.[46] These observations are compatible with the postulate of hexokinase as a metallo ionophoroprotein system with a phospholipid, as well as an acidic nonphospholipid ionophore, as the prosthetic ionophores and Mg^{++} as the dissociable divalent metal. The prosthetic phospholipid appears to have the same chromatographic behavior as lecithin (with admixture by some lysolecithin) in thin-layer chromatography.

As we suspected, we have opened Pandora's box. Energy-coupling systems appear to be a symphony of ionophores. The task of isolating and characterizing the multiple ionophores in bioenergetic systems will not be unlike the task of describing all of the lipids and proteins in a given membrane. We hope we can enlist the cooperation of others in extending and consolidating these initial exploratory efforts.

The Beechey Protein—An Ionophoroprotein

Beechey and his colleagues have isolated, from mitochondria exposed to the minimal concentration of ^{14}C-labeled dicyclohexylcarbodiimide (DCCD) required to suppress oxidative phosphorylation, a chloroform-soluble protein which contains the radioactive label.[47] From the amount of radioactivity, it could be deduced that one molecule of this protein of molecular weight 10,000 reacted with O, H molecules of DCCD. This immobilization of the Beechey protein by DCCD led not only to the loss of oxidative phosphorylation, but also to the loss of coupled ATP hydrolysis. Oligomycin, the specific inhibitor of coupled ATP hydrolysis, acts synergistically with DCCD in the sense that less DCCD is required to suppress oxidative phosphorylation when submaximal amounts of oligomycin had previously been added to the mitochondrion. In other words, both oligomycin and DCCD immobilized the same protein, namely, the chloroform-soluble Beechey protein. Whatever DCCD-accomplished by this immobilization could be duplicated by oligomycin.

For a variety of reasons, we inferred that the Beechey protein was an ionophoroprotein and set about demonstrating that such was the case. The first task was to devise an assay by which to measure ionophoroprotein activity. Since the Beechey protein was soluble in organic solvents and not in water, and since the PMC model predicted that, in coupled ATP synthesis, Pi and ADP would have to be encapsulated within an ionophore, the logical assay for the postulated ionophoroprotein would be the extraction of labeled Pi or ADP from an aqueous to an organic phase. This enhancement by the Beechey protein of the extraction of labeled Pi or ADP into chloroform was indeed observed by Kessler, Tyson, and Vande Zande[56] and it could be shown that these two activities increased progressively from the first crude extract which was the starting point for the preparation to the final purified preparation of the Beechey protein. These activities were by no means marginal. In the most active preparations of the Beechey protein, they approximated to one molecule of Pi or ADP per molecule of protein (assuming a molecular weight of 10,000) (TABLE 3).

TABLE 3

TRANSFER OF LABELED PI AND ADP FROM AN AQUEOUS TO AN ORGANIC PHASE MEDIATED BY THE BEECHEY PROTEIN

Preparation of Beechey Protein	ADP in the Organic Phase (nmol/mg Beechey protein corrected for control without protein)	Pi in the Organic Phase (nmol/mg Beechey protein corrected for control without protein)
I	70.3	
II		63

NOTE: The incorporation of Pi was carried out as follows. The Beechey protein in chloroform-methanol (2:1) (80 μl containing 280 μg) was mixed with 3 ml methanol, 1.5 ml chloroform, 0.04 ml M $MgCl_2$ and 0.4 ml of 10 mM ^{32}Pi (K^+ salt) of pH 7.0 (510 cpm per nmol). The mixture was vortexed to a single phase and allowed to stand for 10 minutes. Chloroform (1 ml) and water (1 ml) were then added and the mixture was vortexed again before spinning in a clinical centrifuge for 5 minutes. The upper phase was removed by aspiration and the lower phase was transferred to a clean test tube and chilled for 5 minutes before a second spinning for 1.5 minutes. The twice-clarified extract was chilled for 30 minutes before an aliquot (0.5 ml) was taken for counting. Protein was determined in another aliquot by the Folin-Lowry method.

The incorporation of ADP was carried out essentially the same way, except for the following differences. No Mg^{++} was added to the donor compartment; the ^{14}C-ADP was 2 mM (820 counts per nmol) and adjusted to pH 6.8. Before phase separation induced by addition of chloroform and water, 0.2 ml of glacial acetic acid was added.

The chilling time was 5 minutes at each of the two stages.

When the Beechey protein fraction in chloroform-methanol was extracted with EDTA before assay, a complete requirement for divalent metal could be shown. Either Mg^{++} or Ca^{++} could satisfy that requirement for the extraction of labeled Pi from an aqueous to an organic phase (TABLE 4). It was not possible to show this requirement consistently. In some preparations of the Beechey protein, not only was divalent metal not required, but the addition

TABLE 4

ESSENTIALITY OF DIVALENT METAL FOR Pi INCORPORATION BY THE BEECHEY PROTEIN

Pretreatment	Metal Added	Pi Incorporated (cpm/mg Beechey protein)
EDTA	0	107
EDTA	$MgCl_2$*	2733
EDTA	$CaCl_2$*	4843

NOTE: The Dyer procedure for generating a one-phase from a two-phase system was used. Pi was determined in the chloroform phase.

* 37.5 mM in the aqueous phase.

of divalent metal was found to diminish the extractability of Pi or ADP.

It occurred to us that preparations of the Beechey protein which showed little or no requirement for divalent metal should contain bound metal, and indeed this was found to be the case (TABLE 5). The metal was Mg^{++} exclusively; no trace of Ca^{++} was found. Moreover, the amount of ADP taken up by this particular preparation corresponded to one mole per atom of Mg^{++}. Why Mg^{++} is tenaciously held in one preparation of the Beechey protein and is labile in another preparation is still an unresolved question. All these observations were consistent with the Beechey protein being an ionophoroprotein, but the direct proof that a bound ionophore was implicated in the protein-mediated transfer of Pi or ADP from an aqueous to an organic phase was still lacking. When the Beechey protein was extracted with a mixture of butanol, acetic acid and water (4:1:0.5), the extracted protein, as well as the extract, was found to be inactive in facilitating the partition of Pi into an organic

TABLE 5

CORRELATION OF ADP BINDING TO THE BOUND Mg^{++} CONTENT OF THE BEECHEY PROTEIN AT VARIOUS STAGES OF PURIFICATION

State of Purification	Nmol ADP Bound per Mg Protein	Nmol Bound Mg^{++} per Mg Protein
Starting chloroform-methanol extract before purification	1.6	54
Standard preparation of the Beechey protein	20.	77
A fraction derived by chromatography of the standard preparation on DEAE Sephadex	83.	78

NOTE: Bound Mg^{++} was determined in a 0.2-ml aliquot of each of the three samples—each sample containing about 1.5 mg protein dissolved in chloroform-methanol 2:1. The samples were evaporated to dryness, charred in a muffler furnace at 650°, cooled to dryness, and extracted with 6 N HCl. Mg^{++} was determined by atomic absorption spectroscopy after addition of 0.1% La $(NO_3)_3$ to the 6 N HCl extract to bind any interfering phosphoric acid.

ADP binding was determined by the protein-mediated transfer from an aqueous phase to the organic phase (chloroform).

phase. Moreover, the extract was found to contain phospholipid and a neutral ionophore of the octadecadienoic acid family (similar in properties to the methyl ester of 9-hydroxyoctadecadienoic acid). At the least, one could conclude that the Beechey protein fraction contained ionophores which could account for the partition function in respect to Pi and ADP. When the Beechey protein was prepared from mitochondria which were treated with DCCD, tripropyl tin and oligomycin, at concentration levels sufficient to abolish oxidative phosphorylation, the final preparation was found to be inactive in the partition of Pi and ADP and no neutral ionophore of the octadecadienoic acid family could be extracted from the protein by the butanol-acetic acid-water mixture. These observations provided an important link between the Beechey protein and the phenomenon of oxidative phosphorylation in that reagents known to inhibit, specifically, oxidative phosphorylation in mitochondria also suppressed the partition function of the Beechey protein when isolated from such treated mitochondria.

A dramatic sequel to these studies of the Beechey protein was the discovery by Maina and Green[41] that lecithin and asolectin could mediate the Mg^{++}-dependent transport of Pi or ADP between two aqueous phases separated by an organic phase. This transport showed all the properties of the ATP synthetase system: replaceability of Pi or ADP by arsenate, inactivity with Ca^{++}, sensitivity to dinitrophenol, picrate and tetraphenyl boron, inactivation by mercurials and, most importantly, extreme sensitivity to DCCD, tripropyl tin and oligomycin. Here was an ionophore (lecithin) that reproduced very closely the properties of ATP synthetase and duplicated the catalytic capability of the Beechey protein. One could not necessarily conclude from these data that the prosthetic ionophore group of the Beechey protein was lecithin, but the results are highly suggestive that a phospholipid either identical or similar to lecithin may play a key role in the ionophoric function of the Beechey protein. The presence of bound lecithin in the purified Beechey protein, releasable only after succinylation of the protein, is the critical supporting evidence for this proposal.

Hormones as Ionophores

The hormones are the molecules that circulate through the body and that control the exercise of physiological function. Since physiological function is the net expression of coupled function, it would be expected from first principles that the control of coupled functions would be operative in such a way that the coupled function could be turned on or off. The simplest way to achieve this switching on and off would be for the hormone to be an ionophore or for the hormone to control the activity of ionophores via some control transition. We have evaluated, by the Pressman cell technique, the first alternative in relation to the steroid hormones and the prostaglandins.[54] The active steroids tested include the sex hormones (progesterone, estrogens), corticosteroids, diethylstilbestrol, etc., and a set of synthetic steroids, all shown to be biologically active. All of them were found to be divalent metal ionophores ranging from ionophores of unusually high activity (progesterone and corticosteroids) to ionophores of moderate activity (estradiol). The prostaglandins tested included the full set of known active prostaglandins. Again, divalent ionophoric activity was found in all species ranging from outstanding activity (prostaglandins A_1, B_1, E_1 and F_1) and modest activity (prostaglandin $F_{1\alpha}$). It clearly cannot be a happenstance that all members of two classes of hormones are uniformly ionophores. The conclusion is inescapable in these two instances, at least that hormones are control molecules by virtue of their intrinsic ionophoric capability. This is not to say that specific proteins are not required for the expression of hormonal action. That such is the case is undeniable. But this essentiality of specific proteins is not incompatible with the exercise of ionophoric activity by hormones. Hormones may stand in the same relation to their specific proteins as does lecithin to β-hydroxybutyrate dehydrogenase. Both are ionophores whose ionophoric function is modulated by specific proteins.

The Control of Mitochondrial Coupling Function

Mercurials and heavy metal ions (Ca^{++}, Cd^{++}, Zn^{++}) suppress one set of mitochondrial coupling functions and induce another set.[49, 50] The set of functions that is suppressed includes oxidative phosphorylation and high-affinity Ca^{++} transport; the set of functions that is induced includes active transport of Mg^{++} and K^+. From these observations, it can be deduced that mitochondria exist in two coupling modes—the mode appropriate for oxidative phosphorylation and the mode appropriate for active transport of Mg^{++} and K^+.

The concentration of mercurials or heavy metal ions required for half-maximal suppression of one set of coupled functions is also half-maximal for the induction of the other set of coupled functions. There is thus an inverse relation between the disappearing and the emerging coupled functions. As one set disappears, the other set appears.

Southard *et al.*[49] have pointed out that this mercurial-induced transition in the coupling modes can be satisfactorily rationalized in terms of three cationic ionophore systems—the systems for active transport of Ca^{++}, Mg^{++} and K^+. In the oxidative phosphorylation coupling mode, the Ca^{++} ionophore is available for active transport; the Mg^{++} and K^+ ionophores, however, are latent. In the other coupling mode, the Mg^{++} and K^+ ionophores are available for active transport; and the Ca^{++} ionophore is latent.

The crucial correlation is the inverse relation between the disappearance of oxidative phosphorylation or Ca^{++} transport and the appearance of K^+ or Mg^{++} transport (FIGURE 13). This inverse relation could be interpreted in either of two ways. Either oxidative phosphorylation depends uniquely on the capacity for Ca^{++} transport, or oxidative phosphorylation is in some way tied in to the latency of the K^+ and Mg^{++} ionophores. When antiport coupling was visualized in terms of a linkage system, the second interpretation was favored, since the two latent ionophores could be assigned to the hypothetical linkage system. But with antiport coupling, now visualized in terms of the negatively charged anion becoming incorporated into a positively charged complex ion as discussed above, the case for relating the latency of the Mg^{++} and K^+ ionophores to oxidative phosphorylation collapsed and the case for correlating oxidative phosphorylation with the high-affinity Ca^{++} ionophore had to be reexamined.

The inverse relationships between the disappearing coupling functions and the emergent coupling functions reflect a transition in the mitochondrion by which one set of ionophores required for a particular set of coupled functions

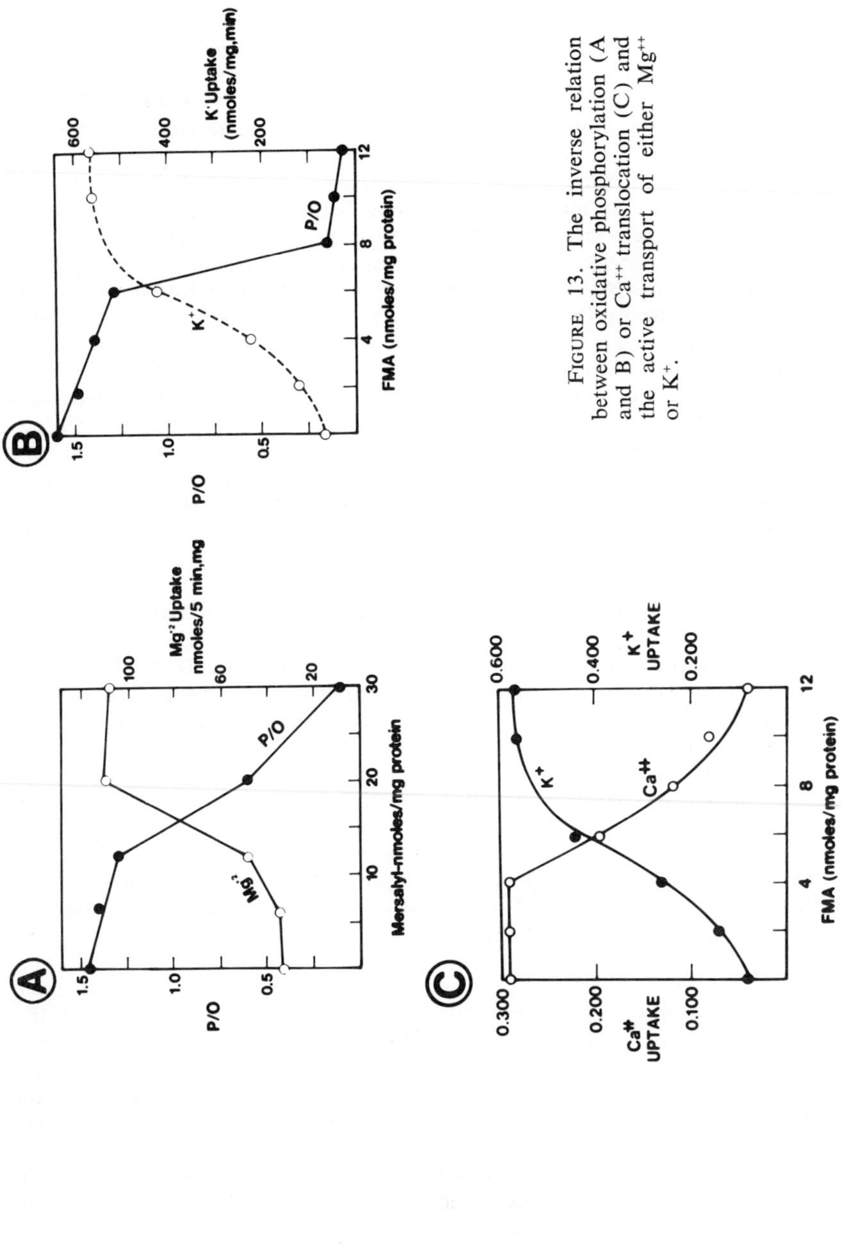

FIGURE 13. The inverse relation between oxidative phosphorylation (A and B) or Ca^{++} translocation (C) and the active transport of either Mg^{++} or K^{+}.

is replaced by another set of ionophores required for an entirely different set of coupled functions. The question is how the physiological regulators (Ca^{++}, Mg^{++}, Pi, fatty acids, etc.) can induce this control transition and how a change in ionophoric capability can result from this transition. The available evidence suggests that two systems play a key role in this control maneuver. These two systems are bilayer lecithin and the ribbon continuum. We have already discussed the way in which lecithin can exist in two states (the Mg^{++} and Ca^{++} states) and how these two states determine the geometric properties of the inner membrane.[41, 45] What we have not considered up to now is the state of the protein in the inner membrane. Evidence is now available that there are two domains in biological membranes—a protein domain and a phospholipid bilayer domain.[55] The protein domains are separated, one from another, by a zone of bilayer phospholipid. The protein domains consist of ribbon structures or continua which carry the projecting elements—the headpiece-stalks. These ribbons contain the complete apparatus for energy coupling—the electron transfer chain, the ATP synthetase and hydrolase, as well as the cation transporting systems. When divalent ions modulate the geometric state of the inner membrane, the protein ribbons undergo a complementary change. In the aggregated configuration, the headpiece-stalk projections are retracted into the membrane. In the orthodox configuration, the headpiece-stalk projections are extended from the membrane. This movement in and out of the headpiece-stalks profoundly alters the geometry of the ribbons and with this change in geometry, the ionophores in the ribbon which can couple to the electron transfer process undergo rearrangement. Hence, Mg^{++} will induce the emergence of one set of ionophores in the ribbon and Ca^{++} another set of ionophores. In this fashion, the end-result of the control of mitochondrial function is the control of ionophoric function.

The transition from the oxidative phosphorylation mode to the mode for active transport of K^{+} and Mg^{++} goes parallel with the transition from the aggregated to the orthodox configurational transition.[41] This configurational transition is activated by a mitochondrial control mechanism. It cannot be a coincidence that Ca^{++} is required to induce the aggregated-to-orthodox configurational transition and Mg^{++} to induce the reverse transition. The high-affinity Ca^{++} ionophore is available in the aggregated configurational state of the mitochondrion, whereas the Mg^{++} ionophore is available in the orthodox configurational state. We thus see the intimate connection between ionophores and the control of mitochondrial function. The ionophores mediate the flow of ions which lead to a configurational transition and the transition in turn leads to a new pattern of ionophores and to a new set of coupling functions.

Summary

If, as we deem inevitable, the principles of energy coupling are universal, then the necessity for charge-separating devices will apply across the board to all bioenergetic systems. Since, apart from the electron transfer chain, ionophores are the only charge-separating devices available in bioenergetic systems, it is predictable that bioenergetics will be a symphony of ionophores. No model of energy coupling that does not feature this central role of ionophores can be taken seriously. The ionophore approach thus opens the royal highway to the ultimate solution of all bioenergetic problems.

References

1. GREEN, D. E. 1974. Biochim. Biophys. Acta **346:** 27.
2. GREEN, D. E. 1974. Ann. N.Y. Acad. Sci. **227:** 6.
3. GREEN, D. E. & S. REIBLE. 1974. Proc. Nat. Acad. Sci. U.S.A. **71:** 4850.
4. GREEN, D. E. & S. REIBLE. 1975. Proc. Nat. Acad. Sci. U.S.A. **72:** 253.
5. GREEN, D. E., G. BLONDIN, R. KESSLER & J. H. SOUTHARD. 1975. Proc. Nat. Acad. Sci. U.S.A. **72:** 896.
6. KEMENY, G. 1974. Proc. Nat. Acad. U.S.A. **71:** 3064.
7. KEMENY, G. 1974. Proc. Nat. Acad. Sci. U.S.A., **71:** 3669.
8. ROY, H. & E. MOUDRIANAKIS. 1971. Proc. Nat. Acad. Sci. U.S.A. **68:** 464; **68:** 2720.
9. HOLLAND, P. C. & H. S. SHERRATT. 1972. Biochem. J. **129:** 39.
10. GREEN, D. E. & R. F. GOLDBERGER. 1967. Molecular Insights into the Living Process. 383–398. Academic Press. New York, N.Y.
11. MACLENNAN, D. H., T. J. OSTWALD & P. S. STEWARD. 1967. Ann. N.Y. Acad. Sci. **227:** 527.
12. BLONDIN, G. A. 1975. This monograph.
13. LOVAAS, E. & G. A. BLONDIN. Unpublished studies.
14. PRESSMAN, B. C. 1974. Ann. N.Y. Acad. Sci. **227:** 380.
15. BLONDIN, G. A., W. J. VAIL & D. E. GREEN. 1969. Arch. Biochem. Biophys. **129:** 158.
16. FEINSTEIN, M. B. & H. FELSENFELD. 1971. Proc. Nat. Acad. Sci. U.S.A. **68:** 2037.
17. PRESSMAN, B. C., E. J. HARRIS, W. S. JAGGER & J. H. JOHNSON. 1967. Proc. Nat. Acad. Sci. U.S.A. **58:** 1949.
18. BLONDIN, G. A. 1974. Ann. N.Y. Acad. Sci. **227:** 392.
19. BLONDIN, G. A. 1974. Biochem. Biophys. Res. Commun. **56:** 97.
20. BLONDIN, G. A. & D. E. GREEN. 1970. Bioenergetics **1:** 479.
21. FLEISCHER, S., H. KLOUWEN & G. BRIERLEY. 1961. J. Biol. Chem. **236:** 2936.
22. TYSON, C. A., H. VANDE ZANDE & D. E. GREEN. J. Biol. Chem. In press.
23. HAYNES, D. H. & B. C. PRESSMAN. 1974. J. Membrane Biol. **16:** 195.
24. LEHNINGER, A. L. & E. CARAFOLI. 1971. Arch. Biochem. Biophys. **143:** 506.
25. SOUTHARD, J. H. & D. E. GREEN. 1974. Biochem. Biophys. Res. Commun. **59:** 30.
26. FLEISCHER, S., H. KLOUWEN & G. BRIERLEY. 1961. J. Biol. Chem. **236:** 2936.
27. SINGER, S. J. 1971. *In* Structure and Function of Biological Membranes. L. I. Rothfield, Ed. 145. Academic Press. New York, N.Y.
28. KORNBERG, R. D. & H. M. MCCONNELL. 1971. Proc. Nat. Acad. Sci. U.S.A. **6:** 2564.
29. WIDMER, C. & F. L. CRANE. 1958. Biochim. Biophys. Acta **27:** 203.
30. GREEN, D. E. & S. FLEISCHER. 1963. Biochim. Biophys. Acta **70:** 554.
31. LESTER, R. L. & S. FLEISCHER. 1961. Biochim. Biophys. Acta **47:** 358.
32. PENNISTON, J. T., H. VANDE ZANDE & D. E. GREEN. 1966. Arch. Biochem. Biophys. **113:** 512.
33. SINGER, M. C., P. R. EHRLICH & L. E. GILBERT. 1971. *Science* **172:** 1339.
34. CRIDDLE, R. S., R. M. BOCK, D. E. GREEN & H. TISDALE. 1962. Biochemistry **1:** 827.
35. CAPALDI, R. A., H. KOMAI & D. R. HUNTER. 1973. Biochem. Biophys. Res. Commun. **55:** 655.
36. KESSLER, R. Unpublished studies.
37. PULLMAN, M. E., H. S. PENEFSKY, A. DATTA & E. RACKER. 1960. J. Biol. Chem. **235:** 3322.
38. ALLMANN, D. W., T. WAKABAYASHI, E. F. KORMAN & D. E. GREEN. 1970. Bioenergetics **1:** 73.

39. Allman, D. W., J. Munroe, T. Wakabayashi, R. A. Harris & D. E. Green. 1970. Bioenergetics 1: 87.
40. Allmann, D. W., J. Munroe, T. Wakabayashi & D. E. Green. 1970. Bioenergetics **1:** 331.
41. Maina, G. & D. E. Green. Unpublished studies.
42. Brand, J. & G. A. Blondin. Unpublished studies.
43. Tyson, C. & G. A. Blondin. Unpublished studies.
44. Silver, S. & G. A. Blondin. Unpublished studies.
45. Vanderkooi, G. 1973. Chemistry and Physics of Lipids **11:** 148.
46. Blondin, G. & E. Løvaas. Unpublished studies.
47. Cattell, K. J., I. G. Knight, C. R. Lindop & R. B. Beechey. 1970. Biochem. J. **117:** 1011.
48. Blondin, G. Unpublished studies.
49. Southard, J. H. & D. E. Green. 1974. Biochem. Biophys. Res. Commun. **61:** 1310.
50. Southard, J. H., P. Nitisewojo & D. E. Green. 1974. Fed. Proc. **33:** 2147.
51. Jurtshuk, P., Jr., I. Sekuzu & D. E. Green. 1963. J. Biol. Chem. **238:** 3595.
52. Menzel, H. M. & G. G. Hammes. 1973. J. Biol. Chem. **248:** 4885.
53. Gazzotti, P., H.-G. Bock & S. Fleischer. 1975. J. Biol. Chem. **250:** 5782.
54. Maina, G., C. A. Tyson & D. E. Green. Unpublished studies.
55. Haworth, R. A., H. Komai & D. E. Green. Unpublished studies.
56. Kessler, R., C. A. Tyson & H. Vande Zande. Unpublished studies.
57. Hunter, D. R., R. A. Haworth & J. H. Southard. Unpublished studies.

Discussion

Dr. Shamoo: It should be pointed out here, and I think Dr. Blondin is involved in this, that you have proposed for the last seven or eight years that ionophores are loose in the membrane and they are not necessarily membrane-bound. Since we have shown that only tryptic digestion or cynogen bromide cleavage of a chemical bond could release those ionophores, you and Dr. Blondin have changed your opinion to include that ionophores are membrane bound. What is your comment on this?

Dr. Green: We did look for monovalent cation ionophore for a considerable time. We did find one such in the lipid, but we discovered the amount of that ionophore was miniscule, and only after we had used tryptic digestion and extraction with the butanol-acetic acid mixture did we increase the yield by a factor of 50 to 100. From that result we realized that they are basically contained. Now it is only within the last two or three years that we've recognized thoroughly the containment aspect. I have to pay tribute to Dr. Shamoo since he was the first to publically state that the ionophores were contained within proteins and had to be liberated by trypsin. At the time we were working with soluble ionophores and so it was only later that we were aware of his very important contributions. But we recognized this containment only as a result of our failure to isolate significant amounts of the monovalent cation-ionophores.

Dr. Mueller: Dr. Green, one of the problems with his whole concept is that these compounds you have may not act in the bilayer because they are constructed in such a way that they are held across the bilayer at the two interfaces. You may find that these compounds transport ions very well across a thick phase in which they dissolve and get turned around and move to the other side. But if you put them in the bilayer, they either stop themselves just on one side or they span the bilayer and are held there fixed and don't do anything. Consequently, the evidence you presented for ionophoric activity doesn't satisfy me, and I don't think it will satisfy just anybody.

DR. GREEN: Yes, we are in agreement.

DR. MUELLER: The lipid story is very clear there. The lipids do not flip-flop, yet they act as ionophores, by your definition, "in those thick phases." The bilayers are held by the interphases.

DR. GREEN: I accept fully what you've said, and as a matter of fact it shows me that I didn't develop my presentation very well. I expressly made the point you made, that when these compounds are fixed they could not act as ionophores; however, I indicated there might be two escapes from that dilemma. One is lipid *c*, where you have a new kind of state of the lipid. A second possibility, which I didn't mention, is that there is bound lipid within many of the proteins and we don't know what its state is. The third point I made, is that it is the very fact that phospholipids are held in the bilayer modality that endows this ability to bind cations with enormous importance because that is probably related to the way in which the membrane changes its state. The ionophore is fixed, if you like, in the membrane because in that way the ionophore can release or take up divalent cations and in that way change the state of membrane.

DR. MUELLER: But Dr. Green, if it gives you the wrong answer in one case, aren't you worried it may also give you the wrong answer in another?

DR. GREEN: That is what is known in philosophy as a *non sequitur*.

DR. MUELLER: Have you considered the possibility that some of your thirty ionophores, which possibly look suspiciously like breakdown products of phospholipids, have been artificially derived by degradation of phospholipids?

DR. GREEN: We can't answer that; obviously at the present stage one could invoke this hypothesis. All we can say is that if the ionophores are meaningful, as we think they are, we should find them intrinsic to certain enzymes. For example, we found one particular ionophoric species in the isolated Beechey protein.

DR. KIRKPATRICK: I've noticed that you have several ionophores that conduct calcium, and you also have several other fatty acid derivatives that appear to conduct phosphate in the presence of a divalent cation. So you appear to have a number of carriers for two of the commonly transported substances, but do not appear to have ionophores out of your total of 16 or more for anything other than phosphates and divalents. Do you have a control for any of these effects in the sense that you have found an amphiphilic charged substance which does not transport calcium, phosphate, or one of the other ions into lipid phases? For instance, do you fail to get results with stearic acid?

DR. GREEN: We find that the capacity for ionophoretic activity is already present in the free fatty acid. It's a hundredth or perhaps a thousandth the activity you would find with the more specialized substituted fatty acids. But this is exactly what you expect; we don't generate a property *de novo*. The ionophoric property is already in the fatty acid, but very weak and as you introduce double bonds, hydroxy groups, keto groups, you enhance it by several orders of magnitude. You will therefore find the traces of these activities all along the line.

I have to make clear that we did not examine systematically the cation specificity of the different ionophoric species we have isolated. We concentrated on particular divalents and monovalents. We were not looking for anything else, so we couldn't very well find it. Now there is no ionophore specific for Pi. The Pi binding that we get is dependent upon the presence of a divalent metal and a protein. That's a completely different thing. The only other binding we're dealing with is the binding of phospholipid to cytochrome *c*, but that's not ionophoric. That is simply due to the fact that cytochrome *c* is surrounded by this lipid sheath.

DR. MUELLER: Do you consider cytochrome *c* an ionophore?

DR. GREEN: I don't think so.

DR. MUELLER: But it works! As you probably know Montal and Crane have developed methods to dissolve cytochrome *c* in hydrocarbons and that system will no doubt act as an ionophore, that is, if you shake it, it will transport ions from one side to the other in the system, and yet nobody in his right mind would consider cytochrome *c* to be an ionophore even though it satisfies all the criteria of your test.

DR. GREEN: I sense that we are talking at cross-purposes, and I suspect that basically, there is no disagreement between us on fundamental issues. A molecule which moves ions between two aqueous phases separated by an immiscible organic phase is, by definition, an ionophore. Therefore, this is not a debatable point. By that criterion, we maintain that we have isolated multiple ionophores from mitochondria. Your point, which is well taken, is that although phospholipid molecules may function as ionophores in a Pressman cell, they may not be capable of such activity when subjected to the constraints of the bilayer in a biological membrane. We agree. However, I pointed out that there may be ways of overcoming these constraints, and in such cases, the ionophoric capability of these phospholipids would be demonstrable, even in a bilayer membrane. Lipid *c* is a polymeric array of phospholipids stabilized by electrostatic interaction with cytochrome *c*. Other basic proteins can replace cytochrome *c* in this stabilization. It is the ionophoric properties of this polymeric system of phospholipids stabilized by cytochrome *c* that I am emphasizing, not the possible ionophoric properties of cytochrome *c*. I mentioned lipid *c* as one tactic by which the constraints of the bilayer could be overcome.

ISOLATION OF IONOPHORES FROM ION-TRANSPORT SYSTEMS*

Adil E. Shamoo and Thomas E. Ryan

Department of Radiation Biology and Biophysics
The University of Rochester
School of Medicine and Dentistry
Rochester, New York 14642

Introduction

There is considerable experimental evidence linking membrane-bound proteins with active and passive transport of ions.[1-20] Primary among these are $Na^+ + K^+$-ATPase and $Ca^{++} + Mg^{++}$-ATPase, both of which have been implicated in the active transport of Na^+ and Ca^{++}, respectively.[1-20]

In 1972, Racker[16] initially demonstrated ATP-dependent Ca^{++} uptake into vesicles reconstituted from sarcoplasmic reticulum $Ca^{++} + Mg^{++}$-ATPase and soybean phospholipids. These experiments were later confirmed by Warren *et al.*[18] using a more defined phospholipid in the reconstituted vesicles. Similar experiments have shown active Na^+ transport in vesicles prepared from the $Na^+ + K^+$-ATPase.[19] This phenomenon has recently been confirmed and elaborated upon by Hokin *et al.*[20] This type of experiment is the first directly indicating that the primary ion "pump" function resides in the various ATPases.

Membrane-bound proteins are also considered responsible for mediating and controlling ion movements and synaptic transmission of nerve membranes. The acetylcholine receptor in some way provides an open channel for monovalent cation movement. These experiments have clearly identified the proteins and controlling factors that mediate various types of transport. Little evidence is available as to the mode of operation of these substances or the nature of the sites where ions interact with or are transported across the membrane.

In this report an ionophore is defined in the broadest sense, that is, material that produces an increase in bilayer lipid membrane (BLM) conductance. Work from this laboratory in the past three years has demonstrated the isolation and characterization of certain cation-dependent and/or -selective ionophores from pure membrane-bound proteins. Dependency is defined as the need for a certain ion for incorporation into the bilayer for the initial increase in conductance. A Na^+-dependent ionophore has been isolated from the $Na^+ + K^+$-ATPase of the eel electric organ.[8, 9, 12] It was further shown that a 2,000 dalton peptide fragment of the $Na^+ + K^+$-ATPase was a cation-selective ionophore. The Na^+-dependent ionophore was also found as part of the 57,000 dalton subunit of $Na^+ + K^+$-ATPase. More recently we have shown that a Ca^{++}-dependent and -selective ionophore is part of the sarcoplasmic reticulum

* This paper was based on work partially performed under contract with the United States Energy Research and Development Administration at the University of Rochester Biomedical and Environmental Research Project and has been assigned Report No. UR-3490-703. This paper was also supported in part by National Institutes of Health Grant No. 1 RO1 AM17571 to A.E.S.

Ca^{++} + Mg^{++}-ATPase obtained from white skeletal muscle of the rabbit. It was demonstrated that solubilized Ca^{++} + Mg^{++}-ATPase and cleaved peptide portions of the ATPase contained similar Ca^{++}-dependent and -selective ionophoric characteristics.

In recent years a variety of attempts have been made to isolate ion-transport mediators from tissues primarily involved in specific ion movements.[8–12, 21–29] Many of these attempts have never clearly identified the protein of origin or purified ionophoric material.[21–29] Furthermore, certain experiments were not reproducible.[27, 29]

In this report we will present evidence for the characterization and isolation of variety of ionophores from various sources. We will also present evidence for the following: (1) the characterization of Na^{+}-dependent ionophore from Na + K-ATPase; (2) the isolation of cation-selective ionophoric peptide fragment from Na^{+} + K^{+}-ATPase; (3) the isolation of K^{+}-dependent ionophore from Na^{+} + K^{+}-ATPase; (4) the demonstration that a cholinergic-agent-sensitive cation-selective ionophore is part of the acetylcholine receptor; (5) the isolation and characterization of Ca^{++}-dependent ionophore as a peptide fragment (<2,000 daltons) from sarcoplasmic reticulum Ca^{++} + Mg^{++}-ATPase.[30]

Conductance Measurements

The lipid-forming solution was oxidized cholesterol prepared according to the published method.[31] The membrane-forming solution consisted of 20 mg oxidized cholesterol per milliliter of *n*-octane. FIGURE 1 depicts the circuit diagram for bilayer conductance measurements. The input impedance of the Keithly electrometer is $10^{14}\Omega$. The conductance of the bilayer lipid membrane (BLM) is measured by imposing a constant voltage pulse and measuring the current through the lipid bilayer, $G = \Delta I/\Delta V$, where G is the condutance in mho/cm^2 and ΔI is the change in current in amperes, due to the imposed voltage ΔV in volts. The current-voltage relationship is obtained by imposing a triangular waveform voltage across the bilayer and simultaneously recording

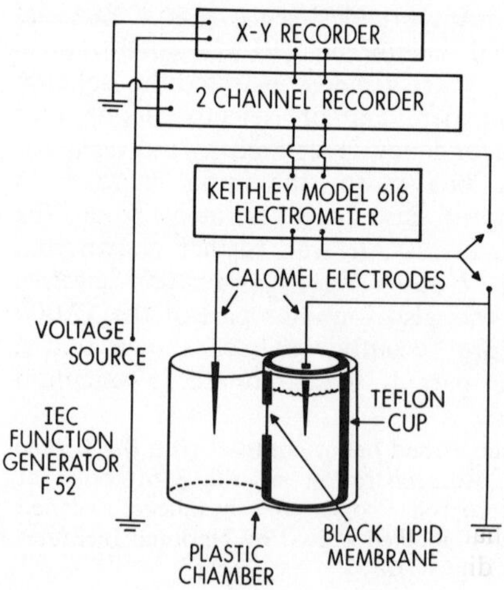

FIGURE 1. Schematic diagram of the measurement of electrical properties of black lipid membranes.

both the voltage and current with an X-Y recorder. The slope of the curve at any section gives the instantaneous conductance at that voltage. If a concentration gradient is placed across the bilayer, a potential across the membrane may develop. This is indicated by the intercept of the current voltage curve on the voltage axis. The sign of the intercept will indicate whether the lipid bilayer is cation- or anion-selective. The magnitude of the potential will be related to the permeability ratio of cations versus anions.

$$V = -\frac{(t_2 - t_1)}{t_2 + t_1} 58 \log \frac{C_2}{C_1} \quad \text{in mV}$$

where t_2 and t_1 are the bilayer transfer numbers for cations and anions, respectively; V is the voltage; and C_2 and C_1 are the concentrations of the salt on the two sides of the bilayer. t_2 and t_1 can be assumed to be proportional to their respective permeabilities. If we have more than one salt solution, we use the generalized Goldman, Hodgkin, and Katz equation[32] for monovalent cations (i.e., NaCl and KCl):

$$V = -\frac{RT}{F} \ln ((Na_i + (P_K/P_{Na})K_i + (P_{Cl}/P_{Na})Cl_o)/(Na_o + (P_K/P_{Na})K_o + (P_{Cl}/P_{Na})Cl_i))$$

where P stands for permeability and i and o represent the two compartments of the bilayer.

The black lipid membrane was formed on a 1-mm^2 hole in a Teflon cup placed in a plastic chamber. The standard electrolyte solution consisted of the indicated salt concentration and 5 mM histidine (pH 7.3, 23°).

Ionophoric Preparations

Due to the use of various tissue and preparative sources of the ionophoric material we will quote those methods published and briefly describe the new methods.

The membrane fraction rich in $Na^+ + K^+$-ATPase was prepared from eel electroplax as has been previously described.[33] The one-hour or the 24-hour tryptic digestion (400 trypsin: 1 protein, w/w) of the membrane fraction and its passage through DEAE column has been described.[10, 13] The two polypeptides of $Na^+ + K^+$-ATPase were obtained by preparative 0.1% SDS (sodium dodecyl sulfate) 5% polyacrylamide electrophoresis.[9]

Sarcoplasmic reticulum $Ca^{++} + Mg^{++}$-ATPase from rabbit skeletal muscles was prepared according to the method of MacLennan.[34] The $Ca^{++} + Mg^{++}$-ATPase was solubilized with 5:1 (w/w) succinic anhydride to $Ca^{++} + Mg^{++}$-ATPase at pH 7.8.[11] This solubilized enzyme contained the Ca^{++}-dependent and-selective ionophore and was termed S-Ca^{++}-ATPase.[11]

Ionophores from $Na^+ + K^+$-ATPase

The initial stage of the conductance change induced in oxidized cholesterol bilayer membrane by a one-hour tryptic digest of eel electroplax membranes rich in $Na^+ + K^+$-ATPase was Na^+-dependent[8] and the voltage dependency of the increase in conductance showed properties similar to the properties of alamethicin-doped bilayer membranes.[8] A 24-hour tryptic digest of the electroplax preparation provided active ionophoric material in greater yield and with similar properties to those of the one-hour tryptic digest with the notable exception that the sodium-dependent incorporation character was lost.[10, 13] The tryptic digests (whether one-hour or 24-hour) were passed through a DEAE-cellulose column equilibrated with 0.2 M ammonium acetate, pH 5.1.[8, 10, 13] The one-hour tryptic digest fragments were apparently inhomogeneous and therefore difficult to purify. Thus, we proceeded to purify the 24-hour tryptic digest with G-25 Sephadex column equilibrated with 0.5 M ammonium acetate at pH 7.0.[10] Each 1-ml fraction from the 32-ml column was tested for protein and ionophoric activity. The column was calibrated with the appropriate molecular weight standards. The elution position of the ionophoric material suggests that a rough estimate for the molecular weight of the ionophore may be approximately 2000 (FIGURE 2).

The $Na^+ + K^+$-ATPase preparation from the electric organ of *Electrophorus* consists of two polypeptides, as resolved by sodium dodecyl sulfate (SDS) preparative gel electrophoresis.[9] The two polypeptides were identified on analytical SDS-gel electrophoresis. The large polypeptide (L) has a molecular weight of 85,200 daltons and the small polypeptide (S) a molecular weight of 57,000 daltons.[9]

FIGURE 3 is a composite of several experiments. Each ionic condition represents an independent experiment on a bilayer. For each ionic condition, the data are representative of at least six different experiments.

The conductance was constant at $\sim 10^{-8}\Omega^{-1}/cm^2$ under all or any of the ionic conditions before the introduction of peptides into the bathing fluid of the bilayer. The conductance in the control period between 0 and 4 minutes was constant and displayed no rectification when voltage was changed from +50 mV to —50 mV. Rectification is defined as an asymmetric change in conductance due to changes in the polarity of the applied voltage. The data represent the experiments in which, at four minute's time, we added about 5×10^{-4} mg/ml of either (a) both polypeptides L and S in weight ratio of 1:1 or (b) small polypeptide alone. Similar data were obtained for both systems.

When the voltage was positive (i.e., inside Teflon cup +), the current flowed from the inside of the Teflon cup toward the outside. The ionophoric material was then placed inside the Teflon cup. When the voltage was positive (between 4 and 6 minutes) on the side where the ionophoric material was located, conductance increased with any cation. When the voltage was negative (between 6 and 8 minutes), not only was there no increase in conductance in the absence of Na^+, but also conductance decreased to the BLM original conductance. When Na was present, the conductance continued to increase slowly. When the voltage was returned to positive (between 8 and 10 minutes), there was again an increase in conductance under all of the ionic conditions. However, when the voltage was negative, we observed that the conductance increased further only in the presence of Na^+. Data on the effect of other fractions or polypeptides of $Na^+ + K^+$-ATPase on BLM conductance can be located in our original paper.[9]

We have encountered several important difficulties in the ionophoric properties of the purified polypeptides of $Na^+ + K^+$-ATPase. We understand some

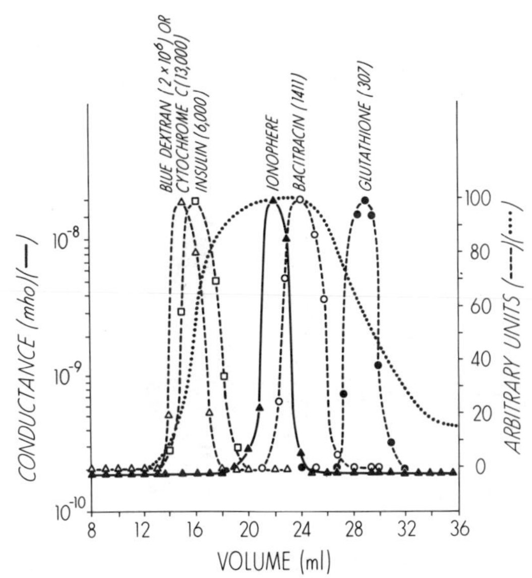

FIGURE 2. Fractionation with G-25 Sephadex column; conductance and concentration of the molecular weight standards versus fraction number. A 32-ml G-25 Sephadex column was equilibrated with 0.5 M ammonium acetate at pH 7.0. The elution pattern of several standards with known molecular weights was determined on the column. The column was then eluted by 50 ml of 0.5 M ammonium acetate, pH 7.0. The presence of the standards in fractions of 1 ml of the eluate was determined by either direct optical density (A_{280}, A_{580}) measurements or by Lowry protein determination. A 1-ml fraction showed maximal amounts of the standard and was taken arbitrarily as 100. Other determinations of the standard were calculated as percentages of the maximal amount. After calibration of the column, 0.6 ml (2 mg/ml) of the eel ionophoric material was added to the column. The column was eluted by 50 ml of 0.5 M ammonium acetate, pH 7.0. One-ml fractions after lyophilization were then tested individually for activity on the conductance of oxidized cholesterol BLM at 5×10^{-4} mg/ml final concentration. The protein concentration of each fraction was determined prior to testing. Then variable volumes (usually from 1 to 10 ml) were introduced to give approximately 5×10^{-4} mg/ml. (*Solid line*): the ionophoric activity (as an increase in conductance) versus the fraction number. (*Broken line*): the location of the standards versus the fraction number. (*Dotted lines*): the location of the protein versus the fraction number.

of the causes for those difficulties, but we still do not understand others. The most important difficulty is that the Na^+-dependency associated with the purified small polypeptide of $Na^+ + K^+$-ATPase was not reproducible for every preparation. In addition, the Na^+-dependency property was lost for several reasons. For example, the number of days the small polypeptide was subjected to a certain treatment in order to remove bound SDS resulted in the loss of Na^+-dependency. If dialysis was not sufficient, a residual SDS affected the bilayer membrane conductance. The treatment of the small polypeptide possessing Na^+-dependent ionophoric activity with 8 M urea to remove most of the bound SDS followed by dialysis to remove the urea results in loss of Na^+-dependency of the ionophoric material present before the urea treatment.

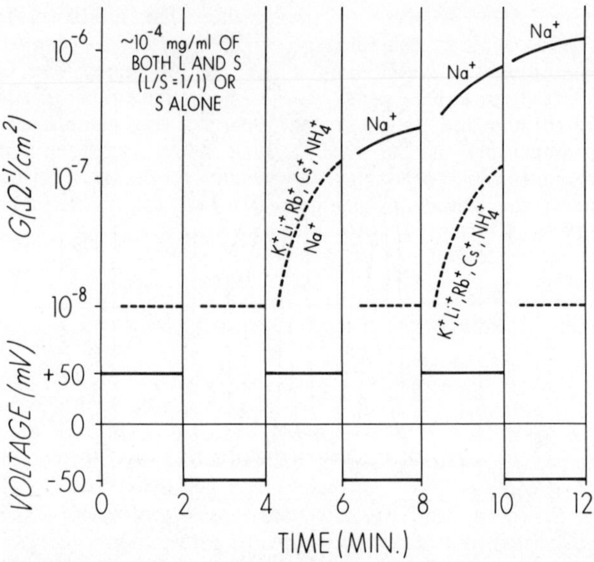

FIGURE 3. Effect of polarity of voltage clamp on a black lipid membrane conductance in the presence of various polypeptides. The vertical axis represents conductance in Ω^{-1}/cm^2 and voltage in mV in the lowest panel. The horizontal axis represents time in minutes. At the period from 0 to 4 min the bilayer bathing solution consisted of 0.1 M salt (either NaCl, LiCl, CsCl, RbCl, or NH_4Cl) on both sides of the membrane and 5 mM histidine, pH 7.4, 19° C. At t = 4 min, we introduced into the bathing fluid inside the Teflon cup either $\sim 5 \times 10^{-4}$ mg/ml of both large and small polypeptide of $Na^+ + K^+$-ATPase together (i.e., weight ratio of 1:1) or $\sim 5 \times 10^{-4}$ mg/ml of the small polypeptide of $Na^+ + K^+$-ATPase. The voltage clamp for every 2 min was either +50 mV or −50 mV as indicated. The +50 mV indicates that the solution inside the Teflon cup is subjected to positive voltage with respect to the solution outside the Teflon cup. The polypeptides were at all times introduced into the inside of the Teflon cup. A positive potential produced a flow of a positive current from the bathing fluid inside the Teflon cup, to the chamber outside the Teflon cup.[9]

Also, succinylation of the small polypeptide results in loss of the Na^+-dependency. The treatment of the small polypeptide with either acetone to remove SDS or with chloroform:methanol (2:1) also resulted in the loss of Na^+-dependency. Another cause for the loss of Na^+-dependency may be the presence of Lubrol with the polypeptides. The presence of Lubrol with the protein causes the formation of a precipitate during Lowry's determination (unpublished data). The small polypeptide after isolation with SDS-preparative gel electrophoresis and prolonged dialysis showed the presence of varied amounts of bound Lubrol as evidenced by the precipitate formed during Lowry protein determinations. However, all of the above preparations maintained monovalent cationic dependency and selectivity. This has led us to conclude that Na^+-dependency depends on the three-dimensional configuration of the small polypeptide and also on the amount of the residual lipid environment.

We now use cyanogen bromide (CNBr) treatment to cleave the intact enzyme and the small polypeptide of $Na^+ + K^+$-ATPase. Cyanogen bromide cleaves peptide bonds adjacent to methionine. The CNBr-treated intact or small polypeptide was passed through G-25 Sephadex column equilibrated with 0.1 M ammonium acetate and eluted by the same. The CNBr-treated $Na^+ + K^+$-ATPase gave two peaks on G-25 Sephadex. One peak appeared near the void column and the other peak near the final volume of the column. The protein appearing at the void volume when tested on BLM showed K-dependent and cation-selective ionophoric properties. Further work is needed to purify the peptide fragments responsible for this K^+-dependent ionophoric activity.

Ionophores from $Ca^{++} + Mg^{++}$-ATPase

In order to generalize the concept of the existence of ion-transport mediators as part of membrane-bound proteins and especially as part of enzymes involved in active transport of ions, we initiated a search for a calcium ionophore in the sarcoplasmic reticulum of white skeletal muscle in the rabbit. Purified $Ca^{++} + Mg^{++}$-ATPase was solubilized by succinylation (succinic anhydride:protein 5:1 w/w, and pH maintained between 7 and 8). This soluble preparation consisted of one protein species of molecular weight 100,000.[11]

FIGURE 4 represents the data on conductance of bimolecular lipid membrane versus time. It is clear that as the concentrations of S-Ca^{++}-ATPase increased, the rate of conductance and the maximal conductance increased. We detected conductance changes with concentration of S-Ca^{++}-ATPase of 1 mg/ml (i.e., 10 pM).

We found that S-Ca^{++}-ATPase caused similar increases in conductance when bilayer membranes were formed from either oxidized cholesterol or soybean

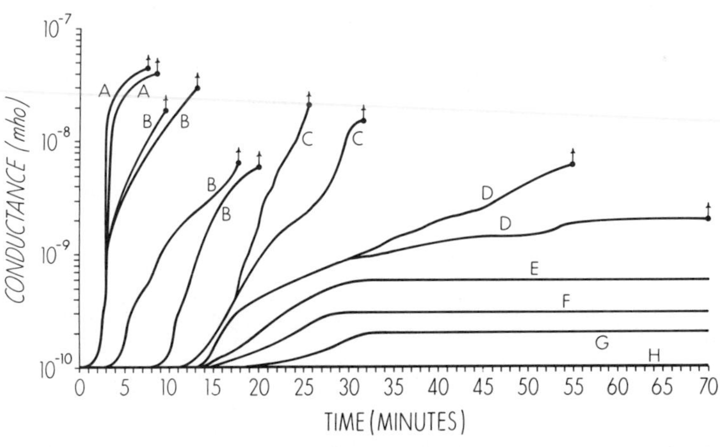

FIGURE 4. The time-response of black lipid membrane conductance in the presence of various amounts of Suc-Ca^{++}-ATPase. The membrane was bathed on both sides with 10 mM CaCl + 5 mM histidine (pH 7.3). The concentrations of Suc-Ca^{++}-ATPase in mg/ml: A = 1.6×10^{-3}; B = 2.5×10^{-4}; C = 5.4×10^{-5}; D = 2.4×10^{-5}; E = 8.0×10^{-6}; F = 5.4×10^{-6}; G = 2.7×10^{-6}; and H = 1×10^{-6} or zero.[11]

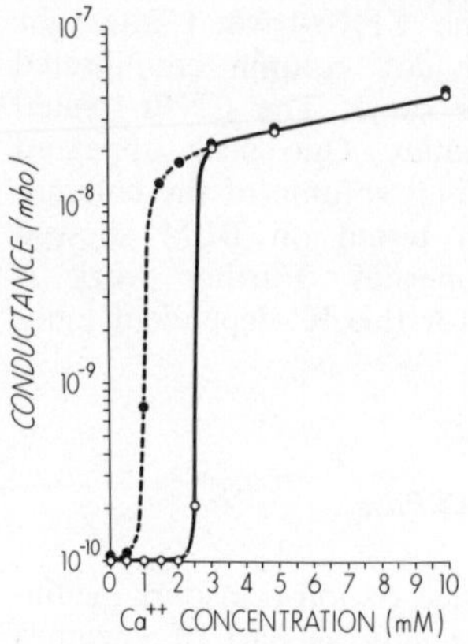

FIGURE 5. Bimolecular lipid membrane conductance against Ca^{++} concentrations in the presence of 2×10^{-4} mg/ml of Suc-Ca^{++}-ATPase (*solid line*) or sonicated Ca^{++}-ATPase (*broken line*). The membrane bathing fluid consisted of the indicated Ca^{++} concentrations with 5 mM histidine (pH 7.3). The conductance was continuously monitored under 50-mV voltage clamp conditions.[11]

phospholipid + cholesterol or soybean phosphatidylcholine + cholesterol (unpublished data).

FIGURE 5 represents the data on conductance versus calcium concentrations. The conductance increased markedly at calcium concentrations beyond 2.5 mM when S-Ca^{++}-ATPase was used and beyond 0.1 mM when Ca^{++}-ATPase was sonicated into the bathing fluid of the bilayer. Sonication of the insoluble Ca^{++}-ATPase into the fluid bathing the membrane proved to be sufficient to incorporate the enzyme into the lipid bilayer and have its ionophoric properties exposed. The lower dosage of Ca^{++} required to increase BLM conductance when sonicated Ca^{++}-ATPase was used instead of S-Ca^{++}-ATPase may be due to the binding of calcium ion with the succinic acid groups within the bilayer in the S-Ca-ATPase. The calcium binding to succinic acid groups may therefore reduce the effective calcium concentration adjacent to the bilayer.

FIGURE 6 represents the data on current-voltage characteristics for bionic potential of 3 mM $CaCl_2$ on one side and 3 mM $ZnCl_2$ on the other side in the presence of S-Ca^{++}-ATPase. The +V indicates that the Ca^{++} side of the membrane was positive and —V indicates that the other side containing $ZnCl_2$ was positive. The intercept with voltage axis (at I = 0) represents the bi-ionic

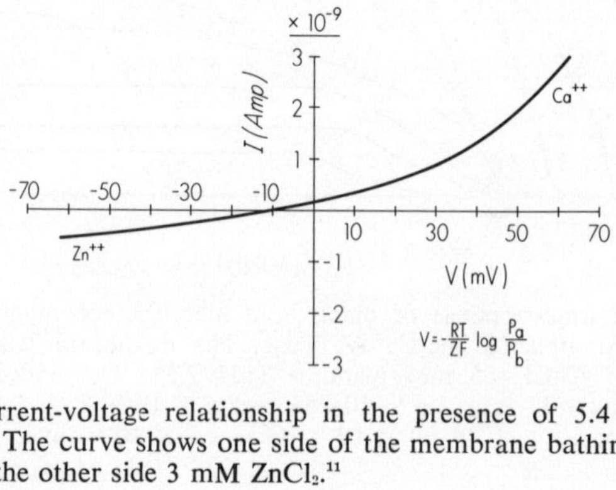

FIGURE 6. Current-voltage relationship in the presence of 5.4×10^{-4} mg/ml of Suc-Ca^{++}-ATPase. The curve shows one side of the membrane bathing fluid containing 3 mM $CaCl_2$ and the other side 3 mM $ZnCl_2$.[11]

potential, which can be related to the permeability of two ions bathing the membrane by the formula

$$V = -\frac{RT}{ZF} \ln \frac{P_a}{P_b},$$

if we assume no anionic permeability. P_a and P_b are the permeabilities of two opposite divalent cations; V is the bi-ionic potential; R, T, Z, and F have the usual meaning. Similar I-V data were obtained for Ca^{++} versus other divalent cations in the presence of S-Ca^{++}-ATPase.[11] The calculated permeabilities with respect to Ca^{++} are shown in TABLE 1. The selectivity sequence is $Ba^{++} > Ca^{++} > Sr^{++} > Mg^{++} > Mn^{++} > Zn^{++}$.

TABLE 1

RELATIVE PERMEABILITY AND INCREASE IN CONDUCTANCE OF BILAYER LIPID MEMBRANE IN THE PRESENCE OF SOLUBILIZED Ca^{++}-ATPASE WITH RESPECT TO OTHER DIVALENT CATIONS*

Ion	$P_{Ca^{++}}/P_{D^{++}}$	Fold Increase in G
Ba^{++}	0.55	912
Ca^{++}	1.00	210
Sr^{++}	1.49	175
Mg^{++}	1.89	66
Mn^{++}	2.04	10
Zn^{++}	2.21	1.0

* The first column represents the relative permeability ratio data calculated from the intercept of current voltage curve under bi-ionic conditions of 3 mM $CaCl_2$ on one side of the membrane and 3 mM salt ($BaCl_2$, $MgCl_2$, etc.) on the other side of the membrane. The Suc-Ca^{++}-ATPase was present on both sides at 5.4×10^{-4} mg/ml. P_D represents the permeability of the other cations. The second column represents the steady-state conductance or simply the maximal conductance achieved with 30 minutes after the formation of the membrane in the presence of 5.4×10^{-4} mg/ml of Suc-Ca^{++}-ATPase and 3 mM chloride salt of the indicated cation.

TABLE 1 also gives the data of the fold increase in conductance in the presence of the divalent cation and S-Ca^{++}-ATPase. The conductance sequence is the same as the selectivity sequence. The selectivity sequence of $Ba^{++} > Ca^{++} > Sr^{++} > Mg^{++}$ fits with the second selectivity sequence predicted by Sherry[35] on theoretical grounds where the free energy of hydration dominates and the site-ion interaction is weak. This simple fit of the selectivity sequence of the ionophore to the simple selectivity sequence of cage-like compound models of the theoretically predicted sequence indicates that the Ca^{++} carrier in the $Ca^{++} + Mg^{++}$-ATPase has a simple structure and is not greatly hindered by the remainder of the $Ca^{++} + Mg^{++}$-ATPase molecule.

An increase in BLM conductance occurred with S-Ca^{++}-ATPase and Ba^{++}, Ca^{++}, Sr^{++}, Mg^{++} and Mn^{++}, but not with Zn^{++}. Furthermore, the presence of 3 mM Zn^{++} with 3 mM Ca^{++} on both sides of the membrane completely inhibited the Ca^{++} conductance. No increase in conductance was ever observed with Na^{+}, or K^{+} alone. If the membrane was formed from the Zn^{++}, Na^{+}, K^{+} side, there was no increase in conductance.[11]

We have found that the inhibitory effects of methylmercury on Ca^{++} transport into sarcoplasmic reticulum and $Ca^{++} + Mg^{++}$-ATPase activity are similar ($K_I = 5 \times 10^{-5}$ M for preincubation).[14, 36] This result indicates that methylmercury acted at a single rate-limiting step in inhibiting both the enzyme activity and the active transport. This is consistent with the theory that the energy source for Ca^{++} transport comes from hydrolysis of ATP by $Ca^{++} + Mg^{++}$-ATPase.

Mercuric chloride also produced parallel dose-effect curves in inhibiting the calcium transport and $Ca^{++} + Mg^{++}$-ATPase. But the K_I for Ca^{++} transport was consistently lower than the K_I for $Ca^{++} + Mg^{++}$-ATPase (8×10^{-5} M and 1.1×10^{-4} M). Also, both K_I's were higher than those for methylmercury inhibition of both parameters. One possible mechanism for the differential potency is that CH_3Hg^+ has one cationic charge and therefore can come closer to the —SH group than the divalent Hg^{++}. This is consistent with the observation that mersalyl inhibits $Ca^{++} + Mg^{++}$-ATPase and Ca^{++} transport in similar fashion to methylmercury,[37] but is not consistent with the observation that *N*-ethylmaleimide (an uncharged —SH blocker) is no more potent than methylmercury.[38] A more reasonable explanation for the higher sensitivity of both functions to methylmercury than to mercuric chloride is that the target —SH groups are in a slightly hydrophobic environment.

We found that $HgCl_2$ blocked the ionophoric activity in the presence of Ca^{++} completely where methylmercury, also a sulfhydryl group blocker, had no effect. Therefore, the $HgCl_2$ inhibition is not due to its binding to a sulfhydryl group, but rather it may be to a replacement of Ca^{++} at the Ca^{++}-binding ionophoric site. This was confirmed since Zn^{++}, Mn^{++}, and La^{+++} all had similar inhibitory effects on the ionophore, as was the case with Hg^{++}. With $HgCl_2$, the active transport of calcium ions was more susceptible to inhibition than the $Ca^{++} + Mg^{++}$-ATPase activity. This is explained in terms of a summation of two independent actions of Hg^{++} on the transport system: the inhibition at the enzyme site and the inhibition at the ionophoric site, first reported here. The evidence that methylmercury does not act on the ionophore is thus both direct (from BLM experiments) and indirect (from differential effect on Ca^{++} transport and ATPase activity). This is the first evidence to indicate that the Ca^{++}-dependent ionophore has no sulfhydryl groups essential for its activity.

Ionophores from Acetylcholine Receptor

Further experiments were performed on acetylcholine receptor to show that ion-transport mediators are part of the membrane-bound proteins. The acetylcholine receptor was purified from the electric organ of the electric ray, *Torpedo californica.* The tryptic digest of the acetylcholine receptor showed consistent carbamylcholine- and acetylcholine sensitivity-induced cation-selective conductance ($P_{Na,K}:P_{Cl} = 4.4$). Curare (10 pM) prevented the action of carbamylcholine.[15]

Purification of the Ca^{++}-Ionophore

The controlled tryptic digest of sarcoplasmic-reticulum-bound $Ca^{++} + Mg^{++}$-ATPase releases four proteins of 55,000, 45,000, 30,000, and 20,000 dalton molecular weight.[11] Those of weight 55,000 and 30,000 are known to come from Ca^{++} + Mg^{++}-ATPase because both are labeled with ^{32}P if they were incubated with γ-^{32}P-ATP under proper ionic conditions prior to tryptic digestion. The 20,000-dalton peptide fragment was shown to crossreact with the 55,000-dalton fragment, but it did not crossreact with the 45,000-dalton fragment.[39] The 20,000-dalton fragment showed a divalent cation-dependent and -selective ionophoric activity.[30]

The 20,000-dalton peptide fragment was digested in a mixture containing 2 mg/ml protein, 70% formic acid, and a 100:1 molar ratio of cyanogen bromide:protein. Following digestion the mixture was lyophilized, redissolved in 1 ml 0.1 M ammonium acetate, and placed on a 1.5 × 45-cm G-25 Sephadex column equilibrated with 0.1 M ammonium acetate. Two peaks, absorbing at 280 nm were resolved: the first, eluting at the void volume (V_0), contained 80% of the total protein eluted; the remaining protein was eluted at the total volume of the column (FIGURE 7). The elution volume (V_e) of the second

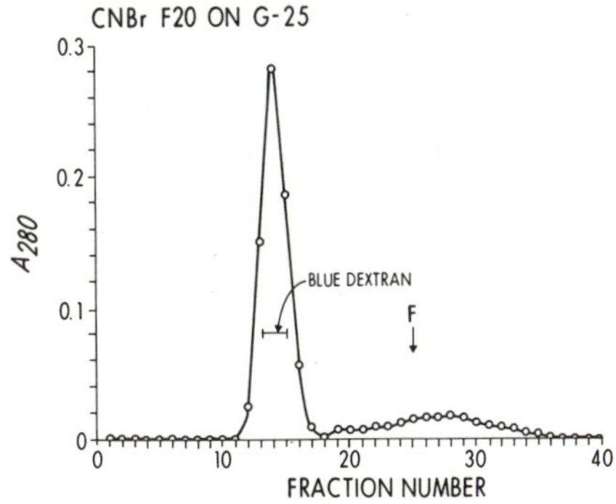

FIGURE 7. Chromatography of the cyanogen bromide digested 20,000-dalton peptide fragment of Ca^{++}-ATPase on Sephadex G-25. The column was equilibrated and developed with 0.1 M ammonium acetate. The volume of each fraction was 3.2 ml. Fractions 12–17 and 20–35 were combined, lyophilized, and tested for ionophoric activity. Molecular weight of the peak eluting at V_T was later estimated by fractionation on Bio-Gel P-2.

peak indicated a species with a molecular weight ≤2,000 daltons ($V_e = V_T$). Each peak was tested for ionophoric activity on the bilayer lipid membrane. Both showed divalent cation-dependent and -selective conductance increases. When analyzed on silica gel thin-layer chromatography each peak showed three and two spots for the respective high and low molecular weight peaks. Two had similar R_f's, indicating that the Sephadex fractionation was incomplete or that aggregation of peptides resulted in elution at V_0.

Model for Active Transport

We propose a hypothetical model for the active transport of ions. The model for active ion transport in FIGURE 8 consists of two main components in series. The first part represents a passive gate as part of the 55,000-dalton head followed by a channel as part of the 45,000-dalton piece. The energetic step for the active transport occurs at the 55,000-dalton fragment. The gate has defined selectivity which is consistent with the overall transport system, as is the case with the selectivity of Ca^{++}-dependent ionophore from sarcoplasmic reticulum Ca^{++} + Mg^{++}-ATPase. The energetic step, whether at the hydrolytic site of ATP or at the channel site, must have a selectivity sequence consistent with the overall transport. The selectivity of the overall transport is greater than the selectivity of each step.

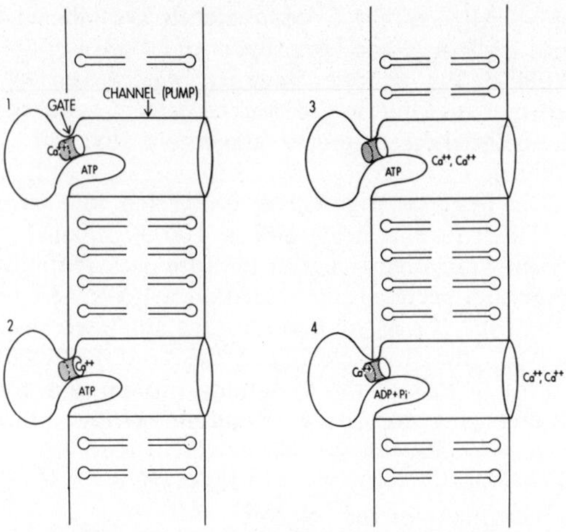

FIGURE 8. Model for active transport of ions (e.g., Ca^{++}).

Evidence supporting this model is as follows:

1. The two functions of Ca^{++} ionophoric activity and $Ca^{++} + Mg^{++}$-ATPase activity were separated by the use of mercurial compounds. Methylmercury inhibited the ATPase activity, but not the ionophoric activity. Furthermore, Ca^{++} transport into SR vesicles was more sensitive to inhibition by mercuric chloride than to the inhibition by methylmercury. Mercuric chloride acts on Ca^{++} transport by inhibiting two series functions: Ca^{++} ionophoric site and the energetic step of ATPase catalytic site. Mercuric chloride inhibited Ca^{++}-dependent ionophore as did Zn^{++}, La^{+++}, and Mn^{++}. The Ca^{++} ionophoric site was inhibited by competition of the divalent cations with Ca^{++} for the Ca^{++} ionophoric site.

2. The selectivity sequence for the Ca^{++} ionophoric activity was consistent with the overall Ca^{++} transport, but not the same as that of the ATPase catalytic function. Similarly, the ATPase hydrolytic function ionic dependency[11] was also consistent with the overall Ca^{++} transport, but not the same as that of the Ca^{++} ionophoric function. The overall selectivity is greater than the selectivity of the two series steps.

3. Zerahn[40] found that the isolated midgut of the larva of the American silkworm actively transports potassium but not Na^+, Li^+, and NH_4^+ from the blood-side to the lumen, when the bathing fluid contains calcium and magnesium. However, when magnesium and calcium were removed from the bathing medium, potassium, sodium, Li^+, and ammonium were transported. This not only implies that the K-selective gate is separate from the channel, but also that Ca^{++} and Mg^{++} concentrations altered the selectivity of the gate.

This model predicts several findings which can be tested experimentally: (1) The Ca^{++} ionophore site which will be associated with the gate can be chemically separated from the hydrolytic site of ATPase. The peptide fragment containing the phosphorylated intermediate or ATPase binding site will be different from the peptide fragment exhibiting the Ca^{++}-ionophoric activity. (2) Selectivity of the pump can be modified through alterations of the gate as long as the agent does not alter the hydrolytic site of the enzyme. (3) The chemically separated gate may act in synthetic bilayer as a carrier and the channel part may act as a channel-former.

Other ways to link ATP hydrolysis to the gate directly may actually exist. For example, the one-ion-dependent ATPases may be an example of the ionophoric and catalytic sites being one and the same.

References

1. SIEGEL, G. J. & R. W. ALBERS. 1970. *In* Handbook of Neurochemistry. Vol. **4:** 13. Plenum Press. New York, N.Y.
2. ALBERS, R. W. 1967. J. Biol. Chem. **239:** 345–350.
3. SKOU, J. C. 1971. Curr. Top. Bioenerg., **4:** 357–398.
4. SHAMOO, A. E. & W. A. BRODSKY. 1972. Biochim. Biophys. Acta **255:** 220–230.
5. SHAMOO, A. E. & W. A. BRODSKY. 1971. Biochim. Biophys. Acta **241:** 846–856.
6. SEN, A. K. & R. L. POST. 1964. J. Biol. Chem. **239:** 345–350.
7. GARRAHAN, P. J. & I. M. GLYNN. 1967. J. Physiol. **192:** 217–235.
8. SHAMOO, A. E. & R. W. ALBERS. 1973. Proc. Nat. Acad. Sci. U.S.A. **70:** 1191.
9. SHAMOO, A. E. & M. MYERS. 1974. J. Membrane Biol. **19:** 163.
10. SHAMOO, A. E., M. MYERS, R. BLUMENTHAL, & R. W. ALBERS. 1974. J. Membrane Biol. **19:** 129.
11. SHAMOO, A. E. & D. H. MACLENNAN. 1974. Proc. Nat. Acad. Sci. U.S.A. **71:** 3522.
12. SHAMOO, A. E. 1974. Ann. N.Y. Acad. Sci. **242:** 389.
13. BLUMENTHAL, R. P. & A. E. SHAMOO, 1974. J. Membrane Biol. **19:** 141.
14. SHAMOO, A. E., D. H. MACLENNAN & M. E. ELDEFRAWI. 1975. Chemico-Biological Interaction. In press.
15. SHAMOO, A. E. & M. E. ELDEFRAWI. 1975. J. Membr. Biol. In press. Biol. **8:** 363–388.
16. RACKER, E. 1972. J. Biol. Chem. **247:** 8198–8200.
17. MACLENNAN, D. H., T. J. OSTWAL & P. S. STEWART. 1974. Ann. N.Y. Acad. Sci. **227:** 527–536.
18. WARREN, G. B., P. A. TOON, N. J. M. BIRDSALL, A. G. LEE & J. C. METCALF. 1974. Proc. Nat. Acad. Sci. U.S.A. **71:** 622–626.
19. GOLDIN, S. M. & S. W. TONG. 1974. J. Biol. Chem. **249:** 5907.
20. HILDEN, S., H. M. RHEE & L. E. HOKIN. 1974. J. Biol. Chem. **249:** 7432.
21. GOODALL, M. C. & G. SACHS. 1972. Nature **237:** 252–253.
22. REDWOOD, W. R., H. MUELDNER & T. W. THOMPSON. 1969. Proc. Nat. Acad. Sci. U.S.A. **64:** 989–996.
23. JAIN, M. K., A. STRICKHOLM & E. J. CORDES. 1969. Nature **222:** 871–872.
24. JAIN, M. K., F. P. WHITE, A. STRICKHOLM & E. H. CORDES. 1972. J. Membrane Biol. **8:** 363–388.
25. KEMP, G., J. P. DOLLY, E. A. BARNARD & C. E. WENNER. 1973. Biochem. Biophys. Res. Commun. **54:** 607.
26. ROMINE, W. O., M. C. GOODALL, J. PETERSON & R. J. BRADLEY. 1974. Biochim. Biophys. Acta **316:** 316.
27. BLONDIN, G. A. 1974. Annals N.Y. Acad. Sci. **227:** 392.
28. BLONDIN, G. A. 1974. Biochem. Biophys. Res. Communn. **56:** 97.
29. GREEN, D. E., G. A. BLONDIN, R. KESSLER & J. H. SOUTHARD. 1975. Proc. Nat. Acad. Sci. U.S.A. In press.
30. SHAMOO, A. E., T. H. RYAN, P. S. STEWART & D. H. MACLENNAN. 1975. J. Biol. Chem. Submitted for publication.
31. TIEN, H. T., S. CARBONE & E. A. DAWIDOWICZ. 1966. Nature **212:** 719–720.
32. HODGKIN, A. L. & B. KATZ. 1949. J. Physiol. **108:** 37.
33. ALBERS, R. W., S. FAHN & G. J. KOVAL. 1963. Proc. Nat. Acad. Sci. U.S.A. **50:** 474–481.

34. MacLennan, D. H. 1970. J. Biol. Chem. **245:** 4508–4518.
35. Sherry, H. S. 1969. *In* Ion Exchange: II. 89–134. Marcel Dekker. New York, N.Y.
36. Shamoo, A. E. & D. H. MacLennan. 1975. J. Membr. Biol. In press.
37. Panet, R. & Z. Selinger. 1970. Eur. J. Biochem. **14:** 440.
38. Hasselbach, W. & K. Seraydarian. 1966. Biochem. Z. **345:** 159.
39. MacLennan, D. H. & P. S. Stewart. Unpublished data.
40. Zerahn, K. 1971. Philos. Trans. Roy. Soc. London, Ser. B **262:** 315.

Discussion

Dr. Fields: Did you do any radioactive flux measurements with calcium, sodium, or potassium to see how it relates to the conductance changes?

Dr. Shamoo: In lipid bilayer it is very difficult to measure calcium flux isotopically; we hope some day to measure fluxes in bubble type bilayer.

Dr. Cohen: Have you done these experiments with phospholipid bilayers? Also, in view of Dr. Haydon's earlier comments on the importance of the lipid species for ionophoric activity, I'd like your comment on the relevance of studies in oxidized cholesterol bilayers to biological function.

Dr. Shamoo: In every paper we state that oxidized cholesterol is the worst model of biological membranes, and we use it simply as an assay tool.

Second, we have tested the calcium-dependent ionophore in other phospholipids and it worked. We have used two others and Dr. Mueller has used another one and it worked similarly. It shows again the calcium-dependency, as we have reported it for oxidized cholesterol bilayers.

Dr. Armstrong (*University of Rochester, Rochester, N.Y.*): You showed a slide in which there was an extremely steep dependence of conductance on calcium.

Dr. Shamoo: What we see in lipid bilayer does not necessarily reflect what was going on *in vivo* in the functional calcium transport system. We believe that the whole field of bilayer is an artifact and we are hoping that what we observe in lipid bilayers is the mirror image part of the function. So I don't even try to explain it; you need several calcium ions to bind and open the channel. You could make such a derivation and fit the data, but we don't think this will necessarily reflect the real system.

Dr. Racker: Dr. Shamoo, we may get into semantic difficulty. You called your material an ionophore, yet in your model, which I like very much, you have pictured your particular fraction from $Ca^{++} + Mg^{++}$-ATPase as a gate. I think it is quite reasonable that by the time you split it up with trypsin it becomes an ionophore. I would like to see how you have pictured as this part being the gate, and perhaps a coupling factor that we have isolated is actually the channel. I don't think there's any conflict in our views on this.

Dr. Shamoo: I agree; there is no conflict with your views.

Dr. Holtz (*National Institutes of Health, Bethesda, Md.*): With your calcium ionophore have you ever looked for single channels?

Dr. Shamoo: This is a good question. The freshly prepared solubilized calcium ATPase causes a smooth increase in conductance, similar to that of valinomycin. Then it appears consistent with the carrier. If you age the preparation for three to eight months the material probably aggregates more and more, and we will observe channels of varying heights. The height of the channel is proportional to the number of months we have stored the material. This is why my answer to Dr. Armstrong was that what you see in lipid bilayer may or may not exactly describe the real system. But seeing those channels and saying that therefore the ionophore is a channel in the real system is beyond me.

DR. PRESSMAN: There is a whole class of electrically silent ionophores. The people using lipid bilayer assay only may be throwing the electrically silent ionophores down the sink.

DR. SHAMOO: I agree that our method only detects those ionophores that cause the transfer of charge eventually.

DR. MUELLER: As you probably know, a lot of us have spent a wasted ten years or so trying to get these various materials into bilayers or something that is planar barrier or something that can be recognized as having physiological significance. One of the big problems we have all run into is the apparent phenomenon that I refer to as a leakage-induced conductance. That is a ligand binding-induced conductance increase. You can take almost any kind of functional protein from an ATPase or trypsin as criteria, or, as has been shown, an antibody-antigen or acetylcholine esterase and acetylcholine. You add the protein if you have a sensitive membrane, and these oxidized cholesterol membranes are very sensitive; you bind the ligand, add the ligand, and the conductance goes up. That is, you have something now that specifically binds sodium and maybe undergoes a configuration change or just a lipid solubility as a response, so it didn't then penetrate the membrane. The effects you then see are not necessarily related to the physiological effects. In the future we have to sort out what is really happening.

DR. SHAMOO: In general, what Dr. Mueller said is correct. But not all the binding proteins show an increase in BLM conductance when their ligand is present. For example, we found that all of the calcium binding proteins from sarcoplasmic reticulum had no effect on BLM conductance in the presence of calcium. Moreover, I have heard that the vitamin-D-sensitive calcium binding protein from intestine had no effect on BLM conductance. Also, the amino acid binding proteins from bacteria had no effect on BLM conductance in the presence of their ligand (Oxender; personal communication). This is why we have started working on the $Ca^{+} + Mg^{+}$-ATPase, where we can obtain "pure" protein. In addition, those ATPases are known to be involved in the active transport of their associated ion. We have shown that a Ca^{++}-dependent and -selective ionophore is part of the intact protein. Moreover, we have shown that a small peptidic fragment of the parent protein contains the Ca^{++} ionophore. Then, when increases in conductances are associated with the transported ion in more than one example we cannot maintain the view that it is a happenstance.

ISOLATION, PROPERTIES, AND STRUCTURAL FEATURES OF DIVALENT CATION IONOPHORES DERIVED FROM BEEF HEART MITOCHONDRIA*

George A. Blondin†

Institute for Enzyme Research
University of Wisconsin
Madison, Wisconsin 53706

The isolation of small molecular weight ion carriers from biological systems has, within the past few years, begun to achieve the firm documentation that has been awaited since the introduction of the concept by Osterhout in 1933.[1] The accumulation of an immense body of data over the past ten years focusing on bacterial and synthetic ionophores and the framework of principles derived from the latter have undoubtedly contributed singularly to this documentation. The documentation itself of biological ion carriers and channels has proceeded on three major fronts and it is worthwhile at the outset to briefly consider the directions that different investigators have taken in order that a clear understanding of the various positions can be seen in perspective.

The pursuit of natural ion carriers in biological systems may first be considered to encompass either (a) the documentation of ionophoretic capacity and function of known chemical species present in biological membranes, or (b) the isolation and identification from transport systems of novel chemical species or molecular complexes which possess demonstrable ionophoretic activity and function. For example, some of the known compounds for which an ionophoretic function has been suggested and partially documented include cardiolipin,[2] lecithin,[3] cytochrome *c*,[4] free fatty acids,[5] and prostaglandin E_1.[6,7] On the other hand, the isolation and identification of novel ionophoretic species have, in general, evolved according to two seemingly diverse methods of approach to the problem. Our own studies have focused on the potentially monodispersed nature of novel natural ion carriers, whereas the predominant feature of the studies of other investigators in this area of inquiry is derived from the recognition of ionophoretic channel activity associated with various lipoprotein preparations.

These two points should be made abundantly clear in order for the relationship between the two experimental approaches to be understood in its proper perspective. The following example may serve to illustrate this relationship more clearly. In 1971, we reported[8] the isolation of a $Na^+ + K^+$-specific dodecapeptide ionophore which was derivable from beef heart mitochondria by simple organic solvent extraction and chromatographic purification. Subsequent studies[9] revealed that this ionophore is not normally monodispersed in mitochondrial membrane bilayer lipid, but that it is, in fact, very tightly associated with membrane protein subunits, wherein lies the modality of its function.

* This work was supported in part by Program Project Grant GM-12847 from the National Institute of General Medical Sciences, United States Public Health Service, Bethesda, Maryland.

† Established Investigator of the American Heart Association.

In other words, the functional unit of $Na^+ + K^+$ ionophoretic activity in mitochondria is the ionophoroprotein complex. However, by the imposition of sufficiently drastic conditions, such as trypsin digestion of lipid-free inner mitochondrial membrane and subsequent extraction with butanol:acetic acid:water, the free ionophore can be displaced from the ionophoroprotein complex and isolated as a classical small molecular weight ionophoretic species. Moreover, when the isolated ionophore is added back to freshly prepared mitochondria, it is able to perform its ionophoretic function in the membrane bilayer.

On the other hand, Goodall and Sachs[10] in 1972, and Shamoo and Albers[11] in 1973, described the isolation of channel-forming ionophoretic material with alkali metal cation specificity from excitable tissue and electroplax tissue, respectively. These ionophoretic materials were described as lipoprotein or protein complexes that possessed documented ionophoretic activity only in artificial lipid bilayer membranes. However, both their manipulation and method of extraction and partial purification would argue against the notion that these preparations are chemically analogous to classical ionophores. Our interpretation of these studies is that these investigators have succeeded in isolating "ionophoroprotein" complexes and we predict that classical ionophores can be displaced from the latter by suitable procedures, as Dr. Ralph Kessler has demonstrated in our laboratory[2] in the case of a specific ionophoroprotein component of the base-piece sector of the mitochondrial oligomycin-sensitive ATPase. In the latter case, the displaceable ionophore component is divalent cation-specific and is a member of a family of structurally related keto- or hydroxy-octadecadienocic acid congeners that we have isolated and characterized from lipid-free mitochondria and that I will now describe.

Isolation of Oxyoctadecadienoic Acid Compounds from Mitochondria

We had previously described[12] a lipid-soluble isolate from a tryptic digest of mitochondrial acetone powder that possessed divalent cation ionophoretic activity. The active material appeared only to be extractable in the presence of sec-butanol:acetic acid:water (4:1:5) and the final active residue obtained after a single chromatographic fractionation was soluble in chloroform and catalyzed $Ca^{++}/2H^+$ exchange across a carbon tetrachloride phase separating two discontinuous aqueous phases, as described by Pressman.[13] However, in the course of our attempts to further resolve the active species, we soon recognized that the activity was referable to multiple, chromatographically different components which were poorly resolvable. We soon discovered that the resolution was enormously simplified by first separating the crude butanol: acetic acid:water (BAW) extracted residue into four subfractions as shown in TABLE 1. For example, by this procedure, all of the subsequently purified neutral congeners were placed in subfraction I, 99% of two conjugated isomeric hydroxy-octadecadienoic acids were placed in subfraction IV, and approximately 95% of a keto acid congener was placed in subfraction III. Subsequent purification of the active species was in each case achieved by chromatographic fractionation on silicic acid and Sephadex LH-20. An example of the results obtained by such procedures is shown in FIGURE 1, which describes the fractionation of the neutral residue (subfraction I) on silicic acid with chloroform: methanol gradients, as shown in A, followed by the refractionation of the activity described under zone III on Sephadex LH-20 in 1,2-dichloroethane:eth-

TABLE 1

SUBFRACTIONATION OF BUTANOL:ACETIC ACID:WATER EXTRACT OF MITOCHONDRIA

BAW residue, 126 ± 11 mg/gm*

1. Solution in 285 ml $CHCl_3$:MEOH:H_2O (1:2:0.8)
2. Phase separation by addition of 75 ml $CHCl_3$ + 75 ml 5% aq. K_2CO_3
3. Re-extraction of lower $CHCl_3$ phase with 285 ml MEOH:H_2O:aq. 5% K CO (2:0.8:1)

Lower phase residue (LP-2R), 69.4 ± 10.3 mg/gm

1. Solution in 100 ml $CHCl_3$
2. Filtration through 40 gm silicic acid (3.0 × 20 cm bed)
3. Continued elution with 300 ml $CHCl_3$:MEOH (9:1)

↓

Filtrate residue (F-1R), 18.6 ± 1.2 mg/gm

1. Chromatography on Sephadex LH-20 (3.0 × 23 cm bed)
2. Elution with 120 ml $CHCl_3$:Hexane (65:35)–Fraction I
3. Elution with 220 ml $CHCl_3$:Hexane:ETOH(40:40:20)–Fraction II

Fraction I: SF-I, 7.39 ± 0.6 mg/gm

Fraction II: SF-II, 10.67 ± 0.85 mg/gm

Combined upper phases (UP-2)

1. Acidified with 56 ml conc. acetic acid
2. Extracted with 300 ml $CHCl_3$
3. Upper phase reextracted with 275 ml $CHCl_3$

↓

Combined lower phase residue (LP-3R),† 36.6 ± 2.3 mg/gm

1. Chromatography on 40 gms silicic acid (3.0 × 20 cm bed)
2. Elution with 600 ml $CHCl_3$–Fraction III
3. Elution with 300 ml $CHCl_3$:MEOH (9:1)–Fraction IV

Fraction III: SF-III, 18.73 ± 1.1 mg/gm

Fraction IV: SF-IV, 3.50 ± 0.3 mg/gm

* Yields expressed as mg dry residue per gm mitochondrial protein starting material ± the average error of four experiments.

† Prior to the evaporation of LP-3, 56 ml toluene was added in order to effect complete removal of residual acetic acid by azeotropic action. The original crude BAW residue was similarly submitted to reevaporation to dryness in the presence of toluene.

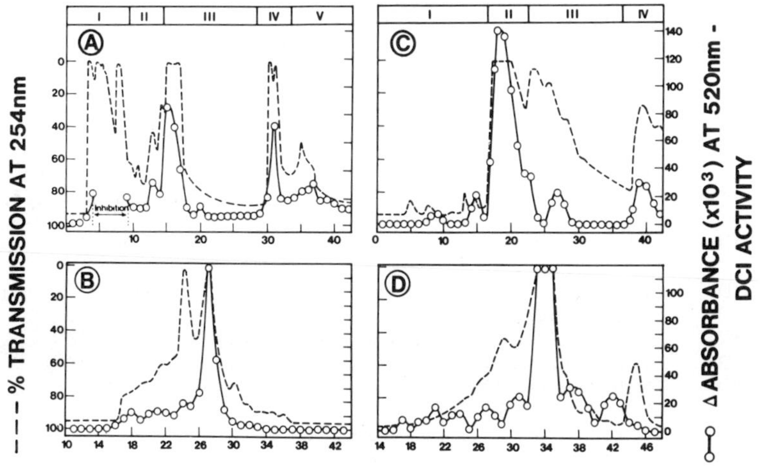

FIGURE 1. Fractionation of the butanol:acetic acid:water subfraction residues. The chromatographic fractionation of SF-I and SF-IV on silicic acid is described by (A) and (C) respectively. The refractionation of fractions described under Zone III in (A) and Zone II in (C) on Sephadex LH-20 is described in (B) and (D), respectively. Silicic acid (34 grams) was slurried in chloroform and packed into a glass column (2.0 cm O.D.) to a bed height of 29 cm. Sephadex LH-20 was preswollen in 1,2-dichloroethane:ethanol (66.5:33.5) and packed into a glass column (2.8 cm O.D.) to a bed height of 105 cm. Column A was eluted with 340 ml chloroform (fractions 1–27) followed by 200 ml (19:1) chloroform/methanol (fractions 28–42). Column C was eluted with 340 ml 1% methanol in chloroform (fractions 1–27) followed by 200 ml (19:1) chloroform/methanol (fractions 28–42). Individual fraction volumes in both A and C varied between 12 and 13 ml each. Columns B and D were both developed with 1,2-dichloroethane:ethanol (66.5:33.5) and individual fractions of 10.5 ml and 11.0 ml each were collected, respectively, in B and D. A continuous recording of the ultraviolet absorption at 254 nm of the effluent was obtained with each chromatogram and individual fractions were assayed for divalent cation ionophoretic (DCI) activity as described elsewhere.[12]

anol (66.5:33.5), as described in B. The ionophoretically active material described in B encompassing fractions 26-29 was additionally purified in this case by fractionation on silicic acid using diethyl ether:hexane gradients and on Sephadex LH-20 in chloroform:hexane (65:35) to yield pure methyl ester of 9-hydroxy-10-*trans*, 12-*cis*-octadecadienoic acid (9-MHOD).

Following a similar protocol, the fractionation of subfraction IV residue on silicic acid is shown in C, and the ionophoretically active material described under zone II was refractionated on Sephadex LH-20 in 1,2-dichloroethane:ethanol, as shown in D. The active material in D between fractions 17 and 23 was additionally purified by recycling chromatography on the same column to yield a mixture of the 9- and 13-hydroxy isomers of octadecadienoic acid. The latter were separated from each other by fractionation on Sephadex LH-20 in chloroform:hexane:ethanol (60:35:5) to yield pure 9-hydroxy-10-*trans*, 12-*cis*-octadecadienoic acid (9-HOD) and 13-hydroxy-9-*cis*, 11-*trans*-octadecadienoic acid (13-HOD).

TABLE 2
SUMMARY OF PURIFIED DIVALENT CATION IONOPHORES DERIVED FROM BEEF-HEART MITOCHONDRIA

Designation	Structure	Sub-fraction Origin	Yield	
			ngm per mg protein	nmol per mg protein
9-HOD	9-hydroxy-*trans*-10-*cis*-12-octadecadienoic acid	IV	419	1.4
13-HOD	13-hydroxy-*cis*-9-*trans*-11-octadecadienoic acid	IV	238	0.8
9-KOD	9-keto-10,12-octadecadienoic acid	III }	198	0.67
13-KOD	13-keto-9,11-octadecadienoic acid	III }		
9-MHOD	Methyl ester of 9-HOD	I	85.4	0.28
9-MKOD	Methyl ester of 9-KOD	I	37.5	0.12
DCI-N_3	Unknown; not related to above acids	I	37	–
DCI-N_2	Unknown; not related to above acids	I	52	–
DCI-$A_2\alpha$	Unknown; related to above acids	III	111	–
DCI-$A_2\beta$	Unknown; related to above acids	III	28	–

The structures of these and other congeners isolated by similar chromatographic procedures were deduced from ultraviolet, infrared, and nmr spectroscopic methods and by mass spectrometry of parent and derivatized compounds; and, the full documentation of these structures will appear elsewhere.[14] TABLE 2 presents a summary of the oxyoctadecadienoate congeners and other divalent cation ionophoretic components which we have isolated from lipid-free beef heart mitochondria, together with their structures and yields, where known. Several key points originate from this Table, including the following: (1) All of the species listed must be considered to have been originally firmly bound to membrane protein, since they are not observed in the original lipid extract. (2) Their release from lipid-free mitochondria requires some form of membrane perturbation, such as trypsin digestion or solubilization with detergents or chaotropic agents. (3) The yields of structurally known compounds appear to fall in the range of the concentration in mitochondria of cytochrome c_1. (4) Certain of the ionophoretically active components are spectroscopically unrelated to the oxyoctadecadienoate family and several have yet even to be purified. Thus, some surprises may be in store as the structures of these species unfold and their ionophoretic activities are more extensively documented.

Ionophoretic Properties of the Neutral Mitochondrial Divalent Cation Ionophores

One would not necessarily predict by looking at the structures of the isolated oxyoctadecadienoate congeners shown in FIGURE 2 that these should possess divalent cation ionophoretic properties which are different from those of ordinary free fatty acids such as oleic or *cis*-linoleic acid. Quite to the contrary, we will show that certain very special features are derivative from the former. For example, the data in TABLE 3 show that the methyl esters of both the

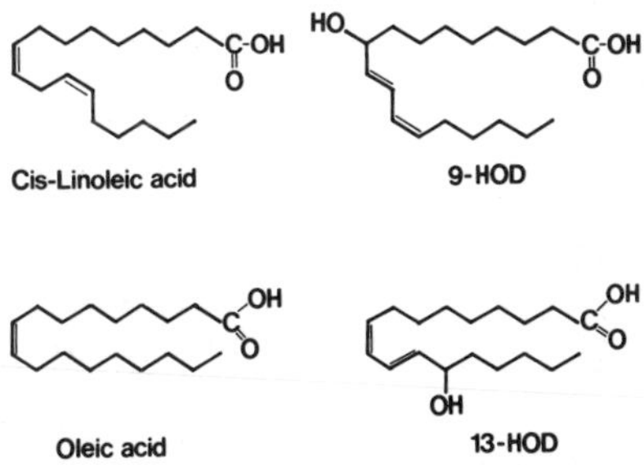

FIGURE 2. Structures of mitochondrial oxyoctadecadienoate congeners and related fatty acids. 13-HOD = 13-hydroxy-9-*cis*, 11-*trans*-octadecadienoic acid and 9-HOD = 9-hydroxy-10-*trans*, 12-*cis*-octadecadienoic acid.

TABLE 3

IONOPHORETIC PROPERTIES OF NEUTRAL MITOCHONDRIAL DIVALENT CATION IONOPHORES

Ionophore	Cpm Calcium-45 (organic) −DNP	Cpm Calcium-45 (organic) +DNP	Mitochondrial Swelling (Δ absorbance)	Reconstitution of Calcium Transport (% of original)	Bulk Phase Mg^{++} Transport (Δ absorbance)
Control	0	0	.020	0	0
9-MHOD	15	368	.225	19.2	.130
9-MKOD	16	827	.195	17.8	–
DCI-N_2	19	486	.120	0	–
DCI-N_3	24	1027	.245	46	–
Me-linoleate	0	0	.020	0	0

NOTE: Equilibrium extraction of Ca^{45} from a 1.0 mM aqueous $CaCl_2$ solution into 70% toluene: 30% *n*-butanol was determined at test ionophore concentrations of 4×10^{-4} M. The aqueous medium was buffered at pH 8.0 with 20 mM Tricine-tetramethyl-ammonium hydroxide. Parallel studies were carried out in the presence or absence of 4×10^{-4} M dinitrophenolate (DNP) anion in the aqueous phase. The measurement of mitochondrial swelling was carried out in 0.1 M magnesium nitrate, 10 mM Tris nitrate, pH 7.4, containing 0.2 mg per ml of mitochondrial protein. The absorbance change at 520 nm was monitored for 5 minutes after the addition of test ionophores and the net decrease in absorbance over this time period recorded. The final concentration of test ionophore was 6.7×10^{-6} M in all cases. The measurement of the recovery of succinate-driven active calcium acetate translocation was determined by monitoring the associated mitochondrial swelling as previously described. The test system contained calcium acetate (1 mM), sucrose (0.25 M), Tris-acetate at pH 7.4 (10 mM), and mitochondrial protein (0.2 mg per ml). Rotenone (2 μg per mg protein) was used to suppress endogenous respiration and the reaction was started by the addition of potassium succinate to a final concentration of 3.0 mM. Ruthenium red (0.1 nmol per mg protein) was used to specifically inhibit active transport of calcium and the attempted reconstitution was accomplished in the presence of 2.7×10^{-5} M test ionophore. Bulk phase transport of Mg^{++} ion (10 mM) was carried out in a "Pressman" cell according to previously published procedures from this laboratory.[12] 9-MHOD and methyl-linoleate at final concentrations of 0.5 mg per ml were present in the organic phase (CCl_4) and the assay was carried out according to the conditions described by Roeske *et al.*[15] for neutral divalent cation ionophores in which the coappearance of picrate anion is measured spectrophotometrically at 356 nm. The data are reported as an increase in absorbance per 8 hours.

9-hydroxy (9-MHOD) and the 9-keto (9-MKOD) octadecadienoic acids are capable of affecting the partition of calcium ions from an aqueous into an organic phase (toluene:butanol), while the same concentration of methyl linoleate is completely without effect. It is also apparent from the data that a lipophilic anion such as dinitrophenolate (DNP) is required in order to maintain electrical neutrality of the expected electrogenic cation complex derived from neutral ionophores.

Isolated mitochondria can be induced to swell coincident with the influx of ions when suspended in isoosmolar magnesium nitrate in the presence of an appropriate ionophore capable of rendering the membrane permeable to Mg^{++} ions. The data of TABLE 3 show that all four neutral species are potent inducers of mitochondrial swelling under the latter conditions, whereas methyl-linoleate is again completely without effect.

Isolated mitochondria accumulate calcium acetate in an energy-dependent fashion and undergo large amplitude swelling during the process. The calcium transport inhibitor, ruthenium red, is able to completely abolish the transport and associated swelling and the data in TABLE 3 present an account of the ability of the various isolated neutral divalent cation ionophores to restore the transport of calcium and associated swelling under the latter conditions. Both MHOD and MKOD possess only marginal effects in this system; however, methyl linoleate is again completely without effect. Finally, the data of TABLE 3 demonstrate the activity of MHOD versus that of methyl linoleate relative to the movement of magnesium ion in the bulk phase transport system. In the two experiments described, the appearance of the organophilic anion, picrate, in the receptor compartment was taken as a measure of the transport of Mg^{++} as documented by Roeske *et al.*[15] The total lack of response in the presence of methyl linoleate is again featured in this assay.

We do not necessarily mean to imply by these results that any or all of the isolated neutral species are involved in the active transport of magnesium or calcium ions by mitochondria. Our purpose at this juncture is only to document the ionophoretic properties of these congeners. Moreover, since the *in vitro* effects of these species may well be a function of the particular protein with which they are complexed, the activities that we have been able to evoke upon their readdition to isolated mitochondria may be unrelated if not qualitatively, at least quantitatively, to their activities *in situ*. For example, the data of Kessler[2] suggest that at least one of these (MHOD) is not involved in the transport of divalent cations, but that it is required for the Mg^{++}-dependent binding or activation of phosphate ion in the membrane sector of the oligomycin-sensitive ATPase.

Ionophoretic Properties of the Carboxylic Divalent Cation Ionophores

The free-acid members of the oxyoctadecadienoate family of divalent cation ionophores occur generally in higher concentrations than that of the esters and possess quite different ionophoretic properties. For example, the data shown in TABLE 4 establish the fact that, unlike the ester series, both hydroxy isomers and the mixed keto isomers are able to extract calcium ions into organic media in the absence of the lipophilic anion DNP. This observation would imply that the organic calcium-ionophore complex is dimeric in order

TABLE 4
IONOPHORETIC PROPERTIES OF CARBOXYLIC MITOCHONDRIAL DIVALENT CATION IONOPHORES

Ionophore	Cpm Calcium-45 (Organic)		Mitochondrial Swelling (Δ absorbance)		Inhibition of Active Transport of Calcium Acetate (%)
	–DNP	+DNP	Calcium	Magnesium	
9-HOD	1,385	6,116	.120	.020	48
13-HOD	1,912	6,336	.140	.000	41
9+13-KOD	3,544	8,360	.150	.015	62
X537A	158,000	166,000	.145	.210	75

NOTE: Equilibrium extraction of calcium-45 into organic media was carried out exactly as described in TABLE 3. Measurement of mitochondrial swelling in 0.1 M calcium or magnesium acetate was carried out essentially as described in TABLE 3. However, the medium was buffered with 10 mM Tris-acetate, pH 6.9, and test ionophores were added to a final concentration of 2.7×10^{-5} M. The change in absorbance in control experiments in the absence of test ionophore has been subtracted from the values that are reported in the table. Measurement of the inhibition of succinate-driven active transport of calcium acetate by mitochondria was carried out as described for the measurement of reconstitution described in TABLE 3 except for the omission in the present protocol of ruthenium red.

to satisfy the requirements of electrical neutrality. However, in a similar fashion to the ester series, DNP does indeed facilitate the complexation of additional calcium ion, which suggests the participation of an electrogenic complex as well. The most outstanding feature of TABLE 4 derives from the observations relating to the calcium ion complexation by the divalent cation ionophore X537A. By comparison with it, the oxyoctadecadienoate congeners possess much less affinity for calcium ion in this test system in which the aqueous Ca^{++} ion concentration was 1 mM. Nevertheless, the results shown in FIGURE 3 support the notion that the oxyoctadecadienoate free acids are efficient translocators of calcium ion under conditions (10 mM $CaCl_2$) favorable to low-affinity transport in a "Pressman cell" system. More importantly, the tremendous increment in activity over that of an ordinary fatty acid (e.g., oleic acid) is documented from these data when attention is drawn to the fact that the calcium transport scale of FIGURE 3 is logarithmic. In addition, it is apparent from the data that 9-HOD possesses a demonstrable preference for calcium over magnesium ion under these conditions. This selectivity for calcium ion is also borne out in studies (TABLE 4) describing the ionophore-induced influx of salt and associated mitochondrial swelling of isolated mitochondria suspended in isoosmotic calcium and magnesium acetate. While the oxyoctadecadienoate congeners display a marked preference for calcium in this system, the classical divalent cation ionophore X537A, as expected, shows little selectivity.

Since the oxyoctadecadienoate free acid congeners appear to induce a massive flux of calcium acetate across the mitochondrial membrane, one might expect that they would be capable of discharging a calcium ion gradient originating from the active transport of calcium acetate into the mitochondria. The data of TABLE 4 also document this prediction and point up the fact that the free acids, as well as X537A, have predictably opposite effects on mitochondrial calcium transport compared with those described earlier for the esters. However, again it must be pointed out that these effects do not necessarily correlate with the role of these ionophores *in situ* where they are bound to their respective ionophoroproteins. Our only intention thus far has been to point out the naked ionophoretic properties of these molecules and to compare them with other ionophores (X537A) and common free fatty acids (e.g., oleate and *cis*-linoleate).

To this end, one of the most startling demonstrations of $Ca^{++} > Mg^{++}$ selectivity is featured from a study of the interaction of the oxyoctadecadienoate congeners with divalent cations and the fluorescent anion 1-anilino-8-naphthalene sulfonate (ANS). The study is based on the observations of Feinstein and Felsenfeld[16] which describe the interaction of ANS with the ionophorous antibiotics, valinomycin, nigericin, and alamethicin, during the binding of alkali cations in aqueous media. The interaction of the ionophore-cation complexes with the fluorescent dye produced enhanced fluorescence emission and a significant blue shift (520 → 460 nm) of the emission maxima of the fluorescence spectrum resulting from the interaction of ANS with low-polarity sites. While the affinity of these ionophores for alkali cations (e.g., K^+) was, as expected, much less in aqueous media (e.g., 430 mM for valinomycin) than in organic solvents (e.g., 49 mM for valinomycin), the relative affinities of the ionophores for various alkali cations were in accord with the data obtained in organic solvents. When the protocol described by these authors

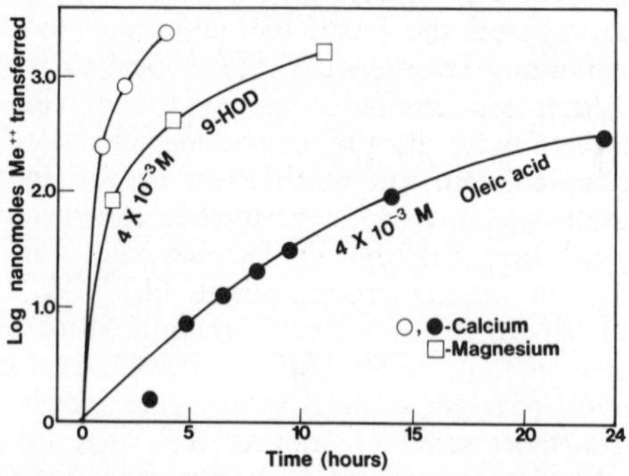

FIGURE 3. Ionophore-mediated transport of calcium and magnesium ions across a bulk phase of carbon tetrachloride. The procedure requires a glass cylinder with a glass partition vertically sealed across the top ⅔ of the cylinder as diagrammed in Reference 13. The bottom half of the vessel is filled with a heavier-than-water organic solvent (6.0 ml CCl_4) containing 4×10^{-3} M 9-HOD or oleic acid, thereby creating two discontinuous compartments above the organic level, which are then filled, one (donor) with 2.0 ml aqueous buffer (25 mM Tricine-tetramethyl ammonium hydroxide, pH 8.5) containing the test cation (10 mM $CaCl_2$ or $MgCl_2$) and the other (receptor) with 2.0 ml buffer alone (50 mM citrate-tetramethyl ammonium hydroxide, pH 5.4) for the measurement of the appearance of the test cation. The result of this arrangement is that two separately exposed aqueous phases communicate with one another only via an organic phase. Aliquots of 50 μl were withdrawn from the aqueous receptor compartment at times indicated in the Figure and assayed for the appearance of calcium or magnesium by atomic absorption spectrophotometry or by assaying the concentration of calcium-45.

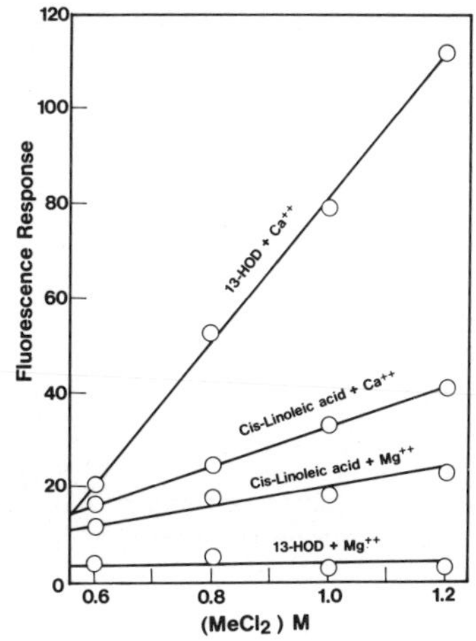

FIGURE 4. Ionophore-dependent fluorescence response of ANS in the presence of divalent cations. The fluorescence emission of ANS at 460 nm due to Me^{++}-ionophore complex formation in aqueous media was measured according to procedures described by Feinstein and Felsenfeld.[16] Test compounds were present at a final concentration of 4×10^{-4} M and the fluorescence response of the total system less the test ionophore has been subtracted from the values plotted.

was applied to the free-acid oxyoctadecadienoate congeners in the presence of calcium or magnesium chloride, the results shown in FIGURE 4 were obtained. The selectivity of these ionophores for calcium over magnesium ions, which had previously been seen in the swelling and bulk phase transport studies, is most dramatically brought out, especially in the case of 13-HOD. Based on the ratios of the slopes of the lines described in FIGURE 4, the Ca^{++}/Mg^{++} selectivity ratio of 13-HOD is 61.5, whereas that for *cis*-linoleic acid is only 1.94. The 9-HOD free acid congener (data not shown) is between these two at 5.5, which is in the range of the selectivity ratio observed for the latter in the bulk-phase transport studies, whereas the one neutral (9-MHOD) oxyoctadecadienoate congener that has been examined in this system exhibited a slight selectivity for magnesium over calcium ions.

13-HOD-Induced Na^+/H^+ Exchange Activity

Isolated beef-heart mitochondria have long been known to possess a sodium-specific electrically silent (Na^+/H^+ exchange) carrier mechanism.[17, 18] Thus, an ionophore specific for Na^+ over K^+ and of the nigericin class of carboxylic ionophores would best typify the kind of carrier activity demonstrable in isolated mitochondria. The fact that isolated mitochondria swell spontaneously in isoosmotic sodium acetate, whereas they are resistent to swelling in isoosmotic potassium acetate may be considered to be an expression of this endogenous activity. Several years ago, we measured the steady-state exchange diffusion kinetics of this carrier in the presence of Na^{22} and K^{42} and estimated that the Na^+/K^+ selectivity ratio should be on the order of 20 (Reference 18). However, our earlier attempts to isolate this postulated carrier by simple organic solvent extraction techniques met with no success and we soon abandoned the project. Since the isolated oxyoctadecadienoate free acids are in principle Me^+/H^+ exchangers, and since at least one of these (13-HOD) has already

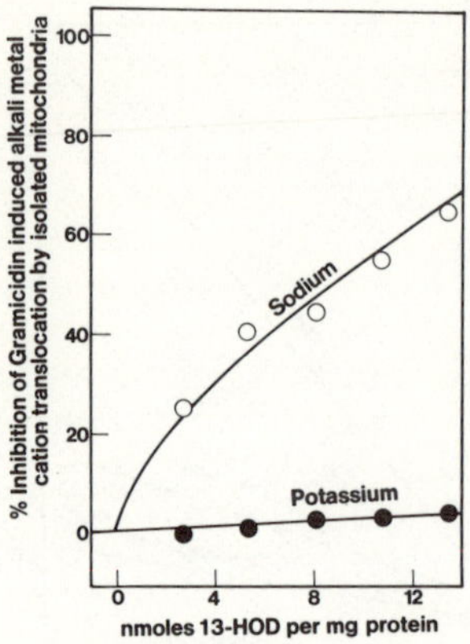

FIGURE 5. Effect of 13-HOD on gramicidin-induced alkali metal ion translocation by isolated mitochondria. The test media (7.3 ml) contained 0.2 M mannitol, 25 mM Tris-acetate (pH 7.5), 4.0 mM sodium acetate, 2 mM Tris-glutamate, 2 mM Tris-malate, and mitochondrial protein at a final concentration of 2.05 mg per ml. After preincubation at 25° for 1.0 minutes, gramicidin (0.2 μg) was added and the uptake of Na^+ ion was monitored with a Beckman No. 39278 sodium ion electrode. A Beckman No. 39137 cationic electrode was used for the measurement of K^+ ion translocation. 13-HOD was added 30 seconds prior to the addition of gramicidin.

demonstrated a remarkable degree of specificity with respect to divalent cations, we decided to test these compounds for Na^+/H^+ exchange activity in isolated mitochondria. However, since there are no known inhibitors of this exchange activity in mitochondria,‡ assays 'that are based on volume changes are of little value, since the control rates of swelling in isoosmotic sodium acetate are too rapid to observe inducible rate changes. The gramicidin-induced active transport of Na^+ and K^+ by isolated mitochondria is, on the other hand, a well-documented phenomenon, as is the competitive inhibition of this translocation by nigericin. An analysis of Na^+/H^+ versus K^+/H^+ exchange activity based on the latter observations and in the presence of 13-HOD is shown in FIGURE 5. It is immediately apparent that 13-HOD is a very potent inhibitor of gramicidin-induced Na^+ ion translocation, while it is practically without effect on K^+ translocation. Assuming a linear slope between 0 and 12 nmol of 13-HOD per milligram of protein, the Na^+/K^+ selectivity ratio obtained is 17.8 to 1, which is in remarkably good agreement with the selectivity ratio estimated from exchange diffusion kinetics of 20. The Na^+/K^+ selectivity ratio obtained in an identical experiment in the presence of nigericin was 0.66, indicating a presence of the latter for K^+, as has previously been documented (e.g., $Na^+/K^+ = 0.5$ in release of ions from erythrocyte ghosts) by Pressman.[20] Even more important is the finding that perceptible activity (25% inhibition) is observed in these studies in the presence of a concentration of 13-HOD (2.5 nmol/mg pro'tein) which is only three-fold greater than the endogenous concentration of 13-HOD based on the yield of purified material. These observations are of very recent origin and we have not yet had the chance to document fully this phenomenon, nor have we examined in any detail the other free-acid oxyoctadecadienoate congeners; nevertheless, the observed specificity and the nature of the inhibition described in FIGURE 5 are to us very encouraging evidence that this class of compounds may be singularly responsible for the well-documented Na^+/H^+ exchange activity of isolated mitochondria.

‡ R. Cockrell *et al.* have presented evidence[19] that suggests that 20 mM $MgCl_2$ effectively inhibits Na^+/H^+ exchange in isolated mitochondria. However, in our hands, the attempted reconstitution of exchange by the measurement of ionophore-induced passive swelling in isoosmotic sodium acetate was severely inhibited in such a system due to the osmotic support afforded by 20 mM $MgCl_2$.

Relationship of Mitochondrial Oxyoctadecadienoate Congeners to Platelet Nonprostanoate Compounds

Hamberg, Samuelsson and colleagues[21, 22] have recently described a series of novel transformation of arachidonic acid in human blood platelets which are depicted in FIGURE 6. The important features that are derivative from this and other work are as follows: (1) Thrombin-induced aggregation of blood platelets is accompanied by a release of calcium ions from this organelle.[23] (2) The divalent cation ionophore X537A is competent on its own in releasing Ca^{++} ions from platelets and causing subsequent aggregation.[24] (3) The thrombin-induced release of calcium ions from platelets is accompanied by a specific metabolic conversion of arachidonic acid to 12L-hydroxy-5,8,10-heptadecatrienoic acid (HHT) and 8-(1-hydroxy-3-oxopropyl)-9,-12L-dihydroxy-5,10-heptadecadienoic acid (PHD).[21, 22] (4) The strcture of HHT is remarkably similar to the structure of the mitochondrial calcium-specific ionophore 13-HOD as shown in FIGURE 7 together with two prostaglandins which possess similar partial structures. Accordingly, these observations taken together suggest that the role of HHT in blood platelets is to induce the release of calcium ions by virtue of its calcium-specific ionophoretic activity, which itself is inferred from its structural similarity to mitochondrial 13-HOD. These predictions are easily open to verification, and such experiments will soon be underway in our laboratory.

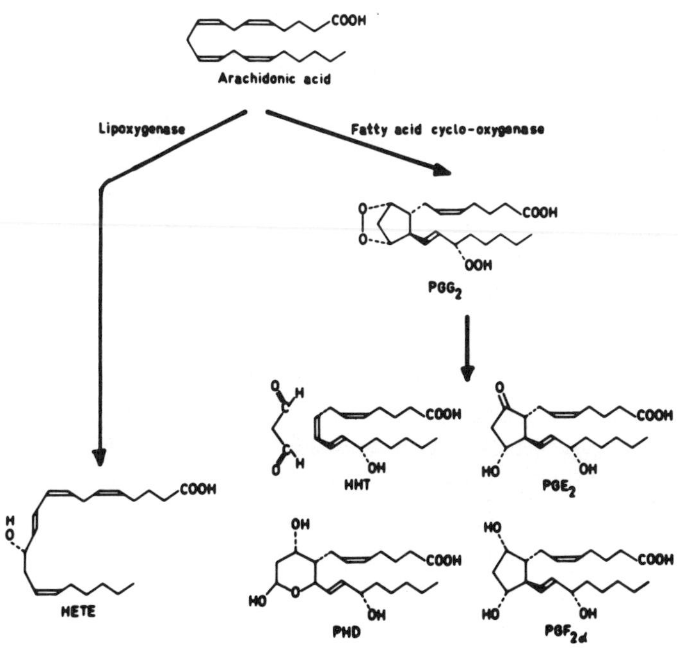

FIGURE 6. Transformations of arachidonic acid in human blood platelets. (After Hamberg *et al.*[22])

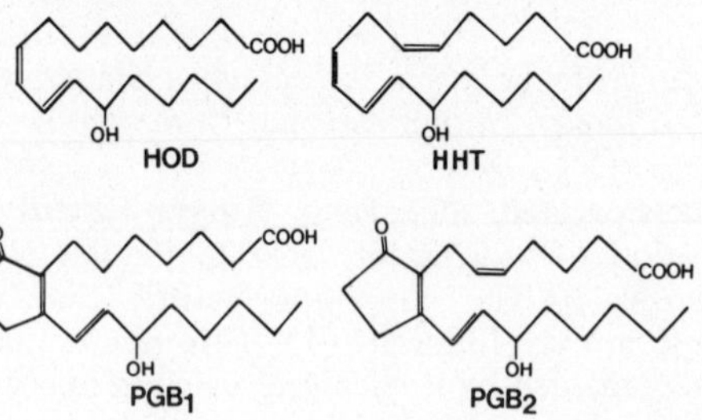

FIGURE 7. Fatty acid derivatives that possess calcium ionophore activity.

Summary

The notion of small molecular weight ion carriers in biological systems is herein documented by a description of the isolation and ionophoretic properties of a family of oxyoctadecadienoate congeners derived from beef-heart mitochondria. Although certain members of this family of compounds have been shown to possess unique ionophoretic properties, one should not lose sight of the fact that the compounds that we have described represent only a portion of the total picture. Other chemically unrelated, yet structurally unknown species have been isolated from beef-heart mitochondria, and compounds similar in both chromatographic and spectroscopic properties to the oxyoctadecadienoate family, as well as other unique structures, have been isolated in our laboratory from sarcoplasmic reticulum[25] and chloroplasts.[26] The important points to be derived from these findings are that there is an apparent abundance of natural ionophores and we should no longer concern should address ourselves to the more relevant task of digging them out and ourselves with the question "are there ionophores in biological systems?" but describing their chemical and physical properties. In view of the apparent abundance of natural ionophores, this is an enormous task, especially when one considers that it only represents half of the problem. The isolation and description of the ionophoroprotein or channel-forming complexes share equally in the overall significance and level of understanding attributable to this area of inquiry and it would appear that many fruitful collaborative ventures are, or should be, on the horizon.

Acknowledgments

I gratefully acknowledge the support and interest of Professor David E. Green and the excellent technical assistance of Mrs. Wanda Fechner. I am indebted to Professor Heinrich Schnoes and Guy Carter of the University of Wisconsin Biochemistry Department for the mass spectrometry results and to James W. Blackbourn of the University of Wisconsin Pharmacy Department for nuclear magnetic resonance spectral determinations.

References

1. OSTERHOUT, W. J. U. 1933. Ergeb. Physiol. **35:** 967.
2. GREEN, D. E. This monograph.

3. Hyman, E. S. 1971. Biophys. Soc. Abstr. **11:** 11a.
4. Margoliash, E., G. H. Barlow & V. Byers. 1970. Nature **228:** 723.
5. Wojtczak, L. 1974. FEBS Lett. **44:** 25.
6. Kirtland, S. J. & H. Baum. 1972. Nature **47:** 236.
7. Carafoli, E. & F. Crovetti. 1973. Arch. Biochem. Biophys. **154:** 40.
8. Blondin, G. A., A. F. DeCastro & A. E. Senior. 1971. Biochem. Biophys. Res. Commun. **43:** 28.
9. Blondin, G. A. 1974. Ann. N.Y. Acad. Sci. **227:** 392.
10. Goodall, K. C. & G. Sachs. 1972. Nature New Biol. **237:** 252.
11. Shamoo, A. E. & R. W. Albers. 1973. Proc. Nat. Acad. Sci. U.S.A. **70:** 1191.
12. Blondin, G. A. 1974. Biochem. Biophys. Res. Commun. **56:** 97.
13. Pressman, B. C. 1973. Fed. Proc. **32:** 1698.
14. Blondin, G., G. T. Carter & H. K. Schnoes. Unpublished studies.
15. Roeske, R. W., S. Isaac, T. E. King & L. K. Steinrauf. 1974. Biophys. Biochem. Res. Commun. **57:** 554.
16. Feinstein, M. B. & H. Felsenfeld. 1971. Proc. Nat. Acad. Sci. U.S.A. **68:** 2037.
17. Blondin, G. A., W. J. Vail & D. E. Green. 1969. Arch. Biochem. Biophys. **129:** 158.
18. Blondin, G. A. & D. E. Green. 1970. Bioenergetics **1:** 479.
19. Douglas, M. G. & R. S. Cockrell. 1974. J. Biol. Chem. **249:** 5464.
20. Pressman, B. C. & M. J. Heeb. 1971. Biophys. Soc. Ann. Meet. Abstr. **11:** 310a.
21. Hamberg, M. & B. Samuelsson. 1974. Proc. Nat. Acad. Sci. U.S.A. **71:** 3400.
22. Hamberg, M., J. Svensson & B. Samuelsson. 1974. Proc. Nat. Acad. Sci. U.S.A. **71:** 3824.
23. Detwiler, T. C. & R. D. Feinman. 1973. Biochemistry **12:** 282.
24. Feinman, R. D. & T. C. Detwiler. 1974. Nature **249:** 172.
25. Blondin, G. & C. Tyson. Unpublished studies.
26. Blondin, G. & J. Brand. Unpublished studies.

THE USE OF ELECTRON PARAMAGNETIC RESONANCE IN STUDIES OF FREE AND BOUND DIVALENT CATION: THE MEASUREMENT OF MEMBRANE POTENTIALS IN MITOCHONDRIA*

Thomas E. Gunter and Jerome S. Puskin

Department of Radiation Biology and Biophysics
The University of Rochester
School of Medicine and Dentistry
Rochester, New York 14642

Introduction

The mechanism by which the mitochondria of higher animals accumulate divalent cation against a concentration gradient has been the subject of considerable work and speculation since the early 1960s. Initially, it was fashionable to describe the transport mechanism as a kind of calcium pump closely coupled to both electron transport and ATP dephosphorylation.[1,2]

The first in a series of observations, which when taken together strongly support the concept that cation accumulation by mitochondria is a carrier-mediated process coupled to electron transport and to ATP dephosphorylation via a membrane potential, was the finding that active mitochondrial uptake of K^+, Rb^+, and Cs^+ can be induced by treating the mitochondria with the antibiotic ionophore valinomycin.[3]

High-affinity external binding to inhibited mitochondria by those divalent cations (Ca^{2+}, Sr^{2+}, and Mn^{2+}) which can be actively accumulated suggests the existence of a carrier.[4] Inhibition of transport by ruthenium red[5] and lanthanides[6] indicates that the number of carriers is $\simeq$0.1 nmol/mg mitochondrial protein. The cooperative character of uptake kinetics and the activating effect of Ca^{2+} on Mn^{2+} accumulation[7,8] strongly suggest two cooperative binding sites for divalent cations on the transport system. Nuclear magnetic resonance studies have contributed some evidence for a highly mobile ioncarrier complex.[9] Finally, a successful attempt at isolating a divalent cation ionophore, with properties similar to those of the bacterial divalent cation ionophores A23187 and X537A, from the mitochondrial membrane system has been recently reported.[10]

The demonstration[11] that valinomycin-induced K^+ efflux can stimulate Ca^{2+} uptake against a concentration gradient showed that Ca^{2+} can be accumulated in response to a membrane potential. The hypothesis that the divalent cation distribution found in energized mitochondria is a steady-state response to a membrane potential is further supported by the observation that the divalent-cation-transporting ionophores A23187 and X537A do not greatly affect the steady-state distribution of Ca^{2+} unless a Ca^{+2} chelating agent such as ethylenedi-

* This paper is based on work performed under contract with the U.S. Energy Research and Development Administration at the University of Rochester Biomedical and Environmental Research Project and has been assigned Report No. UR-3490-722.

aminetetraacetic acid (EDTA) is present in the supporting medium.[12, 13] This hypothesis is further supported by (1) the finding that there are equal Na^+ and K^+ concentration ratios across mitochondrial membranes made permeable to these ions by gramicidin[14] and (2) that the log of the concentration ratios of ^{86}Rb and $^{45}Ca^{2+}$ in the presence of valinomycin differ by a factor of 2, as would be predicted by the Nernst equation.[13]

The magnitudes of the membrane potentials deduced from these Rb^+ and Ca^{2+} concentration ratios differ considerably from those measured earlier by Mitchell and Moyle[15] using a K^+ electrode. Both of these past measurements relied on the highly unlikely assumption that all the internal ions were free.

The use of electron paramagnetic resonance with the paramagnetic Ca^{2+} analog Mn^{2+} allows the quantitative determination of internal and external bound and free fractions of Mn^{2+} (References 16–19), as is described later. This should lead to a more accurate determination of the membrane potential than that previously obtained. Preliminary data of this type are given below.

In addition, it has been claimed[13, 20, 21] that anions of weak acids such as acetate, lactate, butyrate and β-hydroxybutyrate permeate the membrane in a neutral form. These along with $H_2PO_4^-$, which is thought to be transported on an exchange carrier,[22] and H_2CO_3, which can penetrate the membrane as dissolved CO_2, can donate a proton to the mitochondrial matrix. This process could lead to the conversion of a pH gradient into a membrane potential.

The spectral characteristics of intramitochondrial manganese, observed over a range of pH and in the presence of various anions, discussed here and in past publications, allow several inferences to be drawn directly relating to the above hypothesis.

Materials and Methods

Rat liver mitochondria (RLM) were prepared by the technique of Schnaitman and Greenawalt.[23] For most of the experiments described below the RLM were suspended in a medium composed of 88 mM mannitol, 29 mM sucrose, 16 mM Na-Hepes (pH 7.2), 1 mg/ml BSA plus 9.75 mM Na succinate. In addition 50 mM KA was present where A refers to any anion or combination of anions from the group, acetate, bromide, butyrate, chloride, lactate, nitrate, and perchlorate.

Inner membrane preparations were carried out following the method of Schnaitman and Greenawalt.[23] Assays for monoamine oxidase activity[24] and NADH-cytochrome *c* reductase activity[25] were used to show low outer-membrane contamination in the inner-membrane preparation.

Quantitation of epr spectra was carried out as previously reported.[17, 19]

Results

Internal and External Bound Fractions

It has been established that mitochondrial uptake of Mn^{2+} in the absence of permeant anion leads to a condition in which most of the sequestered Mn^{2+} is bound in regions of high local concentration.[16] When acetate is a counter ion, the uptake still leads to the formation of a large bound fraction, but in this case a larger percentage of the Mn^{2+} is held in the "free" form inside the mitochondria.[18] While the nature of this intramitochondrial bound Mn^{2+} has not been fully characterized, it has been shown to involve phosphate (although not in the form of a simple $Mn_3(PO_4)_2$ precipitate) and to be more prominent at high intramitochondrial pH.[19] Competition for binding in this "spin exchange" form between Mn^{2+} and Ca^{2+} has been established. It has been hypothesized that this "spin exchange" fraction represents the first step in the formation of a plaque on the inner side of the inner mitochondrial membrane in which the divalent cation first associated with fixed negative charges on membrane proteins, phosphoproteins, phospholipids, etc. and is then stabilized by binding or precipitation of small anions, such as phosphate, hydroxide, etc.[19] This hypothesis is strengthened by comparing internal and external mitochondrial binding. Reynafarje and Lehninger[4] reported that inhibited mitochondria possesses, both high- and low-affinity binding sites for Ca^{2+} and Mn^{2+}. Epr measurements of external Mn^{2+} binding, carried out in our laboratory, are in essential agreement with this past work. The epr work has also shown that at elevated pH (circa 8.6) spin exchange can be observed (FIGURE 1)

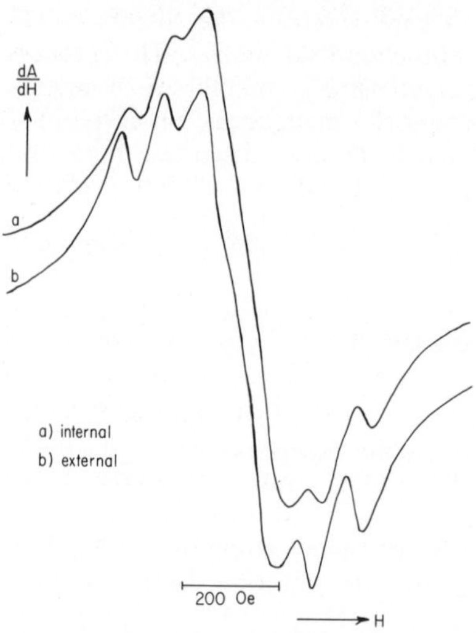

FIGURE 1 (a) Intramitochondrial spin exchange signal. The medium was .25 M sucrose, 1 mg/ml BSA, 20 mM tris (pH 7.4), .1 mM ATP, 10 mM Na succinate. The mitochondrial concentration was 3.35 mg/ml. 269 nmol Mn^{2+}/mg protein was added to the suspension and ten minutes later 300 nmol EDTA/mg protein was added just prior to observing the intramitochondrial Mn^{2+} with epr. (b) External spin exchange signal. The medium was .25 M sucrose, 20 mM tris (pH 8.9), $\simeq 10^{-4}$ M DNP. The mitochondrial concentration was 50 mg/ml. 25 nmoles Mn^{2+}/mg protein was added to the suspension just prior to observing the Mn^{2+} with epr.

in concentrated mitochondrial suspensions inhibited by either dinitrophenol (DNP) or by rotenone plus hydroxyquinoline, and that formation of this signal is unaffected by prior addition of several nmol/mg protein of lanthanum chloride. This "external" spin exchange signal is destroyed by addition of sufficient EDTA to the suspension in contradistinction to the effects of EDTA on the intramitochondrial spin exchange signal.

Further work has shown that such a signal can also be obtained at pH 8.6 in the presence of DNP by addition of Mn^{2+} to a concentrated suspension of mitochondrial inner membranes.

It was therefore concluded that Mn^{2+} could be bound to the outer side of the mitochondrial inner membrane in a fashion somewhat similar to that in which it is bound on the inner surface following uptake through the transport system. Intramitochondrial binding, like external binding, increases with intra-

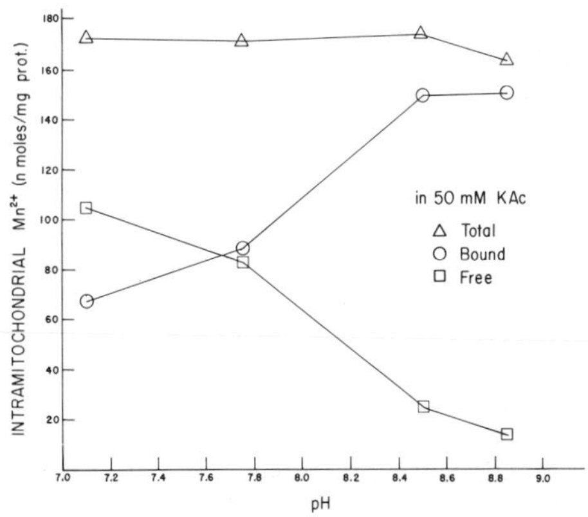

FIGURE 2. Bound, free, and total intramitochondrial Mn^{2+} vs. external pH in the presence of 50 mM [Ac^-]. The suspension medium was the usual one as described in the *Materials and Methods* section; the mitochondrial concentration is 5.3 mg protein/ml.

mitochondrial pH. Where acetate is used to break down the pH gradient, the bound and free fractions of intramitochondrial manganese are plotted from epr data as a function of external $\simeq$ internal pH in FIGURE 2. This supports the view[19] that increased binding and spin exchange averaging, accompanied by decreased internal free Mn^{2+}, as observed earlier in the absence of permeable anions, is the result of intramitochondrial alkalinization occurring during Mn^{2+} uptake (see also References 26–28).

The Internal "Free" Fraction and Measurement of Membrane Potential

Determination of Internal and External Free Manganese. As has been shown for Ca^{2+} by Reed and Lardy,[12] the addition of EDTA to a mitochondrial suspension stimulates a slow rate of loss of Mn^{2+} from the mitochondria (FIGURE 3). Consequently, two techniques were used in the experiments described here, only one of which involves the addition of EDTA to the mitochondrial suspension. These two techniques can then be used as a cross check on one another.

After addition of manganese to a 6 ml mitochondrial aliquot (3–5 mg/ml),

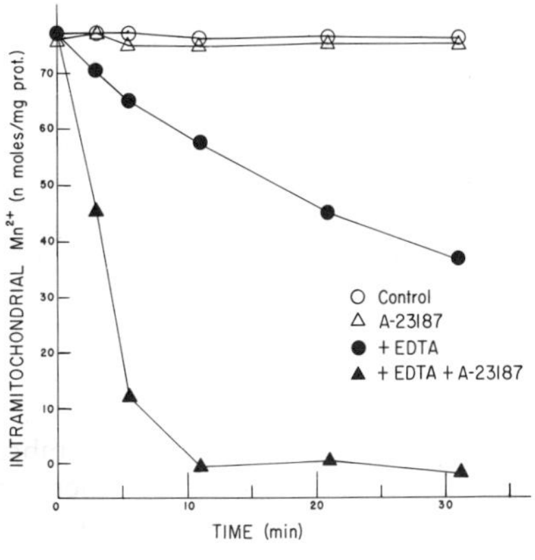

FIGURE 3. Intramitochondrial Mn^{2+} vs. time. The suspending medium was the usual one, as described in the *Materials and Methods* section. 10 mM [Ac^-] is used. At time −10 78 nmol Mn^{2+}/mg protein with a trace of $^{54}Mn^{2+}$ is added to the suspension. At time zero $\simeq$ 69 nmol Na-EDTA/mg protein was added to the samples whose intramitochondrial Mn^{2+} is indicated by the solid symbols and $\simeq$.23 nmol/mg of A23187 was added to the samples whose intramitochondrial Mn^{2+} is indicated by the triangles.

a set time is allowed for uptake, as is described below. At the proper time the suspension is reoxygenated and vortexed to insure homogeneity, a 1-ml sample is withdrawn for centrifugation, and another part of the sample drawn into a quartz flat cell. The flat cell is rapidly mounted in a microwave cavity and the epr spectrum of the first and second (lowest field) lines of the manganese hexahydrate hyperfine sextet are recorded (Spectrum A). Afterwards the original suspension is reoxygenated and an amount of EDTA more than sufficient to chelate all manganese external to the mitochondria is added. Again a sample of the suspension is drawn into the quartz flat cell and another epr spectrum comparable to the first is taken (Spectrum B).

At the same time that the epr spectroscopy is being carried out, the 1-ml sample withdrawn immediately preceding the spectroscopy is centrifuged (8000 rpm in a Beckman JA21 centrifuge head for one minute) to pellet the mitochondria and the supernatant withdrawn. The epr spectrum of this supernatant is then taken (Spectrum C) in the same flat cell used previously.

After quantitation of the hyperfine sextet spectra in a fashion similar to that described previously,[17,19] both $[Mn^{2+}{}_A] - [Mn^{2+}{}_C]$ and $[Mn^{2+}{}_B]$ should be comparable and should be a measure of intramitochondrial free manganese. The validity of this procedure can be judged from the proximity of the points calculated from data obtained through these two methods in the figures that follow.

Concentration ratios can be obtained by correcting the above concentrations for intramitochondrial volume and dividing by $[Mn^{2+}{}_C]$.

Establishment of Steady State: Time-Dependence of External Concentration. The value found for extramitochondrial divalent ion concentration depends upon the time after addition of Mn^{2+} to the suspension at which the sample is withdrawn and centifuged. This time should be long enough so that uptake is complete and a steady-state concentration ratio related to the membrane potential is reached, yet not so long that release or degradation of the mitochondria occurs. FIGURE 4 shows the results of typical measurements yielding values for the amount of manganese in Spectrum C (supernatant) as a function of time after manganese addition. In the experiments described below, spectroscopy is begun 18 minutes after addition of manganese and generally completed within 8 or 10 minutes.

General Characteristics of Measurements in the Presence and Absence of Proton Donating Permeable Anions. For the calculations of internal $[Mn^{++}]$ described here the internal volume was treated as constant (1 μl/mg protein).

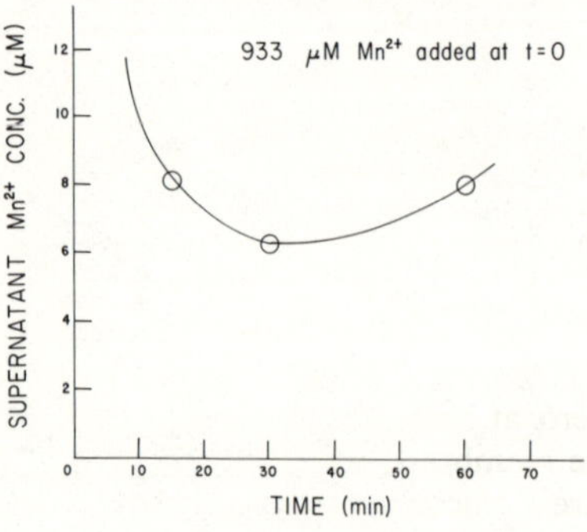

FIGURE 4. Mn^{2+} concentration in the supernatant vs. time. The mitochondria ($\simeq$5 mg protein/ml) were suspended in the usual medium.

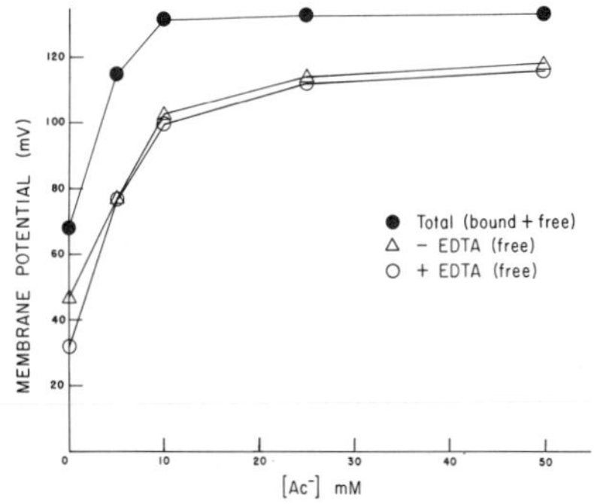

FIGURE 5. Membrane potential vs. [Ac^-]. The mitochondria ($\simeq$3 mg protein/ml) were suspended in the usual medium.

Experimentally, of course, considerable swelling of the mitochondria occurs, particularly under conditions of large manganese uptake at high concentrations of a weak, dissociable acid such as acetate. Under these latter conditions it is almost certain that the internal volume is greater than 1 μl/ mg protein.[13]

While internal volume measurements are currently being carried out in our laboratory, it is felt that the above "rough" value for internal volume is adequate for treatment of the preliminary data reported here because: (1) Due to the logarithmic nature of the dependence of membrane potential on the concentration ratio of divalent cations a twofold error in volume would cause an error of slightly less than 9 mV in the final result, while an error of a factor of 10 in volume would result in an error of slightly less than 30 mV. (2) Even when volume measurements are carried out on the same mitochondria used in the membrane potential measurements, controversy will still ensue over how much of the sucrose inaccessible water is osmotically active.

Membrane potentials across the inner mitochondrial membrane were computed from the Nernst equation:

$$E = \frac{RT}{2F} \ln \frac{[Mn^{2+}{}_{in}]}{[Mn^{2+}{}_{out}]} \qquad (1)$$

where E, R, T, F, and [Mn^{2+}] have their usual meaning, using free ion concentration ratios measured as described above.

FIGURE 5 shows the calculated membrane potential as a function of acetate concentration. In the measurements depicted here A_c^- plus Cl^- were kept constant at 50 mM with potassium as the counter ion. Also shown is the value of membrane potential which would be calculated using Equation 1, assuming that all the divalent ion taken up by mitochondria is held in the "free" form, as was assumed for Ca^{2+} by Rottenberg and Scarpa.[13] While the agreement is fair at high acetate concentration (above, say, 25 mM acetate), it is not good at low concentrations, where in some cases (at zero acetate and high uptake) over 95% of the manganese taken up is bound even in the absence of added ATP.

FIGURE 6 shows the membrane potential as determined by application of Equation 1 as a function of added Mn^{2+}. It is noted that the membrane potential shows no tendency to fall to zero at zero Mn^{2+} concentration, which supports the view that the existence of the membrane potential does not depend on the transport of cations. The apparent increase in membrane potential

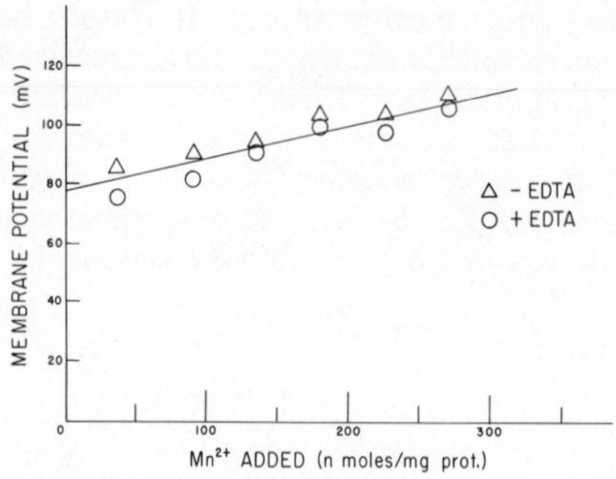

FIGURE 6. Membrane potential vs. Mn^{2+} added. The mitochondria ($\simeq$5 mg protein/ml) were suspended in the usual medium.

with added cation is probably due at least in part to the failure to include mitochondrial swelling in the calculation.

FIGURE 7 is a plot of membrane potential as a function of extramitochondrial pH in the presence of 50 mM acetate, which has been shown to dissipate the pH gradient.[18, 26–28] The fall-off at high pH is taken to be an indication of leakiness or uncoupling induced in these membranes.

TABLE 1 shows a composite of the data on membrane potential and ΔpH

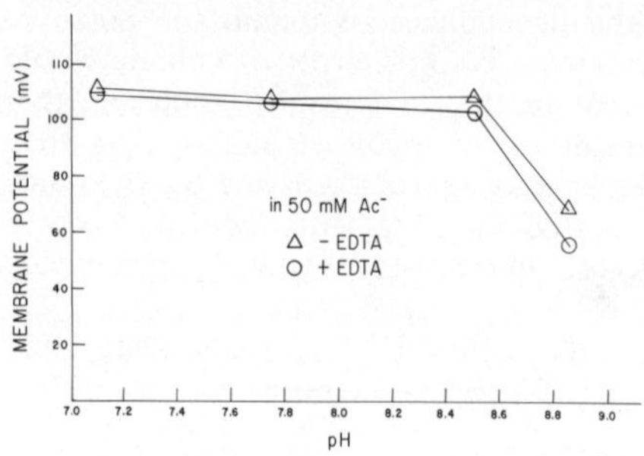

FIGURE 7. Membrane potential vs. pH. The mitochondria ($\simeq$5 mg protein/ml) were suspended in the usual medium, except that for the higher pH points tris buffer was used.

TABLE 1

MITOCHONDRIAL MEMBRANE POTENTIALS AND pH GRADIENTS

	Acetate	Lactate	Butyrate	Nitrate	Perchlorate	Chloride	Bromide	Conditions*
ψ(mV)	109.5	94.7	114.2	16.8	23.2	25.5	25.8	(1)
ψ(mV)	102.4	95.9	89.6	35.3		32.4	49.1	(2)
ψ(mV)	99.4		64.1†	22.3		25.8		(3)
−ΔpH	$\simeq$0					$\simeq$1.5		
ψ − 60 ΔpH	$\simeq$110					$\simeq$115		

* Conditions were as follows: (1) 50 mM A^- concentration, 230 nmol Mn^{2+}/mg protein added; (2) 50 mM A^- concentration, 95 nmol Mn^{2+}/mg protein added; (3) 10 mM A^- concentration, 130 nmol Mn^{2+}/mg protein added.

† There is some indication that a small amount of release occurred.

[*Note added in proof:* Under better uptake conditions, higher values for ψ have been found in the absence of permeable dissociable anions than those quoted in TABLE 1. While still significant, the effect of these anions, therefore, on increasing membrane potential can be considerably less than appears here.]

obtained using a series of permeable and impermeable anions. It should be noted that the data can be divided into sets of anions similar to those described by Lehninger.[21] The largest membrane potentials are found with permeant anions able to exchange a proton. These are precisely the anions found to increase the Ca^{2+}-sequestering capacity of the mitochondria. These data support the hypothesis of Lehninger[21] that through transport of these anions, mitochondria are able to convert the energy of a pH gradient into a membrane potential.

Discussion

Observations reported earlier[18, 19, 26–28] and in this paper on the pH gradient across the inner mitochondrial membrane can be interpreted to imply increasing alkalinization of the mitochondrial matrix during divalent cation transport when such transport takes place in the absence of a permeable counter ion. In addition there is a sizable increase in intramitochondrial binding as pH increases.

These observations also indicate that in the presence of permeable anions, which can dissociate within the mitochondrial matrix so as to contribute a proton, such alkalinization does not occur. Calculations of membrane potential derived from measured free intra- and extramitochondrial Mn^{2+} indicate a lower value of membrane potential in the absence of a permeable, dissociable anion, but gives values up to $\sim$120 mV in the presence of permeable, dissociable anions. The observation that the membrane potential does not approach zero with decreasing $[Mn^{2+}]$ may be taken as evidence that formation of a membrane potential precedes divalent cation uptake. In this light, it is also of interest to note that ADP stimulation of O_2 consumption has been observed after Mn^{2+} uptake ($\simeq$80 nmol/mg protein) in 10 mM A_c^- on an aliquot of the same mitochondrial preparation used for the membrane potential measurements.

These observations lend strong support to the view expressed by Lehninger[21] (which is related qualitatively to the chemiosmotic hypothesis) that an electrochemical gradient of protons can be converted by the inward transport of dissociable anions (phosphate (Pi) *in vivo* in the $H_2PO_4^-$ form) into a membrane potential that can drive Ca^{++} uptake.

It is also interesting that with either an abudance of permeable, dissociable anions (no pH gradient but large membrane potential) or no permeable anions (pH gradient $\simeq$1.5 units but small membrane potential), mitochondria can still sequester quite appreciable amounts of divalent cation. The internal akalinization leading to increased binding in the former case tends to partially compensate for the loss of the membrane potential important in the second case.

The numerical values of the membrane potentials reported here are in rough agreement with those found by Rottenberg and Scarpa[13] and in substantial disagreement with those reported by Mitchell and Moyle.[15] The agreement with the values given by Rottenberg and Scarpa[13] is thought to be at least partly fortuitous since their work was carried out at a low acetate concentration where internal binding of divalent cations is by no means negligible. It is brought about largely through the logarithmic dependence in the Nernst equation.

Sources of Error

It should be pointed out that the data presented above give no indication of the charge state of the carrier. This can be shown very easily in the simple case where the distribution of the free carrier itself is governed by the Nernst equation and where the association constant for the ion-carrier complex is the same on both sides of the membrane. In this case, where n is the number of negative charges on the carrier (n can be positive or negative):

$$k_a = \frac{[MCr^{2-n}]}{[M^{2+}][Cr^{n-}]} \tag{2}$$

$$\frac{[MCr_{in}^{2-n}]}{[MCr_{out}^{2-n}]} = e^{\{(2-n)FE/RT\}} \tag{3}$$

$$\frac{[Cr_{in}^{-n}]}{[Cr_{out}^{-n}]} = e^{\{(-n)FE/RT\}} \tag{4}$$

where E, the membrane potential, F, R, and T have the usual meaning. These equations can be put together to give:

$$e^{\{(2-n)FE/RT\}} = \frac{[MCr_{in}^{2-n}]}{[MCr_{out}^{2-n}]} = \frac{k_{a\ in}[M_{in}^{2+}][Cr_{in}^{n-}]}{k_{a\ out}[M_{out}^{2+}][Cr_{out}^{n-}]} = \frac{[M_{in}^{2+}]}{[M_{out}^{2+}]} e^{\{(-n)FE/RT\}}$$

or

$$\frac{RT}{2F} \ln \left\{ \frac{[M_{in}^{2+}]}{[M_{out}^{2+}]} \right\} = E \tag{5}$$

This simple calculation points to possible sources of error in the above discussed experiments. If, for example, the carrier's distribution is not governed directly by the membrane potential, error could be introduced. Some evidence that this difficulty is not encountered may be found in FIGURE 3, where it is seen that amounts of A23187, an independent Ca^{2+} and Mn^{2+} carrier, sufficient to rapidly deplete the mitochondria of Mn^{2+} in the presence of the chelating agent EDTA, produces no effect in the absence of EDTA. This was interpreted as being an indication that increasing both inward and outward fluxes of divalent ion does not disturb the steady-state concentration ratio and hence that what was being measured from the concentration ratio was indeed a membrane potential.

Another assumption necessary in the derivation of Equation **5** is that the binding affinity for the ion-carrier complex be the same on both sides of the membrane. This is not necessarily the case, particularly where a pH gradient exists. In that case some underestimation of the membrane potential could occur (assuming k_a increases with pH). It is not thought that this should prove important at pH's of 7 or above, however, since H^+ ion competition for binding should be negligible. Finally it should be acknowledged that the "free" signal quantitated with epr is not equivalent to Mn^{2+} chemical activity. While it is much more sensitive to activity than a simple measurement of concentration would be, it tends to be sensitive only to short-range effects (first and second hydration sphere) of counter ions as opposed to long-range interactions which can also influence activity. The necessary activity corrections might be thought to be about as large as the magnitude of the activity coefficient in dilute solutions since in dilute solutions long-range interactions are still important while short-range interactions are not as important. Since monovalent anions are used in the work described here, it is thought that the activity corrections to be used should be more like those for Mn^{2+} with monovalent anions than with divalent or trivalent anions. From simple considerations of Debye-Hückel theory and from observation of measured values of activity coefficients for divalent cations and monovalent anions in the millimolar concentration range, it is felt that 30% would be a conservative estimate of error from activity effects (see, for example, Reference 29). This would lead to an error of about 3.3 mV and probably to an overestimation of membrane potential by this amount.

Acknowledgments

The authors wish to thank Dr. George Kimmich for interesting discussions of the data, and Mr. Tom Hefley, Dr. Karlene Gunter, and Mrs. Phyllis Russell for help with the experiments. A 23187 was the generous gift of R. L. Hamill of Eli Lilly Co.

References

1. Rasmussen, H., B. Chance & E. Ogata. 1965. A mechanism for the reactions of calcium with mitochondria. Proc. Nat. Acad. Sci. U.S.A. **53:** 1069–1076.
2. Chance, B. 1965. The energy-linked reaction of calcium with mitochondria. J. Biol. Chem. **240:** 2729–2748.
3. Pressman, B. C. 1965. Induced active transport of ions in mitochondria. Proc. Nat. Acad. Sci. U.S.A. **53:** 1076–1083.
4. Reynafarje, B. & A. L. Lehninger. 1969. High affinity and low affinity binding of Ca^{++} by rat liver mitochondria. J. Biol. Chem. **244:** 584–593.
5. Vasington, F. D., P. Gazzotti, R. Tiozzo & E. Carafoli. 1972. The effect of ruthenium red on Ca^{2+} transport and respiration in rat liver mitochondria. Biochim. Biophys. Acta **256:** 43–54.
6. Mela, L. 1968. Interactions of La^{+3} and local anesthetic drugs with mitochondrial Ca^{+2} and Mn^{+2} uptake. Arch. Biochem. Biophys. **123:** 286–293.
7. Vanio, H., L. Mela & B. Chance. 1970. Energy dependent bivalent cation translocation in rat liver mitochondria. Eur. J. Biochem. **12:** 387–391.
8. Vinogradov, A. & A. Scarpa. 1973. The initial velocities of calcium uptake by rat liver mitochondria. J. Biol. Chem. **248:** 5527–5531.
9. Case, G. D. 1975. Magnetic resonance studies of the mitochondrial divalent cation carrier. Biochim. Biophys. Acta **375:** 69–86.
10. Blondin, G. A. 1974. Isolation of a divalent cation ionophore from beef heart mitochondria. BBAC **56:** 97–105.
11. Scarpa, A. & G. Azzone. 1970. The mechanism of ion translocation in mitochondria. Eur. J. Biochem. **12:** 328–335.
12. Reed, P. W. & H. A. Lardy. 1972. A23187: A divalent cation ionophore. J. Biol. Chem. **247:** 6970–6977.
13. Rottenberg, H. & A. Scarpa. 1974. Calcium uptake and membrane potential in mitochondria. Biochemistry **13:** 4811–4817.
14. Rottenberg, H. 1973. The mechanism of energy-dependent ion transport in mitochondria. J. Membrane Biol. **11:** 117–137.
15. Mitchell, P. & J. Moyle. 1969. Estimation of membrane potential and pH

difference across the cristae membrane of rat liver mitochondria. Eur. J. Biochem. **7:** 471–484.
16. GUNTER, T. E. & J. S. PUSKIN. 1972. Manganous ion as a spin label on studies of mitochondrial uptake of manganese. Biophys. J. **12:** 625–635.
17. PUSKIN, J. S. & T. E. GUNTER. 1972. Evidence for the transport of manganous ion against an activity gradient by mitochondria. Biochim. Biophys. Acta **275:** 302–307.
18. PUSKIN, J. S. & T. E. GUNTER. 1973. Ion and pH gradients across the transport membrane of mitochondria following Mn^{2+} uptake in the presence of acetate. Biochem. Biophys. Res. Commun. **51:** 797–803.
19. GUNTER, T. E., J. S. PUSKIN & P. R. RUSSELL. 1975. Quantitative magnetic resonance studies of manganese uptake by mitochondria. Biophys. J. **15:** in press.
20. CHAPPELL, J. B. & A. R. CROFTS. 1966. Ion transport and reversible volume changes of isolated mitochondria. *In* Regulation of Metabolic Processes in Mitochondria. J. M. Tager, S. Papa, E. Quagliariello and E. C. Slater, Eds.: 293–316. Elsevier Publishing Co. New York, N.Y.
21. LEHNINGER, A. L. 1974. Role of phosphate and other proton-donating anions in respiration-coupled transport of Ca^{2+} by mitochondria. Proc. Nat. Acad. Sci. U.S.A. **71:** 1520–1524.
22. CHAPPELL, J. B. & K. N. HAARHOFF. 1967. The penetration of the mitochondria membrane by anions and cations. *In* Biochemistry of Mitochondria. E. C. Slater, Z. Kaniuga and L. Wojtezak, Eds.: 75–91. Academic Press. London, England.
23. SCHNAITMAN, C. & J. W. GREENAWALT. 1967. Enzymatic properties of the inner and outer membranes of rat liver mitochondria. J. Cell Biol. **38:** 158–175.
24. SCHNAITMAN, C., V. G. ERWIN & J. W. GREENAWALT. 1967. The submitochondrial localization of monoämine oxidase. J. Cell Biol. **32:** 719–735.
25. SOTTACASA, G. L., B. KUYLENSTIERNA, L. ERNSTER & A. BERGSTRAND. 1967. An electron-transport system associated with the outer membrane of liver mitochondria. J. Cell Biol. **32:** 415–438.
26. GEAR, A. R. L., C. S. ROSSI, B. REYNAFARJE & A. L. LEHNINGER. 1967. Acid-base exchanges in mitochondria and suspending medium during respiration-linked accumulation of bivalent cations. J. Biol. Chem. **242:** 3403–3413.
27. ADDANKI, S., F. D. CAHILL & J. F. SOTOS. 1968. Determination of intramitochondrial pH and intramitochondrial-extramitochondrial pH gradient of isolated heart mitochondria by the use of 5,5-dimethyl-2,4-oxazolidinedione. J. Biol. Chem. **243:** 2337–2348.
28. CHANCE, B. & L. MELA. 1966. A hydrogen ion concentration gradient in a mitochondrial membrane. Nature **212:** 369–372.
29. MOORE, W. J. 1962. Physical Chemistry: 351–357. Prentice-Hall. Englewood Cliffs, N.J.

DISCUSSION

DR. M. BENDER (*Fairleigh Dickinson University, Teaneck, N.J.*): In connection with that relationship of manganese outside versus inside for the membrane potential, does manganese inside take care of all the manganese bound as well as unbound, that is, the total concentration?

DR. GUNTER: No; we're referring to internal free manganese, which is represented by the hyperfine sextet fraction that we're previously identified as manganese hexahydrate. In one slide calculated membrane potentials using total internal manganese were shown. The potentials measured that way are as low as about 12 and as high as 40 millivolts higher than are obtained by using internal free manganese.

DR. BENDER: Would you give me an idea as to the ratio of bound-to-free inside?

DR. GUNTER: It depends on the conditions of uptake. When chloride is the counter ion, there are times when one can get 98% bound; where acetate is the counter ion, sometimes one gets more than half free.

DR. H. FRIEDMAN (*State University of New York, Stony Brook, L.I., N.Y.*): If you have manganous acetate complex in water, don't you still set the sextet of the manganese?

DR. GUNTER: Yes. Manganese and acetate do not complex very strongly, so you're primarily dealing with manganese hexahydrate. Although acetate is going into the mitochondrion as a counter ion and although it affects the activity of the manganese, most of the manganese can be considered free manganese.

DR. FRIEDMAN: I thought the complexing constant was something appreciable, say of the order of ten. Wouldn't that make a substantial error in the membrane potential that you measure?

DR. GUNTER: Yes, it does lead to some error, but not to a large error; what we are looking at is not actually activity. In other words, the actual membrane potential, including activity effects, would be a little lower than what we get. However, an important thing about this technique is that it's sensitive to the same short-range interactions that affect activity such as complex formation. It's not sensitive to the long-range interactions that affect activity, so you do have error from that source.

DR. D. HAYNES (*University of Miami, Coral Gables, Fla.*): Your study was conducted under steady-state conditions, wasn't it? And you still had respiratory control?

DR. GUNTER: Yes; in fact, we measured the respiratory control ratio a couple of times under the conditions in which we measured membrane potential. We did see ADP stimulation of oxygen consumption when we worked at 10 mM [Ac^-] and uptake of about 80 nanomoles of Mn^{2+} per milligram of protein.

DR. HAYNES: Did you then have the mitochondrial energy squiggle or membrane potential or whatever it is that should have been sufficient to drive ATP synthesis?

DR. GUNTER: Yes, it was sufficient under some conditions.

DR. HAYNES: Your paper shows then that under some conditions you have something like the 100-millivolt membrane potential and a 0 pH gradient?

DR. GUNTER: Yes, under certain conditions, such as the presence of acetate, butyrate, or lactate.

DR. HAYNES: Don't some of the people who are arguing for the chemiosmotic hypothesis have something to say about this?

DR. GUNTER: Actually, we find ourselves in no-man's-land because our data look qualitatively very much like the chemiosmotic hypothesis, but quantitatively, the numbers don't come out large enough in terms of membrane potentials to satisfy proponents of this hypothesis. We get shot at from both directions.

Part III. Interaction of Proteins with Lipids

LIPID-LIPID AND LIPID-PROTEIN INTERACTION IN MODEL SYSTEMS AND MEMBRANES*

L. L. M. van Deenen, J. de Gier, R. A. Demel, B. de Kruyff, M. C. Blok, E. C. M. van der Neut-Kok, C. W. M. Haest, P. H. J. Th. Ververgaert, and A. J. Verkleij

Department of Biochemistry
University of Utrecht
Utrecht, The Netherlands

In 1925 Gorter and Grendel[1] proposed that in the erythrocyte membrane the lipid molecules are arranged in a bilayer. They spread the lipids extracted from a given quantity of red blood cells in a Langmuir trough at the air-water interface and observed that the area occupied by the lipids was twice the surface of the erythrocytes. It is not our intention to discuss the merits and pitfalls of this ingenuous approach. After 50 years of intensive research the concept of the presence of a lipid bilayer in membranes appears more generally accepted when compared with the situation some ten years ago. Indeed several membrane properties are understandable on the basis of a lipid bilayer structure with proteins both located at the surfaces and penetrating or spanning the hydrophobic lipid core.

The present paper summarizes some studies from our laboratory on lipid mono- and bilayers (liposomes) in relation to biomembranes.

Action of Polyene Antibiotics on Membranes

Polyene antibiotics can interact with both artificial lipid interfaces and biological membranes, and cause changes in permeability characteristics. It is well established that the presence of sterols in the membrane is a requirement for the action of these antibiotics,[2] which compounds have a macrolide ring structure with a varying number of double bonds and hyroxyl groups (FIGURE 1).

In a collaborative study with Dr. S. C. Kinsky it was demonstrated that the sterol requirement of polyene antibiotics is manifested in the simple system of lipid monolayers.[3, 4] Injection of these antibiotics underneath a lipid monolayer induces at low molar ratios of antibiotic/lipid a significant increase of surface pressure, but only when sterol is present in the lipid film above a given sterol/phospholipid ratio. Recent studies were aimed at understanding the nature of the molecular association between the antibiotics and sterols and at interpreting the permeability alterations occurring in the membrane. In this respect it is appropriate to give first some attention to studies on the interaction between phospholipids and various sterols in monolayers, liposomes, and natural membranes. It was found that only sterols having a 3β-OH group, a planar ring system, and a hydrophobic side chain at C-17 gave a reduction of the

* This paper is dedicated to the memory of Dr. Y. London, who was killed at the Golan Heights in October 1973.

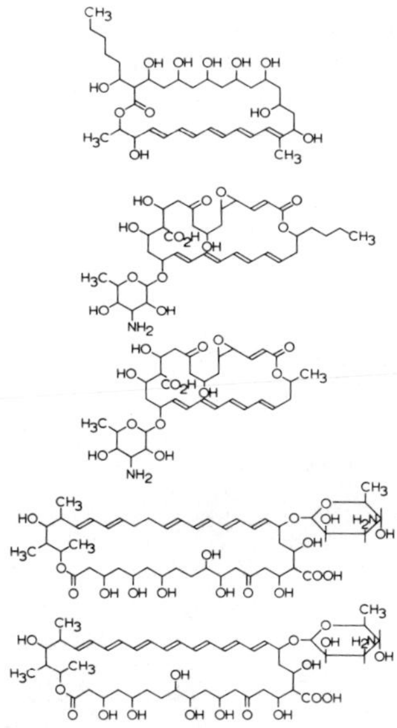

FIGURE 1. Structures proposed for a number of polyene antibiotics.

molecular area of a monolayer of synthetic unsaturated lecithins.[5] The same requirments for the sterol molecule were found in studies on liposomes which demonstrated the importance of these structural parameters to reduce the passive permeability of the lipid bilayers towards nonelectrolytes, such as erythritol, glycerol and glucose, and Rb^+.[6] That the observations on the artificial systems are relevant to biomembranes is supported by experiments that involved the replacement of the cholesterol in the erythrocyte membrane by various sterol analogs.[7] The structural requirement of sterol interaction with other lipid components of biological membranes was studied also after incorporation of various sterols in the cell membrane of *Acholeplasma laidlawii*.[8, 9] The effect of the sterol upon the permeability of this membrane to erythritol was found to follow to the same rules as have been established for liposomes and monolayers.

The relations observed between model systems and natural membranes for lipid-sterol interaction stimulated similar studies on polyene antibiotic-sterol interaction. In monolayers,[10] liposomes,[10–12] and *A. laidlawii* membranes[10–13] it could be demonstrated that the sterol incorporated in the membrane required a planar ring system, a hydrophobic side chain, and a 3β-OH group for polyene action. It is intriguing that there is a clear relationship between the structural requirments involved in (phospho)-lipid-sterol interaction and polyene antibiotic-sterol interaction. The latter association apparently is stronger than the former one. This is demonstrated also by differential scanning calorimetry and spectroscopic studies,[10, 11, 13] which revealed that the presence of polyene antibiotic has withdrawn sterol from its interaction with other membrane lipids so as to form polyene antibiotic-sterol complexes of defined stoichiometry. This complex formation alters the permeability of both artificial and natural membranes in a similar manner. The various polyenes, with the exception of filipin, cause selective permeability changes, suggesting the formation of pores with a diameter of about 8 Å, permitting nonelectrolytes of the size of or smaller than glucose and monovalent cations to pass. Models for these complexes have been proposed by various investigators.[14-16] A space-filling model developed for the amphotericin B-cholesterol complex is visualized as a circular arrangement of eight amphotericin B and eight cholesterol molecules.[16] A schematic representation (FIGURE 2) shows that in such complexes, each forming

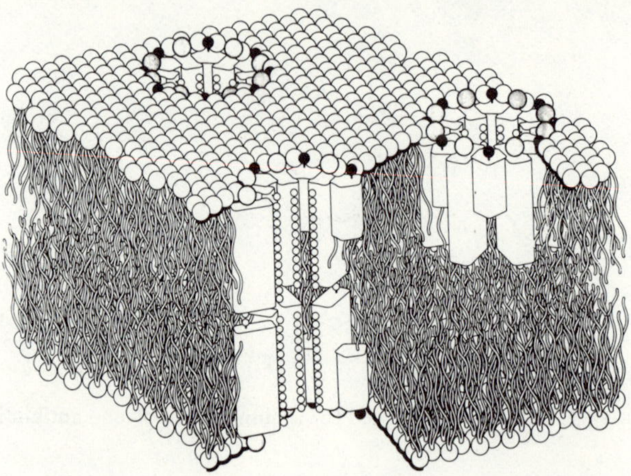

FIGURE 2. Schematic representation of conducting pores formed by amphotericin B and cholesterol in a lipid bilayer.[16]

a half pore, the inside is hydrophilic because of the presence of the hydroxyl groups of the polyene antibiotic. The cholesterol molecules fit in the ring system of the polyene antibiotic and are in close contact with the double bond system without having direct contact with the hydrophilic region(s) of the antibiotic. The hydroxyl group of cholesterol and the charged groups of amphotericin B are believed to be located at the membrane surfaces. Similar structures have been deduced for complexes formed by nystatin and etruscomycin.[16] Furthermore, it was argued that the primaricin-cholesterol complex cannot form a conducting pore because the length of the two half-pores is limited when compared with the thickness of a lipid bilayer. Filipin is known to cause more drastic alterations in both artificial and natural membranes. The permeability changes (e.g., release of the enzyme glucose-6-phosphate dehydrogenase) suggest that significant structural alterations occur in the membrane by the formation of the filipin-cholesterol complex. Freeze-etch electron microscopy demonstrated that filipin induces the formation of aggregates 150–250 Å in diameter in both biological membranes and liposomes which contain cholesterol[17, 18] (FIGURE 3). At the etched face of the membrane "projections" of about 50 Å in height are visible. No change in fracture faces was observed after treatment of cholesterol-rich membranes with amphotericin B.

A hypothetical model for filipin-cholesterol was proposed by De Kruyff and Demel[16] and is reproduced in FIGURE 4. In contrast to the other polyene antibiotics under discussion, filipin does not have a charged carboxyl and mycosamine group. As demonstrated with space-filling models filipin is most

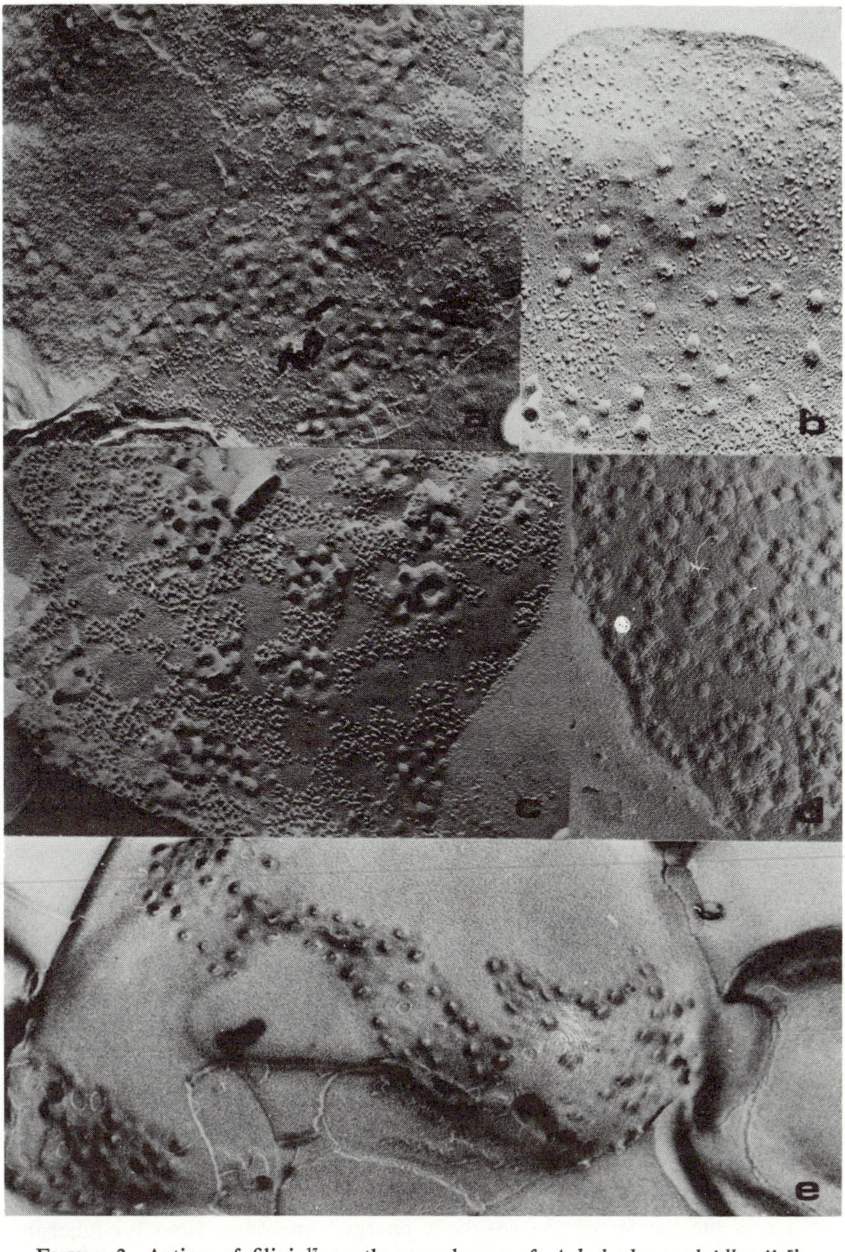

FIGURE 3. Action of filipin[17] on the membrane of *Acholeplasma laidlawii* [inner fracture face and etch face (a), outer fracture face (b)], human erythrocytes [inner fracture face (c), etch face (d)] and liposomes of lecithin and cholesterol (e).

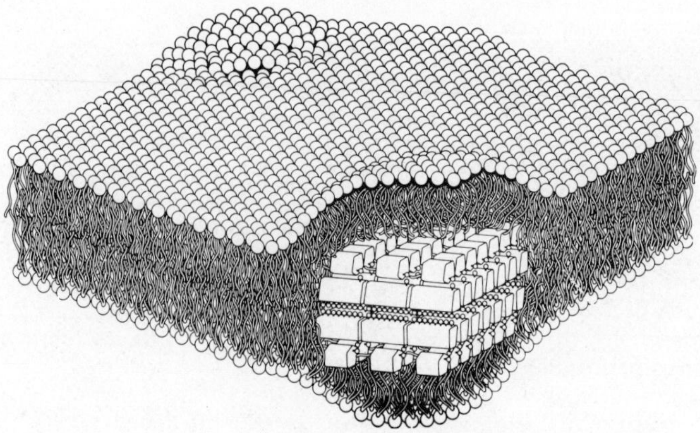

FIGURE 4. Schematic representation of a proposed complex of filipin and cholesterol in a lipid bilayer.[16]

suitably packed in a parallel array. One side of such a planar aggregate is more hydrophobic (double bond system) and the other side hydrophilic (hydroxyl groups). A bilayer of filipin molecules exposing on both sides the double bond system to hydrophobic interaction with cholesterol molecules may give aggregates in the interior of the lipid core of the membrane, thereby leading to membrane deformation and local disruptures.

Effects of Membrane Lipid Composition of Valinomycin-induced Cation Permeability

Excellent correlations have been found between passive permeability of liposomes and natural membranes of different lipid composition. The consequences of induction of chemical variations in lipid structure on membrane behavior could be predicted in many cases on the basis of the properties of the lipids in monomolecular films. As a follow-up of these studies on simple diffusion of nonelectrolytes the valinomycin-mediated cation transport was studied in liposomes and natural membranes which contained: (1) phospholipids with differently charged polar headgroups, (2) hydrophobic chains of a different degree of unsaturation, and (3) a variation in sterol content.

A striking difference in the properties of polar headgroup exists between two closely related (bacterial) phospholipids: the negatively-charged phosphatidylglycerol and positively-charged lysylphosphatidylglycerol. The permeability of Rb^+ and K^+ was found to be higher for liposomes of phosphatidylglycerol than for those of lysylphosphatidylglycerol.[19] Valinomycin was able to increase the transport of these cations in the liposomes of phosphatidylglycerol, but not across the barrier of the positively charged liposomes from lysylphosphatidylglyerol. In order to extrapolate these results from model studies to biological membranes the lipid composition of intact cells of *Staphylococcus aureus* was varied by means of the environmental pH. Indeed it was found that with an increasing lysylphosphatidylglycerol-to-phosphatidylglycerol ratio the valinomycin-mediated exchange of ^{86}Rb across the cell membrane was decreased.[19]

A second lipid parameter which was found to control the rate of valinomycin-mediated transport in black films[20] and liposomes[21] is the nature of the fatty acid constituents of the phospholipids. For instance, it was found that increasing unsaturation of the paraffin chain promotes the valinomycin-induced cation leak. After variation of the fatty acid composition of the membrane lipids in *A. laidlawii* cells the valinomycin-induced K^+ and Rb^+ permeability was measured with three different techniques. In accordance with the experiments on liposomes it was found that the increase in the degree of unsaturation of the lipids causes a higher effectivity of valinomycin.[22] Insertion of cholesterol in liposomes and in *Acholeplasma laidlawii* membranes causes a decrease in permeability.

The observation that the lipid composition, e.g., the degree of unsaturation of fatty acids, influences the rate of valinomycin-mediated transport can be interpreted in different ways. The possibility exists that this ionophore displays a higher affinity for lipid bilayers with higher unsaturation of the phospholipids. Alternatively, the carrier may have a greater mobility within the unsaturated lipid bilayer. The kinetics of valinomycin-induced K^+ leak were studied on liposomes loaded with potassium thiocyanate.[23] The lipophilic anion permits a considerable leakage in the absence of proton conductors, thus making unnecessary the addition of uncouplers.[24] The initial rate of K^+ leakage was found to be proportional to the valinomycin concentration. A rapid partition equilibrium of valinomycin occurs between the liposomal bilayer and the aqueous phase. A high affinity of valinomycin for the lipid bilayer was apparent and the partition constant has a negative temperature coefficient. The kinetic data support a model involving the formation of a ternary complex of valinomycin with K^+ and a thiocyanate ion (FIGURE 5). Using liposomes of phospholipids with a different content of polyunsaturated fatty acids it was demonstrated that the rate of valinomycin-mediated transport increased with increasing unsaturation of the phospholipids; the affinity of the ionophore may even decrease somewhat with increasing unsaturation of the lipid bilayer.[23] In short, it is most likely that the translocation of the complex is the rate-limiting step and that the turnover rate of more complicated transport systems also may

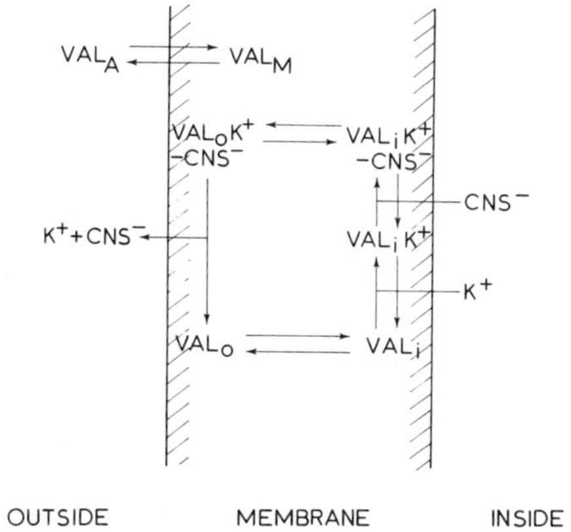

FIGURE 5. Model for the valinomycin-mediated K^+ leak from liposomes containing KCNS.[23]

be controlled by the local lipid composition of the membrane. Insertion of cholesterol in liposomes[21] and *A. laidlawii*[22] was found to bring about a reduction of the valinomycin-mediated transport; current kinetic studies may give further insight in the molecular mechanism involved.

Effects of Lipid-Phase Transition on Membrane Permeability and Lipid-Protein Interaction

In recent years much attention has been given to thermotropic transitions in biological membranes. The transition of lipids from the gel (solid) to the liquid-crystalline phase and vice versa can affect many properties of membranes. In *A. laidlawii* the cell barrier remains intact upon cooling below the transition temperature, although under these conditions the cells are more fragile when exposed to mechanical or osmotic forces. Passive permeation of nonelectrolytes through *A. laidlawii* membranes at temperatures in the range of lipid phase transition appears to occur predominantly in the regions of the membrane that are still in liquid crystalline state.[9] It was reported that in black lipid membranes valinomycin was not capable of increasing the conductivity of the films at temperatures below the transition temperature of the lipids.[25] In order to validate an extension of this conclusion to natural membranes, experiments with *A. laidlawii* were carried out at the extended region of the gel to liquid-crystalline phase transition.[22] It can be seen that the valinomycin-induced efflux of K^+ is (nearly) zero below the temperature of the gel-liquid crystalline phase transition (FIGURE 6). In the temperature range above the phase transition the valinomycin-inducible K^+ leakage gradually increases with increasing temperature. The immobilization of the carrier function of valinomycin below the lipid phase transition could not be demonstrated in *E. coli* because a spontaneous loss of, for example, intracellular K^+ occurs when these cells are rapidly cooled below the transition temperature.[26] Liposomes prepared from given synthetic phospholipids, such as (dimyristoyl)-lecithin, also revealed a release of intracellular K^+ without valinomycin present when the lipid arrangement is transformed from the liquid-crystalline to the gel state.[26] Current experiments showed that this phenomenon may depend

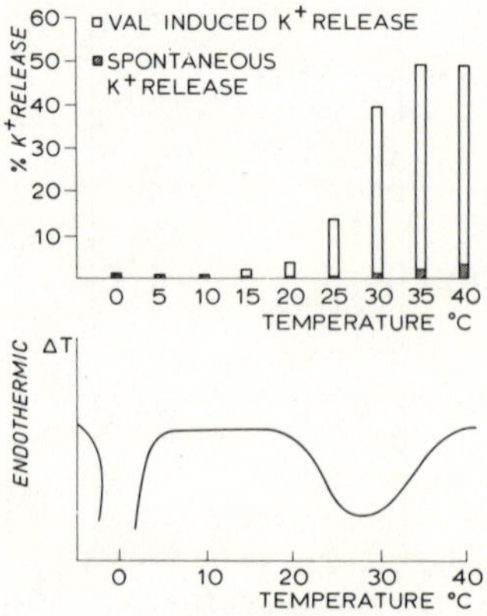

FIGURE 6. Leakage of K^+ from cells of *A. laidlawii* B in relation to the lipid phase transition.[22] Elaidic and palmitic acid added to the growth medium.

to a significant extent on the nature of the lipids making up the bilayer, and the permeation alteration at the phase transition exhibits a certain selectivity even within the range of monovalent cations. Such alterations in ion permeability may find an explanation in the distortions of the lipid conformation during the phase transition.

In this context it is of interest to refer to alterations in the architecture of biological and artificial membranes at lipid phase transition as visualized by freeze-etching electron microscopy.[27] This technique allows a direct observation of the transition from the liquid-crystalline to a gel state of synthetic phosphatidylcholines present in liposomes. Whereas band patterns in the former phase are smooth, quenching from below the transition temperature revealed characteristic band patterns with defined periodicities (FIGURE 7). Liposomes made of two phosphatidylcholine, which have a difference of no more than two CH_2 groups, exhibited a band pattern with a new periodicity[28] (FIGURE 8). These mixtures display one thermotropic transition, and a homogeneous distribution of the phospholipid species below the temperature of the phase transition is likely. Mixtures of phospholipids with a difference in chain length of more than two CH_2 groups or containing one saturated and one unsaturated phospholipid give two thermotropic peaks during differential scanning calorimetry; accordingly at temperatures between, these two phase transition regions of band patterns (phospholipid species in gel phase) and smooth area (phospholipid species in liquid-crystalline state) can be observed (FIGURE 8).[28] At temperatures above both peaks the fracture faces are smooth; and when the mixture is quenched from a temperature below both peaks, the band patterns of both phosphatidylcholine species are visible. Thus the freeze-etching technique makes it possible to visualize the phenomenon of phase separation, which occurs in the temperature range between the phase transition of the two individual phospholipids. A discussion of the effects of cholesterol,[29] different polar headgroups,[30] mixtures of oppositely charged phospholipids,[31] and the presence of cations on the images produced by freeze-etching[32] is beyond the scope of the present contribution. Many of these parameters are important when comparing results obtained on liposomes and biological membranes. Freeze-etching of natural membranes has demonstrated that a phase transition of the lipids is accompanied by an aggregation of the intercalated particles in *A. laidlawii,*[27, 33] *E. coli,*[34, 35] and *Tetrahymena pyriformis.*[36] The onset and the extent of aggregation have been found to be related to the beginning and degree of phase transition (liquid-crystalline to gel), respectively (FIGURE 9). It has been suggested that the aggregation of the protein particles occurs after partial solidification of the lipids, giving regions of lipid in the gel state and a concentration of the protein particles in the lipid regions remaining in the liquid-crystalline state. However, in some organisms (*S. aureus* and *Bacillus subtilis*) particle aggregation was not observed in the fracture face although differential scanning calorimetry[34] and breaks in Arrhenius plots of membrane-bound enzymes indicate that a phase transition of lipids occurred in the temperature region studied.[34] The nonappearance of particle aggregation could be explained by the presence of branched-chain fatty acids in the membrane lipids of these bacteria. Indeed, *A. laidlawii* cells that were grown in the presence of branched-chain fatty acids, and readily incorporated these constituents into membrane lipids, failed to give the particle reorganization induced by the lipid phase transition which was detected by differential scanning calorimetry.[34] A different orientation of the branched-chain fatty acid constituents, giving also in the gel phase a rather loose packing,

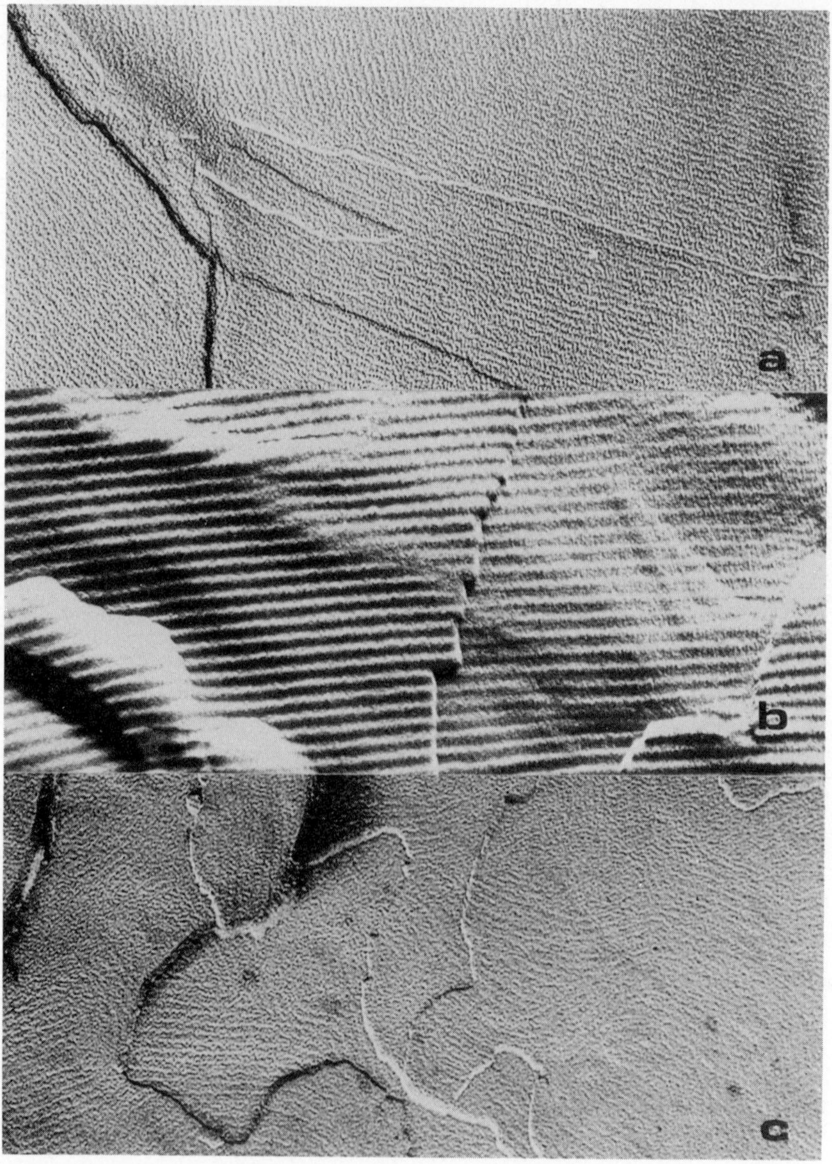

FIGURE 7. Band patterns observed by freeze-etch electron microscopy of synthetic phosphatidylcholines quenched from below the gel-liquid crystalline transition temperature.[27] (a) (dilauroyl)-lecithin; (b) (dimyristoyl)-lecithin; (c) (1-palmitoyl-2-oleoyl)-lecithin.

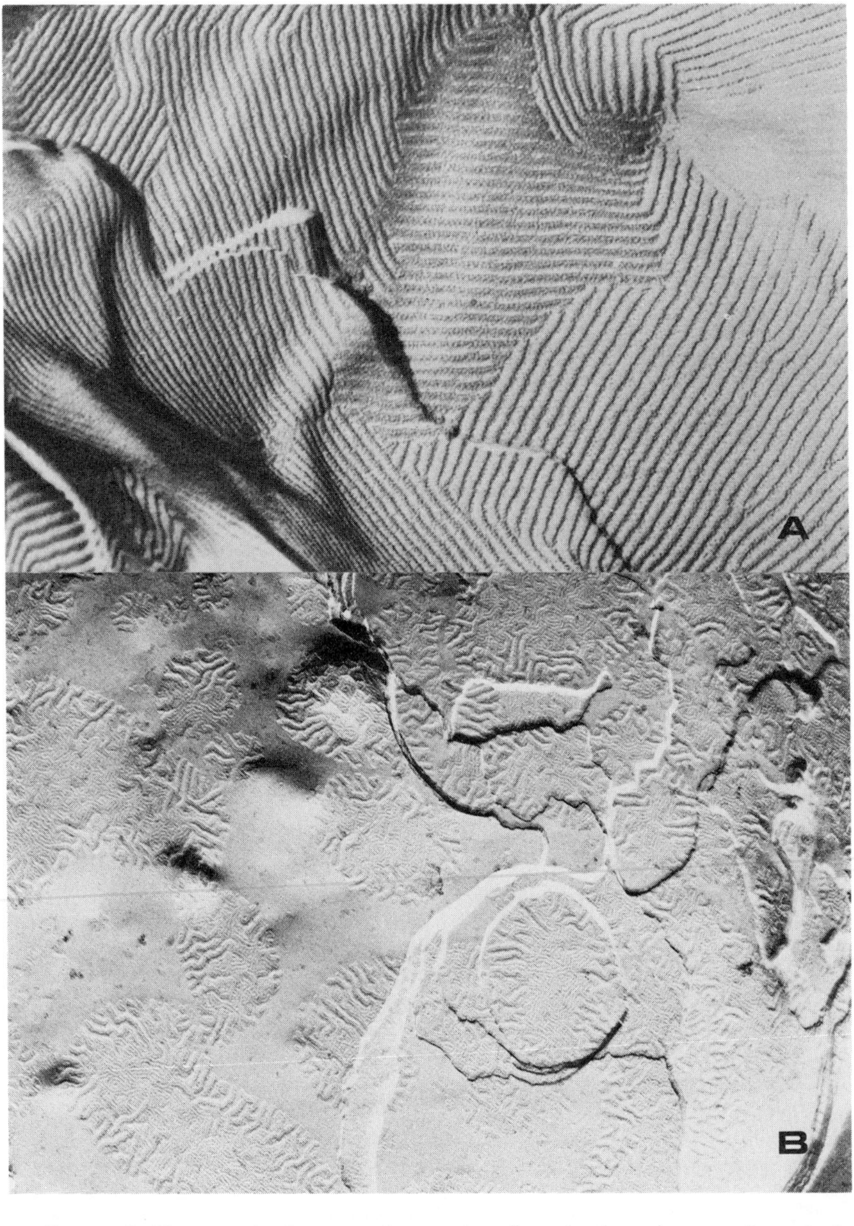

FIGURE 8. Freeze-etch electron micrographs of equimolar mixtures of synthetic phosphatidylcholines.[28] (A) (dilauroyl)-lecithin and (dimyristoyl)-lecithin quenched from below the thermotropic peak. (B) (dipalmitoyl)-lecithin and (1-palmitoyl-2-oleoyl)-lecithin quenched from between the two thermotropic peaks as measured by differential scanning calorimetry.

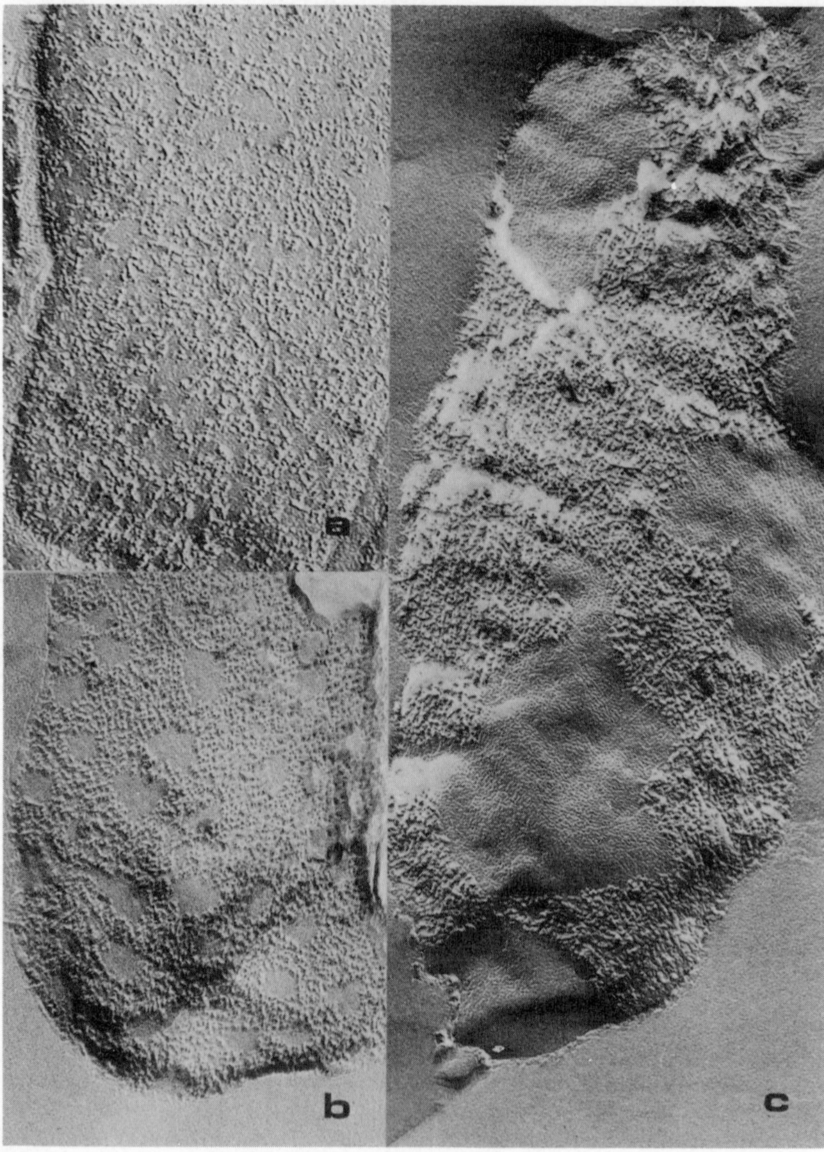

FIGURE 9. Structures of inner faces. *E. coli* cells demonstrated by freeze-fracturing of membranes quenched from above (a), within (b), and below (c) the lipid phase transition.[34]

may explain why the protein particles are not squeezed out and remain homogeneously localized in the hydrocarbon region of the membrane.

In short, lipid phase transition can cause lipid-protein segregation when the lipids attain the gel state, but not when membrane lipids contain branched-chain fatty acids. Oppositely, proteins can affect the transition temperature of lipids[32] and create an ordering effect on the lipid core, particularly when ionic forces are predominantly involved in the interaction between lipids and proteins.

The Nature of Associations between Proteins and Lipids

In biological membranes a number of proteins can be located at either surface of the lipid bilayer, and this lipid-protein interaction may involve principally electrostatic attractions. Proteins that deeply penetrate or even span the lipid bilayer may associate mainly by hydrophobic forces with lipids. It is further realized that protein-lipid associations occur that to a varying degree depend on a subtle interplay between both electrostatic and hydrophobic interaction. For the understanding of membrane architecture and its function it is relevant to investigate the interaction of defined proteins and lipids under well-controlled conditions. The system of monomolecular layers of lipids allows study of the interaction with proteins injected underneath the lipid film by recording changes in surface pressure (at constant surface area) or changes in surface area (at constant pressure). These approaches can be combined with measurements of surface radioactivity by means of labeled protein. Furthermore, it is possible to study the effects of cations on the interaction of protein with lipid as well as the competition between two proteins for a particular lipid. Finally, it is possible to use proteolytic enzymes and phospholipases in combination with lipid-protein monolayers to obtain information about the nature of interaction and to determine which functional groups of both partners are involved. Some examples demonstrating these applications are briefly discussed.

Interaction between A_1 basic protein from CNS myelin membrane and monolayers of various lipid classes demonstrated a pronounced increase of surface pressure (at constant area) with monolayers made of negatively-charged lipids. Of particular interest is the strong reaction between the basic protein and a film of cerebroside sulphate,[37] a lipid that is abundant in this membrane. The results suggested the importance of coulombic interactions for the high affinity between basic protein and negatively-charged lipids, and in this respect it was considered of interest to study the possible effect of charge inhibition by Ca^{2+}. When A_1 protein is bound to the lipid monolayer, no binding of $^{45}Ca^{2+}$ could be detected, but injection of basic protein underneath a lipid-$^{45}Ca^{2+}$ monolayer causes a decrease of surface radioactivity such that $^{45}Ca^{2+}$ is completely expelled from the interface (FIGURE 10). The phenomenon is accompanied by an increase in surface pressure, which is considered to be due to penetration of the protein into the lipid phase. This finding supports the view that electrostatic attractions are effective for the association between lipid and this protein.

Measurement of the surface radioactivity using ^{131}I-labeled A_1 basic protein showed that the amount of basic protein bound to the lipid film correlates well with the increase in surface pressure. That hydrophobic interactions in

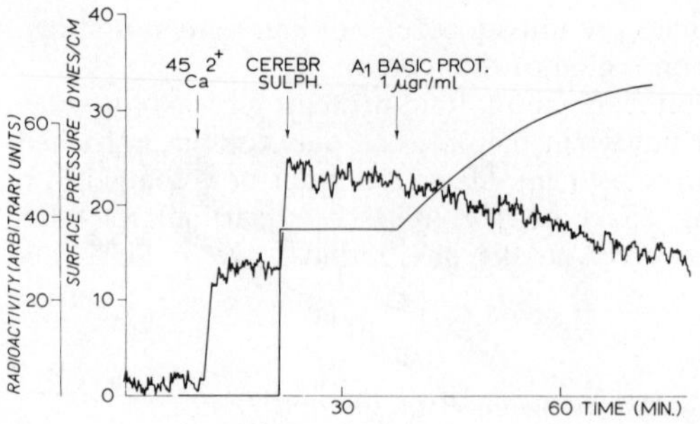

FIGURE 10. The effect of A_1 basic protein on the interaction of Ca^{2+} with a monolayer of cerebroside sulphate. The desorption of $^{45}Ca^{2+}$ was measured by monitoring the surface radioactivity; the penetration of protein was measured by the change in surface pressure.[37]

addition to electrostatic interaction play a part in the formation of this lipid-protein system was suggested by an influence of fatty acid length. Perhaps more conclusive in this respect is the observation that at constant film pressure a most significant increase in surface area (of 200%–400%) was found after injection of A_1 basic protein underneath a monolayer of cerebroside sulphate. This result indicates also that the lipid monolayer is penetrated and that the protein may extend into the apolar region of the film. Experiments were undertaken to elucidate which parts of the protein are particularly involved in the association with the negatively-charged lipids. The lipid-protein layer was treated with proteolytic enzymes such as trypsin, chymotrypsin, subtilopeptidase, or pronase, which were injected underneath the complex.[38] It was found that these enzymes were much less effective in the degradation of the protein after its interaction with lipid monolayers. It could be concluded that certain regions of the protein are protected by the association with lipid. To determine which peptide linkages of the A_1 basic protein were preserved from proteolytic attack, the lipid-protein complex was collected from the interface after the action of trypsin. Peptide maps demonstrated that six peptide bonds located at the N-terminal region between position 23 and 113 of the protein were protected in the protein-lipid complex; a tentative model in which these regions of the protein are considered to penetrate into the lipid phase was postulated (FIGURE 11).

A second myelin protein, the Folch-Lees proteolipid, shows affinity for a variety of lipids and most remarkably for cholesterol as well.[39] This interaction was demonstrated to be dependent on the sterol structure. That these two major myelin proteins have different preferences for associating with lipids was demonstrated by experiments in which both proteins compete for interaction with a lipid monolayer (FIGURE 12). A subsequent injection of A_1 basic protein and Folch-Lees proteolipid underneath a cholesterol monolayer shows a preferential binding of the proteolipid to cholesterol and a desorption of A_1 basic protein from the interface. In similar experiments it was demonstrated that the basic protein had the highest affinity for cerebroside sulphate when compared with Folch-Lees proteolipid. The preferences observed suggest that specific associations of these proteins with particular lipid classes may also occur in the myelin membrane. Extrapolation of these results to the native membrane supports the concept of both lipid and protein asymmetry in this membrane.[39]

An asymmetric distribution of lipids in the erythrocyte membrane has been established on the basis of studies combining the action of phospholipases and freeze-etch electron microscopy.[40] The outer monolayer of the human

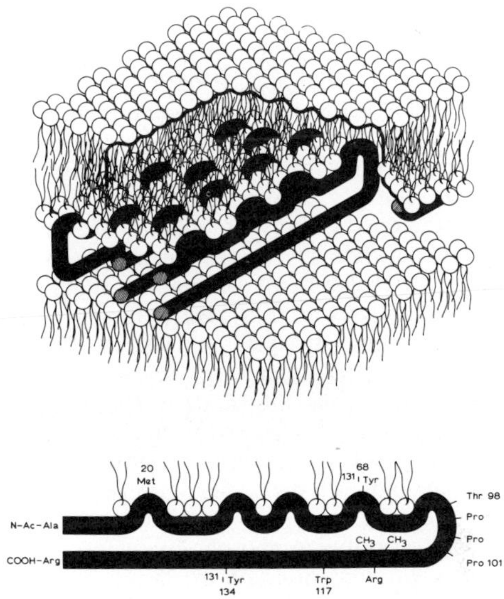

FIGURE 11. Schematic representation of binding sites and conformation of A_1 basic protein after association with myelin lipids.[38]

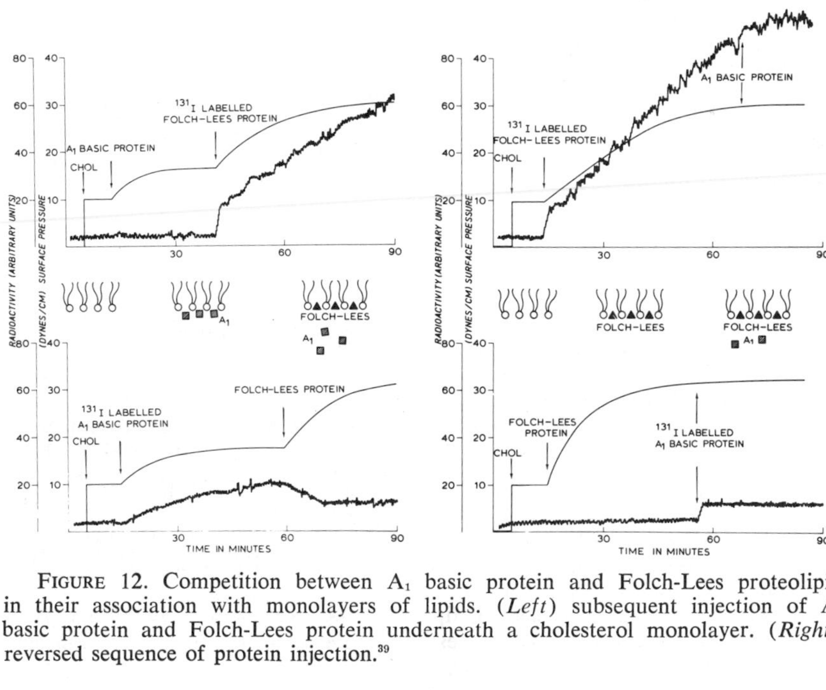

FIGURE 12. Competition between A_1 basic protein and Folch-Lees proteolipid in their association with monolayers of lipids. (*Left*) subsequent injection of A_1 basic protein and Folch-Lees protein underneath a cholesterol monolayer. (*Right*) reversed sequence of protein injection.[39]

erythrocyte is considered to consist mainly of phosphatidylcholine and sphingomyelin and there is an additional 20% of phosphatidylethanolamine. The inner layer is composed of phosphatidylethanolamine, phosphatidylserine, and a small fraction of choline-containing phospholipids. These studies make one aware of the fact that certain phospholipases can act on the phospholipids present in the erythrocyte membrane, whereas others fail to do so. Various interpretations can be made and it was thought to be of interest to make a comparative study of a series of phospholipases on monolayers of phospholipids adjusted at various surface pressures. To date only those phospholipases that are capable of acting on highly compressed phospholipid monolayers (above 30 dynes/cm) were found to be able to attack their substrates in the intact erythrocyte membrane.[41] This relationship permits estimation of the virtual surface pressure of the outer lipid layer of the erythrocyte. This parameter, which gives information about the tightness of the packing of the lipid molecules, is important for the evaluation of the classic experiment of Gorter and Grendel. In addition it seems plausible to conclude that at the erythrocyte surface not all polar headgroups of phospholipids are "occupied" by strong electrostatic interactions with proteins and that not all phospholipid ester linkages susceptible to phospholipase action are shielded off.

A different situation exists for the binding of phospholipid to a pure protein from beef liver, which specifically catalyzes the transfer of phosphatidylcholine between both natural and artificial membranes.[42] This protein was demonstrated to function as a carrier in the transfer of phosphatidylcholine between two separate monomolecular films.[43] The carrier protein contains one mole of phosphatidylcholine per mole of protein. Various phospholipases were unable to hydrolyze this phosphatidylcholine, indicating that at least the polar moiety of the phospholipid molecule is deeply bound into the carrier protein. Only after perturbation with organic solvents or detergents could hydrolysis of phosphatidylcholine occur.[44]

The examples given above may indicate that studies in monomolecular layers in combination with proteolytic and lipolytic enzymes may supply useful information about the molecular segments of proteins and lipids that are involved in their mutual interaction.

References

1. Gorter, E. & F. Grendel. 1925. On bimolecular layers of lipids on the chromocytes of the blood. J. Exp. Med. **41:** 439–443.
2. Kinsky, S. C. 1970. Antibiotic interaction with model membranes. Ann. *Rev.* Pharmacol. **10:** 119–142.
3. Dfmel, R. A., L. L. M. van Deenen & S. C. Kinsky. 1965. Penetration of lipid monolayers by polyene antibiotics. J. Biol. Chem. **240:** 2749–2753.
4. Demel, R. A., F. J. L. Crombrag, L. L. M. van Deenen & S. C. Kinsky. 1968. Interaction of polyene antibiotics with single and mixed lipid monomolecular layers. Biochim. Biophys. Acta **150:** 1–14.
5. Demel, R. A., K. R. Bruckdorfer & L. L. M. van Deenen. 1972. Structural requirements of sterols for the interaction with lecithin at the air-water interface. Biochim. Biophys. Acta **255:** 311–320.
6. Demel, R. A., K. R. Bruckdorfer & L. L. M. van Deenen. 1972. The effect of sterol structure on the permeability of liposomes to glucose, glycerol and Rb^+. Biochim. Biophys. Acta **255:** 321–330.

7. BRUCKDORFER, K. R., R. A. DEMEL, J. DE GIER & L. L. M. VAN DEENEN. 1969. The effect of partial replacements of membrane cholesterol by other sterols on the osmotic fragility and glycerol permeability of erythrocytes. Biochim. Biophys. Acta **181:** 334–345.
8. DEKRUYFF, B., R. A. DEMEL & L. L. M. VAN DEENEN. 1972. The effect of cholesterol and epicholesterol incorporation on the permeability and on the phase transition of intact Acholeplasma laidlawii cell membranes and derived liposomes. Biochim. Biophys. Acta **255:** 331–347.
9. DE KRUYFF, B., W. J. DE GREEF, R. V. M. VAN EYK, R. A. DEMEL & L. L. M. VAN DEENEN. 1973. The effect of different fatty acid and sterol compositions on the erythritol flux through the cell membrane of Acholeplasma laidlawii. Biochim. Biophys. Acta **298:** 479–499.
10. NORMAN, A. W., R. A. DEMEL, B. DE KRUYFF & L. L. M. VAN DEENEN. 1972. Studies on the biological properties of polyene antibiotics; evidence for the direction interaction of filipin with cholesterol J. Biol. Chem. **247:** 1918–1929.
11. NORMAN, A. W., R. A. DEMEL, B. DE KRUYFF, W. S. M. GEURTS VAN KESSEL & L. L. M. VAN DEENEN. 1972. Studies on the biological properties of polyene antibiotics: Comparison of other polyenes with filipin in their ability to interact specifically with sterol. Biochim. Biophys. Acta **290:** 1–14.
12. DE KRUYFF, B., W. J. GERRITSEN, A. OERLEMANS, R. A. DEMEL & L. L. M. VAN DEENEN. 1974. Polyene antibiotic-sterol interactions in membranes of Acholeplasma laidlawii cells and lecithin liposomes. I. Specificity of the membrane permeability changes induced by the polyene antibiotics. Biochim. Biophys. Acta **339:** 30–43.
13. DE KRUYFF, B., W. H. GERRITSEN, A. OERLEMANS, P. W. M. VAN DIJK, R. A. DEMEL & L. L. M. VAN DEENEN. 1974. Polyene antibiotic-sterol interactions in membranes of Acholeplasma laidlawii cells and lecithin liposomes. II. Temperature dependence of the polyene antibiotic-sterol complex formation. Biochim. Biophys. Acta **339:** 44–56.
14. FINKELSTEIN, A. & R. HOLZ. 1973. Aqueous pores created in this lipid membrane by the polyene antibiotics nystatin and amphotericin B. *In* Membranes. G. Eisenman, Ed. Vol. **2:** 377–407. Marcel Dekker, Inc. New York, N.Y.
15. ANDREOLI, T. E. 1973. On the anatomy of amphotericin B-cholesterol pores in lipid bilayer membranes. Kidney Int. **4:** 337–345.
16. DE KRUYFF, B. & R. A. DEMEL. 1974. Polyene antibiotic-sterol interactions in membranes of *Acholeplasma laidlawii* cells and lecithin liposomes. III. Molecular structure of the polyene antibiotic-cholesterol complexes. Biochim. Biophys. Acta **339:** 57–70.
17. VERKLEIJ, A. J., B. DE KRUYFF, W. F. GERRITSEN, R. A. DEMEL, L. L. M. VAN DEENEN & P. H. J. TH. VERVERGAERT. 1973. Freeze-etch electron microscopy of erythrocytes, *Acholeplasma laidlawii* cells and liposomal membranes after the action of filipin and amphotericin B. Biochim. Biophys. Acta **291:** 577–581.
18. TILLACK, T. W. & S. C. KINSKY. 1973. A freeze-etch study of the effects of filipin on liposomes and human erythrocyte membranes. Biochim. Biophys. Acta **323:** 43–54.
19. HAEST, C. W. M., J. DE GIER, J. A. F. OP DEN KAMP, P. BARTELS & L. L. M. VAN DEENEN. 1972. Change in permeability of *Staphylococcus aureus* and derived liposomes with varying lipid composition. Biochim. Biophys. Acta **255:** 720–733.
20. EISENMAN, G., G. SZABO, S. CIANI, S. MC LANGHLIN & S. KRASNE. 1973. Ion binding and ion transport produced by neutral lipid-soluble molecules. *In* Progress in Surface and Membrane Science. Vol. **6:** 139–241. Academic Press. New York, N.Y.
21. DE GIER, J., C. W. M. HAEST, J. G. MANDERSLOOT & L. L. M. VAN DEENEN. 1970. Valinomycin-induced permeation of $^{86}Rb^+$ through the bilayers of liposomes with varying composition. Biochim. Biophys. Acta **211:** 373–375.

22. van der Neut-Kok, E. C. M., J. de Gier, E. J. Middelbeek & L. L. M. van Deenen. Valinomycin-induced potassium and rubidium permeability of intact cells of *Acholeplasma laidlawii* B. Biochim. Biophys. Acta **332:** 97–103.
23. Blok, M. C., J. de Gier & L. L. M. van Deenen. 1974. Kinetics of the valinomycin-induced potassium ion leak from liposomes with potassium thiocyanate enclosed. Biochim. Biophys. Acta **367:** 210–224.
24. Blok, M. C., J. de Gier & L. L. M. van Deenen. 1974. Some factors affecting the valinomycin-induced leak from liposomes. Biochim. Biophys. Acta **367:** 202–209.
25. Krasne, S., G. Eisenman & G. Szabo. 1971. Freezing and melting of lipid bilayers and the mode of action of nonactin, valinomycin and gramicidin. Science **174:** 412–418.
26. Haest, C. W. M., J. de Gier, G. A. van es, A. J. Verkleij, L. L. M. van Deenen. 1972. Fragility of the permeability barrier of Escherichia coli. Biochim. Biophys. Acta **288:** 43–53.
27. Verkleij, A. J., P. H. J. Th. Ververgaert, L. L. M. van Deenen & P. F. Elbers. 1972. Phase transition of phospholipid bilayers and membranes of *Acholeplasma laidlawii* B visualized by freeze fracturing electron microscopy. Biochim. Biophys. Acta **288:** 326–332.
28. Ververgaert, P. H. J. Th., A. J. Verkleij, P. F. Elbers & L. L. M. van Deenen. 1973. Analysis of the crystalization process in lecithin liposomes: A freeze-etch study. Biochim. Biophys. Acta **311:** 320–329.
29. Verkleij, A. J., P. H. J. Th. Ververgaert, B. de Kruyff & L. L. M. van Deenen. 1974. The distribution of cholesterol in bilayers of phosphatidylcholines as visualized by freeze fracturing. Biochim. Biophys. Acta **373:** 495–501.
30. Tocanne, J. F., P. H. J. Th. Ververgaert, A. J. Verkleij & L. L. M. van Deenen. 1974. A monolayer and freeze-etching study of charged phospholipids. I. Effects of ions and pH on the ionic properties of phosphatidylglycerol and lysylphosphatidylglycerol. Chem. Phys. Lipids **12:** 201–219.
31. Tocanne, J. F., P. H. J. Th. Ververgaert, A. J. Verkleij & L. L. M. van Deenen. 1974. A monolayer and freeze-etching study of charged phospholipids. II. Ionic properties of mixtures of phosphatidylglycerol and lysylphosphatidylglycerol. Chem. Phys. Lipids **12:** 220–231.
32. Verkleij, A. J., B. de Kruyff, P. H. J. th Ververgaret, J. F. Tocanne & L. L. M. van Deenen. 1974. The influence of pH, Ga^{2+} and protein on the thermotropic behaviour of the negatively charged phospholipid, phosphatidylglycerol. Biochim. Biophys. Acta **339:** 432–437.
33. James, R. & D. Branton. 1973. Lipid and temperature-dependent structural changes in Acholeplasma laidlawii cell membranes. Biochim. Biophys. Acta **323:** 378–390.
34. Haest, C. W. M., A. J. Verkleij, J. de Gier, R. Scheek, P. H. J. Th. Ververgaert & L. L. M. van Deenen. 1974. The effect of lipid phase transition on the architecture of bacterial membranes. Biochim. Biophys. Acta **356:** 17–26.
35. Kleemann, W. & H. M. McConnell. 1974. Lateral phase separation in *Escherichia coli* membranes. Biochim. Biophys. Acta **345:** 220–230.
36. Speths, V. & F. Wunderlich. 1973. Membranes of Tetrahymena II. Direct visualization of reversible transitions in biomembrane structure induced by temperature. Biochim. Biophys. Acta **291:** 621–628.
37. Demel, R. A., Y. London, W. S. M. Geurts van Kessel, F. G. A. Vossenberg & L. L. M. van Deenen. 1973. The specific interaction of myelin basic protein with lipids at the air-water interface. Biochim. Biophys. Acta **311:** 507–519.
38. London, Y., R. A. Demel, W. S. M. Geurts van Kessel, F. G. A. Vossenberg & L. L. M. van Deenen. 1973. The protection of A_1 myelin basic protein against the action of proteolytic enzymes after interaction of the protein with lipids at the air-water interface. Biochim. Biophys. Acta **311:** 520–530.

39. London, Y., R. A. Demel, W. S. M. Geurts van Kessel, P. Zahler & L. L. M. van Deenen. 1974. The interaction of the "Folch Lees" protein with lipids at the air-water interface. Biochim. Biophys. Acta **332:** 69–84.

40. Verkleij, A. J., R. F. A. Zwaal, B. Roelofsen, P. Comfurius, D. Kastelijn & L. L. M. van Deenen. 1973. The asymmetric distribution of phospholipids in the human red cell membrane. A combined study using phospholipases and freeze-etch electron microscopy. Biochim. Biophys. Acta **323:** 178–193.

41. Demel, R. A., R. F. A. Zwaal, B. Roelfsen, W. S. M. Geurts van Kessel & L. L. M. van Deenen. Unpublished results.

42. Kamp, H. H., K. W. A. Wirtz & L. L. M. van Deenen. 1973. Some properties of phosphatidylcholine exchange protein purified from beef liver. Biochim. Biophys. Acta **318:** 313–325.

43. Demel, R. A., K. W. A. Wirtz, H. H. Kamp, W. S. M. Geurts van Kessel & L. L. M. van Deenen. 1973. Phosphatidylcholine exchange protein from beef liver. Nature New Biol. **246:** 102–105.

44. Kamp, H. H., E. D. Sprengers, J. Westerman, K. W. A. Wirtz & L. L. M. van Deenen. Unpublished results.

MODES OF LIPID-PROTEIN INTERACTIONS IN BIOMEMBRANES*

Donald F. H. Wallach, Vincenz Bieri, Surendra P. Verma, and Rupert Schmidt-Ullrich

Division of Radiobiology
Tufts-New England Medical Center
Boston, Massachusetts 02111

General Introduction

The structures assumed by macromolecules in aqueous media derive in large part from the "hydrophobic effect."[1] Apolar residues, whether parts of lipids, proteins, or nucleic acids, tend to order water molecules about them. This causes a thermodynamically unfavorable decrease in entropy. As a consequence, apolar residues tend to cluster together, excluding water and decreasing the free energy of the system as a whole. The "hydrophobic effect" thus favors formation of structures with apolar cores and polar surfaces. Globular proteins are outstanding examples, as is the bilayer arrangement of amphipathic lipids.

The "hydrophobic effect" predominates in a host of molecular and macromolecular interactions that do not necessarily involve molecular specificity. It thus mediates such diverse associations as of cholesterol with phosphatide acyl chains, heme with certain apolar residues in myoglobin, and α-chain with β-chain in hemoglobin. The "hydrophobic effect" also constitutes a principal mechanism in the association of proteins and lipids in biomembranes. This is not to say that lipid-protein interactions in biomembranes involve apolar interactions exclusively. Polar associations indubitably participate in many instances, and can even predominate in some cases (e.g., cytochrome *c* and mitochondria).

An abundance of evidence indicates that the lipids of biomembranes are primarily in bilamellar array, with their head groups at the two membrane-water interfaces and their apolar residues buried. One must therefore enquire into possible architectural devices that can provide the strong hydrophobic associations known to link membrane proteins to membrane lipid. Although several workers have suggested that these associations could occur via apolar amino acid residues, extending into the hydrocarbon lipid layer from surface-located peptide, it has been shown that such interactions are sterically impossible and energetically improbable.[2, 3] The only alternative is that at least portions of membrane proteins *penetrate into or through the membrane core*. However, this alternative requires that the membrane core is neither all lipid nor all protein, i.e., that the membrane is a *mosaic structure*.

Although Danielli and Davson[4] were the first to present a detailed, molecu-

* Original work was supported by awards CA 12178, CB-43922, and CA 13061 from the United States Public Health Service, CB 32123 from the National Science Foundation, PRA-78 from the American Cancer Society (D.F.H.W.), and a grant from the Max-Planck Gesellschaft zur Foerderung der Wissenschaften (R.S.-U.).

lar model of membrane architecture in 1935, a lipid-protein *mosaic* structure had been suggested already in 1904 by Nathanson.[5] However, the first detailed mosaic hypothesis was proposed by Parpart and Ballentine in 1952.[6] These authors developed their model in an attempt to account for the state of proteins in erythrocyte membranes. They suggested that the erythrocyte membrane consists of a *protein continuum* extending through the membrane thickness, interrupted by lipid-containing cylindrical channels, 50 Å in diameter and lined with polar amino acid residues. They envisioned the head groups of the amphipathic lipids to be oriented toward the polar channel-walls with the hydrocarbon chains lying parallel to the membrane plane to form an apolar core. The authors proposed that binding of lipid to protein would occur through polar associations between phosphatide head groups and polar amino acid residues on the channel walls. They further suggested existence of apolar associations between the apolar residues of the strongly bound phosphatides and weakly bound lipids, including cholesterol. Parpart and Ballentine[6] did not consider apolar lipid-protein associations, and present information does not support the details of the Parpart-Ballentine model. However, as we shall show, the suggestion that membranes contain penetrating lipid "cylinders" constrained by penetrating membrane proteins may approximate the real situation in some membranes.

Modern mosaic models of membrane structure were proposed independently in 1966 by Wallach and Zahler[7] and Lenard and Singer.[8] Both groups based their arguments on evidence indicating extensive apolar interactions between membrane proteins and membrane lipids, experiments indicating high helicity in membrane proteins, and certain newly recognized principles of macromolecular organization. Both groups concluded that some proteins *penetrate* into or through the membrane core and both suggested that apolar amino acids on the penetrating peptide segments were the elements involved in hydrophobic associations with the hydrocarbon chains of membrane lipids. Finally, both groups concurred in the view that penetrating peptide segments were likely to be highly helical and that surface-located peptides, whatever their conformation, could participate in polar interactions with the head groups of phosphatides organized in some form of bilamellar array.

Beyond these points of agreement, the views of the two groups diverge considerably. Thus, Singer and associates[9–11] proposed that lipids form a two-dimensional membrane continuum in which proteins are inserted, as solutes in a solvent. They further argued that, since many phosphatides have fluid hydrocarbon chains at physiologic temperatures, the lipid continuum would be fluid, that membrane lipids and proteins could undergo free translational diffusion parallel to the membrane plane, that membrane lipids and proteins would behave independently and that, except for short-range interactions, the protein distribution parallel to the membrane plane would be random. This "fluidity" concept depends heavily on molecular extrapolations of surface-transposition phenomena involving plasma membrane-associated proteins, the fact that phospholipid hydrocarbon chains can be highly mobile, and evidence that certain membranes exhibit thermotropic lipid-protein phase segregation.

Indubitably many membranes *do* contain fluid lipid regions, but these may be very small domains. Moreover, the plasma membranes of animal cells contain high proportions of cholesterol, which do not favor fluidity. Indubitably, also, *some* membrane proteins undergo rapid *rotational* diffusion,[12, 13] but other proteins are quite restricted in rotational mobility.[14, 15] Finally, whereas electronmicroscopic[16–20] and optical measurements[21] show that the proteins of *some* membranes can undergo extensive and rapid translation parallel or perpendicular to the membrane plane, similar measurements on *plasma membranes* of eukaryotic cells[18, 20, 22] demonstrate severe restrictions on protein mobility.

For these and other reasons we feel that one must view the "fluid mosaic" hypothesis[9–11] with some reserve, and that one must consider other possible modes of lipid-protein interactions in biomembranes, including those proposed in our original mosaic hypothesis.[7, 23, 24] However, before considering these alternatives, we will address some pertinent aspects of protein structure.

Features of Protein Structure that Can Provide for Apolar Lipid-Protein Associations

Low Overall Polarity and/or High Overall Hydrophobicity of Membrane Proteins

Various authors have proposed that "intrinsic" membrane proteins contain disproportionately few polar amino acids and that this low polarity allows for apolar lipid-protein associations (e.g., see Reference 25). Unfortunately, "polarity" has been rather arbitrarily defined in such studies and has not been expressed (as it should be) in terms of ratio (volume of polar residues: volume of nonpolar residues).[26] Indeed this ratio cannot be easily obtained.

It has been suggested that one can obtain a measure of polarity from the polarity index, p, defined as $p = V_e/V_i$, where V_e and V_i are external ("shell") and internal volumes, respectively.[26, 27] Taking known residue volumes,[27] and assuming that the "polar" residues, i.e., Arg, His, Lys, Asp, Glu, Asn, Tyr, Ser and Thr, lie in V_e and all others in V_i, one can calculate p from amino acid compositions. However, x-ray analyses of globular proteins show that, while charged and polar residues are excluded from the interior of these macromolecules, *not all apolar amino acid residues lie buried in the interior*. For this reason, the polarity index can only be considered a useful approximation.

When we apply the polarity-index criterion to membrane proteins that are "elutable" or "nonelutable" by aqueous media (Table 1), we find that both categories of protein, except myelin proteolipid and murein lipoprotein fall in the polarity range of many soluble globular proteins.[27] Interestingly,

Table 1

Polarity Indices (p) and Average Hydrophobicities ($H\phi_{av}$) of Some Membrane Proteins that are Readily Eluted by Aqueous Media and of Some Proteins that Resist Elution*

	Elutable			Nonelutable		
	Human Erythrocyte Spectrin	*S. faecalis* ATPase	Bovine F^1 ATPase	Bovine Retina Rhodopsin	Bovine Myelin Proteolipid	Murein Lipoprotein
p	1.54	1.12	0.96	0.73	0.65	2.04
$H\phi_{av}$	0.96	1.10	1.09	1.23	1.21	0.73

* From amino acid analyses compiled in Reference 3.

murein lipoprotein, which is known to be a penetrating membrane protein, has an exceptionally high *p* value.

The polarity index is a criterion of polarity, but diverse amino acid side chains also differ in *hydrophobicity*. The average hydrophobicity of a protein can be assessed from its amino acid composition and from the hydrophobicities, $H\phi$, of various amino acids, as determined by Tanford[28] from the distribution of free amino acids between water and polar solvents. Values of $H\phi$ range between 0.45 kcal/residue for Thr and 3.00 kcal for Trp.[28] Moreover, Arg and Thr, although polar, are also weakly hydrophobic and Lys even though polar *and* charged, has a high $H\phi$ value, i.e., 1.5 kcal/residue. Knowing the amino acid composition of a protein, one can compute its mean residue hydrophobicity, $H\phi_{av}$ from the individual hydrophobicities of the various residues.[27] The average hydrophobicities, $H\phi_{av}$, of soluble proteins range from 0.440 to 2.020 kcal/residue and 50% of proteins have $H\phi$ values between 1.0 and 1.2 kcal/residue.[27]

We have applied the hydrophobicity criterion to membrane proteins that can be eluted by *ionic manipulations* and those that require *detergents* or *organic* solvents for extraction (TABLE 1). Ionically elutable proteins generally exhibit *lower hydrophobicities* than the more tightly bound category. However, except for murein lipoprotein, the "elutable" and "nonelutable" proteins have $H\phi_{av}$ values in the 0.9–1.2 kcal/residue range typical for most *water-soluble* proteins. Moreover, murein lipoprotein, which is not easily eluted ionically, exhibits an unusually *low* hydrophobicity, 0.73 kcal/residue.

These calculations indicate that one cannot categorize membrane proteins as poorly polar or highly hydrophobic. One must, therefore, search for other features or protein structure, to account for apolar associations of lipids and proteins in biomembranes.

Linear Amphipathic Sequences

The existence in a membrane protein with unremarkable *overall* polarity or hydrophobicity, of highly apolar, hydrophobic sequences (FIGURE 1A) could account for the "anchoring" of such a protein to a biomembrane. A beautiful example of this linear amphipathic device is found in the cytochrome b_5 isolated in rabbit liver endoplasmic reticulum.[29, 30] This enzyme, which has an overall molecular weight of 16,700, consists of an enzymatically active portion, composed of 97 residues and an ~40-residue "tail." The enzymatically active moiety has a very low hydrophobicity, $H\phi_{av}$ — 0.89 kcal/residue, whereas the tail is highly hydrophobic, $H\phi_{av} = 1.7$ kcal/residue. The enzymatically active portion of the protein apparently lies at the membrane surface and is attached to the apolar portions of the membrane by the hydrophobic "tail."

The major glycoprotein of erythrocyte membranes also appears to be linearly amphipathic. Studies initiated by Winzler[31] show that a large portion of this protein can be released by mild proteolysis of intact erythrocytes. The protease-accessible, amino-terminal portion, bearing the sugar moiety contains a large proportion of hydrophilic amino acids,[31, 32] giving it an average $H\phi_{av}$ of only 0.50 kcal/residue, compared with 0.93 kcal/residue for the whole protein. The very high $H\phi_{av}$ of the "residue," 1.33 kcal/residue, is concordant with the concept that this hydrophobic portion penetrates to the

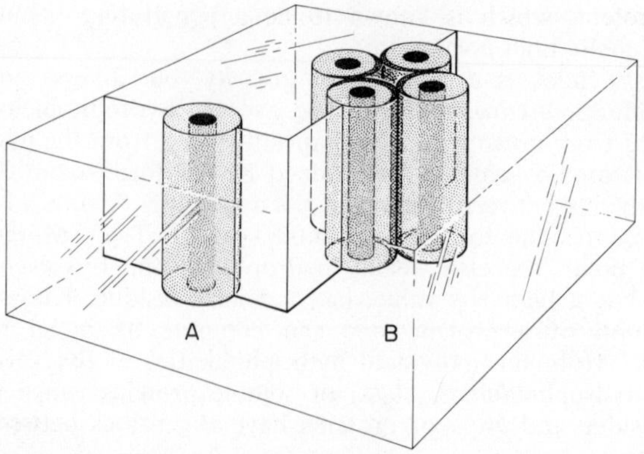

FIGURE 1. Schematic drawing representing two architectural devices allowing polypeptide chains to penetrate into an apolar zone. (A) Peptide segment containing only apolar hydrophobic residues; these surround the peptide backbone and give the whole unit an apolar perimeter. (B) Aggregate of four penetrating, conformationally-amphipathic helices. Polar residues are restricted to the contacts between helices and to a hydrated channel penetrating the assembly.

membrane.[31] The C-terminus appears exposed at the internal membrane surface.[32]

Conformational Segregation of Polar Amino Acid Residues: Amphipathic Helices

X-ray crystallographic analyses of myoglobin,[33] hemoglobin,[34] and many other proteins have revealed an important principle of protein structure: Many helical segments in these proteins contain sequences where apolar residues occur regularly at every third or fourth position. This gives these helices a *polar* face and an *opposite apolar* face; i.e., the helices are *conformationally amphipathic* (FIGURE 1B). The tertiary and/or quaternary structuring of these water-soluble proteins is such as to cause the apolar surfaces of the amphipathic helices to associate with each other within the cores of the macromolecules. The polar surfaces are oriented toward water. In all tetrameric hemoglobins, the quaternary associations also generate a hydrophilic channel running through the center of the protein.

In soluble proteins, amphipathic helices associate via their apolar faces because of entropic factors: minimal free energy and maximal entropy occur when apolar residues are removed from water (as in the formation of bilamellar arrays, or micelles, by amphipathic lipids). It was this fact, and spectroscopic data pointing to high helicity in membrane proteins, that led us[7] to hypothesize that:

(a) Penetrating proteins might contain amphipathic helices, which, being immersed in an *apolar* solvent—phospholipid acyl chains—rather than water, would *associate via their polar faces,* rather than their apolar faces.

(b) Association of amphipathic helices would generate *cylindrical multimers penetrating into or through the thickness of their membranes, with a layer of phospholipid acyl chains bound tightly to their apolar perimeters and with their polar interiors* possibly *forming small aqueous channels penetrating through the membranes* (as in the hemoglobin tetramer).

(c) The apolar amino acids at the perimeters of the multimers would form relatively specific, high-affinity binding sites for the hydrocarbon residues of *tightly* retained membrane lipids, but polar lipid head groups could also participate in polar associations with surface-located protein segments, whatever their conformation.

(d) The association of *tightly* bound lipids with membrane proteins would be analogous to the apolar heme-protein interactions in hemoglobin, but lipids, lying more distant from the protein, e.g., cholesterol, would be bound less tightly and less specifically.

(e) Although in a bilamellar array, the lipid adjacent to the protein would have a composition and organization determined by the vincinal protein and the biologic "ordering" influence of the protein would fall off with distance.

Although the feasibility of our concept was demonstrated by use of space-filling models,[23, 24] information indicating that the hypothesis might actually apply to biomembranes did not emerge until recently, when sufficient data became available to develop meaningful models for high-density plasma lipoproteins and the murein lipoprotein in the outer membrane of *E. coli.* We will summarize this information, before presenting evidence that the organization and mobility of membrane lipids are constrained by membrane proteins.

High-Density Serum Lipoproteins

Recent evidence[35] suggests that high-density serum lipoproteins (HDL) components are important components of erythrocyte membranes. The HDL of human plasma constitute a class of lipoproteins containing the same protein subunits, but differing in lipid composition. Two apoproteins, apoA-I and apoA-II, account for 90% of the peptide. The lipids typically comprise 35–38% of the lipoprotein.

High-density serum lipoproteins are highly α-helical (61–70%) (e.g., see Reference 36). The helical content of the apoproteins increases upon recombination with lipid.[37] Moreover, [^{13}C] nuclear magnetic resonance spectroscopy indicates that the affinity of the apoproteins for phospholipids depends more on the fatty acid chains of the phosphatides than on their head groups.[38, 39]

The amino acid sequence of Apo A-II contains apolar residues at every third to fourth position for a segment at least 20 residues long, with a $H\phi_{av}$ = 1.12 kcal/residue. In the α-helical conformation, such a sequence will produce an apolar band running the length of the helix. Assman and Brewer[40] have proposed that the apolar faces of Apo A-II helices are the sites of apolar *protein-protein* and protein-lipid interactions in HDL. Nuclear magnetic resonance studies[38, 39] are in accord with this notion.

Murein Lipoprotein

The principal lipoprotein in the outer membrane of *E. coli,* murein lipoprotein, constitutes the first documented example of conformational segregation of apolar and polar residues in an "integral" membrane protein. The chemical structure of murein lipoprotein has been determined by Braun and associates.[41–43]

The protein consists of 58 amino acid residues. The N-terminus consists of an unusual amino acid (glycerylcysteine [S-(propane-2′, 3′-diol)-3-thio-2 aminopropanic acid]), which connects to two fatty acids by two ester linkages and to another fatty acid by an amide bond. The protein contains no Pro, Trp, Phe, His, or Gly. The protein has a high polarity ($p = 2.04$) and a very low hydrophobicity ($H\phi_{av} = 0.73$ kcal/residue).

As shown in FIGURE 2, the N-terminus of murein lipoprotein consists

```
Lipid
 |
G — Cys
 |
Ser-Ser-Asn-Ala-Lys-Ile-Asp-Glu-Leu-Ser-Ser-Asp-Val-Gln- - - -
- - Thr-Leu-Asn-Ala-Lys-Val-Asp-Glu-Leu-Ser-Asn-Asp-Val-Asn- - - -
- - Ala-Met-Arg-Ser-Asp-Val-Gln-Ala-Ala-Lys-Asp-Asp-Ala-Ala-Arg- -
- - Ala-Asn-Glu-Arg-Leu-Asp-Asn-Met-Ala-Thr-Lys-Try-Arg-Lys
```

FIGURE 2. Amino acid sequence of murein-lipoprotein. The homologous oligopeptides at the N-terminus have been aligned. (- - -) hydrophobic, (——) hydrophilic and apolar. (Adapted from Hantke and Braun.[43])

of two consecutive oligopeptides (14 and 15 residues) with an almost identical amino acid sequence. Hydrophobic, uncharged residues occur at every third or fourth position along the entire chain.

Murein lipoprotein is more than 70% α-helical.[44] The sequence of repetitive apolar residues thus generates a nonpolar band extending the length of the helix. This also occurs in tropomyosin[45] where the apolar bands of two helices associate hydrophobically to form a coiled-coil structure with a hydrophilic perimeter. Such a coiled-coil structure cannot provide for hydrophobic interaction with membrane lipids. However, as pointed out by us[3] and by Inouye,[46] an alternate arrangement, i.e., polar association of several helices, would generate a multimeric assembly whose perimeter is formed by the apolar bands. Such an assembly could interact hydrophobically with lipids.

Inouye[46] has presented a detailed model of such an array, in which six α-helical lipoprotein units are arranged in a "superhelix" with a 166° turn. The length of the assembly is 76 Å, corresponding to the thickness of the outer membrane and the thickness of a phospholipid bilayer. The superhelix has an apolar perimeter and a central hydrophilic channel 12.5 Å in diameter. Glu-9, Asp-13, Glu-23, and Asp-27 face into the channel interior giving it a fixed negative charge. The central channel is large enough to account for the known movement of many small molecules through the outer membrane of *E. coli*. Inouye[46] accounts satisfactorily for the disposition of the covalently bound fatty acid of the N-terminus and shows that the N-terminus must lie at the external surface of the membrane; the C-terminus is linked covalently to the peptidoglycan layer at the inner surface of the membrane. Inouye[46] also demonstrates that, with greater twist in the superhelix, up to 12 subunits can be accomodated in a 75 Å membrane thickness.

Some data exist therefore, indicating that amphipathic helices may play a role in the apolar interactions between lipids and proteins in lipoproteins and biomembranes. However, as pointed out before, such interaction requires association either between different helices of a single peptide chain (as in myoglobin) and/or association of the multimeric helices of separate peptide chains (as in hemoglobin and as has been proposed for murein lipoprotein).

There is now evidence that certain proteins in plasma membranes can exist as multimers, at least once extracted. Thus, a series of studies demonstrate that the high molecular weight "spectrin" components include, at least in part, multimers of a membrane protein with molecular weight ~50,000D (e.g., see References 47–49). Dimerization of one of the glycoproteins of erythrocyte membranes has also been documented[50] and the concanavalin A-binding protein in the plasma membrane of rabbit thymocytes can exist as monomer (~55,000D) as well as in dimeric and tetrameric form.[51]

Constraints on Lipid Mobility in Biomembranes

Early experiments by Chapman and associates,[52, 53] using proton magnetic resonance spectroscopy and infrared measurements, have suggested that the molecular motions of phospholipid acyl chains in erythrocyte membranes are somehow restrained by apolar interactions with membrane proteins. More recent studies on several different membrane types, employing a variety of probe techniques, support this conclusion.

Thus, Träuble and Overath[54, 55] have used the negatively charged fluorescent probe 1-anilino-8-naphthalene sulfonate (ANS) and the apolar fluorescent probe *N*-phenyl-1-naphthylamine (NPN) to evaluate thermotropic lipid phase transitions in *E. coli* membranes and in aqueous dispersions of lipids extracted from these bacteria. At low concentrations ($< 10^{-5}$ M) NPN monitors the state of lipid hydrocarbon regions, whereas ANS probes more polar regions. A comparison of the limiting fluorescence changes at the phase transition, obtained with membranes and their lipids, suggested that in the membrane 20% of the lipid does not participate in the phase transition. The authors accordingly proposed that approximately 600 phospholipid molecules surround each "integral" protein, of which 130 are closely associated with the protein perimeter.

Turning to animal membranes, Jost *et al.*[56] have examined the interactions of phosphatides with mitochondrial cytochrome oxidase, using 16-nitroxide stearate as a probe. The electron spin resonance spectra indicated that at low phospholipid : protein ratios the probe was immobilized by association with protein. The spectra obtained at higher proportions of phosphatide showed presence of fluid lipid as well as immobilized lipid. This suggested a gradation of lipid fluidity in the vicinity of the protein, with a highly immobilized layer immediately adjacent to the macromolecule.

Studying a different membrane system, Stier and Sackmann[57] evaluated how temperature influences the relative reduction rates of a lipophilic nitroxide (nitroxide stearate) and a water-soluble one (TEMPO-phosphate ester) by microsomal cytochrome P-450-cytochrome P-450-reductase. They interpreted their data to demonstrate a mosaic membrane structure, with the reductase system enclosed in a "rather rigid phospholipid halo" surrounded by more fluid lipid.

More recently,[58] we have examined lipid protein interactions in erythrocyte ghosts (and bovine serum albumin), using 5-nitroxide stearate (5NS) and 16-nitroxide stearate (16NS) as paramagnetic probes and the lipophilic dialdehyde, *o*-phthaladehyde (OPT), as a protein perturbant. We found that when bovine serum albumin, saturated with 5NS or 16NS, was reacted with OPT, *all* the bound spin label became displaced into aqueous solution. OPT produced a similar effect with erythrocyte ghosts equilibrated at equivalent levels of spin label. However, about 5% of the spin label remained membrane bound. Sulphydryl blockers, such as *p*-chloromercuribenzoate, reduced the action of OPT on ghosts by about 70%, but had no effect in the case of serum albumin.

Experiments on liposomes prepared with the total lipids (including OPT phosphatidyl ethanolamine) of OPT-treated ghosts, as well as on phosphatidylethanolamine-lecithin multibilayers reacted with OPT, showed that reaction of OPT with membrane lipids could not account for the displacement of the spin label. We therefore concluded that certain erythrocyte membrane proteins are in close apolar association with lipid acyl chains and that these associations could be perturbed by modifications of protein structure.

We have also examined spin-spin exchange and dipolar interactions of 5NS in erythrocyte membranes.[59] We found that the electron spin resonance spectrum of 5NS bound to erythrocyte membranes varied with the amount of label bound. Moreover, comparative studies on ghosts, lecithin-cholesterol liposomes, and bovine serum albumin, at both ambient temperature and at —160° C, gave clear evidence for spin-spin exchange and dipole-dipole interactions in *ghosts* at high 5NS binding. This indicated that a significant proportion of the label molecules could lie within ~15 Å of each other in the membranes. Significantly, we found that the spectral manifestations due to label clustering could be abolished by reduction of pH and conjoint action of lysolecithin and trypsin, although both perturbations increased 5NS binding. Both perturbations are known to mobilize intramembranous particles by modifying[60] or extracting[61] some membrane proteins. These data again suggested that the lipids and proteins of erythrocyte membranes could exist in a relatively fixed mosaic, and that the mobility of both components could be restricted by some membrane-associated protein framework.

Elucidation of Lipid-Protein Interactions in Erythrocyte and Thymocyte Plasma Membranes by Paramagnetic Quenching of Fluorescence, Raman Spectroscopy, and Resonance-Raman Spectroscopy

Paramagnetic Quenching of Fluorescence

Paramagnetic quenching of fluorescence (PQF) constitutes a new probe approach. When paramagnetic substances, such as nitroxides, closely approach a fluorophore, they reduce its excited-state lifetimes and thereby quench fluorescence. Such paramagnetic quenching involves interaction distances of only 4–6 Å[62] and evaluation of this process in macromolecules and membranes can indicate the position of bound nitroxide radicals and fluorophores relative to each other. We have recently[63] evaluated the effects of temperature on paramagnetic quenching of fluorophore, perylene, incorporated into dipalmitoyllecithin or dipalmitoyllecithin-cholesterol liposomes, using several nitroxide-labeled lipid analogs known to probe different "depths" of membrane hydrophobic regions.[3] We showed that the PQF approach can accurately reflect the state of lipids in model membranes and supported previous spin-label studies,[64] indicating that the cholesterol in cholesterol:phospholipid bilayers tends to form clusters below the phospholipid transition temperature. Our studies also documented that at sterol/phospholipid ratios greater than 0.4, i.e., at ratios typical for plasma membranes,[3] clustering occurs also *above* the transition temperature. We will now present some recent observations on the quenching of the *tryptophan* fluorescence of membrane proteins as a function of temperature and pH, using nitroxide analogs of both cholesterol and fatty acids as quenchers. These experiments constitute an extension of previous studies[58] demonstrating that nitroxide stearates can quench the tryptophan fluorescence of bovine serum albumin and erythrocyte ghosts.

We have measured PQF on hemoglobin-free erythrocyte ghosts prepared as in Reference 65 and plasma membranes isolated from rabbit thymocytes as in Reference 51. The membranes in 50 mM phosphate, 10 mM NaN_3, pH 7.4 (35 μg protein/ml) were equilibrated with buffer solutions plus or minus $10^{-5} - 2 \times 10^{-4}$ M nitroxide-lipid analog, for stated time periods (usually 25 minutes), at stated temperatures and pHs. 5NS, 16NS, and androstane nitroxide (ASL) were purchased from Syva (Palo Alto, California). Tryptophan fluorescence was measured using a Perkin-Elmer MPF3 spectrofluorometer. The exciting wavelength was 285 nm and fluorescence emission was measured at 338 nm (slit widths 7–10 nm). Quenching, Q, is expressed by the Stern-Vollmer expression, i.e., $Q = I_0/I - 1$, where I_0 is the fluorescence intensity without quencher and I is the emission intensity in the presence of quencher. At the concentrations employed, the nitroxides exert no "inner filter" effects.

The quenching efficiency, Q (TABLE 2), of the three nitroxides follows the

TABLE 2

QUENCHING BY THREE PARAMAGNETIC LIPID ANALOGS OF TRYPTOPHAN FLUORESCENCE IN ERYTHROCYTE GHOSTS AND THYMOCYTE PLASMA MEMBRANES

		Q*	
Nitroxide	Concentration ($M \times 10^5$)	Erythrocyte Ghost	Thymocyte Plasma Membrane
5NS	5	0.428	0.316
	10	0.818	0.613
16NS	5	0.254	0.234
	10	0.428	0.408
ASL	5	0.221	0.205
	10	0.282	–

* 25°C; pH 7.3.

order 5NS > 16NS >> ASL in both membrane types and at low as well as at saturating concentrations. With ASL saturation occurs at lower bulk concentration than with 5NS and 16NS.

FIGURE 3 shows the temperature-dependence of PQF by 5NS and ASL (5×10^{-5} M; pH 7.3). Quenching by ASL increases slightly and linearly be-

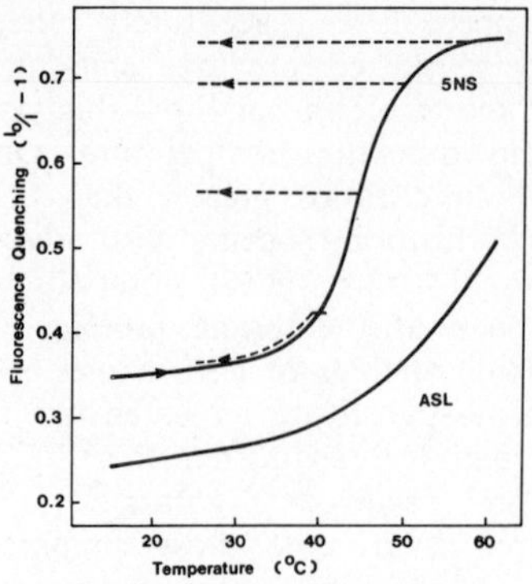

FIGURE 3. Paramagnetic quenching of tryptophan fluorescence of erythrocyte ghosts by 5NS and ASL at different temperatures (---) indicates Q obtained after preequilibration at high temperature and PQF measurements at 25° C (pH 7.3).

tween 15° C and 45° and then climbs abruptly. In contrast, PQF by 5NS rises linearly only up to ~35° C and then jumps dramatically to reach a new stable state at ~60° C. 16NS (not shown) reveals a qualitatively similar pattern and comparable curves have been obtained with thymocyte plasma membranes. Equilibrating the membranes at various temperatures above 25° without 5NS (60 minutes), and reequilibrating at 25° (60 minutes) before PQF measurement, showed that the thermotropic effect depicted in FIGURE 3 *is reversible up to ~42° C,* and that the process does not depend on the presence of the nitroxide. The upswing of PQF by ASL corresponds to the *irreversible* segment of the 5NS curve.

FIGURE 4 demonstrates that PQF by 5NS varies with hydrogen ion concentration between pH 6.0 and 8.0 (50 mM phosphate, 10 mM NaN_3; 35° C). Quenching is minimal between pH 7.0 and 7.5, and increases as the pH is raised or lowered beyond this range.

Paramagnetic quenching of tryptophan fluorescence by 5NS, 16NS, and ASL measures the relative accessibility of these lipid analogs to membrane tryptophans and tyrosines. Although we measure only tryptophan fluorescence, quenching of this fluorescence may also arise from collisions between the nitroxides and tyrosine residues. This is because illumination at 280–290 nm excites tyrosine as well as tryptophan, and, in proteins, the excitation energy of tyrosine is normally transferred to tryptophan.

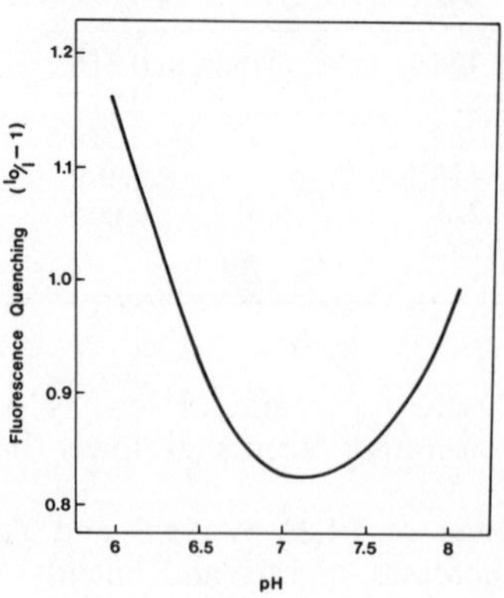

FIGURE 4. PQF in erythrocyte ghosts as a function of pH (35° C).

The fact that 5NS quenches more efficiently than 16NS, even at saturation, suggests that the susceptible fluorophores are more concentrated at a "depth" equivalent to C2 of phosphatide acyl chains than far within the membrane. The paramagnetic residue of ASL should lie at a "depth" equivalent to C8. Its low quenching efficiency at physiological temperatures might thus reflect a relative deficiency of fluorophore at this "depth," or a tendency of this sterol to distribute *away* from membrane protein, possibly in clusters as demonstrated for liposomes.[63, 64]

5NS and 16NS appear to distribute in close vicinity to the proteins. They thereby report that some membrane proteins undergo a thermotropic structure change that is *reversible* within physiological temperatures. This structural transformation becomes *irreversible* above $\sim 42°$ and is then also detected by ASL. Significantly, a *monomeric* apolar, penetrating peptide segment *lacking tertiary structure,* should *not* produce a discontinuous change of PFQ in the 35°–45° range.

We therefore conclude that the *reversible* limb of the thermotropic transition reflects *unfolding* of *globular monomers* and/or *dissociation* of *multimers,* making additional fluorophores accessible to the nitroxide stearates. The *irreversible* curve segment presumably represents denaturation. The enhanced PQF by ASL at the high temperatures may involve decreased sterol clustering as well as excessive protein unfolding.

To what extent does the behavior of ASL and nitroxide stearates resemble that of cholesterol and phosphatide acyl chains? As pointed out in Reference 3, ASL appears a rather good cholesterol analog, and our suggestion that its behavior suggests a low degree of cholesterol-protein interaction fits the easy exchangeability of this sterol.[66] Our proposal of sterol-clustering is consistent with electron microscopic experiments using osmate esters of cholesterol.[67] We further contend that the thermotropic protein transition reported by PQF constitutes a real membrane property, rather than a perturbation effect, since reversibility experiments show that it takes place also in the absence of the nitroxides (FIGURE 3).

The pH effects depicted in FIGURE 4 may reflect the varying ionic interaction of the 5NS carboxyl with basic groups on membrane protein, as diverse protein residues are titrated. More than one ionogenic group must be involved since the pH-PQF curve is biphasic. It is also possible that the noted effects reflect alterations of apolar membrane domains *secondary* to titration of ionogenic groups.

Laser-Raman Spectroscopy (RS) and Resonance-Enhanced Raman Spectroscopy (RERS)

Raman spectra can yield a wealth of information, since any bond-motion that involves a polarizability change may produce a Raman line.[3] We have shown previously[68, 69] that RS can provide information about the state of membrane lipids, the conformation of membrane proteins and the nature of lipid protein interactions. Moreover, our original Raman analyses of erythrocyte ghosts[68] revealed two resonance-enhanced scattering bands, at 1530 cm^{-1} and 1160 cm^{-1} arising from traces of membrane bound β-carotene.[70] In a subsequent report,[70] we documented that the β-carotene can serve as a sensitive, "Raman-active" probe of membrane structure in erythrocyte ghosts and model lipid membranes, and that the β-carotene tends to be excluded from cholesterol-containing acyl chain regions. We will now report on the *thermotropism* of erythrocyte ghosts, using both RS and β-carotene RERS.

Membranes were prepared as already described. Liposomes ± cholesterol and ± β-carotene were prepared as in Reference 70. Raman spectra were recorded with a Ramalog 4 Raman Spectrometer (Spex Industries, Metuchen, N.J.) interfaced to an Interdata (Model 70) computer as detailed in References 68–70. Samples were sealed into 0.9-mm ID Kimex capillaries and these placed in a temperature-controlled Harney-Miller cell. An argon-ion laser (Spectra Physics, 164), tuned at 488 nm, was used as an excitation source with 200 mW power at the sample for RS and 60–70 mW for RERS. Raman scattering was recorded in terms of photons/s. The computer was used to eliminate transients and to integrate net scattering intensities over appropriate frequency intervals. To avoid ambiguities due to photoeffects and/or local heating, multiple scans (2–3 minutes each) were recorded for each sample and multiple 10 μl samples were used for each spectrum.

FIGURE 5 presents pertinent segments of the erythrocyte RS and RERS. The CH stretching vibrations of CH_2 and CH_3 groups occur between 3000 cm^{-1} and 2800 cm^{-1}. As shown previously,[69] a plot-versus-temperature of I_{2890}/I_{2850}, i.e., the integrated intensity of the 2890 cm^{-1} band (asymmetrical CH stretching in CH_2 and CH_3 residues of lipids) relative to the integrated intensity of the 2850 cm^{-1} band (symmetrical CH_2 stretching in lipid acyl chains), can reveal thermotropic lipid phase transitions. The resonance-enhanced carotenoid bands occur at 1530 cm^{-1}, γ (—C═C—), and 1160 cm^{-1}, γ (═C—C═). The CH deformation band, near 1450 cm^{-1}, is nonthermotropic and is used as internal reference. The amide III peptide vibrations occur between 1300 cm^{-1} and 1200 cm^{-1}.

FIGURE 5 shows the variation of I_{2890}/I_{2850} for lecithin:cholesterol liposomes (1:1 molar ratio) and for erythrocyte ghosts. It also shows the temperature-dependence of I_{1530}/I_{1440}, i.e., the integrated intensity of the γ (—C═C—) band of membrane β-carotene relative to the integrated intensity of the CH deformation band. The lecithin:cholesterol liposomes show a very broad discontinuity in the plot of I_{2890}/I_{2850} vs. temperature. This supports RS data by Lippert and Peticolas,[71] indicating that cholesterol does not abolish the gel → liquid-crystalline transition of phosphatide acyl chains, but makes it *poorly cooperative.*

In contrast, erythrocyte ghosts, which have the same overall cholesterol: phospholipid ratio as the cholesterol:phosphatide liposomes, exhibit a sharp discontinuity in the plot of I_{2890}/I_{2850} vs. temperature, suggesting a cooperative

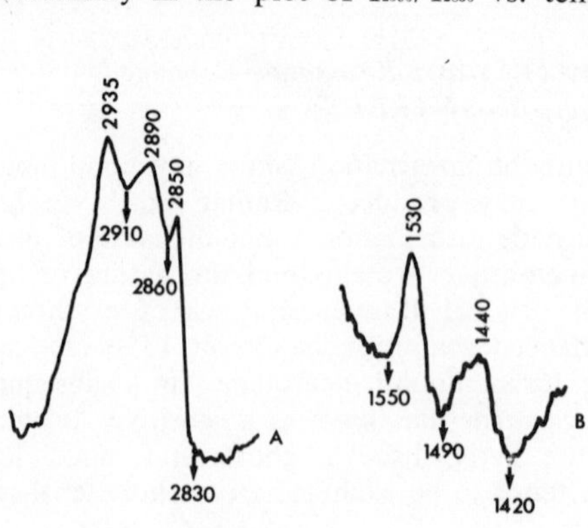

FIGURE 5. (A) Erythrocyte RS in the CH stretching region (3000 cm^{-1}-2800 cm^{-1}) and (B) RERS in the γ (—C═C—) region. Integration is between minima indicated by arrows.

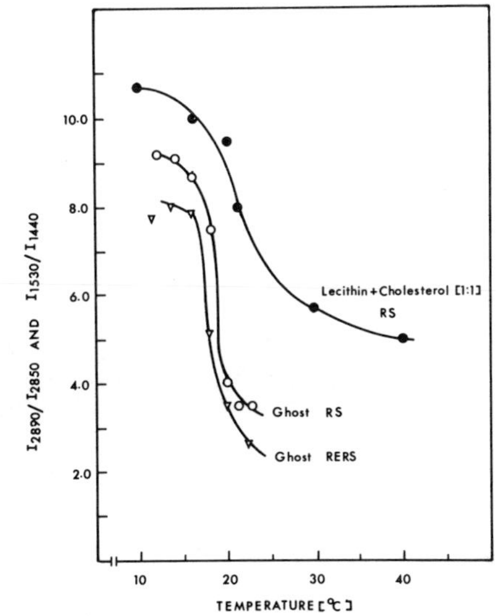

FIGURE 6. Variation of I_{2890}/I_{2850} with temperature in the case of lecithin/cholesterol liposomes, (●) (1:1 molar ratio; egg yolk lecithin) and erythrocyte ghosts (○). A plot of I_{1530}/I_{1440} of ghosts vs. temperature is also given (△).

gel → liquid-crystalline transition. The real transition temperature is probably 18° C, rather than the 16° C indicated in FIGURE 6, due to 2° C local heating.[69] The value of 18° C corresponds to the transition temperature (18–19° C) found by viscometric measurements on ghost sonicates.[72] Thymocyte plasma membranes also show a sharp thermotropic discontinuity of I_{2890}/I_{2850}.[69] This centers near 25° (corrected). The plot of I_{1530}/I_{1450} vs. temperature (FIGURE 6) also indicates a sharp thermotropic transition in ghosts. The transition temperature (corrected for 2° local heating) lies near 19° C.

FIGURE 7 demonstrates extensive changes with temperature in the amide III region. Particularly prominent is the appearance of a band near 1240 cm^{-1} as the temperature is raised from 16° C to 20° C. The emergence of sharp RS bands in this portion of the amide III region might reflect a change of secondary structure in some membrane peptide or an alteration of the local environment of certain peptide segments. In any event, our data indicate protein modification following the thermotropic lipid transition. This information may relate to the thermotropic discontinuity in erythrocyte glucose transport at 19° C[73] and to previous data showing that ATP hydrolysis by erythrocyte ghosts induces infrared signals typical of the antiparallel β-conformation.[74]

That the PQF curves show no discontinuities in a temperature range where RS and RERS reveal a thermotropic transition, is expected: PQF is due to lipid analogs closely associated with membrane protein whereas the discontinuities shown in FIGURE 5, reflect cooperative phase transitions of *unbound* membrane lipid. Moreover, the *sharpness*, i.e., *cooperativity* of the transitions revealed by RS and RERS in high-cholesterol membranes indicate that a substantial proportion of membrane phospholipid is *not* associated with cholesterol, consistent with our proposal that much of the membrane cholesterol lies in clusters. This suggestion is further supported by the β-carotene RERS, since independent data[70] show that β-carotene is excluded from liposomes containing 50 mol% cholesterol.

Finally, that cooperative, thermotropic lipid phase transitions can be re-

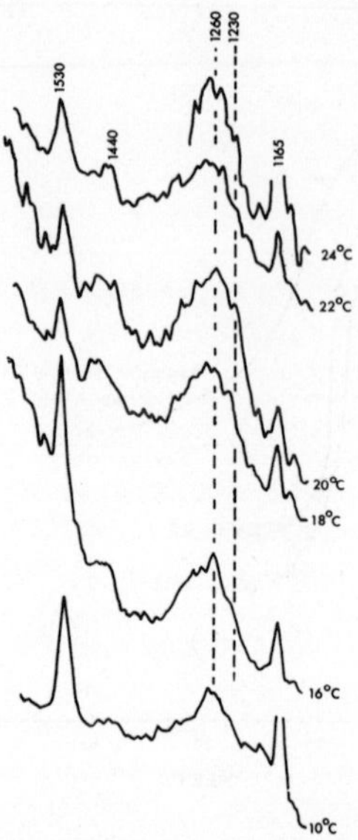

FIGURE 7. Alterations of RS in amide III region as a function of temperature. The γ (—C=C—) region is also shown. Each spectrum was obtained on a separate aliquot of same preparation. Ordinate scale maximum = 4000 counts/s at 10° C–16° C and 25000 counts/s at 18° C–24° C. Only band location is relevant. Note sharpening of bands in region of 1260 cm^{-1} and 1230 cm^{-1} (vertical dashed lines).

vealed in plasma membranes that show (a) no thermotropic *segregation* of protein from lipid by freeze-fracture electron microscopy[20, 60] and (b) no lateral protein mobility by optical methods,[22] indicates that these transitions occur in microscopic domains.

A HYPOTHESIS FOR LIPID-PROTEIN ASSOCIATIONS IN ERYTHROCYTE AND THYMOCYTE PLASMA MEMBRANES

We will now attempt to weld present and other data into a reasonable working hypothesis of lipid-protein interactions in plasma membranes. First, let us summarize four cardinal elements of evidence:

1. PQF data show that a significant proportion of membrane proteins are tertiarily-folded globular monomers and/or multimers with apolar perimeters bearing tightly bound acyl chains. This statement is consistent with evidence that membrane proteins tend to be globular, rather than extended (e.g., see References 3, 7–9), that they can by multimeric,[46–51] and that they are in tight apolar association with some membrane phosphatide.[52–59]

2. PQF also shows that these proteins can fold and unfold reversibly between ~35° C and ~42° C, with an associated increase or decrease, respectively, in lipid-protein association. No other evidence exists bearing directly on this point, but differential thermal calorimetry has shown that membrane proteins denature at temperatures ~50°, and it is known that the initial stages of thermal protein denaturation are generally reversible.

3. Our RS and RERS data indicate that a significant proportion of membrane lipid undergoes a cooperative thermotropic phase transition near 20° C. No techniques as potent as RS or RERS have hitherto been applied to examine the thermotropism of animal-cell plasma membranes. Indeed, in neither erythrocyte nor thymocyte plasma membranes can one detect thermotropic phase

transitions by calorimetric methods, or thermotropic lipid-protein segregation by freeze-fracture electron microscopy.

4. PQF, RS and RERS studies all indicate that cholesterol lies in clusters segregated from membrane protein. These data are consistent with the sterol-clustering documented at high cholesterol/phospholipid ratios in model system (e.g., see References 63, 64) and seen electron-microscopically in erythrocyte membranes.[67]

In FIGURE 8, we have schematized *one* kind of lipid-protein array that is consistent with available experimental data. The scheme shows only one-half of a bilayer membrane and focuses on the distribution of phospholipid and cholesterol within the array. It does not describe polar interactions. The basic assembly consists of a hollow dodecahedral cylinder, comprised of associated cylindrical protein subunits, with the 35 Å-long cylinder axes normal to the membrane plane. The subunits have a molecular weight of ~55,000 daltons, a specific gravity of 1.37, a diameter of ~36 Å, and a molecular volume of 40,000 Å^3. Where not in contact with other subunits, their perimeters comprise apolar amino acid residues. Polar side chains are restricted to the interiors of the subunits (where they may form a hydrated channel as in hemoglobin), and to the membrane-water interface. The membrane lipids are constrained within the protein rings. Moreover, the protein rings are associated with other rings in the membrane plane through protein-protein interactions, to form a complex lattice. Alignment with equivalent lattices in the other half membrane can be speculated to occur via proteins penetrating through the full membrane.

The choice of a 55,000-dalton subunit is somewhat arbitrary, but plasma membranes contain such proteins, some of these are known to form multimers, and the monomer size is compatible with localization in "half membrane"

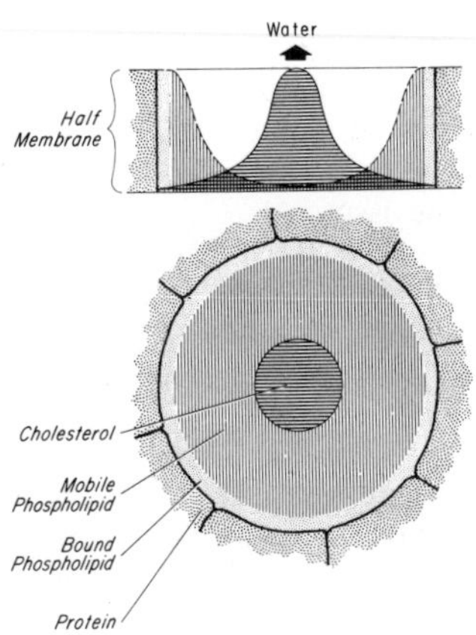

FIGURE 8. A heuristic model of possible lipid-protein associations in plasma membranes, consistent with spectroscopic data. We depict a single lipid-protein assembly within one-half of a bilayer membrane. The assembly consists of a penetrating ring of associated proteins (8–12) and its encompassed membrane lipids. Each ring interacts with other similar assemblies via protein-protein associations (not shown). The schematic drawing deals only with apolar regions of the membrane and depicts both a cross- and a sagittal section. The polar groups of lipids are oriented at the membrane water interface (*top;* not detailed). Polar amino acid residues lie at the membrane-water interface, at protein-protein contacts and/or *within* the protein subunits. The protein surfaces in contact with lipids consist of apolar hydrophobic residues. A layer of bound phospholipids lies immediately adjacent to the inner aspect of the protein ring. A cholesterol cluster occupies the center of the assembly. This is surrounded by a zone of phospholipid whose cholesterol content decreases centripetally. Hypothetical number distributions of phospholipid and cholesterol are depicted. Other geometries might fill the requirements set by the spectroscopic analyses.

The choice of a dodecahedral ring is speculative, but probably approaches a reasonable *lower* limit. It is dictated in part by the assumed subunit-diameter and in part by the need to provide space within the ring for: (a) equimolar amounts of cholesterol and phospholipid; (b) a layer of protein-bound phospholipid; (c) a central, phospholipid-depleted cholesterol cluster; (d) a phospholipid continuum with a number of contiguous phosphatide molecules enough to allow a cooperative gel → liquid-crystal transition. Unfortunately, this number is not known; it lies between ~10 (in *Acholeplasma laidlawii;*) and 200.

With a dodecahedral ring and a subunit diameter of 38 Å, the internal ring radius will be ~50 Å, giving an internal area of ~7,800 Å^2. This can accommodate ~100 phospholipid and cholesterol molecules each, at 50 Å^2/molecules and 37 Å^2/molecules, respectively. Assuming a single layer of bound phosphatides, 7-Å thick, we can account for 40 bound phospholipid molecules. If the cholesterol cluster has a radius ~25 Å, a size smaller than known to occur in cholesterol/phospholipid liposomes[63, 64] and compatible with electron microscopic measurements,[67] this could account for as many as ~53 freely exchangeable molecules.[63] If we assume an outward gradient of cholesterol, approaching a minimum at the bound phospholipid layer, we obtain a cholesterol-depleted phosphatide ring, adjacent to the layer of bound phosphatides, containing ~35 molecules of phospholipid. This should be adequate for the degree of cooperativity observed for the lipid phase transition. Since this transition is localized, it would not produce a segregation of protein and lipid.

We stress that the arrangement depicted in FIGURE 8 is *only one* of a number of diverse arrays that might explain available experimental results. Thus, the numerical values chosen may be quite far off, particularly since the dimensions of cholesterol clusters are not well specified and since we have not allowed for the condensing effects of this sterol, for the different molecular cross sections of phosphatides above and below the transition temperature, nor for the phospholipid heterogeneity in biomembranes. (However, one can speculate that the expansion of phospholipid volume above the phase transition could account for the changes in protein conformation observed in this temperature range). Also, the postulated lipid zones may derive from other geometric protein arrays, e.g., parallel or crossing networks of interacting protein subunits. Finally, our model does not deal with the structuring at the membrane-water interface and we lack the information to speculate about the issue.

The "torus protein" detected in erythrocyte ghosts, as well as certain extracts thereof[75-77] may relate to the lipid-protein array we propose. A single "torus" comprises a hollow, symmetrical ten-unit cylinder. However, the inner diameter, 60 Å, cannot accommodate membrane phospholipid and cholesterol in the manner suggested by spectroscopic measurements.

A more probable morphological correlate of our proposal may be the organizational pattern revealed electron-microscopically after extraction of glutaraldehyde fixed (cross-linking) erythrocyte ghosts with Triton X-100.[78, 79] The extracted, fixed membranes appear as "fenestrated" sheets, perforated by numerous gaps 300–2000 Å in diameter, depending on the degree of cross-linking.[78, 79] At low cross-linking, 100% of the cholesterol and phospholipid, as well as most of the glycoprotein are extracted.[79]

Clearly, the real geometries of the assemblies we postulate (if they exist) can only be established by micromorphologic measurements or x-ray diffraction, but since the latter approach appears remote, we are accordingly striving to develop exact micromorphological correlations to our spectroscopic measurements.

REFERENCES

1. TANFORD, C. 1973. The Hydrophobic Effect. John Wiley. New York, N.Y.
2. HAYDON, D. A. H. & J. T. TAYLOR. 1963. J. Theoret. Biol. **4:** 281.
3. WALLACH, D. F. H. & R. WINZLER. 1974. Evolving Strategies and Tactics in Biomembrane Research. Springer Verlag. New York, N.Y.
4. DANIELLI, J. F. & H. DAVSON. 1935. J. Cell. Comp. Physiol. **5:** 495.
5. NATHANSON, A. 1904. Jahrb. Wiss. Botan. **39:** 607.
6. PARPART, A. K. & R. BALLENTINE. 1952. Trends in Physiology and Biochemistry. E. S. G. Barron, Ed. : 135. Academic Press. New York, N.Y.
7. WALLACH, D. F. H. & H. P. ZAHLER. 1966. Proc. Nat. Acad. Sci. U.S.A. **56:** 1552.
8. LENARD, J. & S. J. SINGER. 1966. Proc. Nat. Acad. Sci. U.S.A. **56:** 1828.
9. SINGER, S. J. & G. NICOLSON. 1972. Science **175:** 720.
10. SINGER, S. J. 1971. Structure and Function of Biological Membranes. L. I. Rothfield, Ed. : 146. Academic Press. New York, N.Y.
11. SINGER, S. J. 1974. Annu. Rev. Biochem. **43:** 805.
12. BROWN, P. K. 1972. Nature New Biol. **236:** 35.
13. CONE, R. A. 1972. Nature New Biol. **236:** 39.
14. JUNGE, W. & A. ECKHOFF. 1973. FEBS Lett. **36:** 207.
15. RAZI-NAQVI, K., J. GONZALEZ-RODRIGUEZ, R. J. CHERRY & D. CHAPMAN. 1973. Nature New Biol. **245:** 249.
16. VERKLEIJ, A. J., P. H. J. VERVERGAERT, L. L. M. VAN DEENEN & E. ELBERS. 1972. Biochim. Biophys. Acta **288:** 326.
17. JAMES, R. & D. BRANTON. 1973. Biochim. Biophys. Acta **323:** 378.
18. SPETH, V. & F. WUNDERLICH. 1973. Biochim. Biophys. Acta **291:** 621.
19. WUNDERLICH, F., W. BATZ, V. SPETH & D. F. H. WALLACH. 1974. J. Cell Biol. **61:** 271.
20. WUNDERLICH, F., W. BATZ, V. SPETH & D. F. H. WALLACH. 1974. Biochim. Biophys. Acta **373:** 34.
21. POO, M. M. & R. A. CONE. 1974. Nature **247:** 438.
22. PETERS, R., J. PETERS, K. H. TEWS & W. BAHR. 1974. Biochim. Biophys. Acta **367:** 282.
23. WALLACH, D. F. H. & A. S. GORDON. 1968. Fed. Proc. **27:** 1263.
24. WALLACH, D. F. H. & A. S. GORDON. 1968. Regulatory Functions of Biological Membranes. J. Jarnefelt, Ed. : 87. Elsevier. Amsterdam, The Netherlands.
25. CAPALDI, R. A. C. & G. V. VANDERKOOI. 1972. Proc. Nat. Acad. Sci. U.S.A. **69:** 930.
26. FISCHER, H. 1964. Proc. Nat. Acad. Sci. U.S.A. **51:** 1285.
27. BIGELOW, C. B. 1967. J. Theoret. Biol. **16:** 187.
28. TANFORD, C. 1962. J. Am. Chem. Soc. **84:** 4240.
29. STRITTMATTER, P., M. J. ROGERS & L. SPATZ. 1972. J. Biol. Chem. **247:** 7188.
30. STRITTMATTER, P. 1971. Proc. Nat. Acad. Sci. U.S.A. **68:** 1042.
31. WINZLER, R. 1970. Int. Rev. Cytol. **29:** 77.
32. SEGREST, J. P., R. L. JACKSON, E. P. ANDREWS & V. T. MARCHESI. 1971. Biochem. Biophys. Res. Commun. **44:** 390.
33. KENDREW, J. D., H. C. WATSON, B. E. STRANDBERG, R. E. DICKERSON, D. PHILLIPS & V. C. SHORE. 1961. Nature **190:** 666.
34. PERUTZ, M. F., J. D. KENDREW & H. C. WATSON. 1965. J. Molec. Biol. **13:** 669.
35. LANGDON, R. G. 1974. Biochim. Biophys. Acta **342:** 213.
36. LUX, S. E., R. HIRZ, R. I. SHRAGER & A. M. GOTTO. 1971. J. Biol. Chem. **247:** 2598.
37. ASSMAN, G. & H. B. BREWER, JR. 1974. Proc. Nat. Acad. Sci. U.S.A. **71:** 989.

38. Stoffel, W., O. Ziesenberg, B. Tungal & E. Schreiber. 1974. Proc. Nat. Acad. Sci. U.S.A. **71:** 3696.
39. Assman, G. & H. B. Brewer, Jr. 1974. Proc. Nat. Acad. Sci. U.S.A. **71:** 3701.
40. Assman, G. & H. B. Brewer, Jr. 1974. Proc. Nat. Acad. Sci. U.S.A. **71:** 1534.
41. Braun, V. & V. Bosch. 1972. Proc. Nat. Acad. Sci. U.S.A. **69:** 970.
42. Braun, V. & V. Bosch. 1969. Eur. J. Biochem. **28:** 51.
43. Hantke, K. & V. Braun. 1973. Eur. J. Biochem. **34:** 284.
44. Braun, V. 1973. J. Infect. Dis. **128:** 59.
45. Sodek, J., L. B. Smillie & L. Jurasek. 1972. Cold Spring Harbor Symp. Quant. Biol. **37:** 299.
46. Inouye, M. 1974. Proc. Nat. Acad. Sci. U.S.A. **71:** 2396.
47. Maddy, A. H. & M. J. Dunn. 1973. FEBS Lett. **36:** 79.
48. Knüfermann, H., S. Bhakdi & D. F. H. Wallach. 1974. Biochim. Biophys. Acta **330:** 356.
49. Mikkelsen, R. & D. F. H. Wallach. 1975. Biochim. Biophys. Acta. In press.
50. Tuech, J. K. & M. Morrison. 1974. Biochem. Biophys. Res. Commun. **59:** 352.
51. Schmidt-Ullrich, R., D. F. H. Wallach & J. Hendricks. 1975. Biochim. Biophys. Acta **382:** 295.
52. Chapman, D., V. B. Kamat, J. de Gier & S. A. Penkett. 1968. J. Mol. Biol. **31:** 101.
53. Chapman, D., V. B. Kamat & R. J. Levene. 1968. Science **160:** 314.
54. Träuble, H. & P. Overath. 1973. Biochim. Biophys. Acta **307:** 491.
55. Overath, P. & H. Träuble. 1973. Biochemistry **12:** 2625.
56. Jost, P. C., O. H. Griffith, R. A. Capaldi & G. Vanderkooi. 1973. Proc. Nat. Acad. Sci. U.S.A. **70:** 480.
57. Stier, A. & E. Sackmann. 1973. Biochim. Biophys. Acta **311:** 400.
58. Wallach, D. F. H., S. P. Verma, E. Weidekamm & V. Bieri. 1974. Biochim. Biophys. Acta **356:** 68.
59. Verma, S. P. & D. F. H. Wallach. 1975. Biochim. Biophys. Acta **382:** 73.
60. Bhakdi, S., V. Speth, H. Knüfermann & H. Fischer. 1974. Biochim. Biophys. Acta **356:** 300.
61. Elgsaeter, A. & D. Branton. 1974. J. Cell Biol. **63:** 1018.
62. Green, J. A., L. A. Singer & F. H. Parks. 1973. J. Chem. Phys. **58:** 2690.
63. Bieri, V. & D. F. H. Wallach. 1975. Biochim. Biophys. Acta **389:** 413.
64. Träuble, H. & E. Sackmann. 1972. J. Am. Chem. Soc. **94:** 4499.
65. Fairbanks, G., T. L. Steck & D. F. H. Wallach. 1971. Biochemistry **10:** 2606.
66. Bruckdorfer, K. R. & J. M. Graham. 1975. Biological Membranes. D. Chapman and D. F. H. Wallach, Eds. Vol. 3. Academic Press, London, England. In press.
67. Higgins, J. A., N. T. Florendo & J. B. Barnett. 1973. J. Ultrastruct. Res. **42:** 66.
68. Wallach, D. F. H. & S. P. Verma. 1975. Biochim. Biophys. Acta **382:** 542.
69. Verma, S. P., D. F. H. Wallach & R. Schmidt-Ullrich. 1975. Biochim. Biophys. Acta **394:** 633.
70. Verma, S. P. & D. F. H. Wallach. 1975. Biochim. Biophys. Acta. In press.
71. Lippert, F. L. & W. L. Peticolas. 1971. Proc. Nat. Acad. Sci. U.S.A. **68:** 1572.
72. Zimmer, G. & H. Schirmer. 1974. Biochim. Biophys. Acta **345:** 314.
73. Lacko, L., B. Wittke & P. Geck. 1973. J. Cell Physiol. **82:** 213.
74. Graham, J. M. & D. F. H. Wallach. Biochim. Biophys. Acta **241:** 249.
75. Harris, J. R. 1969. Biochim. Biophys. Acta **188:** 31.
76. Harris J. R. 1969. J. Mol. Biol. **46:** 329.
77. Harris, J. R. 1971. J. Ultrastruct. Res. **36:** 587.
78. Edwards, H. H., Keyhani, E. & Morrison, M. 1975. Biophys. J. **15:** 19a.
79. Morrison, M. Personal communications.

ION-MEMBRANE INTERACTIONS AS STRUCTURAL FORCES

V. Adrian Parsegian

National Institutes of Health
Division of Computer Research and Technology
Physical Sciences Laboratory
Bethesda, Maryland 20014

I would like to elaborate upon what I see as a inconsistency in the way we look at the passage of an ion across a low dielectric membrane. This process will be discussed in terms of the energy of interaction between ion and membrane. We know that a membrane poses an energetic barrier to the ions in the adjacent solutions. We know that the modulation of this barrier is the primary means by which a membrane controls ionic flow. But we tend to think of the interaction only as one imposed by the membrane on the ion rather than a mutual interaction that might act to perturb the structure of the membrane.

The cause of the ion-membrane interaction is the coulomb field that emanates from the ionic charge. This field acts to polarize the surrounding medium, to attract charge of opposite sign to the ion, to repel like charge and thereby to reduce ionic energy. Polarization is easy in an aqueous medium of high dielectric constant and hard in low dielectric lipid. Hence the hostility of a hydrocarbon medium such as a lipid membrane to ionic charge.

But what do electric fields do to lipid membranes? Voltages of 100 or 200 mV applied to a 30- to 70-Å lipid film will cause it to thin. Potentials of 300 mV or more will cause it to break. A potential of 300 mV across a 50-Å layer corresponds to a breakdown field of 6×10^5 volts/cm.

If one considers a single ionic charge ($e = 4.8 \times 10^{-10}$ statcoulombs) in a lipid medium (dielectric constant $\epsilon \approx 2$) there is an electric field greater than or equal to 6×10^5 volts out to a distance of 35 Å from the ionic charge ($e/\epsilon r^2 \geq 6 \times 10^5$ volts/cm, $r \leq 35$ Å). It turns out that *while an ion is crossing a 50-Å membrane, one or both faces of the membrane suffer an electric field (from the ion) greater than the known breakdown voltage.* The direction of the electric stress is to pinch the membrane at the point of traverse. A thin film differs from a macroscopic medium in that the reach of the electric field from an ion exceeds the dimensions of the membrane. One need not think *a priori* of the membrane as a separate phase.

In the following I will illustrate the magnitude of the electric stress on the membrane faces. I will also review some results for the electrostatic interaction between an ion and different forms of membrane defects created by peptides that facilitate transport. These peptides act to lower the energetic barrier to ion flow.

It appears from some numerical examples that a low dielectric membrane might deform while an ion moves across. Probable energies for creating deformation are comparable to the lowering of ion-lipid interaction energy that would accompany deformation.

Some experiments are suggested for detecting local changes in membrane structure. Several reasons are given for keeping aware of the structural consequences of ion-lipid interactions. Recognizing the symmetry in which ion and membrane influence each other allows us to see that the deformative properties of membrane lipids, as well as membrane "fluidity" or "viscosity," might be critical to the successful design of membrane systems.

Electrostatic Polarization Forces Exerted by an Ion Moving across a Low Dielectric Membrane

One tends to think of the work required to move a charge from water to lipid primarily in terms of removing "water of hydration" from its immediate vicinity and replacing it with hydrocarbon. Neutral cyclic proteins—valinomycin, nonactin, enniatin B, diactin, etc.[3]—are said to act as "carriers" whose polar groups replace the hydration layers. The effective radius of the positively charged complex is much larger than that of the ion. Entry into a lipid is thereby facilitated.

Once within the membrane the positive charge will induce a net negative charge on the interfaces between lipid and water. (Negative charge is pulled more easily to the interface from the water than away from the interface into the lipid.) If the membrane faces remain rigid, the ion must overcome attraction to this interfacial charge (conveniently modeled as an "image" force; see *Appendix A*) in order to move away from the interface. The profile of the corresponding ion-interface interaction energy is shown in FIGURE 1.

The pull on an ion at a distance of 10 Å from the interface is about 3×10^{-6} dynes. The stress on the interface is more than 200 atmospheres nearest to the ion. The local electric field at that part of the membrane is about 14×10^{6} volts/cm.

Incidentally, Eisenman[5] has fitted carrier-transport data to a trapezoid potential which gives essentially the same result. In that mathematical model the ion climbs a hill about 10 kT high over a distance of the order of 10 Å. The net force is $4 \times 10^{-13}/10^{-7} = 4 \times 10^{-6}$ dynes between ion and membrane interface.

At the middle of the membrane the net force on an ion is zero by symmetry. But the inward pressure on either face of a membrane 50 Å thick is greater than one atmosphere over a circular path on the surface 50 Å across.

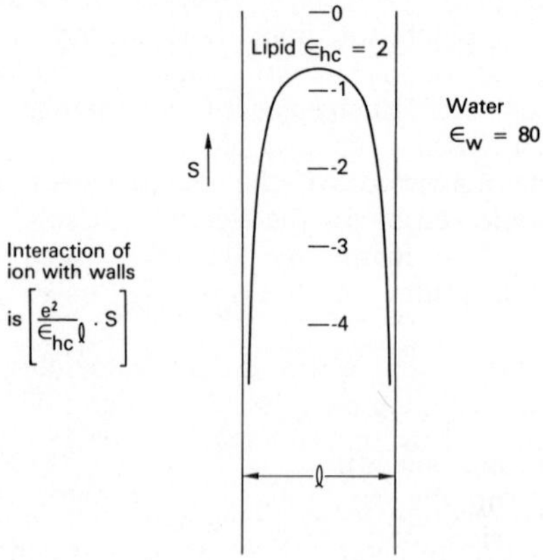

FIGURE 1. The "hill" an ionic charge must climb in order to go across a lipid membrane with fixed interfaces in water. The function S is described in *Appendix A* and has been used as a fixed barrier to ion transport.[4] (Compare Equation A2, *Appendix A*).

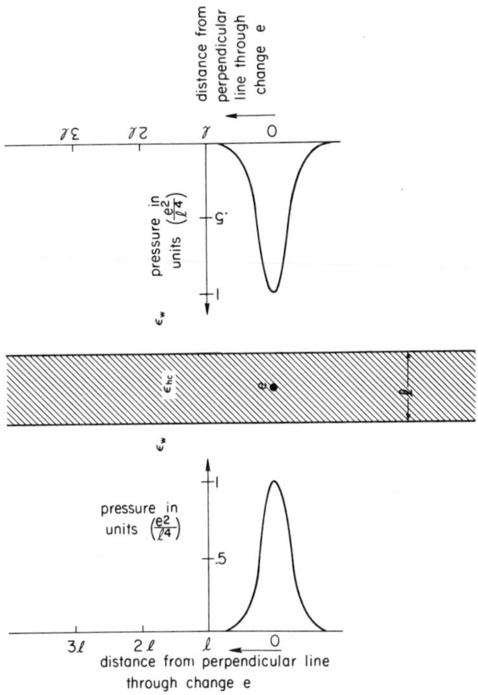

FIGURE 2. Profile of the dielectric stress on lipid-water interfaces due to an ionic charge, e, midway between them.[6] Note that the spread of pressure is over a radius of the order of membrane thickness ℓ and that the magnitude of stress is as the inverse *fourth* power of thickness.

We are told[1,2] that lipid membranes have a static Young's modulus of the order of 10^6 dynes/cm^2 or 1 atm. That is $\Delta\ell/\ell \sim$ compressive pressure (atm). Electrostrictive stress from an ion is apparently significant on this scale. According to FIGURE 2 the stress profile is strongly localized above the point of ionic traverse. It extends over an area comparable in diameter to the distance of the charge from the interface.

For the configuration of an ion in the middle of a 50-Å thick membrane, there will be an electric field greater than the breakdown field, 6×10^5 volts/cm, over a circle of 40-Å radius. This comprises an area much bigger than the cross section of any carrier complex or of a membrane pore.

A bilayer membrane might thin under the stab of the ionic electric field. Is there energy enough to do it?

Electrostatic Interaction Energies between Ionic Charge and Model Membrane

We may refer to three identical configurations of low dielectric membrane, high dielectric medium, and ion for illustrative calculation. These are: (a) an ion in a low dielectric slab of finite thickness (FIGURE A1, *Appendix A*); (b) an ion in the center of a pore drilled through the slab (FIGURE B1, *Appendix B*); (c) an ion surrounded by a spherical jacket whose dielectric polarizability differs from the surrounding medium (FIGURE C1, *Appendix C*). In each case the energy of charging the ion, treated as a conducting sphere, has been found for various dimensions and dielectric values.[6]

(a) The ion's energy differs from that in an infinite medium ϵ by a term

$$-\frac{e^2}{\epsilon\ell}\ln\frac{2\epsilon'}{\epsilon+\epsilon'}$$

where ϵ' is the dielectric outside the slab and ℓ is slab thickness. For typical values of ϵ, ϵ', ℓ and of ionic radius the energy is decreased only a few percent compared to the leading term $e^2/2\epsilon a$ (FIGURE 3). Given present estimates of artificial membrane thickness it is clear that finite width does not substantially change the barrier posed by low ion solubility in a lipid medium.

(b) The barrier for passing through a high dielectric pore is much lower than that of a low dielectric region. The ion energy in a pore of radius b, dielectric ϵ_p is (Equation B5, *Appendix B*)

$$\frac{e^2}{2\epsilon_p a}+\frac{e^2}{\epsilon b}P(R)$$

where R is the ratio of low to high dielectric $R = \epsilon/\epsilon_p$ and P(R) is plotted in FIGURE B2. For water at 37 C, $\epsilon_p = 73$; if $\epsilon = 2$, this energy is 50 kT/b(Å) or about 10 kT = 6.2 kcal/mol for b = 5 A. This is relative to the ion energy in a pore of infinite radius.

We note that these numbers, assuming a water-filled pore, will not be accurate for a gramicidin A channel lined with flexibly mounted carbonyl groups. Judging from the low energetic barrier to ions posed by this pore[7] and arguments regarding pore structure,[8] the effective polarizability of this channel is higher than for one containing only water.

(c) A charge "solvated" by a high dielectric jacket (FIGURE C1) can have a substantially lower charging energy than one "bare" in a low dielectric medium. Reduction by a factor of 70% is reasonable even if the effective jacket dielectric is 20 and that of the medium is 2 for a charge of radius

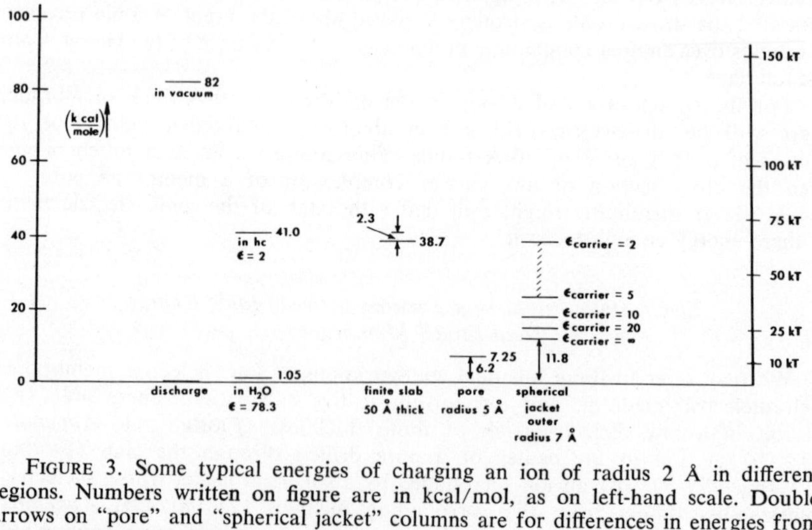

FIGURE 3. Some typical energies of charging an ion of radius 2 Å in different regions. Numbers written on figure are in kcal/mol, as on left-hand scale. Double arrows on "pore" and "spherical jacket" columns are for differences in energies from 1.05 kcal/mole in bulk water. The "pore" is considered to be filled with water $\epsilon = 78.3$ and bounded by hydrocarbon $\epsilon = 2$.

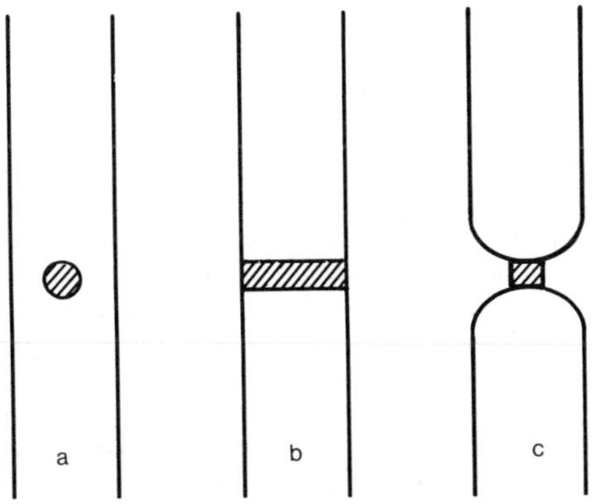

FIGURE 4. Scheme for possible indentation under electric stress (see FIGURE 2). If the shaded regions have the polarizability of water, then the difference in energies of (a) "carrier" and (b) "pore" give a measure of the electrostatic energy decrease in forming (c). Numbers for this case can be taken from FIGURE 3.

1.3 Å and outer jacket radius 7 Å (FIGURE 3). These are sizes described for K^+ ion complexed to valinomycin, which has been considered an ion carrier.[9]

The accuracy of these estimates, based on a macroscopic model, is open to valid question. As in the simple Born theory of ion solvation in water one treats all media as continuum dielectrics about a spherical charge. The ionic charge is given an "effective radius" (greater than its "crystal radius") corresponding to the observed energy of solvation. Further, dielectric saturation (due to intense fields near the ion) is ignored. Although this is not the way to derive a complete description of an ionic charge interacting with its environment, it appears adequate for answers to questions of orders of magnitude required here.

These questions deal with the effect of boundaries between dielectric media introduced a relatively large distance from the charge. One asks for the difference in energy between an ion in infinite medium and within slab walls many angstroms from the ion (FIGURE A1); or between bulk medium and a cylinder-shaped medium where the effect is due to a field induced on the cylinder walls and is weak compared to the field near the ion; or between a charge in a cylinder and in a spherical shape dielectric where the difference in structure occurs several angstroms from the center charge.

The resulting numbers give a rough estimate of the electrostatic energy decrease that would occur if the membrane pinched in at the place where an ion was crossing. We compare the energy of ion in carrier FIGURE 4a with ion in pore FIGURE 4b to get an idea of the energy for indenting as in FIGURE 4c. From the computations of FIGURE 3 one expects energy decreases of the order of 5 to 15 kcal/mol or 10 to 30 kT/ion.

This energy compares favorably with the "cost" of 5.5 kT computed by Haydon[10] to dimple a 48-Å membrane down to 28 Å at the site of gramicidin A or with the work of 1 kcal/mol per angstrom inferred by Urry[8] from the same data. Energies of these magnitudes are also of the order as those interactions responsible for phase transitions in lipid.[11–13]

Comments and Consequences

The pressure pulse exerted by an ion on a membrane will last only for the time of ionic traverse and extend only over a region of hundreds or a few thousand square angstroms. But most of what we know about the response of membranes to applied fields is for indefinitely extended fields lasting for comparatively long times and causing concerted displacements by virtually all the membrane lipid molecules. An indentation that has the shape of the ionic stress (FIGURE 2) and that forms in the ion's wake would not require displacement of more than a few lipid molecules. It would likely be difficult to detect directly. Even the relatively long-lived indentations of lipid near gramicidin A "pores" must be inferred indirectly. Certainly there is enough time for the lipid to respond to the ion. Probable displacements are not necessarily more than what is needed to accommodate a diffusing carrier. (And we do know carrier-coated ions get across membranes!)

If membrane deformations cannot be seen directly, what kind of behavior would suggest its occurrence? The following questions might help:

Is the membrane breakdown voltage ever a function of the ionic strength of the bathing medium or of the amount of assisting protein? (It is obviously impractical to speak of a breakdown current.) If individual ions deform a membrane, then the sum of many deformations may be what leads to breakdown.

Will substances that cannot themselves cross a membrane aid in transport of ions? For example, small fatty alcohols or small amphophiles might stick to the lipid interface to lower its interfacial energy and deformability, but will not change the lipid interior (which determines the barrier posed by an immutable membrane). Facilitation of ion transport would suggest that the interfacial energy is important to ionic transit.

Will substances that act like pores across thin membranes show some carrier properties across a macroscopically thick lipid barrier?[14] For example, the channel-former gramicidin A interacts with ions via carbonyl groups, as does the carrier peptide valinomycin. If the membrane were much thicker than the 38-Å length of gramicidin A, would this drug move as a mobile protein in association with a cation?

Is the probability of an open channel formed by neutral peptides ever a function of salt concentration? The occurrence of alamethicin channels depends strongly on salt in the bathing solution and on applied voltages across the membrane.[3] Formation of gramicidin A pores appears to require some indentation by the host lipid membrane. It may be that presence of an ion in the pore protein creates an electric stress to promote indentation.

The purpose behind asking these questions is to gather information for a physically plausible picture of transport. The lipids might be as important as proteins. Convenient lipid preparations may have anomalous physical properties that make the models qualitatively different from real membranes. On a large scale, solvent-containing phospholipids are probably less deformable than pure bilayers; their ability to conduct "carriers" or accommodate "pores" may also differ. Polar groups which change interfacial energy may be important in the design of real membranes; evidence of this might be seen in model systems.

Rather than thinking of dissolution and diffusion across a membrane, it may be better to think of an instability occurring when an ion gets into a membrane.[15] Phenomenologically these mechanisms might possibly look similar under some model conditions, but physically the processes are very different. Under the actual conditions enjoyed by real membranes this difference may be essential.

For all the progress in making mathematical models and fitting them to data, we still lack a physical theory of transport across lipid membranes. Given the speciality of peptides used to develop our present picture, it seems a bit arbitrary to restrict ourselves to idealized "diffusing carrier" and "channel" models while investigating new compounds. An individual ion can exert stresses known to break membranes, and can effect energy changes sufficient to pinch membranes. It seems reasonable to surmise that the membrane itself be able to change structure to accommodate traversing ions. It should not be surprising if substances that facilitate transport act to guide favorable changes in structure during transport.

Acknowledgments

I am pleased to cite helpful conversations with several colleagues: George H. Weiss for much mathematical advice; Ira Katz for numerous comments on bilayer experiments; Barry Bunow for the initial suggestion that ions might cause membranes to deform; and Harold Lecar for discussions on electrostatics and its consequences in model- and nerve-membrane behavior.

References

1. REQUENA, J., D. A. HAYDON & J. B. HLADKY. 1975. Lenses and the compression of black lipid membranes by an electric field. Biophys. J. **15:** 77.
2. WHITE, S. H. 1974. Comments on electrical breakdown of biomolecular lipid membranes as an electromechanical instability. Biophys. J. **14:** 155.
3. HLADKY, S. B., L. G. M. GORDON & D. A. HAYDON. 1974. Molecular mtchanisms of ion transport in lipid membranes. Ann. Revs. Phys. Chem. **25:** 11.
4. LAUGER, P. & G. STARK. 1970. Kinetics of carrier-mediated ion transport across lipid bilayer membranes. Biochem. Biophys. Acta **211:** 458.
5. EISENMAN, G. 1975. This monograph.
6. PARSEGIAN, V. A. 1969. Energy of an ion crossing a low dielectric membrane: Solutions to four relevant electrostatic problems. Nature **221:** 844.
7. FINKELSTEIN, A. 1974. Aqueous pores created in thin lipid membranes by the antibiotics nystatin, amphotericin B and gramicidin-A: Implications for pores in plasma membranes. *In* Drugs and Transport Properties. B. A. Callingham, Ed.: 241–250. MacMillan. London, England.
8. URRY, D. 1975. This monograph.
9. ANDREOLI, T. E., M. TIFFENBERG & D. TOSTESON. 1967. J. Gen. Physiol. **50:** 2527.
10. HAYDON, D. A. 1975. This monograph.
11. PARSEGIAN, V. A. 1966. Theory of liquid-crystal phase transitions in lipids and water systems. Trans. Farad. Soc. **62:** 848–860.
12. PARSEGIAN, V. A. 1967. Forces between lecithin biomolecular leaflets are due to a disordered surface layer. Science **156:** 939–942.
13. PARSEGIAN, V. A. 1968. An energetic model of ionic lipids in the liquid-crystal state. *In* Membrane Models and the Formation of Biological Membranes. L. Bolis and B. A. Pethica, Eds. North Holland Publishing Co. Amsterdam, The Netherlands.
14. BYRN, S. R. 1974. Biochemistry **13:** 5186.
15. IBERALL, A. S. 1975. Personal communication.
16. SMYTHE, W. R. 1968. Static and Dynamic Electricity, 3rd Ed. McGraw-Hill. New York, N.Y.

Discussion

DR. IBERALL: I would like to add another piece of information from solid-state theory. At extremely high strain rates, that is for rapid processes, there is perhaps some little-known material where in fact the Young's modulus approaches dynamically the order of bulk modulus, which means that the kind of basic resiliances that you face in solid state is on the order of 20,000 atmospheres.

DR. PARSEGIAN: Good point.

DR. FINKELSTEIN: Do you have any estimates of the relaxation times it would take to bring about the pinching down of the membrane that you've mentioned?

DR. PARSEGIAN: Yes, the kinds of deformations that one talks about—because the stress is so localized—aren't that much greater than what would have to be accommodated to get a carrier across anyway. I suspect that the time scale is not that different from what you'd expect from a diffusion, moving the chains apart by 10 or 20 Å. In other words, I don't think that microseconds are too short a time for distances of movement of a few angstroms.

Appendix A: Charge in a Finite Slab

Consider a dielectric medium ϵ which is planar, infinitely extended in the y and z directions but of finite thickness ℓ in the x direction (FIGURE A1). The two faces are put at $x = +\ell/2$ and $x = -\ell/2$. The medium outside this slab has dielectric constant ϵ'.

One wants to know what will be the energy due to a charge, e, inside such a slab compared to the energy of that charge deep in the infinite medium ϵ or compared to its energy when immersed in an infinitely thick slab. It

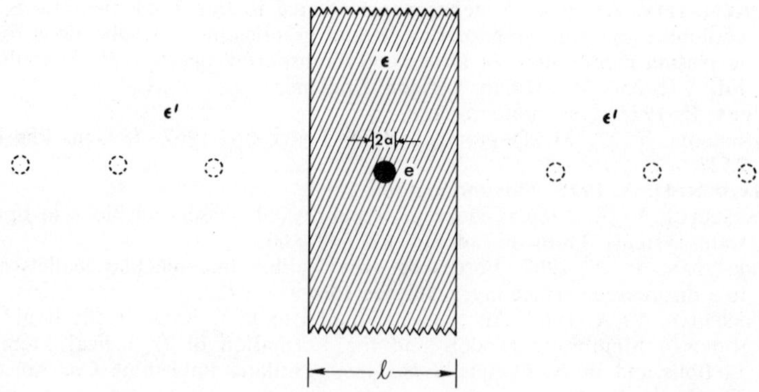

FIGURE A1. Charge e of radius a in a planar slab of dielectric ϵ thickness ℓ. Outside dielectric is ϵ'. The electrostatic potential is given by an infinite series of image charges uniformly spaced at a distance ℓ along a line through the original charge and perpendicular to the plane of the slab.

is necessary to calculate the electrostatic potential about the charge in each of these cases, and then to find the energy of the system in this potential which it has created—that "self-energy."

For a spherical charge in a uniform medium, the self-energy is the well-known Born charging energy

$$e^2/2\epsilon a \tag{A1}$$

The potential due to a point charge in a dielectric slab is the sum of the $e/\epsilon r$ coulomb potential from interaction with the charge itself plus interaction with charge induced on the dielectric boundary surfaces. The latter is conveniently described as a set of image charges at positions which are reflections of the original charge through the dielectric boundary faces.

The change in self-energy of a charge from that in a medium to one in a thin slab is the strength of interaction with the image charges induced (again with reference to a discharged state). When the point charge is a distance, x, from the center of the slab, the free energy of coulombic interaction with images is

$$\frac{e^2}{\epsilon \ell} S = \frac{e^2}{2\epsilon \ell} \sum_{i=1}^{\infty} \left\{ \frac{q^{2i-1}}{(2i - 1 - \alpha)} + \frac{q^{2i}}{i} + \frac{q^{2i-1}}{(2i - 1 + \alpha)} \right\} \tag{A2}$$

where $q = (\epsilon - \epsilon')/(\epsilon + \epsilon')$, $\alpha = 2x/\ell$. (As expected $q = 0$ (no image charge) if $\epsilon = \epsilon'$.)

When the charge is at the slab center $x = 0$, and the extra energy of being in the finite dielectric is concisely

$$-\frac{e^2}{\epsilon \ell} \ln \left(\frac{2\epsilon'}{\epsilon + \epsilon'} \right) \tag{A3}$$

which is zero for $\epsilon' = \epsilon$ (infinite medium) and goes to $-\frac{e^2}{\epsilon \ell} \ln 2$ as $\epsilon' \gg \epsilon$.

In the general case for any α between ± 1 the sum becomes

$$\frac{e^2}{2\,\epsilon \ell} \left[2q \int_0^{\infty} \frac{e^t \cosh \alpha t}{e^{2\epsilon} - q^2} dt - \ln (1 - q^2) \right] \tag{A4}$$

which reduces to Equation A3 when $\alpha = 0$.

The energy is inverse to slab thickness l and to slab dielectric ϵ. If the outside dielectric ϵ' is less than ϵ, $q < 0$ and (A2), (A3) and (A4) are negative. The self-energy is lowered by presence of the medium ϵ' and the $x = 0$ center point (A3) is a maximal energy position for the charge.

The electrostatic stress perpendicular to the interface between slab and medium is the difference

$$\frac{\epsilon E^2 - \epsilon' E'^2}{8\pi} \approx \frac{\epsilon E^2}{8\pi} \tag{A5}$$

when $\epsilon \ll \epsilon'$. The fields, E, are the components perpendicular to the interface and are due to all electric fields emanating from original plus image charges.

Appendix B: Charge in Center of High Dielectric Pore

Again consider a slab of low dielectric material ϵ bounded by semi-infinite regions of dielectric $\epsilon' \gg \epsilon$. Let there be a cylindrical region of radius b, dielectric ϵ_p through the slab (FIGURE B1). What is the charging energy of a charge, e, at the center of the slab and cylinder?

For cylinder radius b much less than slab thickness l the potential in the pore due to a (point) charge e is[16]

$$V(\rho, z) = V_{\text{coulomb}} + V_{\text{induced}}$$

$$= \frac{e}{\epsilon_p}\left[\frac{1}{r} + \frac{2}{\pi}\int_0^\infty \frac{kb(R-1)K_0(kb)K_0'(kb)}{1 - kb(R-1)I_0(kb)K_0'(kb)} I_0(k\rho)\cos kz\, dk\right] \quad \textbf{(B1)}$$

Here r is the distance from the charge; R is the ratio ϵ/ϵ_p; K_0, K_0', I_0 are modified Bessel functions of imaginary argument; ρ and z are cylindrical coordinates measuring distance from the pore axis (ρ) and distance along the axis from the charge (z).

In this potential the ordinary coulomb term $e/\epsilon_p r$ is altered by a term due to the presence of the dielectric boundary at $\rho = b$. It is interaction of the charge with this second term that is the energy of being a distance b from the low-dielectric material outside the pore. Hence one is interested in the magnitude of $V_{\text{induced}}(0)$, which is

$$V_{\text{ind}}(0) = \frac{2e}{\pi\epsilon_p}\int_0^\infty \frac{kb(R-1)K_0(kb)K_0'(kb)}{1 - kb(R-1)I_0(kb)K_0'(kb)}\, dk$$

$$= \frac{1}{b}\frac{2e}{\pi\epsilon_p}\int_0^\infty \frac{y(R-1)K_0(y)K_0'(y)}{1 - y(R-1)I_0(y)K_0'(y)}\, dy \quad \textbf{(B2)}$$

$$= \frac{2e}{b\epsilon_p} f(R).$$

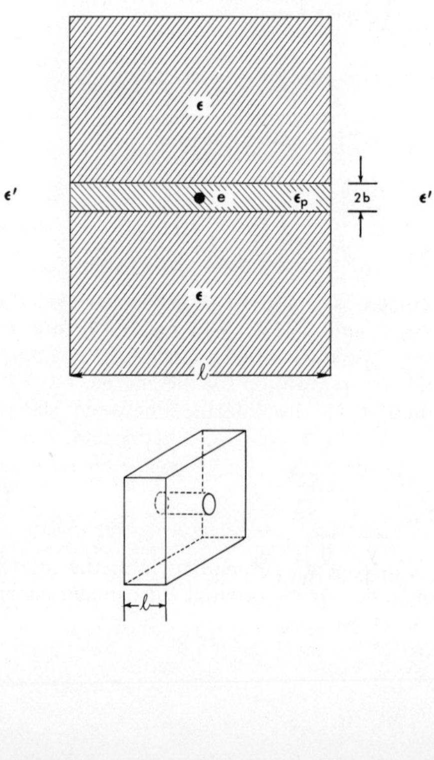

FIGURE B1. Cross section of cylindrical "pore" of dielectric ϵ_p radius b through the low dielectric slab $\epsilon \gg$ thickness ℓ. Treatment in text assumes $b \ll \ell$ and that the ion is not near the ends of the pore.

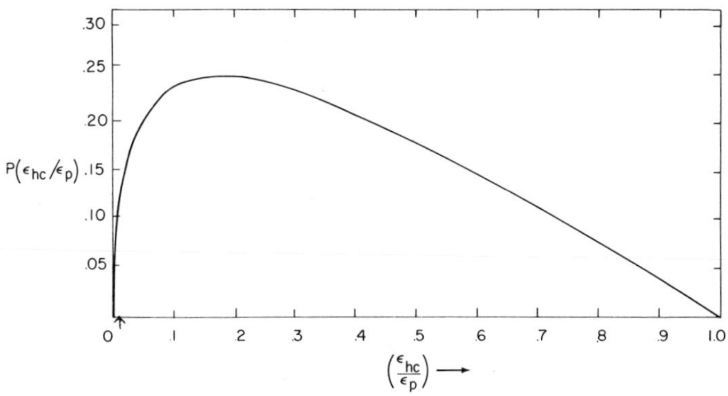

FIGURE B2.

For compactness we define

$$f(R) = \frac{1}{\pi} \int_0^\infty \frac{y(R-1)K_0(y)K_0'(y)}{1 - y(R-1)I_0(y)K_0'(y)\,dy} \tag{B3}$$

This can be evaluated numerically for various $R = \epsilon/\epsilon_p$.* Note that the dependence on pore radius b is a simple factor (1/b). Then the contributions to the energy of charging a particle at point $\rho = 0$, $z = 0$ due to presence of a low dielectric at the radius b is

$$\frac{2f(R)}{\epsilon_p b} \int_0^e e'\,de' = \frac{e^2}{\epsilon_p b} f(R)$$

$$= \frac{e^2}{\epsilon b} P(R) \tag{B4}$$

where we define $P(R) = Rf(R)$.

The total self-energy of a charged sphere at $\rho = 0$, $z = 0$ is then

$$E\text{ (charging in pore)} = E_{born} + E_p$$

$$= \frac{e^2}{2\epsilon_p a} + \frac{e^2}{\epsilon b} P(R)$$

$$= \frac{e^2}{2\epsilon_p a} + \frac{e^2}{\epsilon_p b} f(R) \tag{B5}$$

In the limit where "pore" material and "membrane" material have the same polarizability, $R = 1$, the pore boundary is electrostatically invisible. f(1) and P(1) are zero.

In the limit where ϵ_{pore} is infinitely greater than $\epsilon_{membrane}$, P(0) is again zero.

For the case of $\epsilon_{pore} = \epsilon_{water} = 80$ and $\epsilon_{mem} = 2$, $R = 1/40$ and $P(1/40) = .17$, $f(1/40) = 6.8$.

* The functions P(R) and f(R) ($R = \epsilon_{mem}/\epsilon_{pore}$) are integrated numerically. The logarithmic singularity in the integrand at $y = 0$ is subtracted away and integrated exactly. P(R) is plotted in FIGURE B2 and both functions are tabulated for several values of the ratio R in TABLE 1.

TABLE 1
VALUES OF THE FUNCTIONS f(r), P(R)

$R = \epsilon/\epsilon_p$	f(R), Equation B3	P(R) = Rf(R), Equation B4
.005	18.93	.095
.01	12.35	.123
.015	9.52	.143
.02	6.78	.158
.025	6.82	.170
.03	5.98	.1795
.035	5.37	.188
.04	4.87	.195
.045	4.47	.201
.05	4.13	.207
.055	3.85	.212
.06	3.6	.216
.065	3.39	.220
.07	3.19	.223
.075	3.03	.227
.08	2.87	.230
.085	2.73	.232
.09	2.61	.235
.095	2.49	.237
.1	2.39	.239
.15	1.66	.249
.2	1.24	.249
.25	.977	.244
.3	.785	.235
.35	.641	.224
.4	.529	.212
.45	.438	.197
.5	.364	.182
.55	.302	.166
.6	.248	.149
.65	.202	.131
.7	.162	.114
.75	.129	$.969 \times 10^{-1}$
.8	$.961 \times 10^{-1}$	$.769 \times 10^{-1}$
.85	$.682 \times 10^{-1}$	$.579 \times 10^{-1}$
.9	$.431 \times 10^{-1}$	$.388 \times 10^{-1}$
.95	$.208 \times 10^{-1}$	$.197 \times 10^{-1}$
1.0	0	0

The induction term E_p in the self-energy Equation B5

$$E_p = \frac{e^2}{\epsilon b} P(\epsilon/\epsilon_p) \text{ or } \frac{e^2}{\epsilon_p b} f(\epsilon/\epsilon_p) \quad \textbf{(B6)}$$

can be several kcal/mol ion. This is shown in FIGURE B3, where E_p vs. radius b is plotted for four values of the ratio $R = \epsilon/\epsilon_p$. Values are given both in kcal/mol and kT per ion. In the case of $\epsilon_p = 80$, $\epsilon = 2$ and b in the range 5 to 10 A, the energy of interaction with the pore walls is about 3 to 6 kT per ion or approximately 2 to 4 kcal/mol.

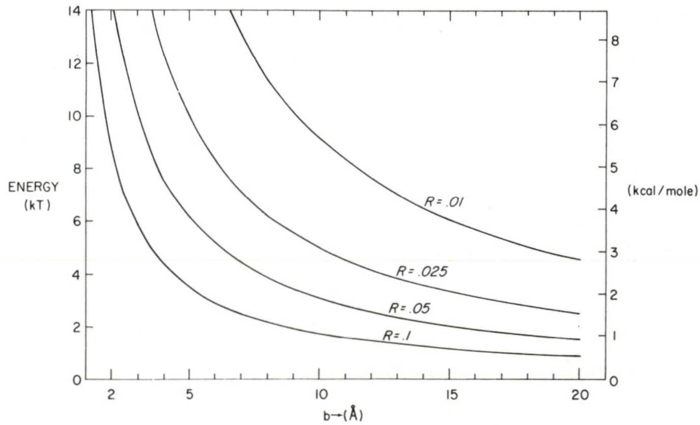

FIGURE B3. Plots of barrier height E_p posed by pore as a function of pore radius, b, and ratio of outside (membrane) dielectric ϵ to inside (pore) dielectric ϵ', $R = \epsilon/\epsilon'$. Energy is in kT units at 37°C and in kcal/mole. $E_p = e^2/\epsilon b\, P(R)$ from Equations B3–B6.

Appendix C: A Spherical Charge Covered by a Spherical Jacket

Let the charge e of radius a be surrounded by a layer of dielectric constant ϵ_c, thickness $(b - a)$, so that the total radius of the sphere is b. The dielectric outside is ϵ_{hc} (FIGURE C1). The electrostatic potential has the form $e/r + d$. In the infinite medium the potential is $e/\epsilon_{hc}r$ for $r > b$. Inside the medium ϵ_c it is $\dfrac{e}{\epsilon_c r} + \dfrac{e}{b}\left(\dfrac{1}{\epsilon_{hc}} - \dfrac{1}{\epsilon_c}\right)$.

The self-energy of the charge is then

$$\frac{e^2}{2}\left[\frac{1}{\epsilon_c a} + \frac{1}{b}\left(\frac{1}{\epsilon_{hc}} - \frac{1}{\epsilon_e}\right)\right] \tag{C1}$$

and subtracting the energy of a charge without jacket, $e^2/2\epsilon_{hc}a$, one gets the difference

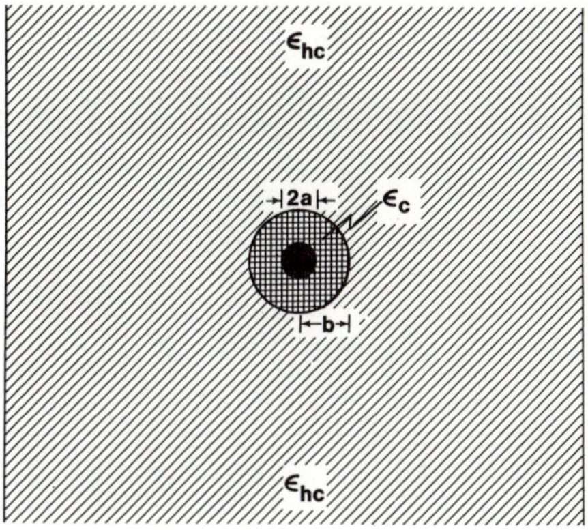

FIGURE C1.

in free energy of charging due to presence of the material

$$\frac{e^2}{2}\left[\left(\frac{1}{b}-\frac{1}{a}\right)\left(\frac{1}{\epsilon_{hc}}-\frac{1}{\epsilon_c}\right)\right]=\frac{e^2}{2\epsilon_{hc}a}\left[\left(\frac{a}{b}-1\right)\left(1-\frac{\epsilon_{hc}}{\epsilon_c}\right)\right] \quad \textbf{(C2)}$$

The factor in square brackets is a convenient form for calculating the energetic effect of the layer ϵ_c relative to the energy of the charge in medium ϵ_{hc} alone.

If ϵ_{hc} is not a constant but varies with distance, r, from the center of the spherical complex, we may write

$$\epsilon_c = \frac{1}{r^2 f(r)} \quad \textbf{(C3)}$$

For any convenient f(r) the self-energy is then

$$\frac{e^2}{2\epsilon_{hc}b}+\frac{e^2}{2}(F(b)-F(a)), \quad \textbf{(C4)}$$

where

$$F(b)-F(a)=\int_a^b f(r)\,dr. \quad \textbf{(C5)}$$

Part IV. Permeation Mechanism across Synaptic Membranes

PURIFICATION OF ACETYLCHOLINE RECEPTOR FROM *TORPEDO CALIFORNICA* AND ITS INCORPORATION INTO PHOSPHOLIPID VESICLES*

Mark G. McNamee, Cheryl L. Weill, and Arthur Karlin

Department of Neurology
College of Physicians and Surgeons
Columbia University
New York, New York 10032

The acetylcholine receptors (AChR) translate the binding of acetylcholine into increases in ion permeabilities of synaptic membranes. The receptor protein can be solubilized from the membrane by nonionic detergents and purified by affinity chromatography. The isolated protein retains the specific binding properties attributed to the AChR *in situ* and is being used to deduce the molecular properties of the AChR. Purification and characterization of the AChR has been the subject of several recent reviews.[1-3]

A complete molecular description of receptor function must include its interactions with other membrane components. Considerable effort is now being dedicated to the reconstitution of cation permeability control in membranes containing AChR[4-6] and our initial efforts in this area are discussed.

Purification of Acetylcholine Receptor

Tissue from the electric organ of *Torpedo californica* is homogenized in 1 mM EDTA (pH 7.4), and the pellet obtained after centrifugation is homogenized in 1 M NaCl–2 mM $NaPO_4$–1 mM EDTA (pH 7.4), sedimented, and resuspended in 1 mM EDTA (pH 7.4), and resedimented. This crude membrane pellet is then extracted for 1 hr at 4° with 3% Triton X-100–50 mM NaCl–10 mM $NaPO_4$–1 mM EDTA–3 mM NaN_3 (pH 8.0). The supernatant after centrifugation contains on the average 16% of the initial protein, 18% of the acetylcholinesterase (AChE), and nearly all of the AChR. The procedure used is identical to that used for extraction of AChR from *Electrophorus electricus* and complete details may be found elsewhere.[3]

The specific activity of the AChR in the detergent extract is approximately 300 pmol of sites per mg of protein, approximately ten times higher than that obtained with *Electrophorus*.

Solubilized AChR is purified by affinity chromatography using *p*-carboxyphenyltrimethylammonium ion, coupled to Affi-Gel 401 (Bio-Rad Lab., Rich-

* This work was supported by Research Grants NS 07065 from the National Institute of Neurological Diseases and Stroke and BMS75-03026 from the National Science Foundation and by a grant-in-aid from the New York Heart Association, Incorporated.

TABLE 1

AFFINITY CHROMATOGRAPHIC PURIFICATION OF RECEPTOR FROM *Torpedo californica*

Elution Sequence	Protein* mg	Protein* %	AChE† nmol	AChE† %	Receptor‡ nmol	Receptor‡ %
Extract§						
Input	1133	100	19.2	100	371	100
Not bound	900	79.4	16.3	85.0	0	0
TNP50¶	6.9	0.6	0.39	2.0	0	0
TNP100	8.5	0.8	0.21	1.1	1.5	0.6
10 mM CARB‖ in TNP 90	26.3	2.3	1.27	6.6	73.5	19.8
Total recovery	942	83	18.2	95	75	20

* Protein was determined by the Lowry method with bovine serum albumin as the standard.[18]

† Acetylcholinesterase activity was measured by Ellman's procedure[19] and is expressed as nmoles of catalytic sites assuming 80,000 daltons per catalytic site and 10 mol acetylthiocholine hydrolyzed per min per g enzyme.

‡ Receptor activity was determined by specific [^{3}H]MBTA labeling (see text).

§ Extract (186 ml) from 600 g of tissue in 3% Triton X-100–50 mM NaCl–10 mM $NaPO_4$–1 mM EDTA–3 mM NaN_3 (pH 7.0).

¶ TNPxx contains 0.2% Triton X-100, xx mM NaCl, 10 mM $NaPO_4$, 1 mM EDTA, 3 mM NaN_3 (pH 7.0).

‖ Carbamylcholine was removed from fractions by dialysis before assay for AChE and receptor.

mond, Calif.), as the adsorbing ligand and carbamylcholine (CARB) as the eluting ligand. TABLE 1 provides a summary of a typical elution pattern from the affinity column.

The AChR can be further purified about 30% by centrifugation in a 5–20% sucrose density gradient containing 0.2% Triton X-100. Typically, the AChR appears as two peaks (FIGURE 1). The faster-moving peak moves as if it were a dimer of the slower peak. The specific activity of AChR is constant across both peaks at about 4 nmol of MBTA sites per mg of protein (see below). Both peaks are resolved from remaining traces of AChE and less than .001% of the pooled AChR fractions is active AChE.

CHARACTERIZATION OF ACHR

Affinity Alkylation

The AChR can be assayed at all stages of purification by affinity alkylation of dithiothreitol-reduced receptor with 4-(*N*-maleimido)benzyltri-[^{3}H]-methylammonium iodide ([^{3}H]MBTA), using a modification of the "quick assay" procedure.[7] The portion of MBTA labeling that is blocked by the principal α-neurotoxin of *Naja naja siamensis*[8] is taken to be specific for AChR; all of the [^{3}H]MBTA labeling of purified AChR is blocked by toxin.

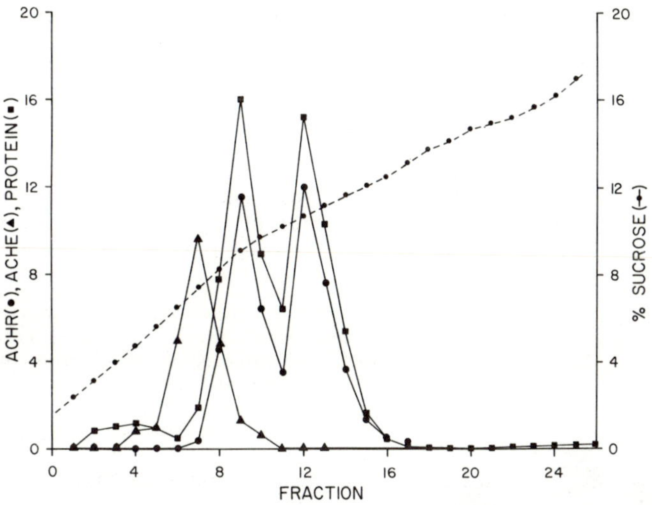

FIGURE 1. Sucrose density gradient centrifugation of *Torpedo* AChR. Acetylcholine receptor, purified by affinity chromatography (1 mg in 1 ml of TNP50 [see footnote (¶) to TABLE 1], 3 nmol MBTA sites/mg) was applied to the top of a 16-ml 5–20% sucrose gradient in TNP150 and centrifuged at 27,000 rpm for 16.5 hr at 5° in the SW27 rotor of an L2-65B Beckman ultracentrifuge. Fractions of 0.5 ml were collected with an ISCO Model 183 Fractionator and analyzed for percentage of sucrose (-●-●-), protein (■, 10 μg/unit),[15] AChR (●, 50 nmol MBTA sites/unit), and AChE (▲, 50 pmol/unit).[6] Approximately 90% of the protein, AChR, and AChE were recovered. The specific activity of the peak AChR fraction was 4.2 nmol MBTA sites/mg.

Toxin Binding

A [^{3}H]methyl derivative of *N. n. siamensis* α-toxin was prepared by reductive alkylation of native toxin with formaldehyde and tritiated sodium borohydride,[3, 9, 10] The toxin is labeled to the extent of about 0.5 mol of methyl groups per mol of toxin, and yields about 2000 cpm per pmol under typical counting conditions. The biological activity of [^{3}H]-toxin is indistinguishable from unlabeled toxin by a bioassay using mice[11] and all of the [^{3}H]-toxin is bound by an excess of receptor.

The binding of [^{3}H]-toxin to receptor is determined by incubating toxin and AChR for one hour at 25° in 0.2% Triton X-100–100 mM NaCl–10 mM $NaPO_4$–1 mM EDTA–3 mM NaN_3 (pH 7.0) and then separating toxin-AChR complex from free toxin on Bio-Gel P-30. The bound radioactivity and the protein of the isolated complex are determined.

We have found that the specific activity estimated by toxin binding is approximately twice that estimated by [^{3}H]MBTA labeling. For example, a preparation of AChR containing 4.2 nmol of sites susceptible to [^{3}H]MBTA labeling per mg of protein was found to bind 8.4 nmol of ^{3}H-toxin per mg. The 2:1 ratio does not appear to be dependent on the degree of purification. Since toxin blocks all of [^{3}H]MBTA labeling of purified AChR, there must be at least one interacting site of toxin and [^{3}H]MBTA labeling. It has not yet been determined whether or not the additional toxin binding represents a second distinct binding site.

Subunit Structure

AChR from *Torpedo* electric tissue purified by affinity chromatography displays two protein components by polyacrylamide gel electrophoresis under nondissociating conditions in the presence of Triton X-100 (FIGURE 2A).

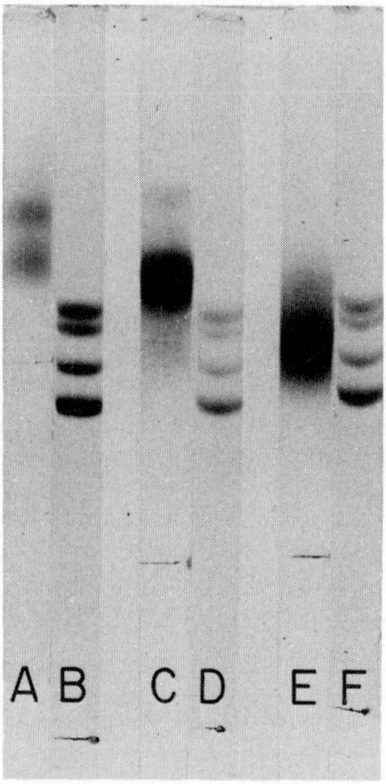

FIGURE 2. Polyacrylamide gel electrophoresis of purified AChR. (A, C, and E) Triton X-100 Gels: To samples of AChR from the affinity column (A) and from a subsequent sucrose density gradient centrifugation, second peak (heavier) C and first peak (lighter) D was added solid sucrose to 10% and bromophenol blue. Samples containing 10–12 μg of proteins were applied to 3.6% gels, and run in 0.2% Triton X-100-100 mM Tris-acetate, pH 8 at 3 ma/gel. Note that gel A was not run as long as gels C and E (ink line marks front). (B, D, and F) Sodium dodecyl sulfate (SDS) gels: Samples of AChR obtained as above were dialyzed against 2% SDS–10 mM Tris acetate–2 mM EDTA–pH 8, precipitated, and washed with dry acetone to remove Triton X-100 and dissolved in an appropriate volume of dialysis buffer containing 20 mM DTT for 0.5–1 hr at 50° C. Samples containing 12–18 μg of protein were applied to 5.6% gels and run in 1% SDS-100 mM Tris-acetate, pH 8 at 8 ma/gel.

Under dissociating conditions, saturated with sodium dodecyl sulfate (SDS) and reduced, this same material gives rise to four polypeptide bands corresponding to apparent molecular weights of 39,000, 48,000, 58,000, and 64,000 daltons (FIGURE 2B).[12] Sucrose density gradient centrifugation results in the separation of two symmetrical peaks of protein and AChR activity, as described above. The protein from each peak yields a single band on Triton X-100 polyacrylamide gels (FIGURE 2C and E). The relative mobilities of the single bands in C and D correspond to the relative mobilities of the bands in A. As expected, the dimer has a smaller electrophoretic mobility. The SDS gel pattern of the protein in each peak from the sucrose density gradient centrifugation (FIGURE 2D and F) is identical to that obtained with AChR before sucrose density gradient centrifugation (FIGURE 2B).

By SDS-gel electrophoresis, it has been shown that the affinity alkylation by [^{3}H]MBTA is directed specifically and exclusively to the subunit of 39,000 daltons.[12] This subunit thus bears all or part of the acetylcholine binding site, as previously demonstrated for the subunit of 40,000 daltons in AChR from *Electrophorus*.[7] The functional significance of the other components has yet to be ascertained.

The specific activity of the most highly purified AChR preparations from

Torpedo (about 4 nmol MBTA sites per mg protein) corresponds to a minimal molecular weight of 250,000 daltons per MBTA binding site. The combined apparent molecular weights of the four putative subunits is 209,000.

Incorporation of AChR into Phospholipid Vesicles

Reconstitution of AChR function in model membrane systems is a principal objective of current receptor research. The mechanism by which the binding of acetylcholine and its congeners to the AChR is transduced into a change in membrane cation permeability is not known, and reconstituted vesicles offer the most attractive system for investigating permeability control at the molecular level.

We incorporate purified AChR into phospholipid vesicles using a dialysis procedure similar to that used by Racker[13] for a variety of membrane systems. The vesicles are formed by dialyzing a solution of AChR and lipids in sodium cholate against detergent-free buffer, under N_2, during a period of 60–72 hr, at either 4° or 25°.

The AChR is purified in Triton X-100 (see above), and this detergent is replaced by sodium cholate before the addition of lipids and dialysis. We do this either by centrifugation of AChR into a 5–20% sucrose gradient containing 0.5% sodium cholate or by repeated dilution with a 0.5% cholate and concentration against 0.5% cholate buffer in a collodion bag ultrafiltration apparatus (Schleicher and Schuell, Inc., Keene, N.H.).

Lipids are isolated from *Torpedo* electric tissue by the procedure of Bligh and Dyer[14] and purified on a silica gel column using chloroform-methanol solutions for elution. The neutral lipids and the phospholipids (mainly phosphatidylethanolamine and phosphatidylcholine) are collected separately. A lipid mixture contaiing 50% egg phosphatidylcholine, 40% *Torpedo* phospholipids, and 10% *Torpedo* neutral lipids is evaporated to dryness under a stream of N_2 and then suspended in buffer containing in different cases 0–3% sodium cholate.

The lipid mixture is added to the AChR solution to give a lipid:protein weight ratio of 10:1 and an AChR concentration of about 1 mg per ml. Routinely, traces of [^{14}C]phosphatidylcholine and of [^{3}H]MBTA-labeled receptor are included in the dialysis mixture to provide markers for the lipids and the AChR. This mixture is then dialyzed. A mixture containing only the lipids was usually dialyzed at the same time. The dialysis buffer is changed every 12 hr and usually contains 200 mM Na^+ (or K^+)–200 mM Cl^- (or 100 mM SO_4^{2-})–10 mM Tris–0.5 mM EDTA–3 mM NaN_3 (pH 7.4); 1–5 mM $CaCl_2$ was included in some preparations.

The binding properties of the receptor are unaffected by the dialysis procedure as judged by specific labeling with [^{3}H]MBTA. We have shown previously[6] that AChR is incorporated into the vesicles by this procedure since the AChR-phospholipid complex can be separated from free AChR and AChR-free phospholipid vesicles by flotation in a sucrose density gradient.

The functional properties of the reconstituted vesicles are tested by measuring the rate of $^{22}Na^+$ influx in the presence and absence of AChR agonists and antagonists. The vesicles are incubated at 25° with $^{22}Na^+$, and at various times, 75 μl aliquots are applied to a 0.8-ml column of Bio-Gel P-6 equilibrated with isosmotic Tris-SO_4 (pH 7.4). A fraction is collected at the void

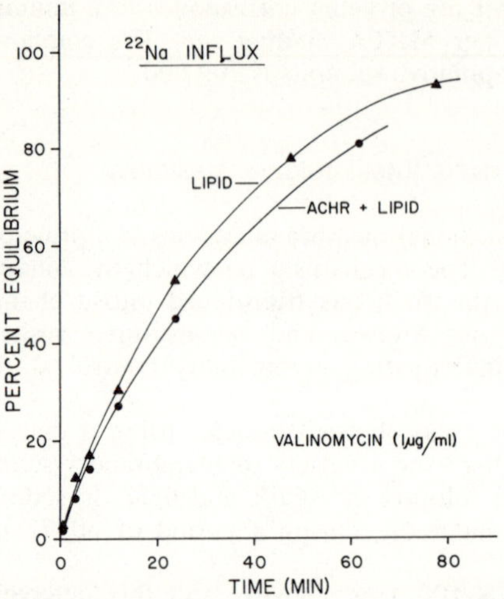

FIGURE 3. Influx of $^{22}Na^+$ into reconstituted vesicles. Vesicles with and without *Torpedo* AChR were prepared as described in the text. To 150 μl of vesicles in 100 mM K_2SO_4–10 mM Na_2SO_4–10 mM Tris SO_4 (pH 7.4) was added 250 μl of the same buffer, 100 μl of $^{22}Na^+$ (50 μCi in 110 mM K_2SO_4) and 5 μl of valinomycin (0.1 mg/ml in ethanol). At time t, 75-μl aliquots were passed through 0.8-ml Bio-Gel P-6 columns equilibrated with isosmotic Tris SO_4 (pH 7.4). The vesicle fraction (0.5 ml) was counted in 5 ml Scintisol.

volume within one minute that contains about 75% of the vesicles and none of the free $^{22}Na^+$. The contained $^{22}Na^+$ counts are divided by the counts due to the [^{14}C]phospholipids, thus giving a normalized value for contained Na^+ per weight of phospholipid.

The vesicles in most cases are relatively impermeable to Na^+, having half-times for equilibration of several hours at 25°. The rate of influx can be increased substantially by adding gramicidin D or valinomycin (1 μg/ml), to the incubation mixture; however, little difference in permeability is seen between vesicles with and without AChR (FIGURE 3).

By using mixtures of Na_2SO_4 and K_2SO_4 in the dialysis and incubation buffers, we have been able to create transmembrane potentials across the vesicles, using valinomycin as a K^+-selective ionophore.[15] From results on intact cells,[16, 17] a potential difference (positive outside) might be expected to enhance receptor activity.

We have not, however, observed any reproducible effect of AChR activators

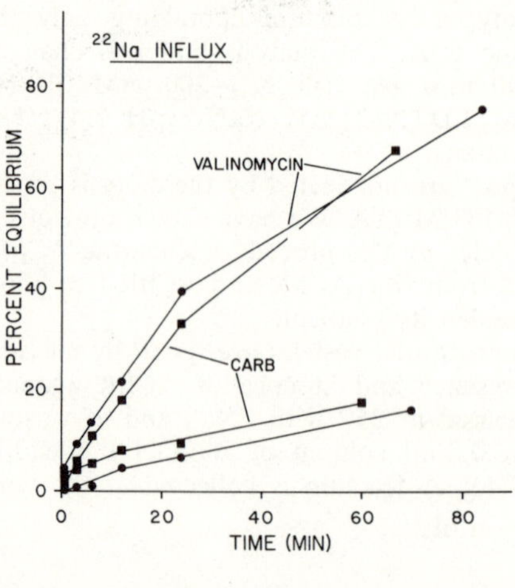

FIGURE 4. Effect of carbamylcoline (CARB) and valinomycin on $^{22}Na^+$ influx into AChR-lipid vesicles. The procedure is as described in FIGURE 3: 5 μl of 10-mM CARB and/or 5 μl of valinomycin (0.1 mg/ml) were added at time zero to samples indicated to give final concentrations of 100 μM and 1 μg/ml, respectively.

or inhibitors on the rate of $^{22}Na^{+}$ influx either in the presence or absence of valinomycin (FIGURE 4).

Hazelbauer and Changeux[5] demonstrated that solubilization of AChR by cholate does not in itself destroy the permeability control properties of the AChR. Recently, Michaelson and Raftery[4] have reported that permeability control could sometimes be reconstituted by procedures similar to those described here, using purified *Torpedo* AChR and lipids.

Our efforts to define the conditions necessary for the reconstitution of the permeability control function of the AChR in phospholipid membranes are continuing. Further characterization of solubilized receptor coupled with biophysical studies of receptor interactions in a membrane environment promise to provide information relevant to the mechanism of permeability control.

REFERENCES

1. KARLIN, A. 1974. Life Sci. **14:** 1385–1415.
2. RANG, H. P. 1974. Quart. Rev. Biophys. **7:** 283–399.
3. KARLIN, A., M. G. MCNAMEE, C. L. WEILL, & R. VALDERRAMA. *In* Methods in Receptor Research. M. Blecher, Ed. Marcel Dekker, Inc. New York, N.Y. In press.
4. MICHAELSON, D. M. & M. A. RAFTERY. 1974. Proc. Nat. Acad. Sci. U.S.A. **71:** 4768–4772.
5. HAZELBAUER, G. L. & J.-P. CHANGEUX. 1974. Proc. Nat. Acad. Sci. U.S.A. **71:** 1479–1483.
6. MCNAMEE, M. G., C. L. WEILL, & A. KARLIN. 1975. *In* Protein-Ligand Interactions. H. Sund & G. Blauer, Eds. Verlag Walter de Gruyter. Berlin, Federal Republic of Germany.
7. KARLIN, A. & D. A. COWBURN. 1973. Proc. Nat. Acad. Sci. U.S.A. **12:** 3636–3640.
8. KARLSSON, E., H. ARNBERG & D. EAKER. 1971. Eur. J. Biochem. **21:** 1–16.
9. BIESECKER, G. 1973. Biochemistry **12:** 4403–4409.
10. RICE, R. H. & G. E. MEANS. 1971. J. Biol. Chem. **246:** 831–832.
11. BOROFF, D. A. & U. FLECK. 1966. J. Bact. **92:** 1580–1581.
12. WEILL, C. L., M. G. MCNAMEE, & A. KARLIN. 1974. Biochem. Biophys. Res. Commun. **61:** 997–1003.
13. RACKER, E. 1972. J. Biol. Chem. **247:** 8198–8200.
14. BLIGH, E. G. & W. J. DYER. 1959. Can. J. Biochem. Physiol. **37:** 911–917.
15. KORNBERG, R. D., M. G. MCNAMEE, & H. M. MCCONNELL. 1972. Proc. Nat. Acad. Sci. U.S.A. **69:** 1508–1513.
16. RUIZ-MANRESA, F. & H. GRUNDFEST. 1971. J. Gen. Physiol. **57:** 71–92.
17. MAGLEBY, K. L. & C. F. STEVENS. 1972. J. Physiol. Lond. **223:** 173–197.
18. LOWRY, O. H., N. J. ROSEBROUGH, A. L. FARR, & R. J. RANDALL. 1951. J. Biol. Chem. **193:** 265–275.
19. ELLMAN, G. L., K. D. COURTNEY, V. ANDRES, & R. M. FEATHERSTONE. 1961. Biochem. Pharm. **7:** 88–95.

DISCUSSION

DR. ELDEFRAWI: I think the experiments that you did with the ACh receptor of *Torpedo* are essentially a repetition of the work published by Michaelson and Raftery. They reported an increase in sodium efflux from the vesicles that were carrying receptor in response to applied carbamylcholine. Was your attempt similar to theirs then? Did your attempt fail, while they claimed at least partial success?

DR. KARLIN: Yes, that is the case. Michaelson and Raftery have reported success, which was at least intermittent; very few details are given so it's hard to evaluate whether they got a specific or a general cation permeability. But if they succeeded, it's very important to know whether this can work, even if it doesn't succeed brilliantly or show a high increase in permeability, because we really want to know whether these receptors are complete and/or functional because none of the binding studies are assaying for the functional properties of the receptor.

DR. TOSTESON: In the experiment in which valinomycin increased the sodium influx, were the K concentrations equal on the two sides?

DR. KARLIN: Yes.

DR. TOSTESON: Are the results the same?

DR. KARLIN: Yes, but the kinetics are somewhat different. Interestingly as the field is dissipated, we see the influx reaching a peak and then coming down as if we are dissipating the gradient. In the absence of a gradient the influx approaches the same final value.

DR. GOLDEN: By your gradient technique, when did you replace the Triton with cholate? Is there also a phospholipid equilibrated along the gradient?

DR. KARLIN: No.

DR. GOLDEN: I mention this because in recent work reported by Metcalfe in the *Proceedings of the National Academy of Sciences* they tried to change the lipid environment before reconstituting calcium ATPase. They found that in order to maintain pumping activity for subsequent reconstitution, phospholipid has to be equilibrated down the gradient. So perhaps if you have phospholipid when you're replacing a detergent you might get some results.

DR. KARLIN: We plan to do that.

MOLECULAR AND FUNCTIONAL PROPERTIES OF THE ACETYLCHOLINE-RECEPTOR

Mohyee E. Eldefrawi and Amira T. Eldefrawi

Section of Neurobiology and Behavior
Cornell University
Ithaca, New York 14853

Adil E. Shamoo

Department of Radiation Biology and Biophysics
University of Rochester School of Medicine and Dentistry
Rochester, New York 14642

Interest in the electric organs of fish dates back to the pioneering studies of the seventeenth century on the electric rays, *Torpedo* sp., of the Mediterranean.[1] This was enhanced by the discovery of bioelectric potentials in the electric eel, *Electrophorus electricus* by Faraday[1] in 1758, in *Torpedo* by Walsh[2] in 1773 and in the electric sheath fish, *Malapterurus elecricus* by Bilharz[1] in 1857. The electric organs of *Torpedo* and *Electrophorus* were found to be embryonically derived from skeletal muscle.[3] The ensuing physiological, pharmacological, and biochemical studies proved that these electric organs were identical to skeletal muscles in having acetylcholine (ACh) as their neurotransmitter, and they established them as excellent sources, probably the richest, for the macromolecules involved in cholinergic transmission.[4]

At cholinergic synapses, ACh produces an increase in the sodium and potassium permeabilities (or conductances) of the postsynaptic membrane[5, 6] as a result of its binding to its receptor in the postsynaptic membrane. Until very recently what we knew about the ACh-receptor was that it must exist, but we could only speculate on its physicochemical character or the nature of the coupling between the receptor and the ion-conductance modulators (carriers or channels). The isolation of the ACh-receptor proteins from the electric organs of *Torpedo* and *Electrophorus* proved the earlier hypothesis of their existence. Recently we, as well as others, have obtained results suggesting that part or all of the ACh-receptor molecule acts as the ion-conductance modulator. In this presentation, we shall discuss our work with the ACh-receptor, mainly of *Torpedo californica,* and compare the data with those of others whenever feasible.

Identification of the Acetylcholine Receptor in Noncellular Fractions

A major problem in the isolation of the ACh-receptor was how to identify the molecules in noncellular fractions. The biochemical approach was based

* The work reported herein was supported in part by Grants AM 17571 and NS 09144 from the National Institutes of Health. It was also based in part on work performed under contract with the United States Energy Research and Development Administration at the University of Rochester Biomedical and Environmental Research Project and has been assigned Report No. UR-3490-710.

on the assumption that *in vitro,* the ACh-receptor retained its affinity for binding of cholinergic activators and inhibitors (ligands). This was confirmed, but other cellular components were also found to bind one or more of the cholinergic ligands used.[7] Therefore, only if the pharmacological profile of the molecule *in vitro* mimicked that *in situ* was the identity confirmed (TABLE 1). Another difficulty facing the researchers was the nature of the binding

TABLE 1

BLOCKADE OF BINDING OF [ACETYL-^{3}H]ACETYLCHOLINE (AT 0.1 μM) TO THE LUBROL-SOLUBILIZED ACH-RECEPTOR FROM *Torpedo* ELECTROPLAX BY DIFFERENT DRUGS*

Choline Esters, Nicotinic Drugs and Neurotoxins (10 μM)	% Blockade ± % Error (Significance†)	Anticholinesterases and Muscarinic Drugs (10 μM)	% Blockade ± % Error (Significance†)
Butyrylcholine	91 ± 37 (.01)	Eserine‡	30 ± 6 (.01)
Carbamylcholine	74 ± 16 (.01)	Neostigmine‡	82 ± 20 (.01)
Benzoylcholine	12 ± 3 (.05)	Pyridostigmine‡	20 ± 2 (.01)
Choline	6 ± 1 (NS)	Edrophonium‡	100 ± 25 (.01)
Succinylcholine	75 ± 19 (.01)	Paraoxon‡	15 ± 2 (.01)
Decamethonium	54 ± 8 (.01)	Muscarine	4 ± 2 (NS)
Hexamethonium	12 ± 1 (.01)	Areocholine	3 ± 1 (NS)
d-Tubocurarine	60 ± 6 (.01)	Pilocarpine	12 ± 1 (.05)
Gallamine	32 ± 5 (.01)	Atropine	10 ± 1 (.05)
Nicotine	41 ± 7 (.01)	Scopolamine	1 ± 0.2 (NS)
Anabasine	19 ± 3 (.01)	Acetyl β-methylcholine	9 ± 2 (.05)
Hemicholinium-3	19 ± 2 (.01)		
α-Bungarotoxin§	99 ± 12 (.01)		
Cobra toxin¶	98 ± 10 (.01)		

* Binding was determined by equilibrium dialysis with [acetyl-^{3}H]acetylcholine alone or in the presence of a drug or neurotoxin.[12]

† Blockade was not significant (NS) when P was >.05.

‡ Drug used at 1 mM.

§ Neurotoxin used at 0.1 μM.

¶ Neurotoxin used at 1 μM.

NOTE: Among the stimulants, strychnine (at 10 μM) blocked 25 ± 3% of ACh binding, but amphetamine, metamphetamine, caffeine, and picrotoxin had no effect. Dibenamine was the only adrenergic drug that blocked ACh binding significantly (14 ± 2). Ineffective adrenergic ligands (at 10 μM) were epinephrine, norepinephrine, dopamine, L-dopa, and ergotamine. No blocking was observed by local anesthetics (procaine, piperocaine, tetracaine, codeine; at 10 μM), hypnotics (sodium diethyl barbital) or hallucinogens (*d*-LSD-25, mescaline hemisulfate). Other ineffective compounds were serotonin, histamine, glutamate, glutamine, and γ-aminobutyrate.

for, unlike the case in an enzyme whose activity was easily detectable *in vivo* or *in vitro* by virtue of its catalytic nature, the only means of identifying the receptor was by its noncatalytic binding of specific ligands. This necessitated the use of large quantities of tissue or the finding of tissues very rich in cholinergic innervation, such as the electric organs of fish. Earlier attempts to isolate the ACh-receptor were unsuccessful mainly because of incorrect identification,[7-9] but they served to emphasize the difficulties involved and led the way to later successes.

Three kinds of ligands are used to identify ACh-receptors in noncellular fractions and to monitor the receptor as purification proceeds: (1) Reversibly binding cholinergic ligands, such as ACh, nicotine, decamethonium, and curare, which are radio-labeled and their binding determined mainly by equilibrium dialysis.[10–14] (2) Affinity label compounds that covalently bind to, among other groups, the reduced disulfide bond close to the active site of the receptor, such as [^{3}H]-4-(*N*-maleimido)-α-5-benzyltrimethylammonium iodide.[15] (3). Relatively "irreversibly" binding radio-labeled α-neurotoxins from snake venoms, such as α-bungarotoxin (α-BGT) from the krait, *Bungarus multicinctus* or the α-neurotoxin from the cobra, *Naja naja*. Their use was introduced by Changeux and coworkers[16] and used by Miledi *et al.*,[17] and they are now the most popular tool in the identification of nicotinic ACh-receptors. The use of the three kinds of ligands gave comparable estimates for the concentration of ACh-receptor binding sites in electric organs, 0.02–0.03 nmol/g tissue in *Electrophorus*[10, 15, 18] and about 1 nmol/g tissue in *Torpedo*.[12, 13, 17, 19]

Isolation and Purification of the Acetylcholine Receptor

The ACh-receptor is a membrane-bound protein that is not solubilized by hypotonic or strong salt solutions or by sonication. It was successfully solubilized by a variety of ionic and nonionic detergents, the most popular of which is Triton X-100.[12, 14, 17] Conventional biochemical techniques failed to purify the ACh-receptor to any significant degree without loss of its activity. Column chromatography gave twofold purification[12] and isoelectrofocusing purified it about sixfold.[20]

A major advance in ACh-receptor research was heralded by the application for receptor purification of the technique of affinity chromatography (or affinity adsorption), which was introduced by Cuatrecasas[21] and successfully used to purify ACh-esterase.[22] In this technique, a ligand that is specific for the receptor is covalently attached to a solid support, such as Sepharose 4B, and is either packed into a column and the protein mixture chromatographed through it, or is added directly to the protein solution then filtered. In either case, the affinity gel selectively adsorbs the specific protein, and the nonadsorbed or nonspecifically adsorbed proteins are easily washed away by salt solutions. The desired protein is then desorbed by a solution of a high concentration of a specific ligand.

Affinity adsorption is a rapid technique and has been successfully applied to the purification of the ACh-receptors of electric organs of *Electrophorus* and *Torpedo*[23–37] (Table 2). In the studies presently reported, 100 g of fresh or frozen electric organ of *Torpedo californica* were homogenized in a solution of 0.2 M NaCl and 0.05 M Na_2HPO_4, pH 7.4 at 4° C. The 16,000 $\times$ g, 1-hr centrifugation pellet was resuspended and then shaken for 30 minutes in the same buffer containing 1% Triton X-100. The 100,000 $\times$ g, 1-hr supernatant was then stirred for 2 hours at 23° C with 20 g of affinity gel of Sepharose 4B. The covalently bound ligand was the α-neurotoxin purified from the venom of *Naja naja siamensis* by ion-exchange chromatography on Sephadex CM 50 with a gradient of ammonium acetate, pH 6.6 (0.05–1.0 M). The toxin concentration used was 0.25 mg/g gel. The affinity gel, with the ACh-receptor adsorbed, we filtered on a glass filter and then transferred to a beaker, and 50 ml of the same buffer containing 0.1% Triton X-100 added.

TABLE 2

SUMMARY OF THE AFFINITY CHROMATOGRAPHY METHODS USED FOR PURIFICATION OF ACH-RECEPTORS FROM THE ELECTRIC ORGANS OF THREE FISH SPECIES AND CERTAIN PROPERTIES OF THESE RECEPTORS*

Immobilized Ligand	Ligand (mM) for Desorption of Receptor	Specific Binding†		ACh-Receptor/ ACh-Esterase	Percentage of Detergent in Purified Receptor Solution	Reference
		α-Toxin	ACh			
Flaxedil analog	Flaxedil-1	2.0–3.3	–	100	1% Triton X-100	24
Flaxedil analog	Flaxedil-2.5	5.5–6.5	2.2–2.6‡	100	1% Triton X-100	33
α-Toxin, *Naja naja*	Hexamethonium-50	6.6	–	2,000	0.5% Tween 80	26
Phenyltrimethylammonium	Carbamylcholine-50	2–4§	–	50	0.2% Triton X-100	31
Phenyltrimethylammonium	Decamethonium-1	4.5	–	140	1.5% Triton X-100	28
α-Toxin, *Naja naja*	Decamethonium-10	4.1	–	–	0.5% Triton X-100	32
Phenyltrimethylammonium	Bis-Q¶-0.003	5.0–6.5	2.5–3.3	2,000–13,000	0.01% Brij	30
α-Toxin, *Naja naja*	Carbamylcholine-500	2.3	–	–	1% Triton X-100	23
α-Toxin, *Naja naja*	Carbamylcholine-1000	–	7.8†–10.4‖	20,000	0.0007% Triton X-100	27
Quaternary ammonium	NaCl-500	6	3**	–	0.1% Triton X-100	25
Quaternary ammonium	NaCl-500	10	5**	–	–††	37
α-Toxin, *Naja naja*	Carbamylcholine-1000	10	10	20,000	0.0007% Triton X-100	35
Phenyltrimethylammonium	Carbamylcholine-50	6	–	2,000	0.2% Triton X-100	79
α-Toxin, *Naja naja*	Carbamylcholine-100	12.2‖	–	20,000	0.1% Triton X-100	80

* The first seven receptors are from *E. electricus*, the next two from *T. marmorata*, the following three from *T. californica*, and the last one from *T. nobiliana*.

† Expressed in nmol/mg protein, which is based on analysis by the method of Lowry *et al.*

‡ Binding in presence of 1% Triton X-100.

§ Binding of 4-(*N*-maleimido)-benzyltrimethylammonium iodide.

¶ 3,3′-bis[α-(trimethylammonium)methyl]-azobenzene bromide.

‖ Based on protein value obtained by amino acid analysis.

** Binding of ACh in the presence of 0.1% Triton.

†† For reconstitution experiments, concentration of Triton was 0.0105 mg/mg protein.

After stirring at 23° for 20 minutes, the gel was filtered and this step repeated so as to remove as much of the lightly adsorbed or trapped material as possible. To remove nonspecifically bound proteins, the gel-receptor was stirred two more times, each with 50 ml of .05 M Na_2HPO_5, 1 M NaCl and 0.1% Triton X-100 and was then filtered. To desorb the ACh-receptor from the affinity gel 50 ml of 1 M carbamylcholine in 0.05 M Na_2HPO_4 solution, pH 7.4, (without any Triton) was added to the gel, and the mixture stirred for 16 hours at 23° C. The mixture was filtered on a glass filter to remove the gel, and the filtrate passed through a Millipore 0.45 μm HAWP filter to remove any bacterial contamination. This carbamylcholine solution, containing the ACh-receptor, was dialyzed extensively against seven changes of 2.5 liters, each of 5 mM Na_2HPO_4 (pH 7.4), until the concentration of carbamylcholine dropped below 0.1 nM. The concentration of protein in the final preparation ranged from 0.05–0.12 mg/ml as determined by Lowry *et al.*[38] Recovery of receptor from the Triton extract was 70 ± 20%. To reduce oxidation of the receptor, all solutions used were deaerated, and the final dialysis was performed under nitrogen at 4° C.

ACh-receptor preparations from various laboratories have different specific binding (TABLE 2). Our *T. californica* receptor binds 10 nmol/mg protein of either ACh, determined by equilibrium dialysis, or α-BGT, determined by ion-exchange gel filtration.[35] In three other purified receptor preparations,[30, 33, 37] the ratio of α-neurotoxin to ACh binding sites was about 2. This contrasts with the equal number of sites per g electroplax found for ACh and α-BGT in the membrane-bound ACh-receptor of either *Torpedo*[12, 13, 17, 19] or *Electrophorus.*[10, 15, 18] The reason for this discrepancy is still unknown. However, it is possible that during some purification procedures, the receptor molecule is exposed to conditions that reduce its ability to bind ACh without affecting its α-neurotoxin binding. We found that if we used 100 mM benzoquinonium for desorption of the ACh-receptor of *T. marmorata* instead of carbamylcholine, the purified receptor bound [^{125}I]-α-BGT, but failed to bind [acetyl-^{3}H]acetylcholine.[27] Also, the presence of 0.4 mM Zn^{++} reduced ACh binding (at 10^{-8} M) of the Triton-solubilized ACh-receptor by 70% after 12 hours at 4° C, but did not affect binding of α-BGT.[39] Furthermore, we recently observed that freezing and thawing of a Triton extract of *T. californica* electric organ caused total loss of ACh-binding, while the ability to bind α-BGT was reduced by only 30%.

Relative purity of the various ACh-receptor preparations is obtained by comparison of their specific binding, although that may be affected by denaturation of the active site. Our purified *T. californica* ACh-receptor behaves as a single homogeneous species in isoelectrofocusing (FIGURE 1), disc gel electrophoresis (FIGURE 2a), and column chromatography.[35] A minor contaminant is ACh-esterase, which is present at the ratio of 1 esterase site for every 20,000 receptor sites. Also present is Triton X-100 at a concentration equivalent to 0.113 mg Triton per mg protein, as will be discussed later.

Binding of Acetylcholine to Its Receptor

Among the various reversible cholinergic ligands, ACh proved to be quite specific for the ACh-receptor of electric organs (TABLE 1). We studied its binding by equilibrium dialysis after inhibiting ACh-esterase activity by a

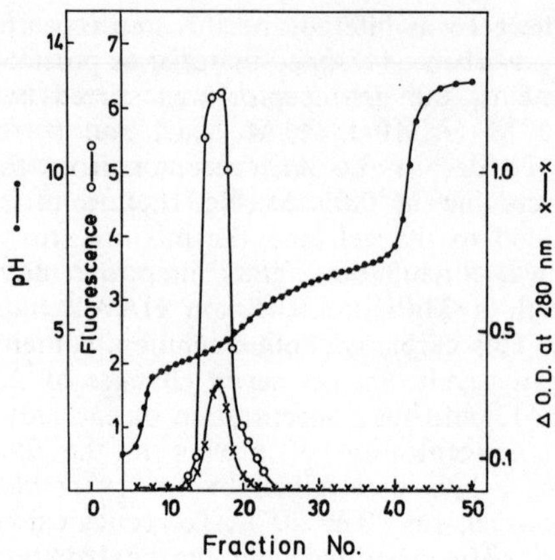

FIGURE 1. Electrofocusing of the purified *T. californica* ACh-receptor in an LKB Produkter column (110 ml capacity) using 1% Ampholine, pH range 3–8.

concentration (up to 0.1 mM) of tetram(*O,O*-diethyl *S*-diethylaminoethyl phosphorothiolate) that did not block receptor binding.[13] The *T. californica* membrane-bound or solubilized (with 1% Triton X-100) ACh-receptors bound [acetyl-^{3}H]acetylcholine with a single high affinity (K_d = 0.02 μM), and half the membrane-bound receptor sites (1.1 nmol/g electroplax) were solubilized with Triton (FIGURE 3a). Identical binding was also observed in the

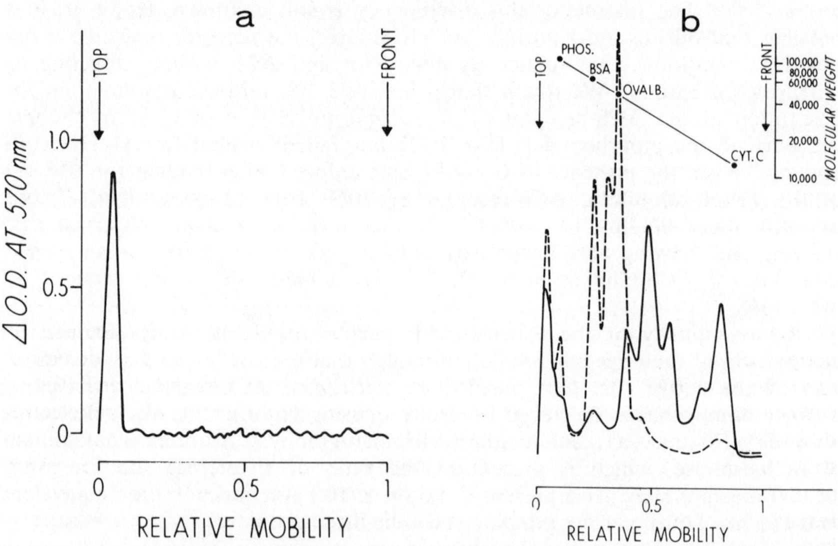

FIGURE 2. Scan of electrophoresis gels of the tryptic digested ACh-receptor purified from *T. californica*. Solid lines are of receptor after one-hour incubation at 23° C with trypsin (50:1, w/w). Broken lines are of untreated receptor (a) Polyacrylamide gel electrophoresis. The untreated receptor gives an identical scan except that the protein peak is larger by about 5%. (b) SDS-gel electrophoresis. Standards are run on a separate gel. Phos, phosphorylase *a* (94,000); BSA, bovine serum albumin (66,000); Oval, ovalbumin (43,000); Cyt. C; cytochrome *c* (11,700).

receptors of the electric organs of fresh *T. marmorata* and *T. ocellata* (M. Eldefrawi; unpublished material). In addition to this high-affinity binding, low-affinity binding sites (K_d in the μM range) were found in the membrane-bound or solubilized ACh-receptors of the electric organs of *T. marmorata* that were lyophilized and stored at —20° C.[12, 13, 20] The appearance of low-affinity binding was attributed by O'Brien and Gibson[40] to oxidation of some of the high-affinity binding sites.

It was most interesting to find upward curvature in the Scatchard plots of ACh-binding to the membrane-bound or solubilized receptors of the electric organs of the three *Torpedo* species at low ACh concentrations (FIGURE 3a),

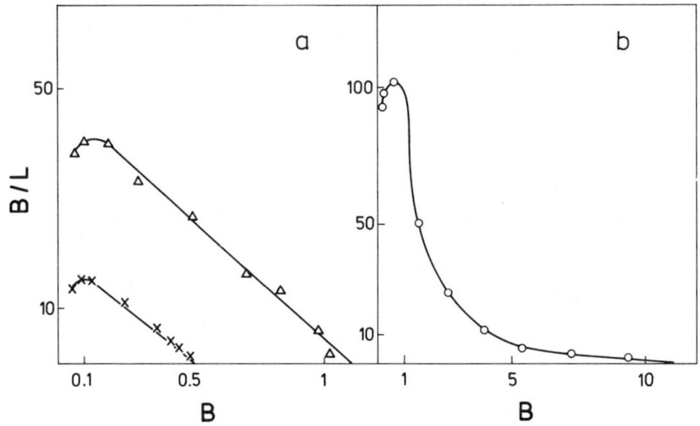

FIGURE 3. Scatchard plot of the binding of [acetyl-^{3}H]acetylcholine to the *T. californica* ACh-receptor, measured by equilibrium dialysis at 4° C for 16 hours. (a) Binding to crude receptor preparations; membrane-bound (△); 1% Triton extract of the membranes (×). (b) Binding to the purified receptor. B, nmol bound per g tissue in (a) and per mg protein in (b). L, concentration of [acetyl-^{3}H]-acetylcholine in μM.

which was also reflected in Hill coefficients of 1.5 to 1.8. These data indicate positive cooperativity in the binding of ACh to its receptors. Similar positive cooperativity was also reported for the binding of ACh to the purified *T. marmorata* receptor[41] and to membrane-bound (but not solubilized) ACh-receptors of *T. marmorata,* measured either by centrifugation[42] or by displacement of a bound fluorescent ligand.[43] Electrophysiological data had previously demonstrated positive cooperativity in the response to activators of neuromuscular[44] or electroplax[45] preparations. Therefore we may speculate that the cooperative response is a result of the receptor-ACh binding step, and that in some solubilized or purified preparations this character is preserved.

In contrast to the membrane-bound ACh-receptors, the preparations purified by others exhibited low-affinity binding for ACh (K_d in the μM range).[33, 37] However, the ACh-receptor we purified by the method described above, exhibited both high- and low-affinity binding of ACh, whether its source was *T. californica* (FIGURE 3b), *T. marmorata,*[27] or *T. ocellata* (M. Eldefrawi; unpublished material). The lower affinities observed in the purified *marmorata* receptor at higher ligand concentrations may be due to a change in the constraints on the molecule in its new environment. However, since only high-affinity binding can now be obtained in membrane-bound receptors from fresh electric organs, this low affinity that appears after purification may be related to that appearing in the aged *T. marmorata* membranes.[13] It is possible that alterations take place in the receptor molecule during purification as a result of the activities of proteases and esterases that may be present in the electroplax or because of oxidation or reduction. Such activities may lead to lowered affinity. Another contributing factor may be the change that occurs in the susceptibility of the receptor to detergents after its purification. We found that addition of 0.1% Triton X-100 to the purified *californica* receptor converted its high-affinity binding to low affinity.[46] Many purified ACh-receptor preparations contained high concentrations of detergent and sometimes Triton was even added to the dialysis media at concentrations of 0.1[36] or 1%[33] (TABLE 2).

Molecular Weight and Subunit Structure

The molecular weight of the ACh-receptor was determined by several techniques, which gave widely varying estimates. A major difficulty was that some of the detergent used for solubilization remained bound to the receptor molecule, adding to its weight and at the same time increasing its buoyancy. Gel filtration of crude ACh-receptors gave an apparent molecular weight value of approximately 500,000[47] and so did the purified receptor.[48] Values obtained for the crude receptor preparation by sucrose gradient centrifugation in H_2O and D_2O, after correction for the contribution of the detergent, were 360,000 and 320,000 in Triton X-100 and deoxycholate, respectively.[49] In another approach, the molecular weight of the purified *Electrophorus* ACh-receptor was obtained by sodium dodecylsulfate (SDS-) gel electrophoresis after crosslinking the receptor with suberimidate[49] or glutaraldehyde,[28] and the values obtained were 230,000 and 260,000, respectively. Since this method could give 20% underestimations,[50] the former value was raised to 275,000.[49]

The molecular weight of our purified ACh-receptor of *T. californica* was determined by sedimentation equilibrium in a Beckman Model E ultracentrifuge equipped with a scanner and on-line computer system.[51, 52] Using [^{3}H]Triton X-100 for solubilization and purification of this receptor preparation, the endogenous Triton X-100 was calculated to be 0.113 mg per mg protein. The sedimentation equilibrium data[46] indicated the presence of two molecular species (FIGURE 4), whose average apparent molecular weight was 510,000. The apparent molecular weights of the two species were calculated to be 330,000 and 660,000. The addition of Triton X-100 to this preparation reduced the average apparent molecular weight down to 330,000 at 0.1% Triton (FIGURE 5).

Estimates of the molecular weight of the receptor protomer (i.e., the subunit carrying one ACh-binding site) are calculated from the specific binding, which, as discussed above, depend on the purity of the prepartion as well as the denaturation that may occur. The smallest calculated unit ranges from 83,000 daltons[23] to 90,000 daltons[26, 48] in molecular weight, which indicates that the ACh-receptor molecule is composed of several protomers. In fact, electron micrographs of negatively stained synaptic membranes of *T. marmorata* showed doughnut-shaped particles with electron-dense cores, 8–9 nm in diameter, and

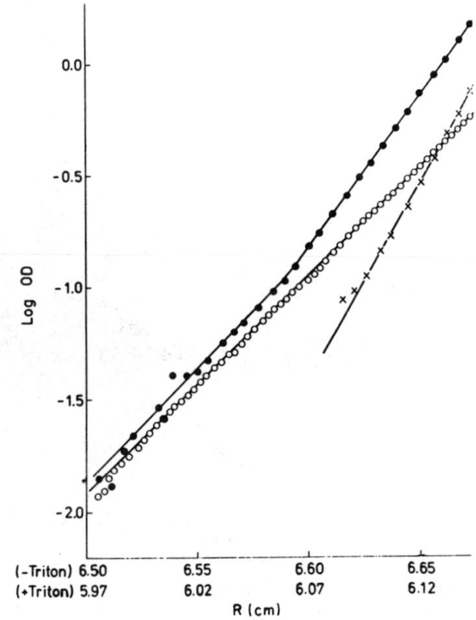

FIGURE 4. Sedimentation equilibrium results on the purified ACh-receptor of *T. californica.* Data are presented as log OD versus R rather than R^2 since R is effectively proportional to R^2 over the limited range of values encountered. Experiments are described for solutions with no added Triton X-100 (●) and with 0.2% Triton X-100 (○). Data for the heavier species in the absence of added Triton X-100 (×) are obtained by subtracting extrapolated values of the concentration of the lighter species from the concentrations at corresponding radial positions beyond the break in the curve.

each apparently was made up of 5–6 subunits.[53, 54] Electron micrographs of the purified ACh-receptor show it to be similar in appearance[54] (FIGURE 6).

SDS-gel electrophoresis has been used by several laboratories to determine the molecular weights of the receptor subunits. In one study[26] a single subunit appeared. However, in several of the other highly purified preparations, there were two subunits, one having a molecular weight of 42,000[32] or 45,000,[49] and the other of 54,000.[32, 49] In another case[37] the weights were 40,000 and 50,000. The difficulty in accurately determining these molecular weights is underscored by the calculation of different values of 43,000 and 48,000 or 45,000 and 54,000 for the two subunits of the same preparation by two different techniques

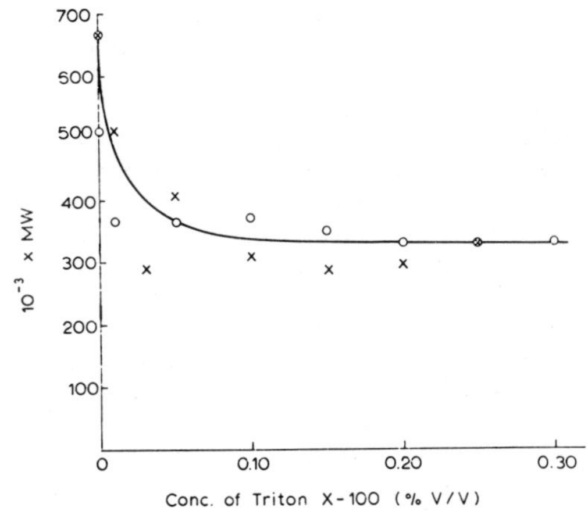

FIGURE 5. Dependence of apparent molecular weight on concentration of Triton X-100 for purified receptors from *T. californica* (○) and *T. marmorata* (×).

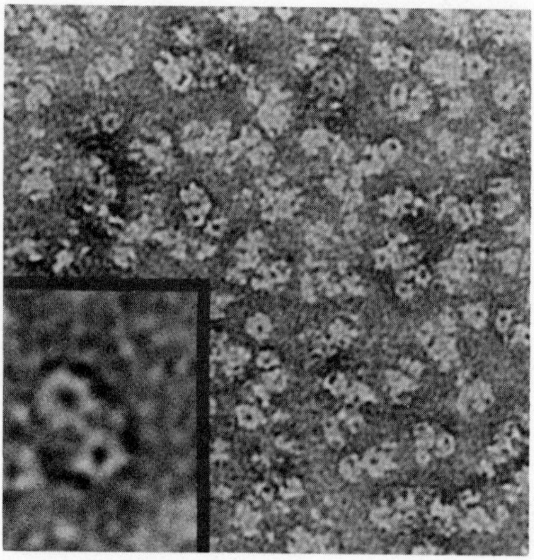

FIGURE 6. Electron micrographs of the purified *T. californica* ACh-receptor negatively stained with 2% phosphotungstate (437,000 ×). Receptor molecule appears as a doughnut with an electron-dense core. Inset (1,093,000 ×) shows the subunit rosette arrangement of the receptor. (Courtesy of Dr. John N. Telford).

of SDS-gel electrophoresis, that of Weber and Osborn and that of Davis and Stark, respectively.[33] Sometimes, in addition, larger subunits appear in the SDS-gel electrophoresis, but these are apparently receptor molecules only partially dissociated by the SDS and mercaptoethanol treatment[33, 55] (FIGURE 2b). Usually of the two major subunits, only the smaller one bound α-neurotoxin[55] or maleimide compound.[33] In view of these data, the receptor molecule of 330,000 may be composed of unidentical subunits, possibly three of each of the two subunits.

Kinetic and binding studies of reversible ligands or α-BGT to the membrane-bound ACh-receptors of electric organs have led to suggestions that sites for binding activators might be different from those binding inhibitors[56, 57] and that half-of-the-sites reactivity occurred.[58] If so, it is not yet known whether these sites would be on the same or different subunits.

Amino Acid Composition

The isolated ACh-receptor of electric organs is a protein and its amino acid composition has been determined (TABLE 3) and found to be remarkably similar to that of ACh-esterase.[27] The amino acid composition of the receptor of *T. californica* was determined and found to be similar to the composition reported for ACh-receptors purified from other *Torpedo* species and from *Electrophorus* (TABLE 3). The basic amino acids make up 10.7–13 mol percent, while the acidic amino acids comprise 18.8–22.5 mol percent, and this is reflected in an acidic isoelectric point of 4.5–4.8[27, 28] (FIGURE 1). The percentage of polar residues of the ACh-receptor range from 44 to 48 (TABLE 3), which are higher than most membrane proteins.

We found that the ACh-receptor of *T. californica,* like *T. marmorata,* contained tryptophan (>2.0 mol percent). When excited at 290 nm, the receptor solution fluoresced with a maximum at 336 nm. Using this native fluorescence of the protein we could detect as little as 1 μg/ml (FIGURE 1), which was undectable by uv absorption or Lowry analysis. Unfortunately, there was no change in fluorescence emission spectra upon binding of carbamylcholine, *d*-tubocurarine, ACh, or decamethonium.[35] This eliminates the utilization of the protein's native fluorescence in kinetic studies of its binding of ligands.

The ACh-receptor protein contains 0.9–2.8 mol percent cysteic acid. In *T. californica,* we found that approximately 18% of the total (equivalent to 20 nmol of free SH groups per mg protein) was present as cysteine residues.[35] The majority of these free SH groups were oxidized during purification of the receptor unless deaerated solutions were used, and the final dialysis was performed under nitrogen.[35]

There is a good possibility that the ACh-receptor is a glycoprotein since hexosamines were found in some of the purified preparations.[33, 37, 59] Based on the data of the interactions of the purified ACh-receptor with plant lectins, it was concluded that the receptor contained a carbohydrate moiety of at least D-mannose and N-acetyl-D-galactosamine.[33] *N*-acetyl-D-glucosamine was detected in another preparation[37] and by means of ion-exchange chromatography mannose, galactose and D-glucose were also found.

TABLE 3

AMINO ACID COMPOSITION (MOL %) OF ACH-RECEPTORS FROM THE ELECTRIC ORGANS OF SEVERAL FISH

Amino Acid	*T. californica* (35)	(37)	*T. marmorata* (27)	(34)	*T. nobiliana* (59)	*E. electricus* (26)	(33)
Lysine	5.4	6.1	6.1	5.0	4.5	4.6	6.3
Histidine	2.4	2.7	2.1	2.5	2.5	2.2	2.5
Arginine	3.9	4.1	3.5	3.3	3.7	4.2	4.2
Aspartic acid	11.6	11.9	11.8	12.4	12.2	11.4	9.8
Threonine	6.4	6.3	6.3	6.2	6.8	5.6	6.0
Serine	7.9	6.6	7.1	8.1	6.4	6.2	8.2
Glutamic acid	10.0	10.2	10.7	8.7	9.7	10.2	9.0
Proline	5.9	5.9	6.2	5.6	7.1	5.7	6.7
Glycine	4.6	4.9	6.4	5.0	5.0	5.9	4.8
Alanine	5.0	5.1	6.0	5.0	4.5	5.8	5.4
Cysteic acid	1.2	0.9	2.0	–	2.8	2.0	1.7
Valine	7.1	7.0	5.5	7.3	6.2	8.6	6.9
Methionine	2.0	1.8	1.7	2.5	1.6	2.0	3.4
Isoleucine	8.2	7.5	5.2	7.4	6.2	6.4	8.1
Leucine	9.5	9.7	9.3	10.1	10.2	10.5	10.7
Tyrosine	3.7	3.8	3.6	3.5	4.2	4.0	3.8
Phenylalanine	4.5	4.6	4.4	4.5	4.2	5.7	5.1
Tryptophan	2.4	–	2.1	–	1.5	0	2.4
Glucosamine	–	0.9	–	–	2.0	–	+*
Percentage of polar residues†	47	48	48	46	46	44	46

* On the basis of the binding of the ACh-receptor to several plant lectins, it was suggested that the receptor carried a carbohydrate moiety containing at least D-mannose and *N*-acetyl-D-galactosamine.

† The sum of Asp, Glu, Lys, Ser, Arg, Thr and His as classified by Capaldi and Vanderkooi.[81]

Immunological Identification

When the ACh-receptors, purified from electric organs of *Electrophorus*[60, 61] and *T. marmorata*,[34] were injected into rabbits, antibodies were produced. Flaccid paralysis always appeared in the immunized rabbits, accompanied by abnormal electromyographs characteristic of neuromuscular blockade, eventually leading to death. The antisera precipitated the ACh-receptor, and it was suggested that the antireceptor antibodies produced reacted with the rabbit's own ACh-receptors.[60]

The intramuscular injection of 25 μg per rabbit or 12.5 μg per rat of the *T. californica* ACh-receptor was effective in stimulating antibody production. A sharp and strong immunoprecipitin line formed between pure receptor and its antisera from rabbit (FIGURE 7) or rat (N. Norcross and M. Eldefrawi; unpublished material).

Three weeks after inoculation, most of the immunized rabbits became paralyzed and died within 48 hours. However, the immunized rats showed only mild transient muscular weakness, even though they developed a high titer of antibodies equal to that found in the fatally paralyzed rabbits. Antisera from immunized rabbits and rats precipitated the crude and pure *T. californica* ACh-receptor, but the precipitates retained 50 and 10% of their ACh-binding capacity, respectively. Evoked potential electromyography of the forearm flexors of rats, immunized with the pure ACh-receptor, showed a decrement in response, which disappeared after injection of neostigmine (FIGURE 8). These data suggest that neuromuscular ACh-receptor of the rat may be impaired by the antibodies produced against the ACh-receptor of *T. californica.*

Calcium Content

When the ACh-receptor protein of *T. californica* was analyzed for the presence of cations by atomic absorption, it was discovered that 4.7% of its weight was due to bound Ca^{++}, which could not be removed even by extensive dialysis.[62] When Ca^{++}-free solutions (containing 1 mM EDTA) were used during

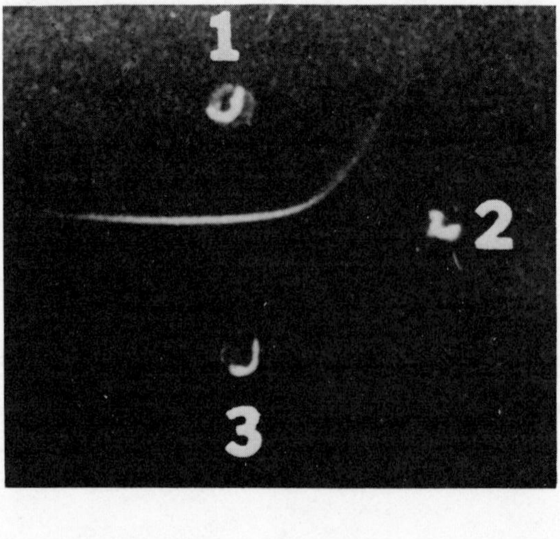

FIGURE 7. Immunodiffusion assay of the ACh-receptor purified from *T. californica* against its antisera from rabbits. *Well 1,* ACh-receptor; *well 2,* rabbit antiserum 7 days postinoculum; *well 3,* rabbit antiserum at 14 days postinoculum. (Courtesy of Dr. Neil L. Norcross.)

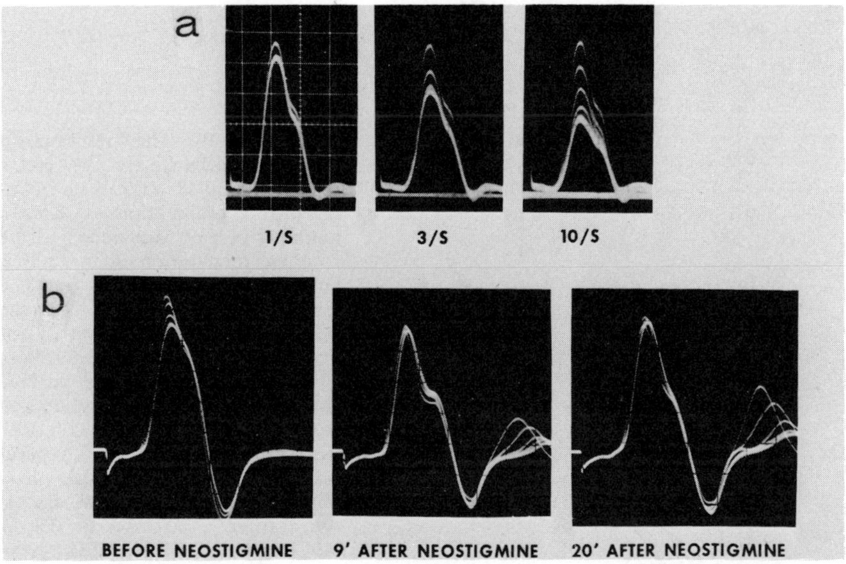

FIGURE 8. Evoked EMG of the forearm flexor of rats immunized with ACh-receptor purified from *T. californica;* superimposed potentials. (a) Various stimulations per second (s). Decrement is present at 1/s stimulation and increases with frequency of stimulation. Rat, 11 days after initial injection with receptor. (b) 10/s stimulation before and after neostigmine injection. The decrement disappears at 9 minutes and does not reappear by 20 minutes. Screen amplification is less after neostigmine. Rat, eight days after initial injection. (Courtesy of Dr. Donald B. Sanders).

purification, 0.6% of the molecular weight of the receptor was still due to bound Ca^{++} (FIGURE 9). This was equivalent to 15 mol of Ca^{++} for each mol of ACh bound at saturation. It is evident that the ACh-receptor has a very high affinity for Ca^{++}, and it is possible that this Ca^{++} is originally bound to the receptor *in situ.*

Cation-Selective Ionophore as Part of the Acetylcholine Receptor

It is generally accepted that binding of ACh to its receptor in the postsynaptic membrane causes specific cation fluxes across that membrane. It was of interest to determine whether or not a cation-selective ionophore† was a part of the ACh-receptor molecule.‡ Therefore, we added the ACh-receptor, purified from *T. californica,* to black lipid membranes (BLM) and studied the effect of its presence and activators on ion conductance.

Conductance increases in BLM have been observed upon addition of substrates to the bathing solution containing their respective enzymes (trypsin,

† Any material that causes an increase of conductance of lipid membranes is considered to be an ionophore. Although the word means "ion carrier," we are using it in the general sense without implication as to whether it is a carrier or a channel.

‡ The ACh-receptor molecule is defined as the noncatalytic protein that binds ACh, α-BGT, and cholinergic activators and inhibitors with high affinities.

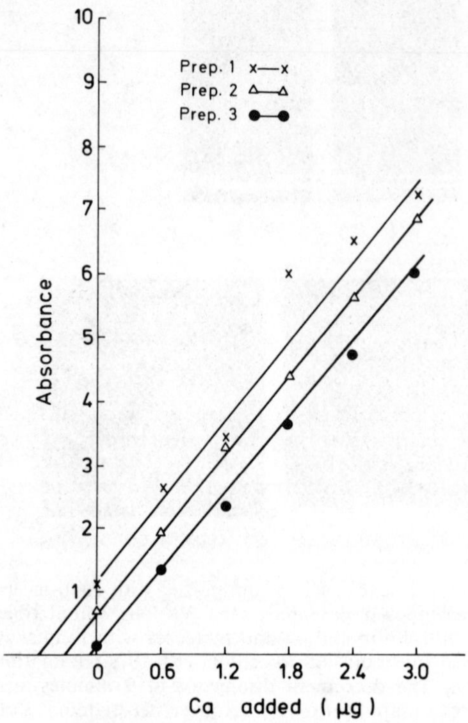

FIGURE 9. The relationship between absorbance and Ca^{++} added to three purified *californica* ACh-receptor preparations. Known amounts of Ca^{++} are added to the receptor preparation, then Ca^{++} is analyzed by atomic absorption. The intercept of each line with the x-axis indicates the amount of endogenous Ca^{++} in each preparation. The three preparations are purified similarly by affinity adsorption, except for differences in the solutions used during purification. Preparation 1, original-Krebs Ringer phosphate solution (0.67 mM Ca^{++}); preparation 2, solution of 0.2 M NaCl and 0.05 M Na_2HPO_4; preparation 3, 0.2 M NaCl, 0.05 M Na_2HPO_4 and 1 mM EDTA.

chymotrypsin, lactic dehydrogenase, and ACh-esterase[63, 64]). Such changes contrast with the selective ionophoric activities recently observed, such as Na^+-dependent and Ca^{++}-dependent and -selective ionophores from $Na^+ + K^+$-ATPase and $Ca^{++} + Mg^{++}$-ATPase, respectively.[65–68] Increases in BLM conductances were reported by several researchers using impure receptor protein.[69–72] In some of these studies transient increases in conductance were observed either in the presence of the protein alone[72] or in the presence of cholinergic ligands as well.[73] Very recently, ACh-receptors purified from *T. californica* were incorporated into vesicles made from *Torpedo* phospholipids, and the presence of activators, ACh or carbamylcholine, accelerated $[^{22}Na]$ efflux of some ACh-receptor preparations.[36, 37] However, the ion selectivity of this flux was not determined.

In this study, we used oxidized cholesterol prepared by the method of Tien *et al.*[74] to form a bilayer of 1-mm surface area separating two compartments. The membrane was formed after the addition into each compartment of 5 ml of the desired electrolyte solution, 5 mM histidine (pH 7.4), and the ACh-receptor purified from *T. californica.* All measurements were made at 23° C. The circuitry was as previously described[66] with a digital Keithly Electrometer Model No. 616 as ammeter/voltmeter. The electrometer was connected directly either in series (ammeter) or in parallel (voltmeter) with the bimolecular lipid membrane. The electrometer output was connected simultaneously to a two-channel recorder and an X-Y recorder. The voltage source was a function generator IEC F52A. Conductance of the bilayer was continu-

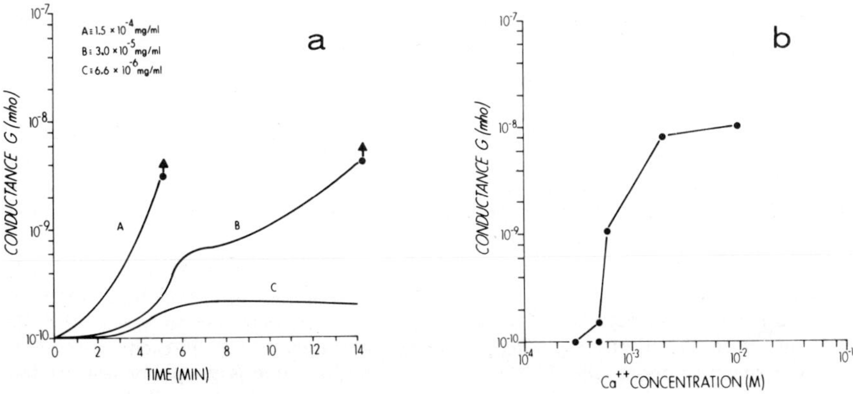

FIGURE 10. Conductance of BLM in presence of Ca^{++}. Data are representative of six experiments. (a) Time-response relationship when $CaCl_2$ concentration is 10 mM and the ACh-receptor purified from *T. californica* is studied at three concentrations, A, B, and C. The arrow indicates breaking of the membrane. (b) Relationship between conductance and $CaCl_2$ concentration. Receptor concentration is 0.14 μg/ml.

ously monitored under the condition of triangular waveform of ±75 mV maximal voltage clamp and at a frequency of 0.015 Hz.

We observed increases in BLM conductance in the presence of Ca^{++} (FIGURE 10a), whose magnitude and rate were increased as the concentrations of receptor or Ca^{++} were raised (FIGURE 10a and b). The same results were obtained (FIGURE 11a) when the receptor was first incubated at 23° C for one hour with trypsin (50:1, w/w), a soybean trypsin inhibitor was added, and the

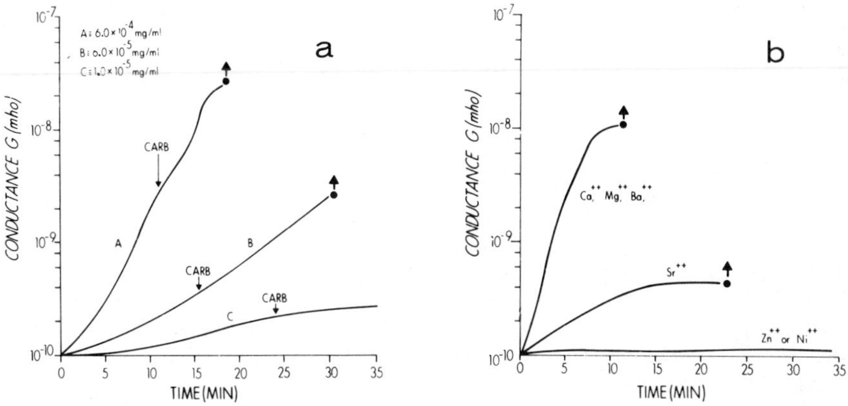

FIGURE 11. The time response of BLM conductance in the presence of the TD-CMC-treated *californica* receptor and the chloride salt of a divalent cation. (a) The salt is $CaCl_2$ (3 mM) and the receptor is used at three concentrations A, B, and C. The arrow next to carbamylcholine indicates its addition at 10 μM final concentration. Data are representative of six experiments. (b) Various divalent cations are used at 3 mM and the TD-CMC-treated receptor at 0.3 μg/ml. Data are representative of four experiments.

positively charged peptides formed were removed by mixing with Sephadex CM 50 and then filtered. This treatment removed about 5–7% of the total protein, but did not affect the binding of ACh by its receptor. The increase in Ca^{++} conductance in the presence of the untreated ACh-receptor or the TD-CMC-treated one was not affected in a reproducible or consistent manner by the addition of ACh or carbamylcholine at a final concentration of 0.1 to 100 μM. When chloride salts of various divalent cations were studied, the TD-CMC-treated ACh-receptor showed a preference for Ca^{++}, Mg^{++} and Ba^{++} over Sr^{++}, Zn^{++} and Ni^{++} (FIGURE 11b).

In measurements of diffusion potential in the presence of either the untreated or TD-CMC-treated ACh-receptor, there was no selectivity between Ca^{++} and Cl^-. This suggests that the receptor, in the presence of Ca^{++}, interacts strongly with the lipid bilayer to produce large holes, not within, but surrounding the receptor molecule. This increase in conductance is probably not related to the cholinergic-activator-induced cation-selective conductance. It is consistent with the fact that the conductance increase does not require the presence of a cholinergic activator (FIGURE 11a). Thus, the conductance increase may indicate a chemical event occurring due to interaction of Ca^{++}-receptor with lipids. What has been observed in conductance with Ca^{++} and the receptor is similar to the recent report that a mitochondrial glycoprotein, which has a very high affinity for Ca^{++}, causes a dramatic nonselective increase in conductance.[75] It is possible that the glycoprotein and the ACh-receptor do not easily become associated with the bilayer unless Ca^{++} is present. On the other hand, Ca^{++} may play a role in a cellular membrane as a modulator of selectivity among monovalent cations, as was found in the transport of alkali metals by the isolated midgut of the American silkworm.[76] The increased conductance of BLM in the presence of Ca^{++} and receptor may be related to the high affinity the receptor has for Ca^{++} (FIGURE 9). These data with Ca^{++} suggest a physiological role for it, as in a proposed model that emphasizes a regulatory role for membrane-bound Ca^{++} and a close relationship between it and the ACh-receptor.[77]

When Na^+ was the cation present in the bathing fluid, the bilayer conductance was also sensitive to the presence of the untreated ACh-receptor (FIGURE 12a). However, the concentration of receptor required to cause an equal increase in BLM conductance to NaCl was 20- to 100-fold greater than that to $CaCl_2$. Addition of ACh or carbamylcholine (10 μM) sometimes increased Na^+ conductance, as in C (FIGURE 12a), but the effect was neither reproducible nor specific. In the case of the TD-CMC-teated ACh-receptor, the concentration required to induce an increase in conductance without carbamylcholine varied within tenfold from batch to batch of ACh-receptor, but was somewhat constant for the same batch. For each batch was first determined the TD-CMC-treated ACh-receptor dosage, which caused no increase in conductance itself. Once that was established in five experiments, we then proceeded to add carbamylcholine; we observed increased conductance (FIGURE 12b). The amount of conductance change due to carbamylcholine varied from a tenfold increase to a more than 100-fold increase. If curare was present (at 10 μM) with salt and receptor, and then carbamylcholine was added after the formation of bilayer, no increase in conductance was observed. When the ACh-receptor was left at room temperature for two days, causing it to lose capacity to bind ACh, no ionophoric properties could be detected, whether with Ca^{++} or Na^+ and with or without carbamylcholine.

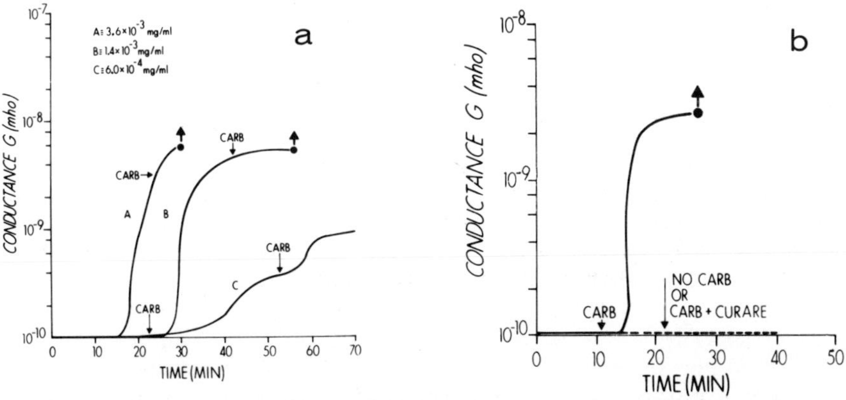

FIGURE 12. The time-response of BLM conductance in the presence of 0.1 M NaCl and ACh-receptor purified from *T. californica* (untreated or TD-CMC-treated). The arrow next to carbamylcholine indicates its addition at 10 μM final concentration. Data are representative of five experiments. (a) Untreated ACh-receptor is used at the three concentrations indicated, A, B and C. (b) TD-CMC-treated ACh-receptor is used at 0.8 μg/ml, and curare at a final concentration of 10 μM.

When the experiments were repeated with 0.1 M KCl instead of NaCl, data similar to those in FIGURE 12b were obtained. There was no selectivity difference between Na^+ and K^+. However, in experiments measuring diffusion potential it was found that P_{Na}/P_{Cl} was 4.4, where P stands for permeability. The cation selectivity is consistent with the *in vivo* responses of the receptor to activators.[6]

Since only the TD-CMC-receptor responds to ACh or carbamylcholine in a consistent manner, it was of interest to determine what tryptic digestion did to the receptor. It is known that trypsin preferentially catalyzes the hydrolysis of peptide bonds between the carboxy group of arginine or lysine and the amino group of another amino acid. It also acts as an esterase and amidase. Exposure of the purified ACh-receptor to a low concentration of trypsin for one hour definitely results in hydrolysis of the receptor into smaller subunits (FIGURE 2b). Nevertheless, the molecule retains its capacity to bind ACh and apparently remains intact (FIGURE 2a), with its subunits attached possibly through disulfide, hydrogen, and hydrophobic bonds. What may also be helpful is its highly folded nature, which is suggested by its resistance to the treatment with guanidine hydrochloride[78] and by studies of its far ultraviolet circular dichroism spectrum.[59] It is possible that the new side chains, exposed by trypsin digestion, permit the incorporation of the receptor in BLM in a functional conformation that allows interaction between the active site and the ionophoric activities.

The data demonstrate the presence of monovalent cation-selective ionophores as part of the ACh-receptor molecule. Further studies are needed to elucidate the role of Ca^{++}, to reduce the variabilities observed, to test different synthetic lipids, and to study the dose-response relationship for various agonists. It is also important to determine whether the whole receptor molecule acts as the ionic channel or whether a small part is the ionophore, and if the latter is the case, to isolate it.

Acknowledgments

We thank Roger C. Carrol and Susan Carson for their performance of the electrophoresis experiments, and Mr. Steven Levy and Mrs. Y. P. Essig for their expert technical help. we are grateful to Dr. Neil L. Norcross (Veterinary College, Cornell University), Dr. John N. Telford (Section of Biochemistry, Molecular and Cell Biology, Cornell University) and Dr. Donald B. Sanders, (University of Virginia School of Medicine, Charlottesville, Virginia) for providing FIGURES 7, 6 and 8, respectively, which are based on studies performed with the ACh-receptor we purified from *T. californica*.

References

1. DuBois-Raymond, E. 1887. Translations of Foreign Biological Memoirs. I. On the Physiology of Nerve, of Muscle and of the Electric Organ. J. Burdon-Sanderson, Ed.: 367–413. Oxford University Press. London, England.
2. Nachmansohn, D. 1955. Harvey Lectures **49:** 57–99.
3. Keynes, R. D. & H. Martins-Ferreira. 1953. J. Physiol. **119:** 315–351.
4. Bennett, M. V. L. 1970. Ann. Rev. Physiol. **32:** 471–527.
5. Takeuchi, A. & N. Takeuchi. 1960. J. Physiol. **154:** 52–67.
6. Ruiz-Manresa, F. & H. Grundfest, 1971. J. Gen. Physiol. **57:** 71–92.
7. O'Brien, R. D., M. E. Eldefrawi & A. T. Eldefrawi. 1972. Annu. Rev. Pharmacol. **12:** 19–34.
8. Chagas, C., E. Penna-Franca, K. Nishie, & E. J. Garcia. 1958. Arch. Biochem. Biophys. **75:** 251–259.
9. Ehrenpreis, S. 1961. Bioelectrogenesis. C. Chagas and A. P. DeCarvalho, Eds.: 379–396. Elsevier. Amsterdam, The Netherlands.
10. Eldefrawi, M. E., A. T. Eldefrawi & R. D. O'Brien. 1971. Proc. Nat. Acad. Sci. U.S.A. **68:** 1047–1050.
11. Eldefrawi, M. E., A. T. Eldefrawi & R. D. O'Brien. 1971. Mol. Pharmacol. **7:** 104–110.
12. Eldefrawi, M. E., A. T. Eldefrawi & R. D. O'Brien. 1972. Arch. Biochem. Biophys. **150:** 210–218.
13. Eldefrawi, M. E., A. G. Britten & A. T. Eldefrawi. 1971. Science **173:** 338–340.
14. Changeux, J.-P., J.-C. Meunier & M. Huchet. 1971. Mol. Pharmacol. **7:** 538–553.
15. Karlin, A., J. Prives, W. Deal & M. Winnik. 1970. *In* Ciba Foundation Symposium on Molecular Properties of Drug Receptors. R. Porter and M. O'Connor, Eds.: 247–261. Churchill. London, England.
16. Changeux, J.-P., M. Kasai & C. Y. Lee. 1970. Proc. Nat. Acad. Sci. U.S.A. **67:** 1241–1247.
17. Miledi, R., P. Molinoff & L. T. Potter. 1971. Nature **229:** 554–557.
18. Suszkiw, J. B. 1973. Biochim. Biophys. Acta **318:** 69–77.
19. Raftery, M. A., J. Schmidt & D. G. Clark. 1972. Arch. Biochem. Biophys. **152:** 882–886.
20. Eldefrawi, M. E. & A. T. Eldefrawi. 1971. Proc. Nat. Acad. Sci. U.S.A. **69:** 1776–1780.
21. Cuatrecasas, P. 1970. J. Biol. Chem. **245:** 3059–3065.
22. Berman, J. B. & M. Young. 1971. Proc. Nat. Acad. Sci. U.S.A. **68:** 395–398.
23. Karlsson, E., E. Heilbronn & L. Widlund. 1972. FEBS Lett. **28:** 107–111.
24. Olsen, R. W., J.-C. Meunier & J.-P. Changeux. 1972. FEBS Lett. **28:** 96–100.
25. Schmidt, J. & M. A. Raftery. 1972. Biochem. Biophys. Res. Commun. **49:** 572–578.

26. Klett, R. P., B. W. Fulpius, D. Cooper, M. Smith, E. Reich & L. D. Possani. 1973. J. Biol. Chem. **248:** 6841–6853.
27. Eldefrawi, M. E. & A. T. Eldefrawi. 1973. Arch. Biochem. Biophys. **159:** 362–373.
28. Biesecker, G. 1973. Biochemistry **12:** 4403–4409.
29. Raftery, M. A. 1973. Arch. Biochem. Biophys. **154:** 270–276.
30. Chang, H. W. 1974. Proc. Nat. Acad. Sci. U.S.A. **71:** 2113–2117.
31. Karlin, A. & D. Cowburn. 1973. Proc. Nat. Acad. Sci. U.S.A. **70:** 3636–3640.
32. Lindstrom, J. & J. Patrick. 1974. *In* Synaptic Transmission and Neuronal Interaction. M. V. L. Bennett, Ed.: 191–216. Raven Press. New York, N.Y.
33. Meunier, J.-C., R. Sealock, R. Olsen & J.-P. Changeux. 1974. Eur. J. Biochem. **45:** 371–394.
34. Heilbronn, E. & C. Mattson. 1974. J. Neurochem. **22:** 315–317.
35. Eldefrawi, M. E., A. T. Eldefrawi & D. B. Wilson. Biochemistry. In press.
36. Michaelson, D. M. & M. A. Raftery. 1974. Proc. Nat. Acad. Sci. U.S.A. **71:** 4768–4772.
37. Michaelson, D., R. Vandlen, J. Bode, T. Moody, J. Schmidt & M. A. Raferty 1974. Arch. Biochem. Biophys. **165:** 796–804.
38. Lowry, O. H., N. J. Rosebrough, A. L. Farr & R. J. Randall. 1951. J. Biol. Chem. **193:** 265–275.
39. O'Brien, R. D., W. R. Thompson & R. E. Gibson. 1974. *In* Neurochemistry of Cholinergic Receptors. E. DeRobertis & J. Schact, Eds.: 49–62. Raven Press. New York, N.Y.
40. O'Brien, R. D. & R. E. Gibson. Arch. Biochem. Biophys. In press.
41. Eldefrawi, M. E. & A. T. Eldefrawi. 1973. Biochem. Pharmacol. **22:** 3145–3150.
42. Weber, M. & J.-P. Changeux. 1971. Mol. Pharmacol. **10:** 15–34.
43. Cohen, J. B. & J.-P. Changeux. 1973. Biochemistry **12:** 4855–4864.
44. Ariens, E. J., A. M. Simonis & J. M. vanRossum. 1964. *In* Molecular Pharmacology. E. J. Ariens, E.: 119–286. Academic Press. New York, N.Y.
45. Changeux, J.-P. & T. Podleski. 1968. Proc. Nat. Acad. Sci. U.S.A. **59:** 944–950.
46. Edelstein, S. J., W. B. Beyer, A. T. Eldefrawi & M. E. Eldefrawi. J. Biol. Chem. In press.
47. Meunier, J.-C., R. W. Olsen & J.-P. Changeux. 1972. FEBS Lett. **24:** 63–68.
48. Raftery, M. A., J. Schmidt, M. Martinez-Carrion, T. Moody, R. Vandlen & J. Duguid. 1973. J. Supramolec. Struct. **1:** 360–367.
49. Hucho, F. & J.-P. Changeux. 1973. FEBS Lett. **38:** 11–15.
50. Spatz, L. & P. Strittmatter. 1973. J. Biol. Chem. **248:** 793–799.
51. Crepeau, R. H., S. J. Edelstein & M. J. Rehmer. 1972. Anal. Biochem. **50:** 213–233.
52. Crepeau, R. H., C. P. Hensley & S. J. Edelstein. 1974. Biochemistry **13:** 4860–4865.
53. Nickel, E. & L. T. Potter. 1973. Brain Res. **57:** 508–517.
54. Cartaud, J., E. L. Benedetti, J. B. Cohen, J.-C. Meunier & J.-P. Changeux. 1973. FEBS Lett. **33:** 109–113.
55. Gordon, A., G. Bandini & F. Hucho. 1974. FEBS Lett. **47:** 204–208.
56. Bulger, J. E. & G. P. Hess. 1973. Biochem. Biophys. Res. Commun. **54:** 677–684.
57. Eldefrawi, M. E. 1974. *In* The Peripheral Nervous System. J. I. Hubbard, Ed.: 181–200. Plenum Press. New York, N.Y.
58. Fu, J. L., D. B. Donner & G. P. Hess. 1974. Biochem. Biophys. Res. Commun. **60:** 1072–1080.
59. Moore, W. M., L. A. Holladay, D. Puett & R. N. Brady. 1974. FEBS Lett. **45:** 145–149.
60. Patrick, J. & J. Lindstrom. 1973. Science **180:** 871–872.
61. Sugiyama, H., P. Benda, J.-C. Meunier & J.-P. Changeux. 1973. FEBS Lett. **35:** 124–128.

62. ELDEFRAWI, M. E., A. T. ELDEFRAWI, L. A. PENFIELD, R. D. O'BRIEN & D. VAN CAMPEN. 1975. Life Sci. **35:** 925–936.
63. DEL CASTILLO, J., A. RODRIGUEZ & C. A. ROMERO. 1967. Ann. N.Y. Acad. Sci. **144:** 803–815.
64. LEUZINGER, W. & M. SCHNEIDER. 1972. Experientia **28:** 256–257.
65. SHAMOO, A. E. & R. W. ALBERS. 1973. Proc. Nat. Acad. Sci. U.S.A. **70:** 1191–1194.
66. SHAMOO, A. E. & M. M. MYERS. 1974. J. Membrane Biol. **19:** 163–178.
67. SHAMOO, A. E. 1974. Ann. N.Y. Acad. Sci. **242:** 389–405.
68. SHAMOO, A. E. & D. H. MACLENNAN. 1974. Proc. Nat. Acad. Sci. U.S.A. **71:** 3522–3526.
69. KEMP, G., J. P. DOLLY, E. A. BARNARD & C. E. WENNER. 1973. Biochem. Biophys. Res. Commun. **54:** 607–613.
70. JAIN, M. K. 1974. Arch. Biochem. Biophys. **164:** 20–29.
71. GOODALL, M. C., R. J. BRADLEY, G. SACCOMANI, & W. O. ROMINE. 1974. Nature **250:** 68–70.
72. ROMINE, W. O., M. C. GOODALL, J. PETERSON & R. J. BRADLEY. 1974. Biochem. Biophys. Acta **316:** 316–325.
73. PARISI, M., E. RIVAS & E. DE ROBERTIS. 1971. Science **172:** 56–57.
74. TIEN, H. T., S. CARBONE & W. A. DAVIDOWICZ. 1966. Nature **212:** 719–720.
75. PRESTIPINO, G., D. CECCARELLI, F. CONTI & E. CARAFOLI. 1974. FEBS Lett. **45:** 99–103.
76. ZERAHN, K. 1971. Phil. Trans. Roy. Soc. Lond. Ser. B. **262:** 315–321.
77. CHANG, K. J. & D. J. TRIGGLE. 1973. J. Theoret. Biol. **40:** 125–154.
78. CARROLL, R. C., M. E. ELDEFRAWI & S. J. EDELSTEIN. 1973. Biochem. Biophys. Res. Commun. **55:** 864–872.
79. WEILL, C. L., M. G. MCNAMEE & A. KARLIN. 1974. Biochem. Biophys. Res. Commun. **61:** 997–1003.
80. ONG, D. E. & R. N. BRADY. 1974. Biochemistry **13:** 2822–2827.
81. CAPALDI, R. A. & G. VANDERKOOI. 1972. Proc. Nat. Acad. Sci. U.S.A. **69:** 930–932.

PART V. MOLECULAR MECHANISM OF ION-IONOPHORE COMPLEXATION

CONFORMATION AND MOLECULAR MECHANISMS OF CARRIERS AND CHANNELS*

D. W. Urry, M. M. Long, M. Jacobs, and R. D. Harris

Laboratory of Molecular Biophysics; and the Cardiovascular Research and Training Center University of Alabama Medical Center Birmingham, Alabama 35294

INTRODUCTION

A schematic representation of the action of carriers and channels on the lipid layer of lipid bilayers is given in FIGURE 1. The carrier is viewed as

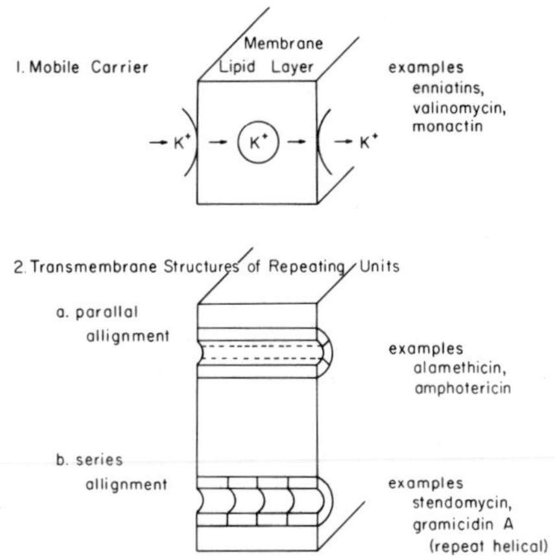

FIGURE 1. Molecular mechanisms of cation permeation of membranes. (From Urry.[1] By permission of *International Journal of Quantum Chemistry*.)

capable of residing at the aqueous-lipid interface where it chelates an ion, enters the lipid milieu, and then reverses the process when the ion-charged carrier next collides with an interface. Cation selectivity of carriers derives in part from the conformational energy required to form a carbonyl coordination sphere of the required dimensions. The conformations of two carriers, enniatin B and valinomycin, extrapolate readily to structures that are capable of spanning the lipid layer to form channels. Also as depicted in FIGURE 1, transmembrane structures capable of forming channels are commonly constructed of repeating units. These repeats may be arranged in series or parallel (barrel stave) fashion. Depending on the structure of the repeating units

* This work was supported by Grant No. HL-11310 from the National Institutes of Health.

both series- and parallel-generated structures could be expected to exhibit a dependence on the electric field across the lipid bilayer which could adhere to the Hodgkin-Huxley kinetics.[1, 2]

The usual approach in determining conformations and molecular mechanisms of peptide carriers and channels is to extrapolate from physical studies on solutions and to infer structure at the lipid bilayer with reliance on descriptions of black lipid membrane studies. The conformation of valinomycin in dimethyl sulfoxide with a lipophilic side and a polar side and the conformaton and stability of the valinomycin-potassium ion complex with a polar core and a lipophilic exterior favor extrapolation to a carrier mechanism. Since channels need not form stable complexes, inferences from solution data are less direct. The solution conformation of gramicidin A is dependent on solvent, concentration, temperature, and pressure. At higher concentration and temperature ranges, hydrogenated gramicidin A exhibits a circular dichroism pattern that is characteristic of right-handed helices whereas at lower concentration and temperature, the pattern is characteristic of left-handed helices. Effects of pressure depend on solvent and indicate that differences between conformers at high and low concentrations are more than helix handedness. In ethanol increased pressure favors the conformer observed at higher concentration and temperature. By combining these four variables the means is available to approximate the dimensions of the channels in solution. These data are then compared to the black lipid membrane data.

Carrier Mechanisms

There are several properties that a polypeptide should have in order to be an effective carrier; it should be capable of residing at an interface and at this site be capable of rapid binding and release of cations; its ion complex should be small enough to exhibit sufficient mobility; its complex should be lipophilic, and commonly a carrier is desired that exhibits a substantial degree of ion selectivity.

In order to effectively bind cations a lipophilic polypeptide must, in general, be capable of generating a carbonyl coordination sphere. The dimensions of this sphere and the energy of the conformation required to generate it are critical factors in achieving ion selectivity and affinity. This was implied in the name given by this laboratory to the conformation of the valinomycin-potassium ion complex, i.e., the "core" conformation which emphasized the polar core formed by the carbonyls.[3] Subsequently conformational energy calculations demonstrated a striking correlation between the energy to form a polar core of a given size and the size of the preferentially bound ions.[4] The results of the conformational energy calculations were in satisfying agreement with the later kinetic studies of Grell and Funck (which demonstrated that little ion-selective interaction occurs as the first several carbonyls interacted with the cation in methanol solution, but dramatic selectivity arose once conformational closure had occurred[5]) once the polar core (the carbonyl coordination sphere) was formed.

A second lipophilic conformation of valinomycin has previously been noted, which was referred to as the "pore" conformation[3]; it has the same secondary structure as the "core" conformation, that of a series of β-turns, but the direction of the chain is reversed and the carbonyls point outward rather than inward (see FIGURE 9). The energy of the "pore" conformation in the absence of cation is close to that of the "core" conformation.[4] This conformation, which has a 4.5 Å pore, should continue to be kept in mind as a possible structure. A serial stacking of the "pore" conformations could conceivably form a transmembrane structure capable of passing ions. Also, if the depth of the pore structure were extended, utilizing the cross-β-conformation, additional transmembrane structures could be considered (see FIGURE 10).

The enniatin B molecule with its alternating L-D sequence,[6] (-L-Methyl-Val-D-HyV-)$_3$, could be considered a cyclic analog of the linear gramicidin A peptide with its alternating L-D sequence.[7] When complexed with cation, the carbonyls of enniatin B alternately point above and below the plane of the complex with component directions alternately parallel and antiparallel to the symmetry axis.[8] The linear analog of this structure is the single-stranded β^6-helix proposed for gramicidin A (Reference 9), which has 6.3 residues per turn and which alternately has its carbonyls pointing parallel and antiparallel to the helix axis. It has been proposed that the carbonyls of a single-stranded β^6-helix rotate toward the helix axis as the cation passes through.[10] Again there is the analogy to the enniatin B-K^+ complex. Thus it is perhaps not unwise to consider the structure and conformation of cyclic carriers as having more information than that involved in the carrier mechanism alone, for the cyclic carriers may provide insights into possible channel correlates. This relationship has been called the concept of cyclic conformations with linear conformational coorrelates.[11]

CHANNEL MECHANISMS

The poblem of deducing the conformation of a channel, as it functions within a lipid bilayer, from studies in solution, is more difficult than deducing the conformation of a carrier complex from solution data because the channel does not form as stable a complex with the permeant ion, because the channel conformation itself may be destabilized in a solvent that cannot enter the channel, and because both polar and lipophilic portions of the channel must be satisfied by the solvent. It is possible, however, to deduce channel structures in solution and to correlate these findings with the data obtained from black lipid membrane studies.

The Gramicidin A Channel

In 1971 this laboratory proposed a new set of polypeptide conformations considered to be relevant to the gramicidin A transmembrane channel.[9, 12] This set of new conformations was called $\pi_{L,D}$-helices because the first member in the set had the same number of residues per turn as the π-helix of Low and Baybutt.[13] Since it was more descriptive, we subsequently changed the name to β-helix because the hydrogen bonding pattern between turns in the single-stranded helix was the same as between chains in the parallel β-pleated sheet conformation.[10, 11] It was considered that the most likely conformation for the gramicidin A channel was the β^6-helix with 6.3 residues per turn.[9–11] In 1972 Ramachandran and Chandrasekaran presented calculations on peptides with an alternating L-D sequence, also arriving at the new conformations and also suggesting the β^6-helix to be the gramicidin A channel. Very recently Veatch *et al.*[15] have extended the concept of a β-helix to that of a double-stranded β-helix. (See FIGURE 7 for the generation of β-helices.) The solution studies that follow are intended to clarify the existence and possible relationship between single-stranded and double-stranded β-helices in solution.

Solution Studies

In our previous solution conformational studies on gramicidin A′ (References 16 and 17)† we utilized the hydrogenated derivative because it removed the absorption of the tryptophan side chains and left only the peptide backbone as the source of absorbance and ellipticity. Our approach was first to determine the peptide absorbance in a host of solvents and to choose that solvent system for further study which was most likely to contain gramicidin A′ in a helical conformation. In doing so, it was found that at 1 mg/ml a marked hypochromism occurred in trifluoroethanol (TFE) with an ϵ^{190} of 4500 implying helicity in this solvent. Next the circular dichroism spectrum in TFE at 1 mg/ml was determined and found to be indicative of a left-handed helix, which was further stabilized at lowered temperatures.[16] Proton magnetic resonance data were then employed for the purpose of evaluating the αCH-NH dihedral angle which discounted the α-helix and was "consistent with the recently proposed $\pi_{L,D}$ (β)-helices." It was concluded that "these results are what are to be expected for $\pi_{L,D}$ (β)-helices."[16] These conclusions remain. What is now possible with the fine work of the Blout laboratory[15, 18, 19] is to extend our understanding of the gramicidin A molecule in solution.

Concentration Effect in Several Solvents. In trifluoroethanol (TFE) the residue molar extinction coefficient at 189 nm for a 20 mg/ml solution is 7000, whereas it drops to 4700 l./mol-cm at 0.2 mg/ml. There is also a hypochromism on dilution from $\epsilon^{189} \simeq 6300$ at 20 mg/ml to $\epsilon^{189} \simeq 5700$ at 0.2 mg/ml in ethanol and from $\epsilon^{189} \simeq 6100$ at 20 mg/ml to $\epsilon^{189} \simeq 5500$ at 0.2 mg/ml in ethanol-water (9:1 v/v). The usually observed extinction coefficients for polypeptides are about 4000 for the α-helix, 8,000 for the β-pleated sheet and 7,000 for the random coil. The hypochromism is indicative of an increasingly helical state on dilution in all three solvents.

The dependence of the circular dichroism pattern on concentration is given for four solvents in FIGURE 2. The effect common to all solvents is a shift toward positive ellipticities on dilution. The effect is most pronounced in FIGURE 2A for TFE, in correspondence with the larger hypochromism exhibited in this solvent on dilution. The very large negative ellipticities at high concentration also indicate an ordered aggregated state. This concentration effect followed at 295 nm for gramicidin A′ was interpreted by Veatch and Blout to reflect a monomer-dimer equilibrium.[18] Thus the large negative ellipticity of an aggregated ordered state and the move toward positive ellipticities and the hypochromism on dilution are evidence for an ordered state at dilute concentrations.

† The commercial preparations contain mixtures of gramicidins A, B and C, which differ in the residue in position 11 with gramicidin A being the primary species, commonly about 70%. Gramicidin A has L-Trp, B has L-Phe and C has L-Tyr in position 11.

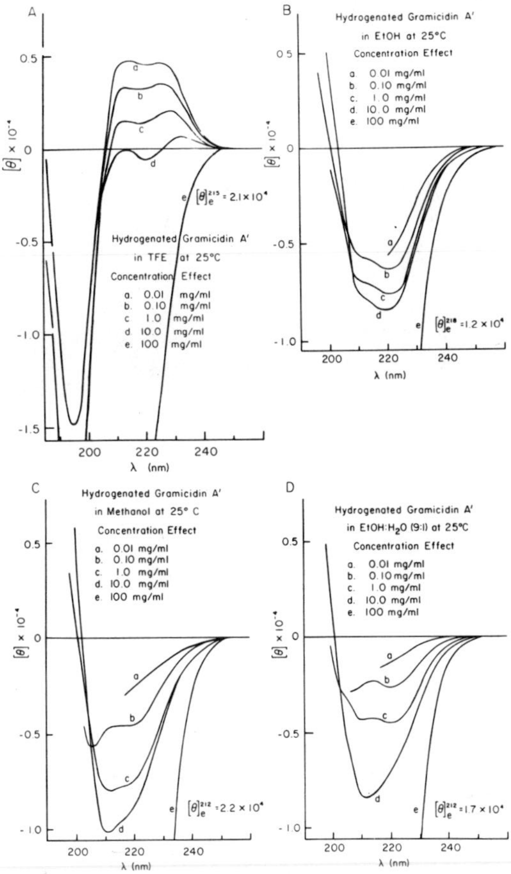

FIGURE 2. Concentration study of circular dichroism patterns of hydrogenated gramicidin A′ in four different solvents: (A) trifluoroethanol (TFE); (B) absolute ethanol; (C) methanol; (D) ethanol:water (9:1). In all solvents it is seen that the circular dichroism pattern becomes more negative with increasing concentration (and reciprocally decreasing path length). The large negative ellipticities are therefore indicative of an ordered aggregated state.

Temperature Effect in Several Solvents. The temperature curves for a 1 mg/ml solution in the same four solvent systems are given in FIGURE 3. Decreasing the temperature shifts the CD patterns toward positive ellipticities; increasing temperature shifts the CD pattern more negatively. In the concentration studies it was shown that the high concentration state, the aggregated state, exhibited negative ellipticity. Normally therefore one would expect increased temperature to drive the equilibrium toward the less aggregated state observed at dilute concentrations, whereas the inverse is observed.

This apparent contradiction can be resolved if the aggregated system has many accessible states and the low concentration state is an ordered structure with a more limited number of states. By the Boltzmann relation we have

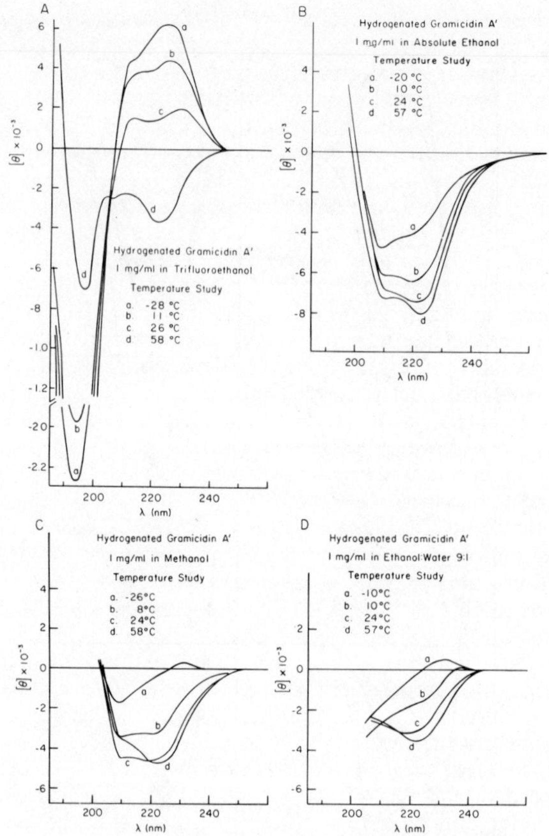

FIGURE 3. Temperature study of the circular dichroism patterns of hydrogenated gramicidin A′ in four different solvents: (A) trifluoroethanol; (B) absolute ethanol; (C) methanol; (D) ethanol:water (9:1). The concentration in each case was 1 mg/ml. Regardless of the starting point and solvent, an increase in temperature results in a shift in the CD pattern of the long-wavelength bands in a negative direction. This indicates that the state represented by a large negative ellipticity near 226 nm is a state of lower entropy, even though the concentration effect indicates this state to be of an ordered aggregate. (See text for further discussion.)

that the entropy, S, of a system is proportional to the natural logarithm of the "number of *a priori* equally probable states accessible to the system,"[20] i.e.,

$$S \equiv R \ln W \tag{1}$$

where proportionality, R, is the gas constant. The temperature results indicate that the entropy for the system at high concentration, S_{hc}, is greater than the entropy for the system at low concentration, S_{lc}, i.e.,

$$S_{hc} > S_{lc} \tag{2}$$

and

$$\Delta S = S_{hc} - S_{lc} = \text{positive}. \tag{3}$$

An aggregated system which would have a large number of equally probable states could indeed be the set of double-stranded β-helices proposed by Veatch *et al.*[15] where, for example, the parallel and antiparallel double-stranded β-helices have several essentially equally probable states of hydrogen bonding between chains. There is additionally the possibility of a mixture of right- and left-handed double helices. This gives a large number of accessible states at high concentration. Now if the dilute system has a more uniquely determined ordered state such as the single-stranded β^6-helix, it is apparent that Equation 2 would obtain, and both the concentration and temperature studies are understandable.

Partial Specific Volume and Refractive Index Data. A very important set of information in understanding the conformations of gramicidin A′ in solution has been provided by ultracentrifugation data of Derechin *et al.*,[21] where it was reported that the partial specific volume of gramicidin A′ in absolute ethanol increased dramatically at concentrations below 1 mg/ml, but that this anomolous behavior did not occur in an ethanol:water system (9:1, v/v). As an independent means of assessing the occurrence of a conformational change on dilution in ethanol that resulted in a marked increase in volume of the monomer,[22] Derechin and Jordan[23] have determined a decrease in the specific refractive index increment, dn/dc, in ethanol from approximately 0.2 ml/gm at concentrations greater than 2 mg/ml to 0.165 ml/gm at high dilution. In ethanol-water (9:1, v/v) dn/dc remained essentially constant, with concentration giving a value of about 0.193 ml/gm.

If we recognize that in FIGURE 2B and D and in the absorption studies referred to above, the conformational change occurs both in ethanol and in the ethanol-water solvent system, the set of results are explicable if the channel conformation formed on dilution is too small to contain the ethanol molecule, but is of a size to accommodate water. A channel conformation occurring at high concentrations, such as in the double-stranded β-helices, would have to be large enough to accommodate ethanol. The concentration and temperature data and the qualitative aspects of the partial specific volume and refractive index data appear to be consistent with an equilibrium between a single-stranded (ss) β^6-helix favored at low concentration and temperature and a multistranded β-helix, such as the double-stranded (ds) β^7-helix. The ss-β^6-helix has a channel diameter of close to 4.0 Å, whereas the ds-β^7-helices have channel diameters greater than 4.3 Å. The ds-β^6-helices have channel diameters of about 2.3 Å.

Effect of Pressure on the CD Patterns. If the above analysis is to be correct, then the application of sufficient pressure to the ethanol solution should cause a collapse of the channel conformation seen at low concentration, thereby shifting the equilibrium toward the conformation found at high concentration. This would mean an increase of negative ellipticity for the ethanol solution with a significant increase in pressure. Preliminary data at 2500 psi have demonstrated an increase in positive ellipticity.

The implication that ethanol cannot enter the channel formed at dilute concentrations, but that water can fixes the diameter of the channel at 3 Å to 4 Å. Further information on the volume of the channel can be obtained from an evaluation of the equilibrium constants at two pressures by the equation[24]

$$\Delta V = 2.303\ RT \log [(K_{P_2}/K_{P_1})]/(P_2 - P_1) \tag{4}$$

This result, when obtained, can be compared with the channel volume of the single-stranded β^6-helix.

Analysis of the Temperature Data. A plot of the mean residue ellipticity of hydrogenated gramicidin A′ at 100 mg/ml in TFE is given in FIGURE 4a. The plateau is not clearly reached at either end of the temperature scale. If we use a value of 1×10^4 for the low temperature end, which is the low temperature value reached with a concentration of 10 mg/ml, and approximate -3.4×10^4 for the value reached at high temperature, where the aggregated conformation begins competing with denaturation, then plots of log K vs. T^{-1} (°K) can be made. The plot, if we assume a dimer-dimer equilibrium

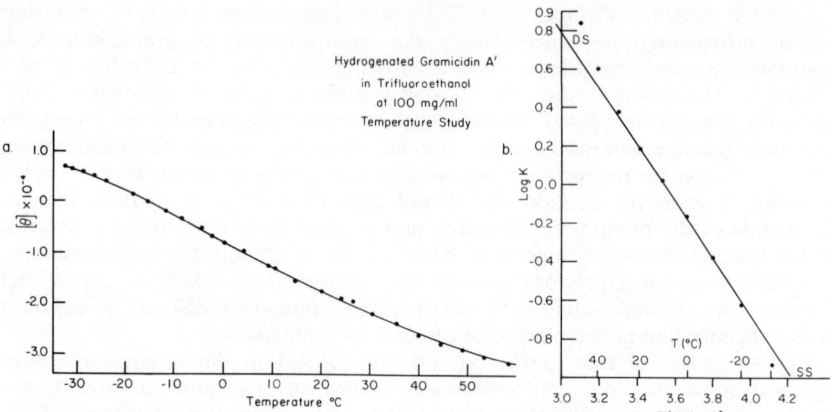

FIGURE 4. Temperature study of hydrogenated gramicidin A′ at 100 mg/ml in trifluoroethanol. (a) Ellipticity at 223 nm. (b) Plot of log K vs. T^{-1} where K has been calculated assuming the equilibrium $D_{ss} = 2M_{ss} = D_{ds}$. (See text for discussion.)

at this concentration, fits the data better than a plot obtained on the assumption of a monomer-dimer equilibrium. Consistent with this, it is not unreasonable to expect at the higher concentration that a single-stranded β-helix would dimerize, and indeed the low temperature plateaus obtained at concentrations of 10 mg/ml, 1 mg/ml, 0.1 mg/ml, and 0.01 mg/ml change in a way expected for a monomer-dimer equilibrium occurring with the low concentration conformation. Accordingly for the plot in FIGURE 4b we have assumed the equilibrium of the sort

$$D_{ss} \rightleftharpoons 2\,M_{ss} \rightleftharpoons D_{ds} \tag{5}$$

where, for example, D_{ss} and M_{ss} stand for the single-stranded dimer and monomer, respectively, and D_{ds} stands for the double-stranded dimer. It should also be appreciated that both M_{ss} and D_{ds} are in equilibrium with a disordered monomer, particularly at elevated temperature. This would explain the greater deviation from a straight line at elevated temperatures in FIGURE 4b. The slope and intercept of FIGURE 4b give a ΔH of 6.7 kcal/mol and a ΔS of 23.6 cal/mol-degree. This means that at room temperature in trifluoroethanol ΔF is slightly negative, i.e.,

$$\Delta F = \Delta H - T\Delta S = 6.7 - 7.0 = -0.3 \text{ kcal/mol} \tag{6}$$

slightly favoring the double-stranded conformation at 100 mg/ml with $K = [D_{ds}]/[D_{ss}] = 1.7$. Again the greater entropy of the D_{ds} conformation would derive from the many states accessible to the double-stranded conformations (see Equation 1).

Thus our analysis of the data is that in solution there exists an equilibrium between two ordered conformational systems. One is favored at dilute concentrations, < 1 mg/ml, which in trifluoroethanol is likely dominated by a left-handed helical structure capable of existing both as a monomer and a dimer and which contains a channel too small for ethanol to enter but large enough for water to fill, i.e., a channel 3 Å to 4 Å in diameter. The second conformational system, favored at high concentration and elevated temperatures, is an aggregate with many nearly equally probable states. If any of the latter contain channels, then the channel of the dominant specials has a diameter large enough for ethanol to enter. The left handed single-stranded β^6-helix satisfies the requirements of the conformation that is dominant in dilute solutions (see FIGURE 8), and the set of double-stranded β^7-helices (see FIGURE 12) would satisfy the requirements of the high-concentration conformations.

Correlation of Solution and Black Lipid Membrane Studies

Chemical Modifications. In our efforts to test the proposal of a single-stranded β-helix which appeared capable of a unique head-to-head hydrogen-bonded dimerization, we chemically modified gramicidin A′ (see footnote on p. 206) and the derivatives were examined by M. C. Goodall for their activities on black lipid membranes. First the gramicidin A′ was deformylated; Goodall found this derivative to be essentially inactive. Next the *N*-acetyl derivative was made, and Goodall found it to be about 1% as active as gramicidin A′. Then the amino ends of two deformyl gramicidin A′ monomers were covalently connected by a malonyl,

$$—\overset{\overset{\displaystyle O}{\|}}{C}—CH_2—\overset{\overset{\displaystyle O}{\|}}{C}—,$$

bridging group; Goodall found this malonyl dimer to be, on a concentration basis, more active than normal gramicidin A′ with activities being observed at a 10^{-15} M concentration in the aqueous bathing solution, but most importantly he found the malonyl dimer to exhibit a first-order concentration-dependence rather than the second-order dependence exhibited by gramicidin A′. Subsequently we have made *N*-t-butyloxycarbonyl gramicidin A′ and Goodall reports that the activity of the preparation is 0.1% that of gramicidin A′. In agreement with the results of Veatch and Blout,[18] who reported that *N*-acetyl-deformyl gramicidin "has almost identical dimerization constant and species mole fraction as gramicidin in ethanol," we find similar solution equilibria for *N*-t-butyloxycarbonyl deformyl gramicidin A′ and for gramicidin A′. This suggests that the dominant aggregation seen at higher concentrations and elevated temperatures is not directly relevant to channel formation in the black lipid membrane studies.

Principle of Microscopic Reversibility. The evidence that channel formation is a second-order reaction has been made more conclusive by the voltage step relaxation and temperature studies of Bamberg and Läuger.[25, 26] These workers and J. A. Hall have independently studied the malonyl dimer by the voltage step relaxation approach (private communications), and they observed first-order kinetics for channel formation for the synthetic dimer. This

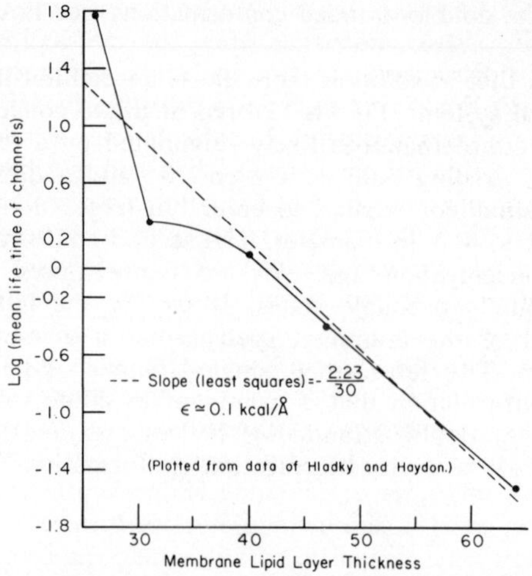

FIGURE 5. Plot of the data of Hladky and Haydon[27] of membrane lipid layer thickness vs. mean lifetime of gramicidin A channels. From the least squares slope an energy of ~0.1 kcal/Å is approximated. This is interpreted as the energy of dimpling (see FIGURE 6 and the text).

further substantiates dimerization and more specifically head-to-head dimerization as the rate-limiting step for channel formation.

Hladky and Haydon[27] have demonstrated that, while channel conductances are essentially independent of lipid layer thickness, the channel mean lifetime is dependent on membrane thickness. FIGURE 5 contains a plot of the Hladky and Haydon data, log (mean lifetime of channels) vs. membrane lipid layer thickness. Since there is a membrane deformation energy component, $\epsilon\delta l$, to the free energy of a channel where ϵ is the energy per unit length, the slope in FIGURE 5 is $\epsilon/2.3$ RT. This gives a deformation energy value of approximately 0.1 kcal/Å for the dimpling process required for the channel to span from side to side. A helical structure spanning a lipid layer is depicted in FIGURE 6.

The dependence of channel mean lifetime on the membrane thickness requires, by the principle of microscopic reversibility, that the process of dimerization also be integrally associated with membrane thickness, that is, if the rate of dissociation for channel termination is membrane-thickness-dependent, then so too is the association of monomers in the dimerization that results in channel formation. The latter is reflected in the membrane-thickness-dependence of the frequency of channel formation.[27]

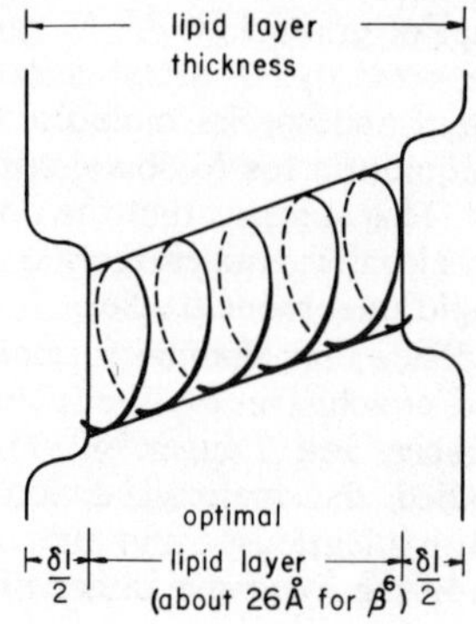

FIGURE 6. Schematic representation of two helices spanning the membrane lipid layer. Since the length of the two helices is too short there is a dimpling of the lipid layer. An energy for this dimpling distortion of the lipid layer is approximated to be 0.1 kcal/Å in FIGURE 5. (Adapted from Urry.[10])

In other words, if the rate-limiting process, the molecular mechanism for channel formation, is the association of two molecules, then the rate-limiting process for channel disruption is the dissociation of two molecules. Therefore if the channel lifetime is a function of lipid layer thickness, then the dimerization process must be a function of lipid layer thickness. For the double stranded β-helix to be the channel one either has the prospect that folding and unfolding occurs within the lipid layer with each end in contact with each surface, or the less attractive hypothesis that the rate of dimerization at the surface on one side of the membrane depends on whether the other side is 60 Å or 30 Å away. The former possibility could conceivably occur by a double-helical zipper mechanism, but it would seemingly require the presence of the monomers in the membrane in an already helical structure such as the single-stranded β-helix. Thus the formation of a double-stranded transmembrane structure would seem to require as a first step the end-to-end association of single-stranded helices, for example, the head-to-head dimerization of two ss-β^6 helices. If there is a preferential solubility of the ends of the ss β helices, as is possible because of the more polar ethanolamine hydroxyl moiety at the carboxyl end, then the preferred product would be an antiparallel-ds-β-helix which is clearly not an essential structure by the demonstrated activity of the malonyl dimer. By this analysis it would seem that the formation of channels by the head-to-head association of two single-stranded β-helices is the more probable event. And the dilute concentrations that favor the single-stranded structure would seem to be more relevant to the black lipid membrane studies.

Channels Constructed of Repeating Units

With the conformational versatility exhibited by gramicidin A′ it is apparent that there is reason to discuss the generation of transmembrane channel structures based on what has been learned from characterized carriers and channels and from our general knowledge of protein conformations. A major reason for considering channels constructed of repeating units is that such channels can provide molecular mechanisms for voltage-dependence of channel formation. As indicated in FIGURE 1, channels constructed of repeating units may be derived by series or parallel alignment of repeats.

Series Alignment of Units

Single-Stranded β-helices. The single-stranded β-helices, particularly the β^6-helix with alternating L and D amino acid residues, can readily be generated from a polypeptide chain in the β-pleated sheet conformation. At the top of FIGURE 7, indicated by the solid-lined structure, is a schematic representation of an edge view of a polypeptide chain in the β-pleated sheet conformation. All residues are of the L-configuration and the R group is seen oriented perpendicular to the chain direction. The process of converting an L-residue to a D-residue is simply one of interchanging the backbone and the R group at a given center. Interchanging backbone and R group at L_2 converts the residue to D_2 with the chain continuing along as the dashed structure. Again interchanging R group and backbone at residue 4 converts it to a D-residue and causes

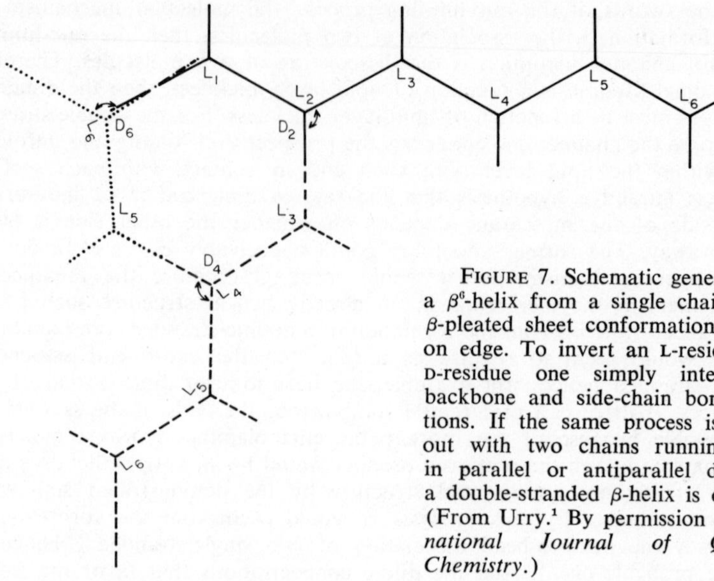

FIGURE 7. Schematic generation of a β^6-helix from a single chain in the β-pleated sheet conformation as seen on edge. To invert an L-residue to a D-residue one simply interchanges backbone and side-chain bond directions. If the same process is carried out with two chains running either in parallel or antiparallel directions a double-stranded β-helix is obtained. (From Urry.[1] By permission of *International Journal of Quantum Chemistry*.)

the remainder of the chain to continue as the dotted structure. A repeat converting L_6 to D_6 causes the next dash-dot backbone segment to overlie the initial segment leading to center L_1. The overlay is such that hydrogen-bonding occurs between the segments. The result is a helical structure with slightly more than six residues per turn. The helical structure can be built either left- or right-handed, but a preference for one or the other handedness can occur depending on which R group is energetically favored and whether the polypeptide has more residues of one configuration than the other. For example, gramicidin A has eight L-residues, six D-residues and one glycine residue in the sequence position of a D-residue. If the L-residues are more energetically favored in the left-handed helical structure than the D-residues, then the left-handed helix will preferentially occur.

The single-stranded β-helices are such that they can hydrogen bond end-to-end, resulting in a series alignment of repeating units. In the specific case of gramicidin A, a head-to-head (amino end to amino end) alignment appears more favorable because the formyl blocking group at the amino end is not sterically crowded. This does not preclude the possible occurrence, particularly in general, of head-to-tail or tail-to-tail (carboxyl end to carboxyl end) alignment.

Models of gramicidin A in the left-handed β^6-helix are shown in FIGURE 8 in an axis (channel) view (A), in a side view of the head-to-head dimer (B), and in a side view of the wire model of the head-to-head dimer (C). In this structure (FIGURE 8A) entrance of the cation to the 4-Å channel would be assisted by the three carbonyl oxygens and the ethanolamine hydroxyl moiety, which would facilitate partial displacement of the cation hydration sphere. The carbonyl oxygens are capable of librating toward the helix axis in a manner mimicking the enniatin B carbonyl orientations in the potassium ion complex of that carrier. At the center of the structure in FIGURE 8B are the formyl moieties. Replacing the two formyl protons by a CH_2 bridging group covalently attaches the two molecules to allow the formation of an uninterrupted channel by a single molecule. The wire model in FIGURE 8C shows the hydrogen-bonding pattern between turns of a monomer to be that of the parallel β-pleated sheet structure, and the hydrogen-bonding pattern between each monomeric unit is that of the antiparallel β-pleated sheet structure.

By a series of arguments leading from the conformation of cyclic hexapeptides to linear peptides, a variant of the single stranded β^6-helix was derived.[11] Instead of having peptide carbonyls pointing alternately parallel and antiparallel to the helix axis, the variant had two carbonyls pointing parallel and the third pointing antiparallel to the helix axis; i.e., for each six residues four carbonyls pointed in one direction and two carbonyls pointed in the other. To delineate the two helices the variant was indicated as a $\beta^6_{2,4}$-helix and the structure in FIGURE 8A was called a $\beta^6_{3,3}$-helix. The interesting feature of a $\beta^6_{2,4}$-helix is that it required a L-L-D or a L-L-Gly amino acid sequence. Several syntheses of (Ala-Ala-Gly)$_n$ peptides were carried out with n = 4 and 5, and the synthetic peptides were found to form channels in a black lipid membrane study.[29]

Stacked Valinomycin "Pore" Structure. As indicated above in the discussion on carrier mechanisms, a conformation of valinomycin that has the same secondary structure as the cation complex, but that has the chain direction

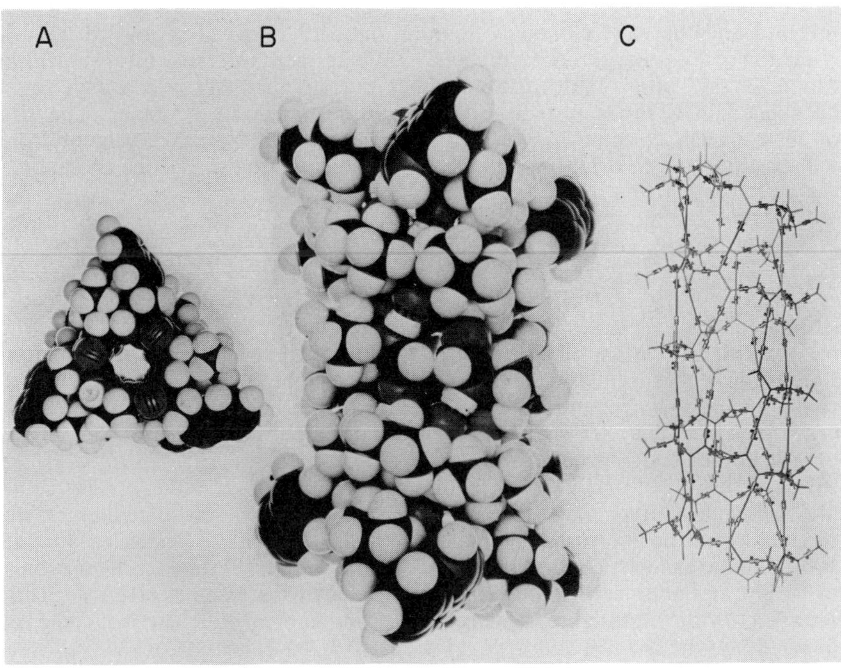

FIGURE 8. The single-stranded-β^6-helix of two molecules of gramicidin A hydrogen bonded head to head. (A) Channel view; (B) side view; (C) side view of wire model.

FIGURE 9. Pore conformation of valinomycin. This is a relatively low energy conformation that differs from that of the valinomycin-K^+ complex in that the chain direction is reversed and the ester carbonyls are pointing outward rather than inward. Conceivably a stack of such conformations or a molecule in which the distance between β-turns was extended could form a transmembrane structure (see text and FIGURE 10). (From Urry and Ohnishi.[3] By permission of the publisher.)

reversed and the valyl carbonyls pointing outward rather than inward, is also a relatively low-energy conformation.[3, 4] We termed this structure shown in FIGURE 9 the "pore" conformation. It has a 4.5-Å diameter pore which could be aligned in a series fashion to form a transmembrane channel. The size of such a channel could vary with the number of repeats. Enlargement from a dodecamer to a hexadecamer could possibly enlarge the size of the channel.

Parallel Alignment of Units

A Cross-β-Channel. The unrolled perspective of the valinomycin "pore" structure shows a series of β-turns.[3] Elongation of the chains between the turns gives, in the unrolled perspective, a cross-β-structure as indicated in FIGURE 10a. Accordingly, a cross-β-channel can be viewed as an elongated valinomycin pore structure. In this structure the carbonyls are hydrogen-bonding between antiparallel chain segments with the carbonyls oriented approximately perpendicular to the chain direction and to the channel axis. If the structure were not made of a continuous sequence, then the chains could also be aligned in parallel directions. In the latter case the diameter of the channel could be increased by one chain at a time as indicated in FIGURE 10b. With a channel constructed of six chains alternate residues would have to be glycine in order that the R group not block the channel or be sterically crowded. With a structure constructed of more chains, R groups could be directed into the channel and impart selectivity to the channel.

The Helical Rack Model. Instead of the repeating unit being a single chain in an extended β-conformation, the unit could be cyclic with one stretch of peptide being in the extended β-conformation and the other in a helical conformation, an α- or 3_{10}-helix for example. If the helical segment were

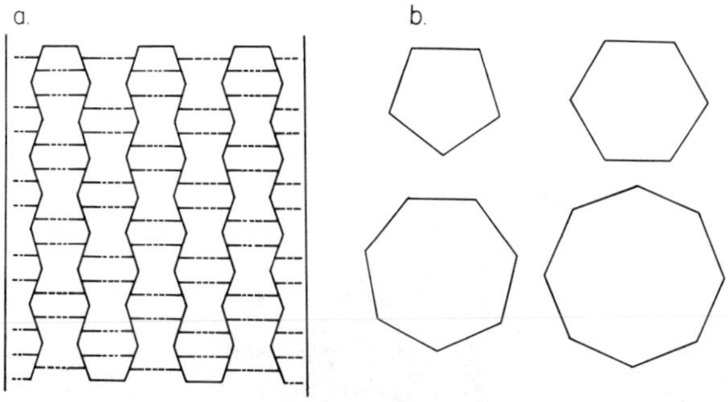

FIGURE 10. Extensions of the valinomycin pore-type conformation by extending the length of chain between β-turns (a) and by varying the number of sides (b). (From Urry.[1] By permission of *International Journal of Quantum Chemistry.*)

comprised predominantly of lipophilic residues, then it would reside at the interface of the membrane as indicated in FIGURE 11a with the β-chain at the aqueous interface and the helix in the lipid layer. This is a stylization of the conformation of alamethicin proposed by Bauman and Mueller.[2] Such structures could associate with the β-chain's hydrogen-bonding much as in the structure depicted in FIGURE 10a, and as will be discussed below, the stability of the structures could be a function of the electric field across the membrane.

Multistranded β-Helices

Looking back at FIGURE 7, if the same set of steps were taken with two (or three) β-chains running either parallel or antiparallel, a set of double- (or triple-) stranded helices could be formed in which the pitch of the helix is approximately twice (or three times) that of the single-stranded β-helix. FIGURE 12 contains the molecular model of one of the interchain hydrogen-bonded registers of the right-handed, antiparallel, double-stranded β^{7}-helix pro-

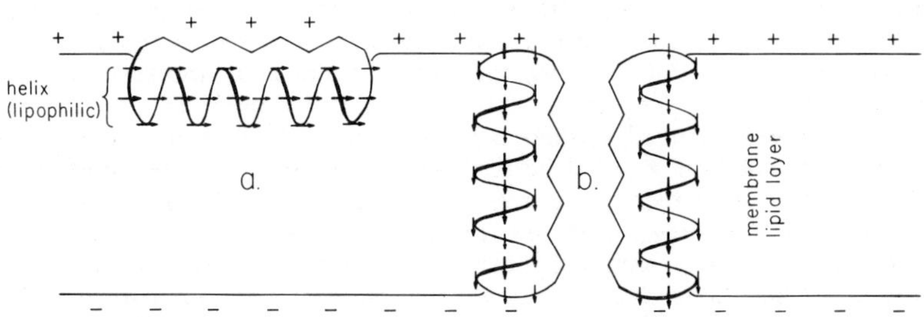

FIGURE 11. The helical rack model of voltage-dependent channel formation. In the classical helical conformations, α-helix, 3_{10} helix, etc., the peptide dipole moments sum to make one large dipole moment with its direction along the helix axis. This means that the transmembrane orientation of the helix is stabilized by the presence of an electric field across the membrane. See text for discussion. (From Urry.[1] By permission of *International Journal of Quantum Chemistry.*)

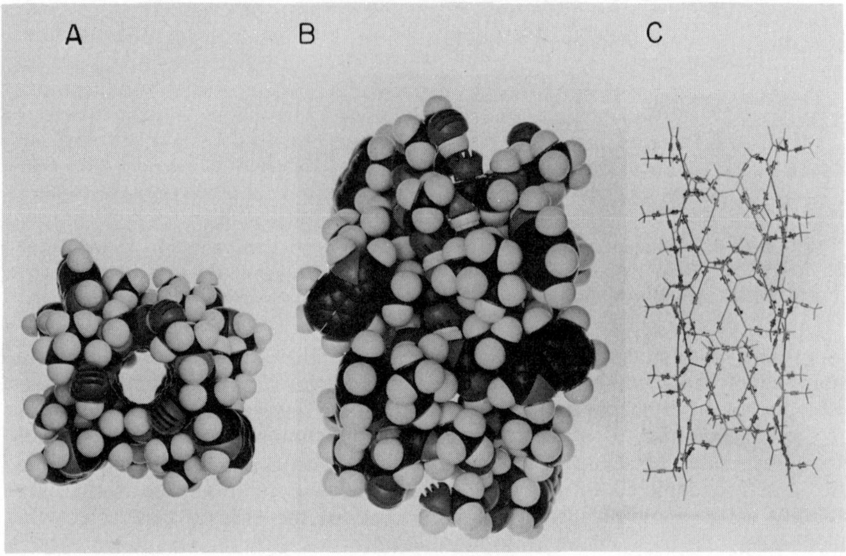

FIGURE 12. The antiparallel, double-stranded-β-helix of gramicidin A. The channel view is given in (A) and the side view in (B), both as CPK models; the side view of the wire model is given in (C).

posed by Veatch *et al.*[15] The structure in channel view (FIGURE 12A) appears to be almost square rather than triangular (as in FIGURE 8A) and the channel is larger, about 4.6 Å in diameter. In side view (FIGURE 12B) the structure has a pseudo twofold symmetry axis. In FIGURE 12C the wire model shows the antiparallel hydrogen bonding pattern between chains.

This structure could also be generated from two right-handed single-stranded β-helices attached head-to-head, as seen in FIGURE 8C by holding one monomer fixed while turning the other in the helix direction with the formyl moieties each dovetailing between the turns of helix of the other. This is what was referred to above as a double-helical zipper mechanism. The double-stranded helix could be generated by a zipper-like series of hydrogen-bond shifts associated with a rotation on the helix axis of one monomer with respect to the other, with the direction of rotation being that of the helix screw sense. This could be achieved without a significant change in the length of the helix and with the helix ends at the dimpled surface of the membrane (FIGURE 6). This would adhere to the above discussed principle of microscopic reversibility for the gramicidin A channel formation as related to the mean channel lifetime dependence on lipid layer thickness. What is unclear is the rate at which such a process might occur. Clearly the process would be slower than the initial head-to-head association. It is also possible that the process could generate a structure that would be nonconducting.

Peptide Dipole Moment Model for Electric Field Dependence of Channel Formation

The peptide moiety has a large permanent dipole moment, about 4×10^{-18} esu cm. If these moments were sufficiently ordered and if structures existed that could utilize these dipoles in coupling with an electric field to increase or decrease the stability of a channel, then transport of ions could be coupled to the transmembrane electric field. Below are briefly described two channels, one serially aligned and the other aligned in parallel, in which the peptide dipole moments could be utilized to stabilize or destabilize the channel structure.[1, 11] The free energy, ΔF, of orientation of a dipole in an electric field is

$$\Delta F = -\frac{\mu}{e} FE \cos\theta$$

23.6 kcal/volt; E the electric field in volts/cm; and $\cos\theta$ is the orientation factor, with θ being the angle between the field direction and the direction of the permanent dipole moment. With an angle as large as 30°, $\cos\theta$ decreases to 0.866; at 20° $\cos\theta$ is 0.94.

The serially aligned conformations are the single-stranded $\beta^6_{2,4}$-helix which contains a channel and the syn-β_2^6-spiral in which the channel has been closed by the 90° rotation of four of the six peptides of the single-stranded $\beta^6_{2,4}$-helix. The change in dipole moments is approximately 15×10^{-18} esu cm per 4.4 Å along the transmembrane axis. For a 30-Å lipid layer thickness and a 30-Å transmembrane structure this sums to be about 100×10^{-18} esu cm. With a 100-mv potential E is 3.3×10^5 volts/cm, giving a difference in free energy between the two structures of about 1.6 kcal/mole. The application or removal of a field shifts the equilibrium by a factor of 10.

For the helical rack model in FIGURE 11, the dipole moments of the β-chain cancel such that there is approximately a permanent dipole moment of 3.5×10^{-18} esu cm per 1.5 Å of an α-helix, i.e., for each peptide in the helical structure. This gives a value of 70×10^{-18} esu cm for the summed dipoles and a free energy of orientation in an electric field of about 1.0 kcal/mol per helix. If the channel is made of six oriented helices, then the channel structure would be stabilized by about 6 kcal/mol of channels. This would alter the equilibrium by about a factor of 10^4. One could view the mechanism as the association of several units at the membrane surface and which, as a cluster, align across the membrane without exposure of the polar β-chains to the lipid. This helical rack model is structurally analogous to the mechanism proposed by Bauman and Mueller[2] for alamethicin, except that they utilized a bound cation at one end of the structure to achieve orientation. With a linear peptide a polar side chain rather than a polar β-chain would provide the polar ridge for the pore side of the structure and several associated units would again bury the polar ridge on going from surface to transmembrane orientation.

ACKNOWLEDGMENTS

The authors acknowledge Dr. Kouji Okamoto for synthesis of hydrogenated gramicidin A and Boc-gramicidin A and thank M. Derechin for allowing reference to work prior to publication.

REFERENCES

1. URRY, D. W. 1975. Int. J. Quantum Chem. **9:** in press.
2. MUELLER, P. This monograph.
3. URRY, D. W. & M. OHNISHI. 1970. *In* Spectroscopic Approaches to Biomolecular Conformation. D. W. Urry, Ed. : 263–300. American Medical Association Press. Chicago, Ill.
4. MAYERS, D. F. & D. W. URRY. 1972. J. Am. Chem. Soc. **94:** 77–81.
5. GRELL, E. & TH. FUNCK. 1973. J. Supramol. Struct. **1:** 307–335.
6. PLATTNER, PL. A., K. VOGLER, R. O. STUDER, PL. QUITT & W. KELLER-SCHIERLEIN. 1963. Helv. Chim. Acta **46:** 927–935.
7. SARGES, R. & B. WITKOP. 1964. J. Am. Chem. Soc. **86:** 1861–1862.
8. DOBLER, M., J. D. DUNITZ & J. KRAJEWSKI. 1969. J. Mol. Biol. **42:** 603–606.
9. URRY, D. W., M. C. GOODALL, J. D. GLICKSON & D. F. MAYERS. 1971. Proc. Nat. Acad. Sci. U.S.A. **68:** 1907–1911.
10. URRY, D. W. 1973. *In* Conformation of Biological Molecules and Polymers. The Jerusalem Symposia on Quantum Chemistry and Biochemistry, V. : 723. The Israel Academy of Sciences and Humanities. Jerusalem, Israel.
11. URRY, D. W. 1972. Proc. Nat. Acad. Sci. U.S.A. **69:** 1610–1614.
12. URRY, D. W. 1971. Proc. Nat. Acad. Sci. U.S.A. **68:** 672–676.
13. LOW, B. W. & R. B. BAYBUTT. 1952. J. Am. Chem. Soc. **74:** 5806–5807.
14. RAMACHANDRAN, G. N. & R. CHANDRASEKARAN. 1972. Indian J. Biochem. Biophys. **9:** 1–11.
15. VEATCH, W. R., E. T. FOSSEL & E. R. BLOUT. 1974. Biochemistry **13:** 5249–5256.
16. URRY, D. W., J. D. GLICKSON, D. F. MAYERS & J. HAIDER. 1972. Biochemistry **11:** 487–493.
17. GLICKSON, J. D., D. F. MAYERS, J. M. SETTINE & D. W. URRY. 1972. Biochemistry **11:** 477–486.
18. VEATCH, W. R. & E. R. BLOUT. 1974. Biochemistry **13:** 5257–5264.
19. FOSSEL, E. T., W. R. VEATCH, YU. A. OVCHINNIKOV & E. R. BLOUT. 1974. Biochemistry **13:** 5264–5275.
20. EYRING, H., D. HENDERSON, B. J. STOVER & E. M. EYRING. 1964. *In* Statistical Mechanics and Dynamics. : 91 Wiley. New York, N.Y.
21. DERECHIN, M., D. M. HAYASHI & B. E. JORDAN. 1974. Life Sci. **15:** 403–413.
22. DERECHIN, M. 1974. Abstract No. 113 of the 9th Middle Atlantic Regional Meeting of the American Chemical Society, Wilkes-Barre, Pa., April 23–26.
23. DERECHIN, M. & B. E. JORDAN. Private communication from M. Derechin.
24. JOHNSON, F. H., H. EYRING & M. J. POLISSAR. 1954. *In* The Kinetic Basics of Molecular Biology. : 305. Wiley. New York, N.Y.
25. BAMBERG, E. & P. LAUGER. 1973. J. Membr. Biol. **11:** 177–194.
26. BAMBERG, E. & P. LAUGER. 1974. Biochim. Biophys. Acta **367:** 127–133.
27. HLADKY, S. B. & D. A. HAYDON. 1972. Biochim. Biophys. Acta **274:** 294–312.
28. TOLMAN, R. C. 1962. *In* The Principles of Statistical Mechanics. : 163. Oxford University Press. London, England.
29. GOODALL, M. C. & D. W. URRY. 1973. Biochim. Biophys. Acta **291:** 317–320.

"SANDWICH" COMPLEXATION IN CYCLOPEPTIDES AND ITS IMPLICATIONS IN MEMBRANE PROCESSES

V. T. Ivanov

Shemyakin Institute of Bioorganic Chemistry
USSR Academy of Sciences
Moscow, USSR

In the mid-1960s Pressman and Lardy discovered the singular ability of certain antibiotics, subsequently called ionophores, to induce alkali metal ion transport across biological membranes. This new class of bioregulators naturally attracted considerable interest, and little time passed before they had been subjected to thorough physicochemical and biological investigation. However, despite the mass of information that had accumulated in the course of the decade since the appearance of the first papers,[1, 2] many aspects of the induced transmembrane ion-transport phenomenon, involving the complicated interplay of ion, ionophore, and membrane, still remain obscure, and numerous papers on this subject are continuously descending upon the editorial boards of the various biological and chemical journals.

In general it can be considered to be firmly established that the underlying principle of the membrane-affecting properties of the ionophores is their selective affinity for metal ions (particularly of the alkali and alkaline earth groups) with which they form lipophilic complexes in solution, capable of permeating the lipid zones of the membranes. It stands to reason, therefore, that any model describing the ionophore-mediated transport of an ion from one side of a membrane to the other must be in complete accord with the available data on the structures and properties of the free and the complexed ionophores.

From this aspect, of considerable interest is the comparatively recent finding in our laboratory[3–5] that the enniatin ionophores are capable of forming "sandwich" complexes and apparently more complicated structures (stacks) with macrocycle:cation ratios of 2:1 and 3:2, as well as the long known[6–8] equimolar 1:1 complexes. It is just these newly discovered associates that have turned out to be responsible for the transport of potassium and cesium ions across artificial bilayers.[3, 5]

The present report deals with an extension of our stoichiometric studies of the complexing reaction in solution to the metal-binding cyclopeptides and their synthetic analogs. We begin with the simplest of these compounds we have investigated, viz., cyclo[-(L-Val-Sar)$_3$-], a peptide analog of enniatin B wherein, besides substitution of the three ester groups by amide groups, three isopropyl radicals have been deleted and the asymmetric centers are inverted.

As with the enniatins, the interaction of cyclo[-(L-Val-Sar)$_3$-] with alkali metal ions was followed spectrometrically after gradually adding a solution of the corresponding salt to a solution of the ionophore. Thus, titration with 0.97 M $NaClO_4$ in 96% aqueous alcohol is accompanied by strong changes in the dichroic absorption up to limiting values corresponding to complete complexation (FIGURE 1). The dependence of the degree of complexation (α) on the salt:cyclopeptide concentration ratio (b_0/a_0) is represented by a

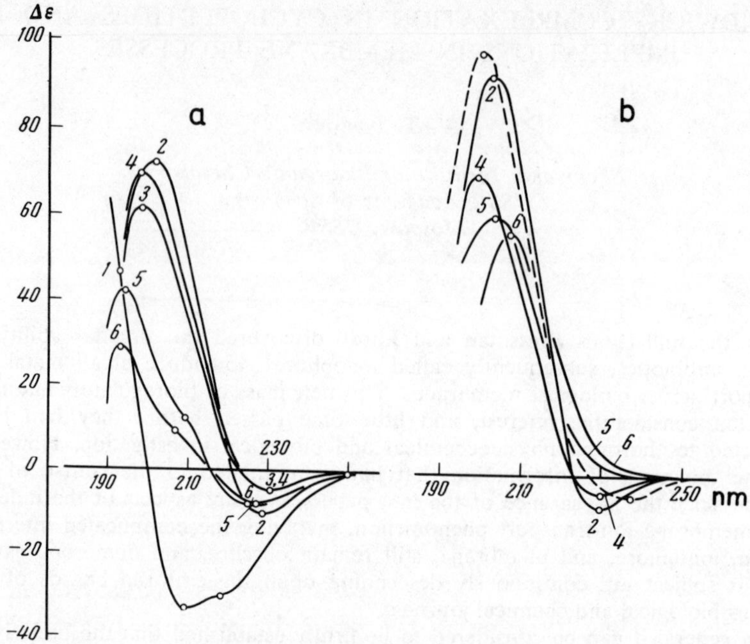

FIGURE 1. CD curves of cyclo[-(L-Val-Sar)$_3$-] (a) Experimental curves; (b) curves calculated for 100% complexation. Concentrations of the cyclopeptide $(0.6–0.9) \cdot 10^{-4}$ M. 1 = EtOH; 2 = $7.02 \cdot 10^{-2}$ M LiCl/EtOH; 3 = 0.1 M $NaClO_4$/EtOH; 4 = 0.97 M $NaClO_4$/EtOH; 5 = $0.72 \cdot 10^{-2}$ M KCl/EtOH; 6 = $1.73 \cdot 10^{-2}$ M CsCl/EtOH. The dashed line refers to the enniatin B potassium complex.

smooth curve leveling off at large excesses of the salt, as illustrated by FIGURE 2a for titration of cyclo[-(L-Val-Sar)$_3$-] with $NaClO_4$.

The results point to the ultimate formation of an equimolar (i.e., 1:1) macrocycle:salt complex. Total complexing could not be attained on titrating cyclo[-(L-Val-Sar)$_3$-] with lithium, potassium, or cesium salts because of low

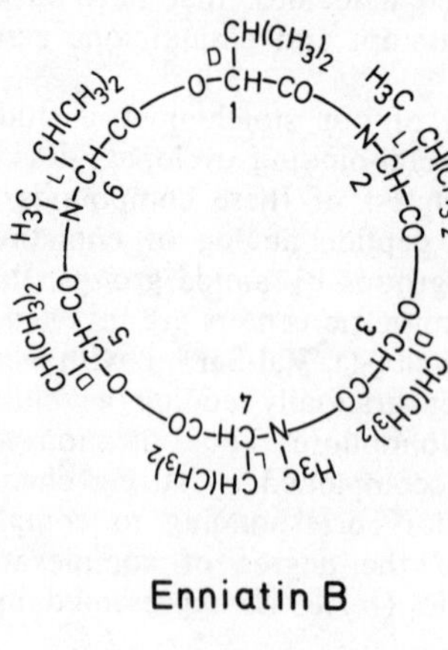

Enniatin B

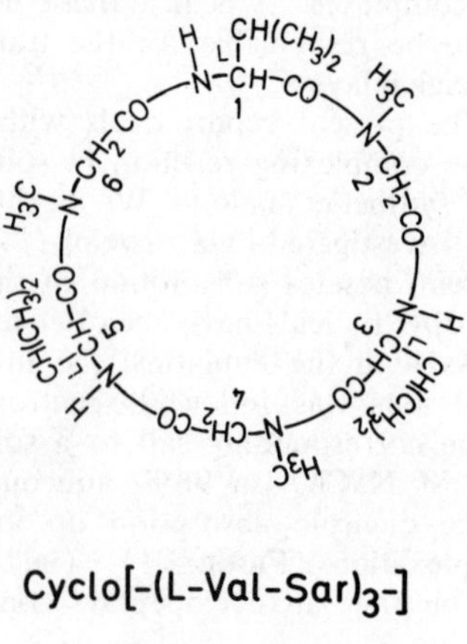

Cyclo[-(L-Val-Sar)$_3$-]

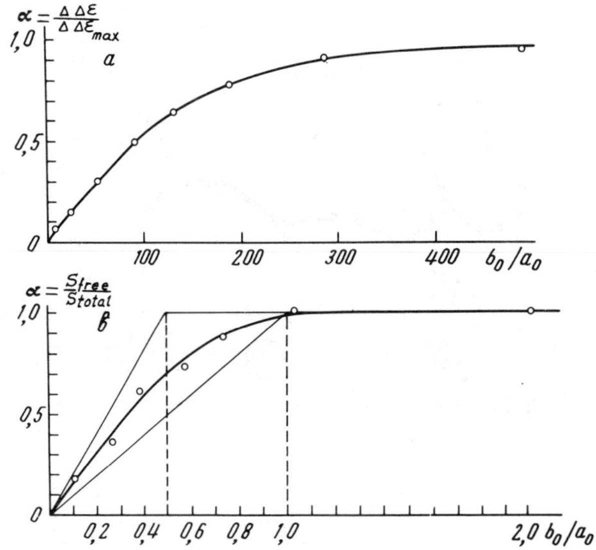

FIGURE 2. Titration curves of cyclo[-(L-Val-Sar)$_3$-]: $NaClO_4$ in EtOH, control by CD, $a_0 = 0.72 \cdot 10^{-4}$ M; (b) NaSCN in CD_3CN, control by nmr, $a_0 = 0.63 \cdot 10^{-1}$ M. S_{free} = total area under the N-Me signals of the noncomplexed cyclo[-(L-Val-Sar)$_3$-] as determined from the area of high field (δ 2.74 ppm) signal (see FIGURE 5); S_{total} = overall area under the N-Me signals.

complex stability or limited salt solubility. The stability constants of equimolar complexes of cyclo[-(L-Val-Sar)$_3$-] in 96% aqueous ethanol at 25° (K_1, M^{-1}) are as follows:

Li	Na^+	K^+	Cs^+
60	280	150	50

These are only approximate values, but they leave no doubt as to the order of complexing selectivity:

$$Na^+ > K^+ > Cs^+, Li^+$$

Comparison of the CD curves of cyclo[-(L-Val-Sar)$_3$-] and enniatin B (to be more exact enantiomeric enniatin) complexes shows them to be very similar in both the region of $n \rightarrow \pi^*$ transition (220–240 nm) and long wavelength π–π^* transitions (205 nm) of the tertiary amide groups. This is illustrated by the curve for enantio-enniatin B represented in FIGURE 1. Accordingly the conclusion thrusts itself upon one that the cyclo[-(L-Val-Sar)$_3$-] and the enniatin complexes are built along the same principles. The structure of the Na^+·cyclo[-(L-Val-Sar)$_3$-] complex based on these principles is shown in FIGURE 3. It is of threefold symmetry and the amide bonds are "all-*trans*." The centrally situated cation interacts with all six of the amide carbonyls. Of these, three belonging to the Sar residues are oriented downwards from the average plane of the ring, whereas the three belonging to the L-Val

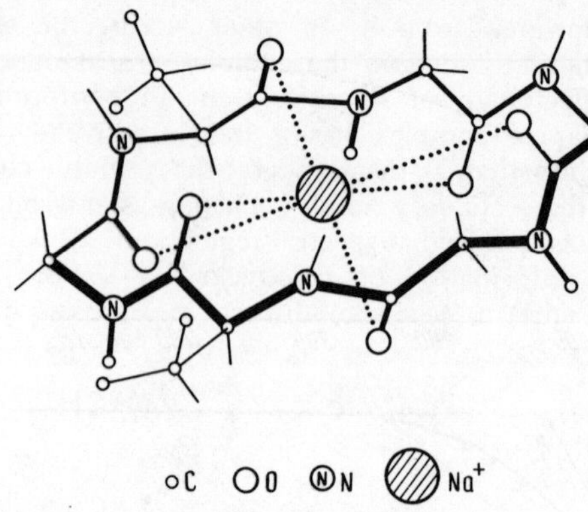

FIGURE 3. Conformation of the sodium complex of cyclo[-(L-Val-Sar)$_3$-].

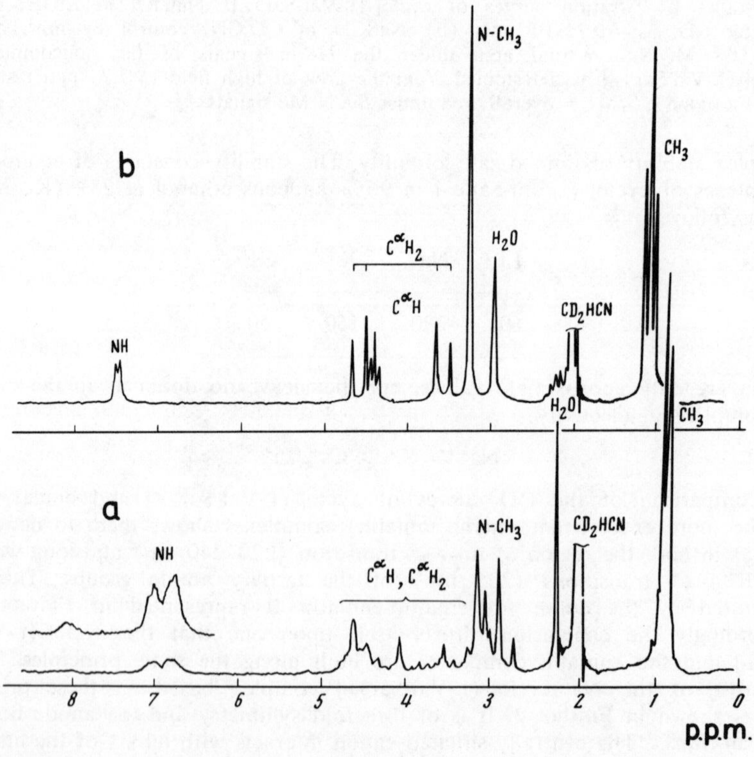

FIGURE 4. Proton magnetic resonance spectra of cyclo[-(L-Val-Sar)$_3$-] (a) and its Na^+ complex (b) in CD_3CN.

residues are oriented upwards from it. In other words, the secondary and tertiary amide chromophoric systems in the complex are of opposite chirality, i.e., their chiroptical effects are of opposite sign. In conformity with this, the Na^+·cyclo[-(.-Val-Sar)$_3$-] complex has a less intensive $\pi - \pi^*$ dichroic absorption band at 205 nm than is displayed by the enniatin complex, which is quite logical because the secondary amide-replacing ester groups of enniatin B undergo no $\pi - \pi^*$ transition in that spectral region.

Important structural information on the cyclo [-(L-Val-Sar)$_3$-] complexes have yielded their nmr spectra. One can discern in FIGURES 4a and 5 (see

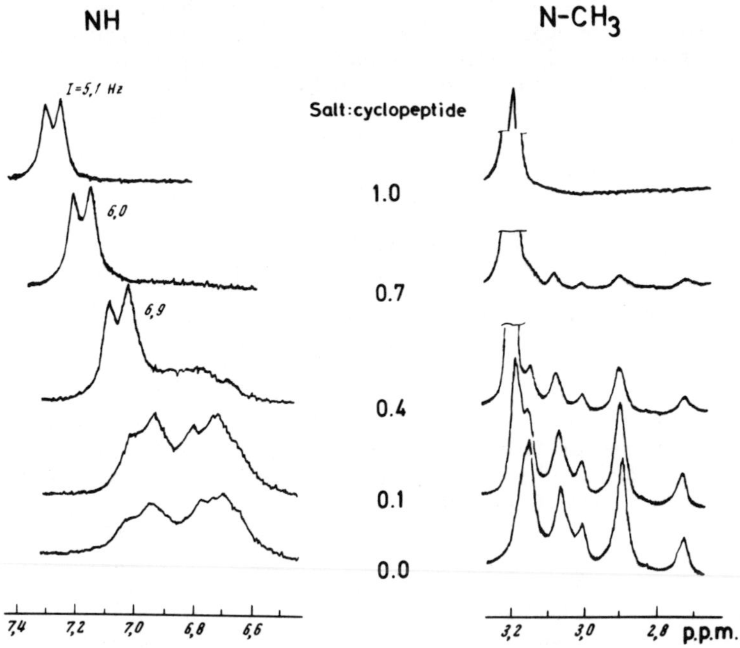

FIGURE 5. NH and N-CH_3 nmr spectral regions of cyclo[-(L-Val-Sar)$_3$-] (*lower spectrum*), its Na† complex (*upper spectrum*), and their equilibrium mixtures in CD_3CN at different cyclopeptide:NaSCN ratios.

also Reference 9), that in CD_3CN solution the cyclopeptide exists as a mixture of three conformers separated by quite considerable energy barriers. Analysis of the *N*-methylamide region shows that two of the conformers comprising 52% and 25% of the overall mixture have no symmetry elements, whereas the third ($\delta_{N\text{-}Me}$ = 3.14 ppm; 23%) possesses, as one can see from the formula, a symmetry axis (at least in the nmr time scale). On adding NaSCN the signal intensities of the unsymmetric forms diminish, and the signals of the symmetric form increase besides being shifted to lower field. Hence the ion oscillates at high frequency (in the nmr scale) between a molecule of the complex and the symmetric molecule of the free cyclopeptide, a process that is possible only for identical "all *cis*" or "all *trans*" tertiary amide configurations in the complex and the symmetric form. This is in harmony with the structure shown in FIGURE 3, and with the earlier-demonstrated[9] "all *trans*" amide configuration of the symmetric form.

The curve depicting the dependence of the complexation on excess salt as calculated from the intensities of the *N*-methyl signals is higher than the theoretical curve for 1:1 complexes of infinitely large stability constants, indicating that in the case of salt deficiency the complexes are of a greater than 1:1 structure (FIGURE 2b). From the initial slope of the curve we ascribe a sandwich form to them with a 2:1 macrocycle:cation ratio. As could be expected from general considerations, maximal sandwich concentration is observed at $b_0/a_0 \approx 0.5$, 1:1 complexing taking place on addition of more salt. The process can be followed from the changes in the $^3J_{NH-CH}$ constant of the complex, which is 6.9 Hz at $b_0/a_0 = 0.4$, but gradually falls to 5.1 Hz at $b_0/a_0 = 1.0$, changing no further on passing to $b_0/a_0 \approx 2.0$. We shall presently go into the reason for the decrease in the $^3J_{NH-CH}$ constant in the equimolar complex relative to the sandwich. The fact that the 2:1 complex is almost absent already at $b_0/a_0 = 1$ indicates that K_2 is at least by two orders of magnitude lower than K_1.

$$\text{Cyclopeptide} + \text{Na}^+ \rightleftarrows (\text{cyclopeptide} \cdot \text{Na})^+$$

$$K_1 = \frac{[(\text{cyclopeptide} \cdot \text{Na})^+]}{[\text{cyclopeptide}] \cdot [\text{Na}^+]} \qquad (1)$$

$$\text{Cyclopeptide} + (\text{cyclopeptide} \cdot \text{Na})^+ \rightleftarrows (\text{cyclopeptide})_2 \cdot \text{Na}^+$$

$$K_2 = \frac{[(\text{cyclopeptide})_2 \cdot \text{Na}^+]}{[\text{cyclopeptide}] \cdot [(\text{cyclopeptide} \cdot \text{Na})^+]} \qquad (2)$$

Thus, owing to the low values of K_2, it is only natural that practically no 2:1 complexes are formed in dilute solutions. Apparently this is why they were not detected in the course of the CD studies, which were performed at cyclopeptide concentrations of 10^{-4} M.

When cyclo[-(L-Val-Sar)$_3$-] was titrated with KSCN and CsSCN in CD_3CN, crystalline complexes (m.p. 325° and 316–317°) with K^+ and Cs^+ contents exactly corresponding to 1:1 stoichiometry (as determined by flame spectrophotometry) precipitated out even at low salt concentrations ($b_0/a_0 \approx 0.2$). A picture similar to that for CD_3CN was observed in the titration with NaSCN in $(CD_3)_2SO$ except that in the initial solution the unsymmetric structure was the sole form present and large excess of salt was required for complete complexation ($b_0/a_0 > 5$); in the case of K^+ complex 20% of the peptide remained free even at $b_0/a_0 \approx 7$. The $^3J_{NH-CH}$ constants are 5.4 Hz in the sodium complex and 6.5 Hz in the potassium complex, reflecting the change in the θ angle* caused by rotation of the amide plane due to increased size of the ion (FIGURE 6). A similar effect (increase in the $^3J_{NH-CH}$ constant from 6.4 to 7.9 Hz when passing from Na^+ to K^+) was observed earlier in the (tri-*N*-desmethyl)-enniatin B complexes.[4, 11]

The cyclo[-(L-Val-Sar)$_3$-] sandwiched ion can interact with six *N*-methylamide or six desmethyl amide groups, or with three *N*-methylamide and three amide groups. Analysis of molecular models shows that in the last two cases at least some of the *N*-methyl group must lie within the sandwich, giving rise to steric hindrance, so that preference has been given to the first possibility. It is also the more likely form when account is taken of the somewhat higher

* For the angular dependence of the $^3J_{NH-CH}$ constant see Reference 10.

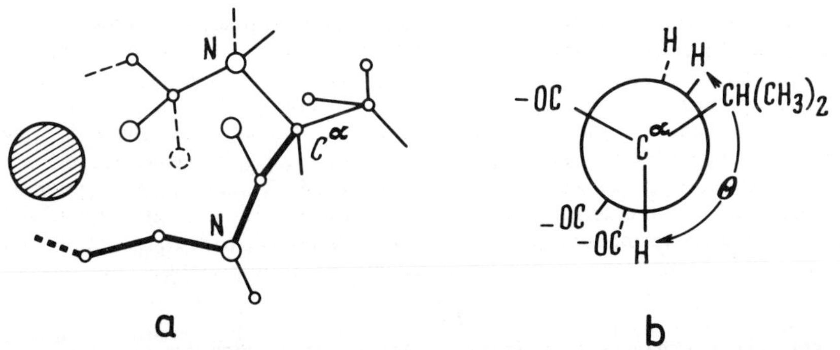

FIGURE 6. Schematic representation of the conformational rearrangement accompanying the size increase of the bound ion in equimolar cyclo[-(L-Val-Sar)$_3$-] complexes. (a) Side view; (b) Newman projection along the C^{α}-N bond.

electrostatic field intensity of the tertiary amides than secondary amides.[12] In line with the above mentioned $^3J_{NH-CH}$ constants (5.1 Hz in the equimolar complex and 6.7 Hz in the case of equilibrium mixture of the equimolar and sandwich complexes; FIGURE 5) in the preferred 2:1 complex the electrostatic repulsion of the secondary carbonyl oxygens confers to the amide planes a tendency to turn further away from the symmetry axis than in the 1:1 complex where they are attracted somewhat inwards by the cation in the molecular cavity. This effect is seen in FIGURE 7.

Thus in its interaction with alkali metal ions cyclo[-(L-Val-Sar)$_3$-] in many respects resembles enniatin B, but differs by a higher Na^+ complexing selectivity, apparently due to the specific electronic structure of its amide ligands.[12] Also, cyclo[-(L-Val-Sar)$_3$-] complexes are considerably less stable and less lipophilic than those of the enniatins. Apparently because of this the cyclopeptide exhibits no ionophoric effect on bilayers nor ability to inhibit microbial growth.

The significance of the cyclo[-(L-Val-Sar)$_3$-] study lies in the more direct proof it provided of the sandwiching capacity of 18-membered rings than was possible with compounds previously investigated by measuring the areas under the nmr signals for the free and bound forms. We have thereby for the first time been able to obtain experimental data on the nature of the conformational reorganizations accompanying the transition from equimolar to sandwich complexes. Properties similar to cyclo[-(L-Val-Sar)$_3$-] are apparently exhibited by cyclo[-(L-Pro-Gly)$_3$-], investigated by Madison and coworkers.[13]

FIGURE 7. Schematic representation of the 2:1 (a) and 1:1 (b) alkali metal ion complexes of cyclo[-(L-Val-Sar)$_3$-]. 1-*N*-methylamide (Val) oxygens; 2-secondary amide (Sar) oxygens.

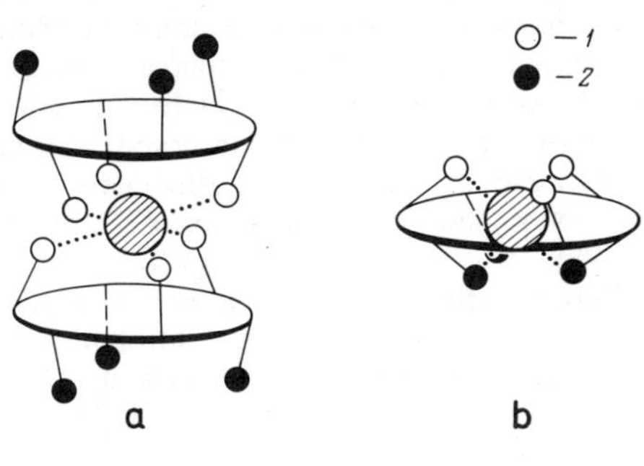

Since we were now on sure ground as to the existence of sandwich complexes of the enniatins and their analogs in solution, our next objective was to ascertain their role in ion transport across simple model systems. Using a spectrophotometric technique, we studied the ability of enniatin B to carry Li^+, Na^+, K^+ and Cs^+ picrates from aqueous solution into methylene chloride. The experiments were performed at constant, high salt concentrations and varying enniatin B concentrations. In other words, conditions were provided for maximal cation entering into methylene chloride, but with varying probability of sandwich formation. Although the data obtained, which are summarized in TABLE 1, do not lend themselves to quantitative interpretation, the consider-

TABLE 1

METHYLENE CHLORIDE EXTRACTION OF ALKALI METAL PICRATES (MPI) FROM AQUEOUS SOLUTIONS IN THE PRESENCE OF ENNIATIN B (EN)*

	Concentration of Enniatin B (c_{EN}, M)					
	10^{-3}		10^{-4}		10^{-5}	
M	λ_{max}^{Pi}(nm)	c_{Pi}/c_{EN}	λ_{max}^{Pi}(nm)	c_{Pi}/c_{EN}	λ_{max}^{Pi}(nm)	c_{Pi}/c_{EN}
Li^+	363	0.15	366	0.15	364	0.32
Na^+	362	0.79	367	0.83	362	0.97
K^+	370	0.55	369	0.53	367	0.86
Cs^+	369	0.22	371	0.37	363	0.45

* The tabulated results are the average of three independent experiments. The extraction was carried out by shaking equal volumes of an enniatin B solution of methylene chloride and a 1-M aqueous solution of the respective salt containing $(3–5) \cdot 10^{-3}$ M of picric acid. The solution pH was maintained at 7.5–8 by addition of the corresponding alkali. Blank experiments showed that, alone, enniatin B is not carried over into the aqueous phase, nor are the picrates, in the absence of enniatin B, extracted into the organic phase. In calculating c_{Pi}, use was made of the molar extinctions of the picrate anion $\xi_{366} = 16{,}100$ given by Roeske *et al.*[14]

able changes in $c_{Pi}/c_{EN} = c_M/c_{EN}$, accompanying the change in antibiotic concentration in the extraction of any of the ions, leave no doubt as to the correctness of the basic conclusion; namely, enniatin B transports alkali metal ions into the organic phase with varying stoichiometries, the relative content of the various forms depending upon the experimental conditions. Moreover, clearly at 10^{-3}–10^{-4} M of enniatin B potassium is transported predominantly in the sandwich form, whereas at 10^{-5} M, it is predominantly transported in the form of an equimolar complex. Sodium ions, on the other hand form predominantly equimolar complexes, whatever the cyclodepsipeptide concentration, although their greatest relative amount is at low antibiotic concentrations. The small c_{Pi}/c_{EN} values observed for all concentrations of the antibiotic in the case of Li^+ can be attributed to low stability of the corresponding complexes, and, in the case of Cs^+, to their lower solubilities in methylene chloride.

Comparison of the λ_{max}^{Pi} values in TABLE 2 with data in the literature shows that the picrate ion always forms ion pairs with the complex cation which are more stable than had been reported by Wong *et al.*[15, 16] for the separated ion pairs of a number of cyclopolyethers (λ_{max}^{Pi} = 369 nm for Na^+ and 378 nm for K^+ in $CHCl_3$),

TABLE 2
EXTRACTION OF SODIUM PICRATE FROM AQUEOUS SOLUTION BY METHYLENE CHLORIDE CONTAINING ANTAMANIDE OR PERHYDROANTAMANIDE*

Cyclopeptide	$c_{cyclopeptide}$(M)	λ_{max}^{Pi}(nm)	c_{Pi}/cyclopeptide
Antamanide	$1.01 \cdot 10^{-4}$	357	0.65
	$0.97 \cdot 10^{-5}$	359	0.77
Perhydroantamanide	$0.91 \cdot 10^{-4}$	358	0.79
	$0.91 \cdot 10^{-5}$	357	0.73

* For the experimental conditions see the footnote to Table 1.

but less stable than the contact ion pairs (λ_{max}^{Pi} = 356 nm for Na^+ and 360 nm for K^+ in $CHCl_3$). Noteworthy is the shift of λ_{max}^{Pi} from 370 nm to 367 nm when in the extraction of K^+ the enniatin B concentration was lowered—a reflection of the more effective screening of the cation from the anion in the sandwich as compared with the equimolar complex.

From the aforementioned it follows that whereas in homogeneous solutions the K/Na enniatin B-selectivity is quite low,[4, 6] in bulk membranes sandwiching by this antibiotic occurs more easily in the case of potassium than sodium. This property apparently lies at the basis of the marked K/Na selectivity of enniatin B-induced ion transport across bilayers.[3, 7]

Of particular interest is the discovery of valinomycin bicomplexes since up till now this antibiotic has been regarded as the classic monocarrier of potassium ions across biomembranes.[1, 2] In the course of investigations into the stability of complexes of valinomycin analogs by CD-controlled metal salt titration, we made use of valinomycin as reference. It was found, however, that sometimes the behavior of this compound did not conform to equimolar complexing. This is illustrated by FIGURE 8, which shows the change in the CD curves of $\sim 10^{-3}$ M solution of valinomycin in ethanol in the course of titration with KCl as well as the titration curve calculated from the results

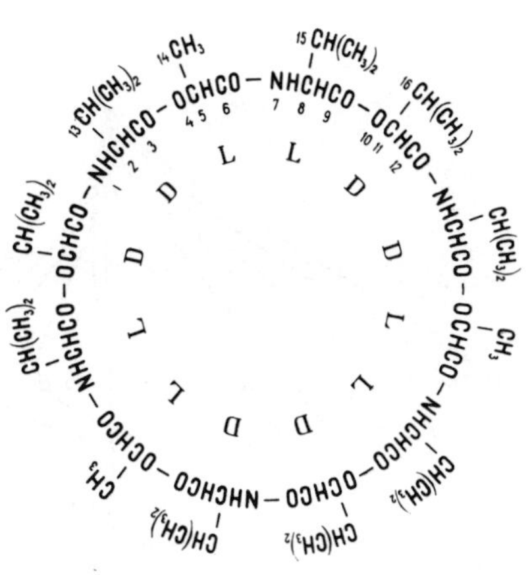

VALINOMYCIN

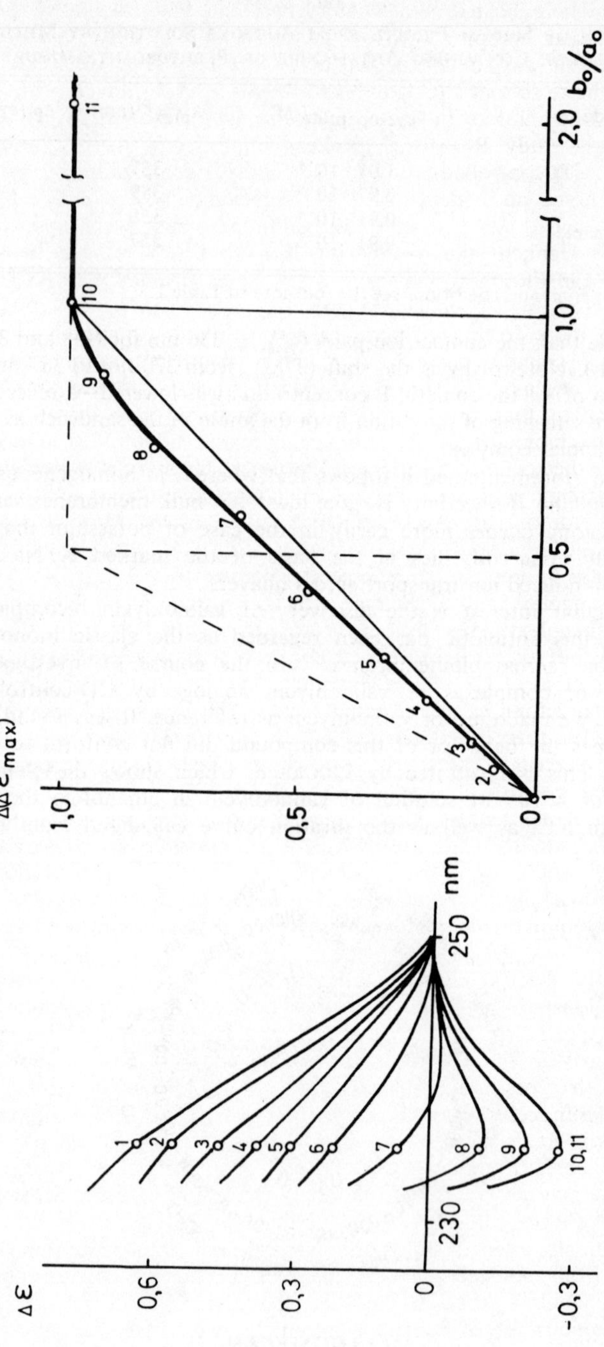

FIGURE 8. Titration of the valinomycin with KCl in 96% aqueous ethanol at 25°. (*Left*) experimental CD curves; (*right*) titration curve.

obtained, which is located somewhat higher than that calculated theoretically for an equimolar complex with an infinitely high stability constant.

Such a result was so unexpected that for a long time we treated it as an artefact, the more so that analogous experiments with cesium chloride in ethanol or potassium perchlorate in acetonitrile displayed no anomaly. In the course of the present study Rose and Henkens published a paper[17] that was also devoted to a CD-controlled study of Na^+ and K^+ complexing by valinomycin, and they observed no evidence of 2:1 complexing in ethanol. We have found that with $a_0 \approx 10^{-4}$ M, i.e., in dilute solutions which were apparently used by Rose and Henkens the complexing is in fact 1:1, as can be seen from titration curve in FIGURE 9. However, when the experiment with KCl in ethanol was repeated at higher concentrations ($a_0 \approx 4 \cdot 10^{-3}$ M) the titration curve deviated still more from the theoretical curve for an equimolar complex than it did at 10^{-3} M; similar curves were obtained when the titration was controlled spectrophotometrically (FIGURE 9). With Ba^{2+}, the ionic radius of which (1.43 Å) is similar to that for K^+ (1.33 Å), titration yields direct evidence of a varying stoichiometry of its complexing with valinomycin. During initial salt additions the value of $\Delta\epsilon$ increases, after which it stops growing (maximal sandwich content) and then begins to fall, changes sign and levels off at $b_0/a_0 > 45$ (FIGURE 10). According to E. Grell of the Max Planck Institute of Biophysical Chemistry (private communication), the titration of valinomycin by calcium salts in methanol gives a similar picture.

All the foregoing leads to the following conclusions:

(i) In ethanol at 25° valinomycin interacts with potassium ions to form not only 1:1, but also 2:1 cyclopeptide:cation complexes. Since the 2:1 sandwich content at $a_0 = 10^{-4}$ M is very low, one can assume that the constant K_2 in Equation 2 for this case does not exceed 10^3, i.e., is by about three orders of magnitude less than K_1 (Equation 1).

(ii) The electronic spectra and the optical activity of the equimolar and sandwich complexes of valinomycin are quite similar.

In the case of ϵ and $\Delta\epsilon$ values of the sandwich intermediate between the free molecule and the 1:1 complex, the titration curves of the two types of complexes would be indistinguishable and CD or uv could not be used to detect sandwiching. An example of large difference between $\Delta\epsilon$ of the sandwich and the equimolar complex can serve the Ba^{2+} complexes of valinomycin (FIGURE 10).

Because the spectral range investigated (220–250 nm) corresponds to electronic transitions of the amide (and not ester) groups, the similarity of the uv and CD spectra of the 1:1 and 2:1 potassium complexes bears evidence of the similarity in the electronic structures and mutual orientations of these groups in both types of complexes. This circumstance is a strong argument in favor of the bracelet system of amide groups with their six strong intramolecular hydrogen bonds firmly established in the equimolar complex[18–20] to be also present in both valinomycin molecules of the sandwich complex. On that basis one may propose a scheme for the structure of the sandwich complex as shown in FIGURE 11. As in the equimolar complex the sandwiched ion interacts with six ester ligands (three from each valinomycin molecule). It can be seen from molecular models that such a sandwich structure is stabilized by van der Waals interaction between the alkyl side chains, the sterically most advantageous site of contact between the two molecules apparently being the lactyl-containing ends of the "cylinders."

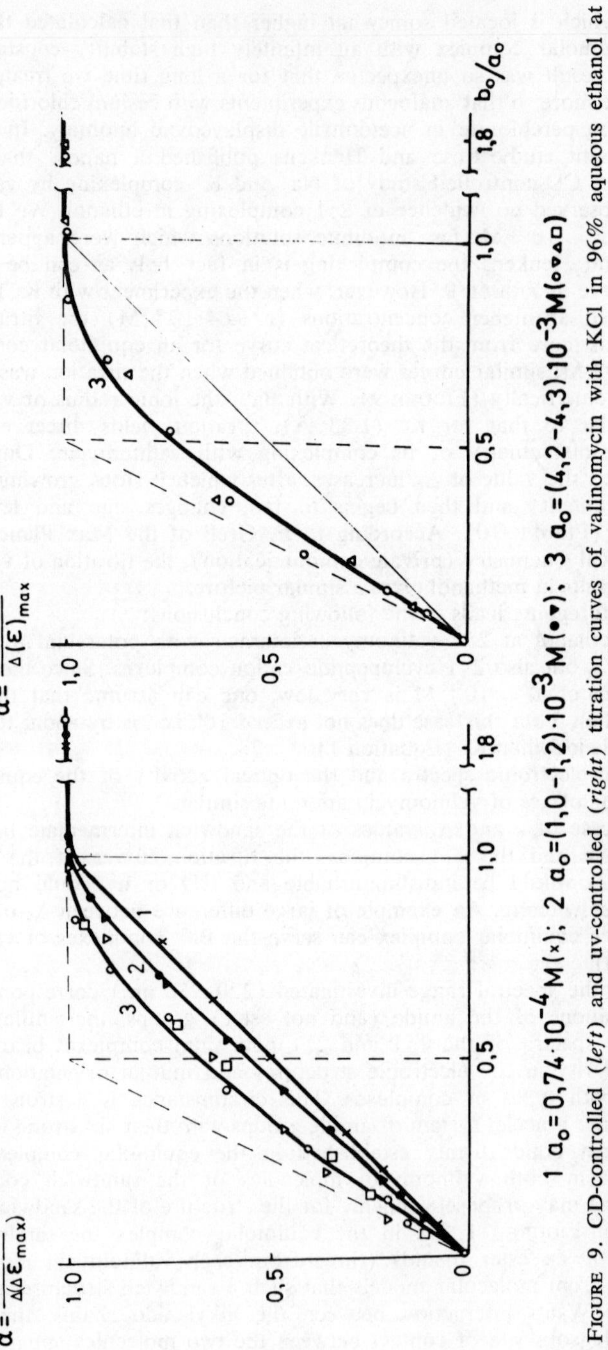

FIGURE 9. CD-controlled (*left*) and uv-controlled (*right*) titration curves of valinomycin with KCl in 96% aqueous ethanol at different antibiotic concentrations. The different designations refer to different runs.

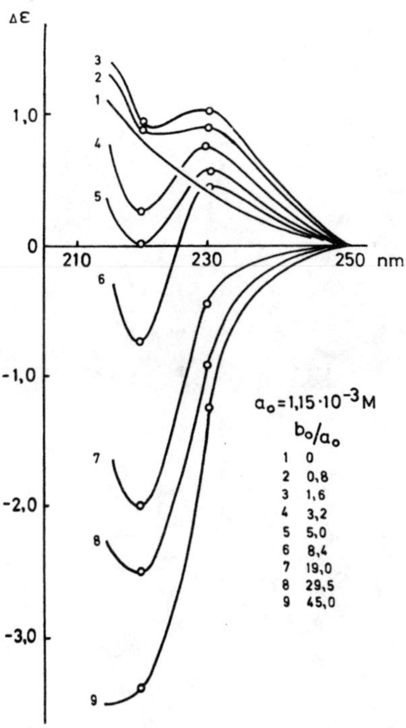

FIGURE 10. Titration of valinomycin with $Ba(ClO_4)_2$ in acetonitrile at 25°.

The results obtained give rise to the conjecture of possible participation of sandwich complexes in transmembrane potassium ion transport. For instance, under certain conditions the concentration-dependence of the potassium permeability of lipid bilayers induced by valinomycin or its analogs is of a superlinear nature, a fact that was regarded as indicating that one of the stages in the transport process is the transition of a potassium ion from one valinomycin molecule to another (Reference 2, p. 290). Now these results can be interpreted by assuming the formation of sandwiches.

In his review[22] Grell discusses the possibility of valinomycin being "incorporated" into the phospholipid membrane, as shown in FIGURE 12, rather than simply "floating" about within it (FIGURE 13). The plausibility of the first

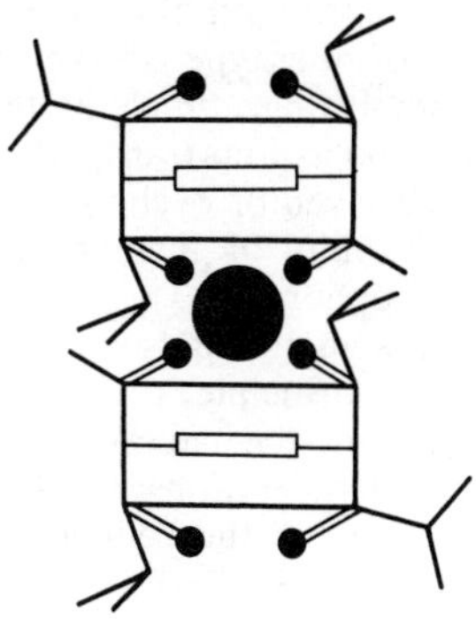

FIGURE 11. Hypothetical structure of the valinomycin sandwich.

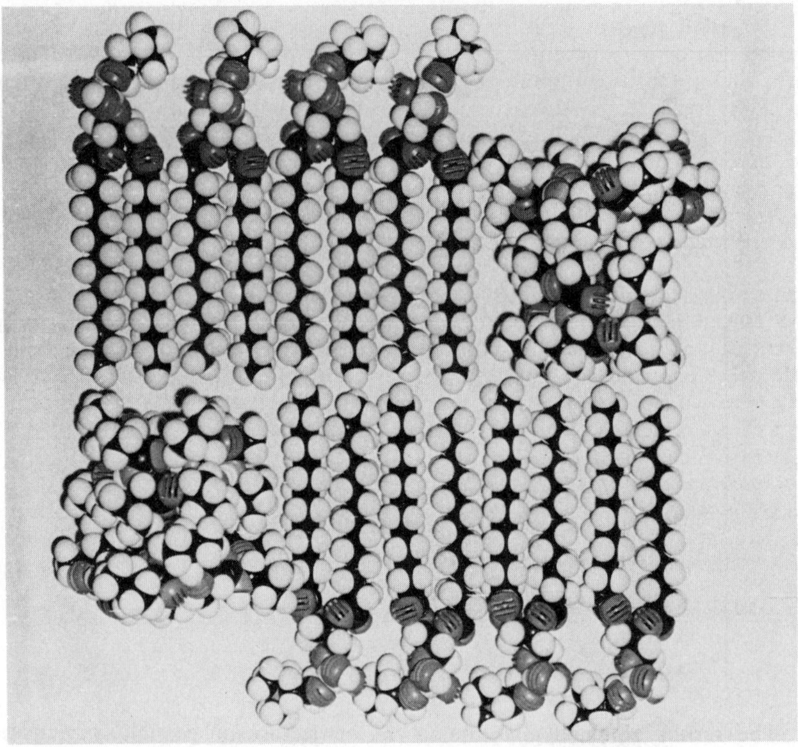

FIGURE 12. "Floating" valinomycin molecule in a dimyristoyllecithin bilayer.

variant can be seen for instance from the lesser "kinking" (according to Träuble[23]) of the alkyl groups, which results from permeation of valinomycin into the membrane. The slight changes in the mobility of the hydrocarbon chains found by Hsu and Chan[24] in their nmr studies and the increase in mobility of the phosphate groups on incorporation of valinomycin into nonsonicated liposomes are also in agreement with the viewpoint expressed here, although the above authors themselves had offered another, and in our opinion less convincing, explanation (sorption of valinomycin on the membrane surface). The presence of valinomycin clusters of the type shown in FIGURE 12 should considerably facilitate sandwich complexing, which in the limiting case could even assume the key role in the potassium-transport process.

In the concluding section a discussion will be given of cyclic peptide antamanide, which while not affecting the ion permeability of membranes, still exhibits a number of properties relating it to the ionophores.

Antamanide is found in very small amounts in the poisonous mushroom *Amanita phalloides* and is an antitoxin to its principle toxic components, namely, phalloidin and the amanitines,[25] which are the causes of irreversible pathological changes in the mammal liver.[26–28] Wieland *et al.* ascribe the antitoxic effect of antamanide to its specific ability to "tighten" the cellular membrane, lowering its permeability to the toxin molecules.[29, 30]

In an independent series of experiments it has been shown that antamanide and several of its synthetic analogs are capable of binding alkali and alkaline earth metal ions in solution, the most stable complexes being with Na^+ and Ca^{2+} (Reference 2, pp. 35–41 and the references cited therein). A comparison of the complexing and antitoxic properties of antamanide and its analogs led to the conclusion that the ability to complex Na^+ (or Ca^{2+}) ions is a necessary but insufficient condition for the manifestation of biological activity in this series of compounds. The reason for the interdependence of these two phenomena still remains unclear.

Considerable effort was spent in elucidating the molecular structures of antamanide complexes in the crystalline state[31–33] and in solution.[34–36] It can now be considered to be established that under all conditions equimolar complexes of antamanide with Li^+, Na^+, K^+ and Ca^{2+} assume a common type of saddle-like conformation, the cation interacting with four amide carbonyl oxygens situated approximately in the apices of a square. This is illustrated in FIGURE 14, which shows an example of the structure of crystalline Li^+·antamanide determined in Karle's laboratory.[31, 32] A distinguishing feature of the equimolar complexes of antamanide, in comparison with those of enniatin B or valinomycin, is that in the former only part of the solvent sheath of the ion is supplanted by the ligands of the macrocycle. For instance, Li^+ bound to antamanide retains a solvate molecule of acetonitrile (FIGURE 14), and

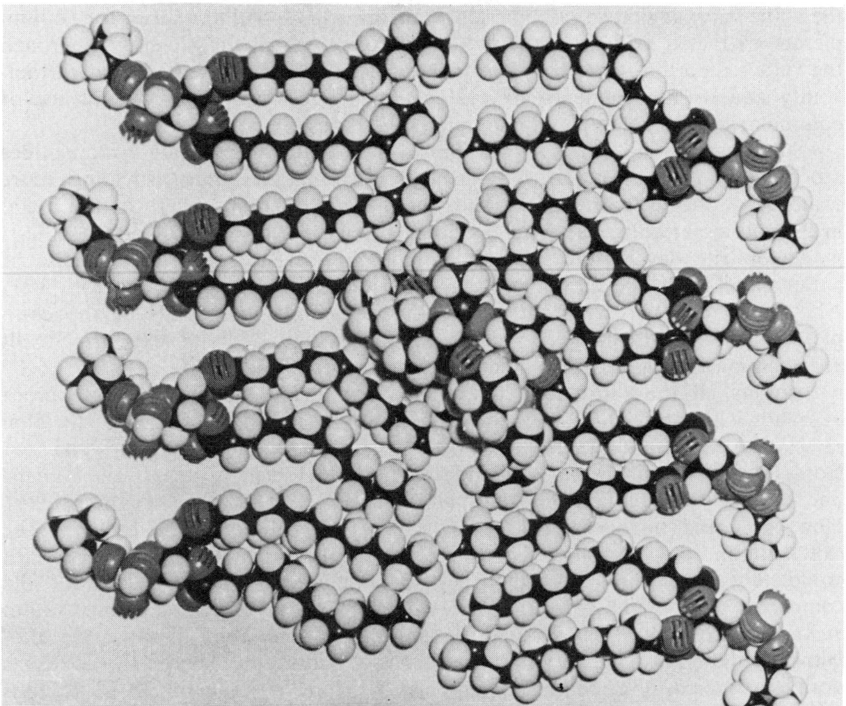

FIGURE 13. Valinomycin stacks in a dimyristoyllecithin bilayer.

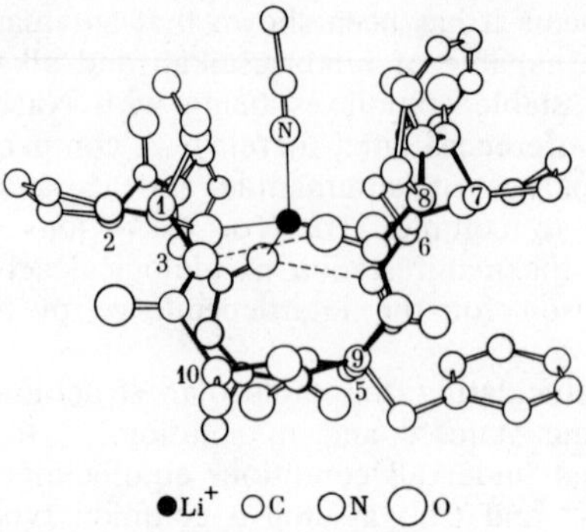

FIGURE 14. Conformation of the Li^+ complex of antamanide (after Karle[32]). Side chains are partially omitted for clarity. Dashed lines indicate the CO···HN hydrogen bonds.

Na^+ complexed with [Phe^4,Val^6]antamanide is in contact with a molecule of ethanol.[31, 33]

The relative accessibility of the cation to the environment in complexes of the antamanide series of cyclopeptides is also reflected in their ability to form ion pairs, as we have found on extracting an aqueous solution of sodium picrate with methylene chloride containing antamanide or perhydroantamanide (an analog with the phenyl groups replaced by cyclohexyls). As one can see from TABLE 2 in this case $\lambda^{Pi}_{max} \approx 358$ nm is considerably shifted towards the short-wave region in comparison with the corresponding values for sodium picrate extracted with enniatin B (~362 nm), which very closely approach the values reported for contact ion pairs (~356 nm[15, 16]). The high Na^+ extractability and its independence of cyclopeptide concentration bears evidence of equimolarity of the complex under such conditions.

Subsequent studies have shown, however, that the antamanide cyclopeptides are capable as well of forming sandwich complexes, thus providing more evidence of the unsaturated character of the coordination sphere of the metal in the equimolar complex. Whereas titration curves with $NaClO_4$ and $Ca(ClO_4)_2$ in acetonitrile deviate only very slightly from the theoretical in the case of antamanide itself, bordering on the experimental error (FIGURE 15), [Val^6, Ala^9]antamanide with its considerably higher stability constants of the complexes[37] gives rise to curves (FIGURE 16) the shape of which leaves no doubt as to the marked presence of sandwich complexes at b_0/a_0 from 0.2 to 0.7.

Among all the substances investigated heretofore the most definite proof of sandwich complexing has been obtained in the nmr-controlled titration of [Val^6,Ala^9]antamanide with $Ca(ClO_4)_2$ in acetonitrile. As one can see from FIGURE 17, on initial salt addition each NH proton gives three signals, one due to the free complexone, and one to each of the two complexes. Now, as could have been expected, with every subsequent addition of the salt solution, the intensity of the free complexone signal diminishes until it completely disappears at large excess of salt. The signal intensities of the complexes bear witness to the following processes: the predominant form at low salt concentrations gradually assumes the minor role, disappearing altogether at the end of the titration; the formerly minor complex, on the contrary, gradually increases in content, eventually attaining a value of 100%. It thus follows that in the early predominating form the macrocycle:cation ratio is 2:1 (very probably not larger, although this possibility cannot be strictly ruled out), whereas the second form has the ratio 1:1. The closeness of

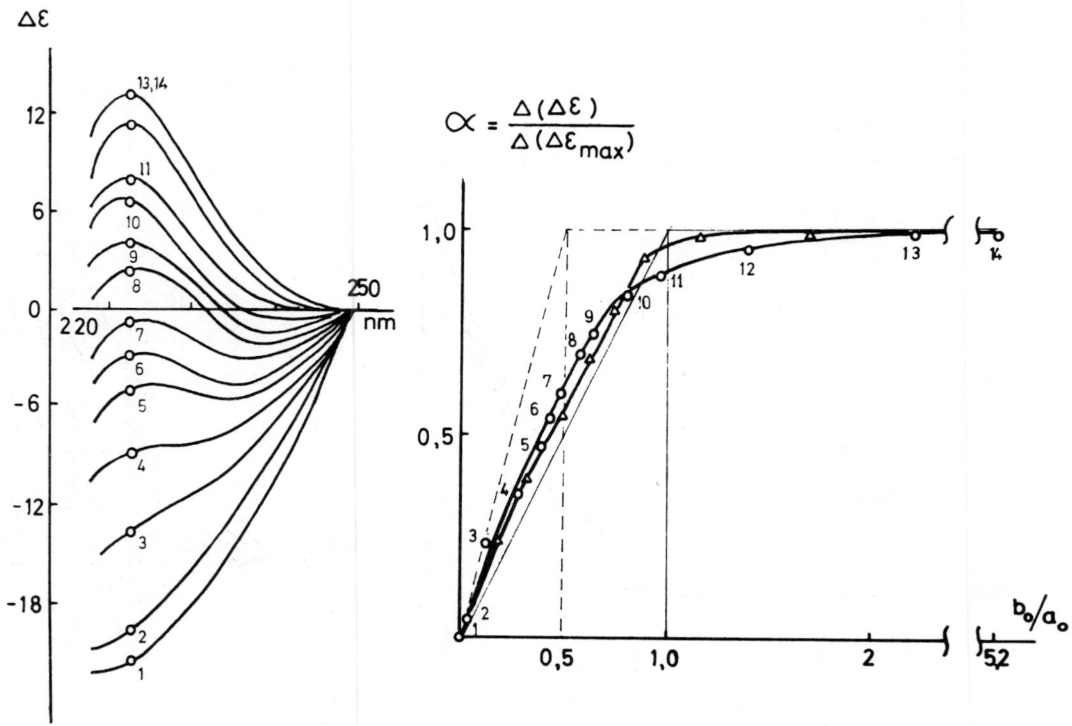

FIGURE 15. Titration of antamanide in acetonitrile at 25°. (*left*) CD spectra on titration with $NaClO_4$; (*right*) titration curves for $NaClO_4$ (○, $a_0 = 4.8 \cdot 10^{-3}$ M), and $Ca(ClO_4)_2$ (Δ, $a_0 = 5.0 \cdot 10^{-3}$ M).

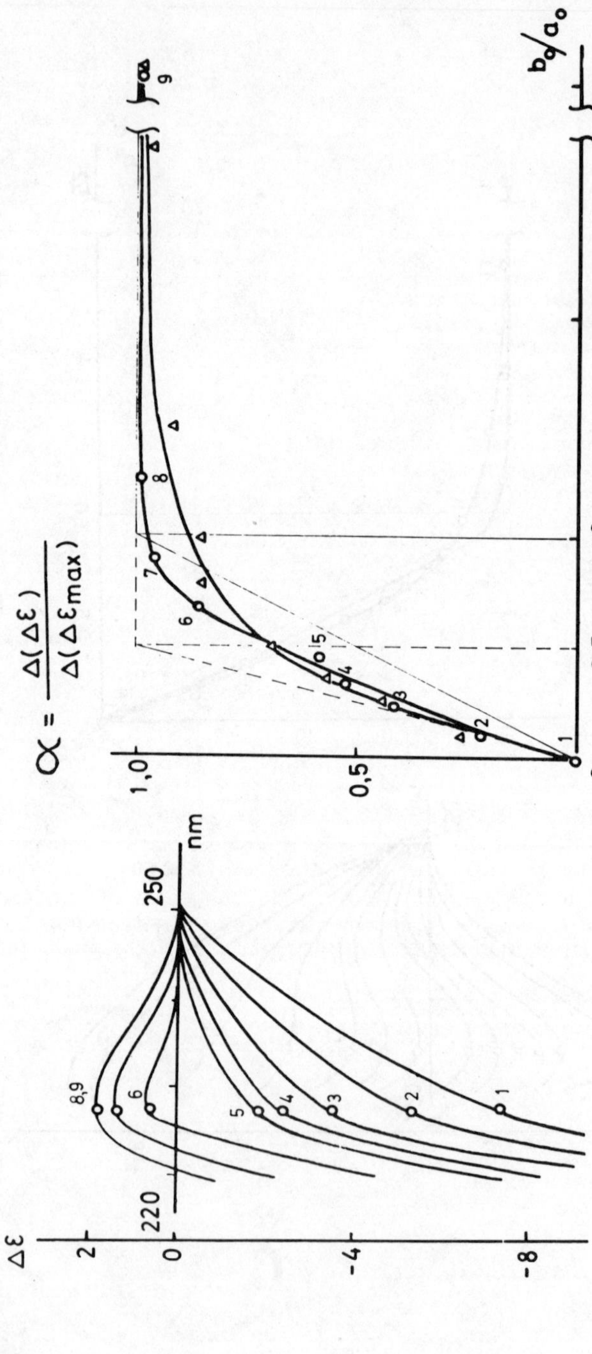

FIGURE 16. Titration of [Val^6, Ala^9] antamanide in CH_3CN at 25°. (*Left*) CD spectra on titration with $NaClO_4$; (*right*) titration curves for $NaClO_4$ (○, $a_0 = 8.8 \cdot 10^{-4}$ M) and $Ca(ClO_4)_2$ (Δ, $a_0 = 8.8 \cdot 10^{-4}$ M).

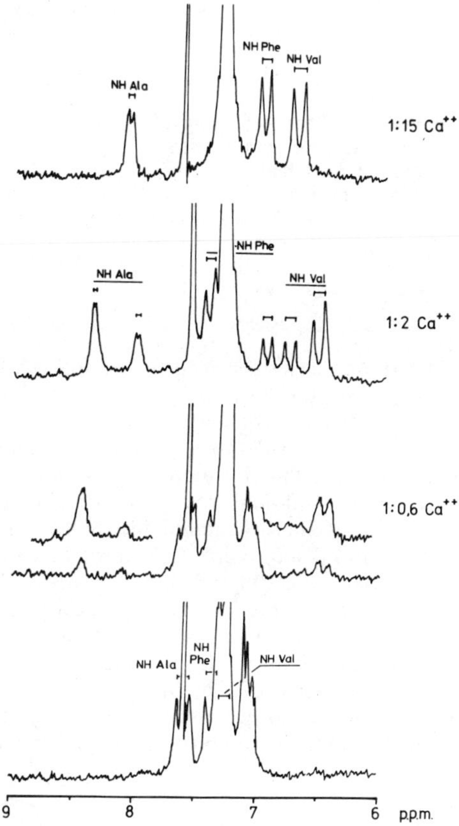

FIGURE 17. The low-field (6–9 ppm) region of the nmr spectra of [Val[6],Ala[9]]antamanide and its Ca^{2+} complexes in acetonitrile. The lowest spectrum refers to the antibiotic in salt-free solution, and each following spectrum refers to increasing concentrations of $Ca(ClO_4)_2$. For assignments of the signals see References 38 and 39.

the chemical shifts and of the $^3J_{NH-CH}$ constants of the like protons in both types of complexes is evidence of the conformational similarity of their peptide chains, whence it follows that the sandwich complexes of antamanide and its analogs have a common type of conformation, represented in FIGURE 14. The underlying structural principle of the 2:1 complexes is shown in FIGURE 18.

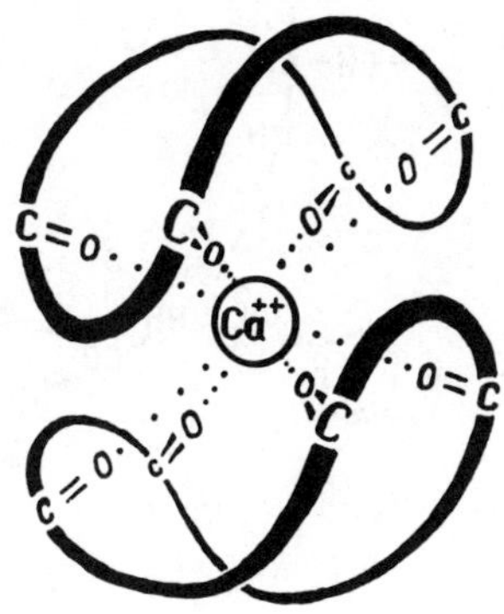

FIGURE 18. Possible structure of antamanide sandwich complexes.

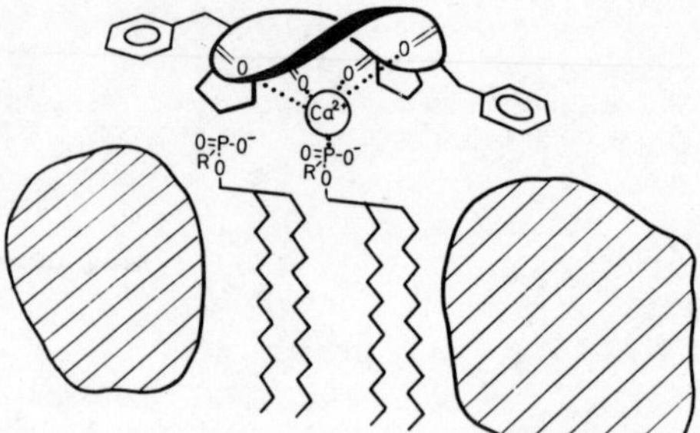

FIGURE 19. Proposed principle of antamanide interaction with a biomembrane (hatched areas refer to protein globules).

The results described lead to the following proposal for the mode of action of antamanide, the basic principle of which is schematically shown in FIGURE 19. Binding the Ca^{2+} or Na^{+} ions sorbed on the membrane surface, antamanide "covers" quite considerable areas of the latter (250–300 Å^2 according to molecular models), thereby modifying its properties, including its permeability to the *Amanita phalloides* toxins. Since the antitoxic action of antamanide is stereospecific (enantio-antamanide is less active by an order of magnitude than the natural antitoxin[40]), the protein components of the membrane apparently also participate in the binding process. The most probable type of interaction is stacking of the aromatic groups of the protein with the phenyl groups of antamanide, inasmuch as substitution of the latter by cyclohexyls, as in perhydroantamanide, while practically without effect on the complex stability, causes complete loss of the antitoxic potency.[41] It is interesting that ionophorous properties concurrently appear so that, like the naturally occurring ionophores, perhydroantamanide inhibits microbial growth and stimulates the outflow of

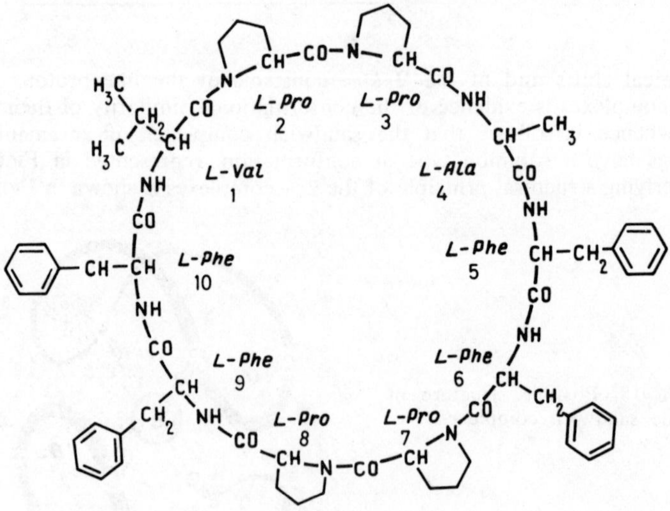

Antamanide

Na^+ or K^+ ions from lecithin liposomes; antamanide itself lacks these properties.[42] Thus the intimate study of the properties of two outwardly biologically very different classes of compounds (ionophores and antitoxins) has dramatically disclosed a very much related type of action.

In summary, it may be said that the present study has shown that the interaction of peptide ionophores, antamanide and their analogs with metal ions is not limited only to the formation of equimolar complexes. When analyzing their behavior in solutions and in membranes, one must take into account the possibility of the formation of more complicated complexes and also of ion-pairing with the counterions. The study has posed a number of important questions still awaiting their solutions, such as the optimal conditions for sandwich complexing, the stabilities of the sandwiches, and their spatial structures. The ability to form composite complexes as a function of the primary structure of the cyclopeptide has also only barely been touched. We hope to have an answer to them in the very near future.

Acknowledgments

The investigations were carried out with the participation of L. A. Fonina, N. N. Uvarova, L. V. Sumskaya, A. V. Evstratov, I. A. Lavrinovitch, T. B. Kropotnitskaya, N. M. Chekhlyaeva, T. N. Deshko, S. L. Portnova, T. A. Balashova, and S. A. Kozmin. The author acknowledges with pleasure numerous helpful discussions with Prof. Yu. A. Ovchinnikov and Dr. V. F. Bystrov.

References

1. DUNITZ, J. D., P. HEMMERICH, J. A. IBERS, C. K. JORGENSEN, J. B. NEILANDS, D. REINEN & R. J. P. WILLIAMS, Eds. 1973. Structure and Bonding. Vol. 16. Springer Verlag. Berlin, Germany.
2. OVCHINNIKOV, YU. A., V. T. IVANOV & A. M. SHKROB. 1974. Membrane active complexones. Elsevier. Amsterdam, The Netherlands.
3. IVANOV, V. T., A. V. EVSTRATOV, L. V. SUMSKAYA, E. I. MELNIK, T. S. CHUMBURIDZE, S. L. PORTNOVA, T. A. BALASHOVA & YU. A. OVCHINNIKOV. 1973 FEBS Lett. **36:** 65.
4. OVCHINNIKOV, YU. A., V. T. IVANOV, A. V. EVSTRATOV, I. I. MIKHALEVA, V. F. BYSTROV, S. L. PORTNOVA, T. A. BALASHOVA, E. N. MESHCHERYAKOVA & V. M. TULCHINSKY. 1974. Int. J. Pept. Protein Res. **6:** 465.
5. OVCHINNIKOV, YU. A. 1974. FEBS Lett. **44:** 1.
6. SHEMYAKIN, M. M., YU. A. OVCHINNIKOV, V. T. IVANOV, V. K. ANTONOV, A. M. SHKROB, I. I. MIKHALEVA, A. V. EVSTRATOV & G. G. MALENKOV. 1967. Biochem. Biophys. Res. Commun. **29:** 834.
7. SHEMYAKIN, M. M., YU. A. OVCHINNIKOV, V. T. IVANOV, V. K. ANTONOV, E. I. VINOGRADOVA, A. M. SHKROB, G. G. MALENKOV, A. V. EVSTRATOV, I. D. RYABOVA, I. A. LAINE & E. A. MELNIK. 1969. J. Membr. Biol. **1:** 402.
8. WIPF, H.-K., L. A. R. PIODA, Z. STEFANAK & W. SIMON. 1968. Helv. Chim. Acta **51:** 377.
9. IVANOV, V. T., I. A. LAVRINOVITCH, S. L. PORTNOVA, V. V. LAPSHIN, P. V. KOSTETSKY & YU. A. OVCHINNIKOV. 1974. Izv. Akad. Nauk SSSR, Ser. Khim. : 2328.
10. BYSTROV, V. F., V. T. IVANOV, S. L. PORTNOVA, T. A. BALASHOVA & YU. OVCHINNIKOV. 1973. Tetrahedron. **29:** 873.

11. Ovchinnikov, Yu. A., V. T. Ivanov, A. V. Evstratov, V. F. Bystrov, N. D. Abdullaev, E. M. Popov, G. M. Lipkind, S. F. Arkhipova, E. S. Efremov & M. M. Shemyakin. 1969. Biochem. Biophys. Res. Commun. **37:** 668.
12. Eisenman, G. & S. J. Krasne. 1973. *In* MTP International Review of Science, Biochemistry Series. Vol. 2. Butterworths. London, England.
13. Madison V., M. Atreyi, C. M. Deber & E. R. Blout. 1974. J. Am. Chem. Soc. **96:** 6725.
14. Roeske R. W., S. Isaak, T. E. King & L. K. Steinrauf. 1974. Biochem. Biophys. Res. Commun. **57:** 554.
15. Wong, D. H., M. Bourgoin & J. Smid 1974. Chem. Commun. :715.
16. Wong, K. H., K. Yagi & J. Smid. 1974. J. Membr. Biol. **18:** 379.
17. Rose, M. C. & R. W. Henkes. 1974. Biochem. Biophys. Acta **372:** 426.
18. Ivanov V. T., I. A. Laine, N. D. Abdullaev, L. B. Senyavina, E. M. Popov, Yu. A. Ovchinnikov & M. M. Shemyakin. 1969. Biochem. Biophys. Res. Commun. **34:** 803.
19. Pinkerton, M., L. K. Steinrauf & P. Dawkins. 1969. Biochem. Biophys. Res. Commun. **35:** 512.
20. Ivanov, V. T., I. A. Laine, N. D. Abdullaev, V. Z. Pletnev, G. M. Lipkind, S. F. Arkhipova, L. B. Senyavina, E. N. Meshcheryakova, E. M. Popov, V. F. Bystrov & Yu. A. Ovchinnikov. 1971. Khim. Prir. Soedin. :221.
21. Ovchinnikov, Yu. A. & V. T. Ivanov. 1974. Tetrahedron **30:** 1871.
22. Grell, E. 1974. *In* Membranes. Vol. 3. Marcel Dekker. New York, N.Y.
23. Träuble, H., 1971. J. Membr. Biol. **4:** 193.
24. Hsu M. & S. I. Chan. 1973. Biochemistry **12:** 3872.
25. Wieland Th., G., Lüben, H. Ottenheym, J. Faesel, J. X. De Vries, W. Konz, A. Prox & J. Schmid. 1968. Angew. Chem. **80:** 209.
26. Wieland, Th. 1972. Naturwissenschaften **59:** 225.
27. Wieland, Th. & V. M. Govindan. 1974. FEBS Lett. **46:** 531.
28. Frimmer, M., R. Kroker & J. Postendörfer. 1974. Naunyn-Schmiedebergs Arch. Pharmakol. **284:** 395.
29. Wieland, Th., H. Faulstich, W. Jahn, V. Govindan, H. Puchinger, Z. Kopitar, H. Schmaus & A. Schmitz. 1972. Hoppe-Seyler's Z. Physiol. Chem. **353:** 1337.
30. Faulctich, H., Th. Wieland, A. Walli & K. Birkmann. 1974. Hoppe-Seyler's Z. Physiol. Chem. **355:** 1162.
31. Karle, I. L., J. Karle, Th. Wieland, W. Burgermeister, H. Faulstich & B. Witkop. 1973. Proc. Nat. Acad. Sci. U.S.A. **70:** 1836.
32. Karle, I. L. 1974. J. Am. Chem. Soc. **96:** 4000.
33. Karle, I. L. 1974. Biochemistry **13:** 2155.
34. Ivanov, V. T., A. M. Miroshnikov, N. D. Abbulaev, L. B. Senyavina, S. F. Arkhipova, N. N. Uvarova, K. Kh. Khalilulina, V. F. Bystrov & Yu. A. Ovchinnikov. 1971. Biochem. Biophys. Res. Commun. **42:** 654.
35. Patel, D. J. 1973. Biochemistry **12:** 677.
36. Patel, D. J. & A. E. Tonelli. 1974. Biochemistry **13:** 788.
37. Miroshnikov A. I., K. Kh. Khalilulina, N. N. Uvarova, V. T. Ivanov & Yu. A. Ovchinnikov. 1973. Khim. Prir. Soedin. : 214.
38. Bystrov V. F., S. L. Portnova, T. A. Balashova, S. A. Kozmin, Yu. D. Gavrilov & V. A. Afanasev. 1973. Pure Appl. Chem. **36:** 19.
39. Ivanov, V. T., A. I. Miroshnikov, S. A. Kozmin, E. A. Meshcheryakova, L. B. Senyavina, N. N. Uvarova, K. Kh. Khalilulina, V. F. Bystrov & Yu. A. Ovchinnikov. 1973. Khim. Prir. Soedin. 378.
40. Wieland, Th., C. Birr, W. Burgermeister, P. Trietsch & G. Rohr. 1972. Liebig's Ann. Chem. **759:** 71.
41. Wieland, Th., M. Faulstich & W. Bürgermeister. 1972. Biochem. Biophys. Res. Commun. **47:** 984.
42. Ovchinnikov, Yu. A., V. T. Ivanov, L. I. Barsukov, E. I. Melnik, N. I.

Oreshnikova, N. D. Bogolyubova, I. D. Ryabova, A. I. Miroshnikov & V. A. Rimskaya. 1972. Experientia **28:** 399.

Discussion

Dr. C. Deber (*University of Wisconsin, Madison, Wisc.*): I'm really intrigued by the clusters of valinomycin molecules and by the fact that apparently cations reside in more than one place with respect to the ionophore. It is fascinating to picture a stack of valinomycin molecules one on top of the other, thereby doing the ultimate step of turning the carrier into a channel.

Dr. Ivanov: At the moment this would be too much to picture—I'm not insisting that this is the only mechanism by which valinomycin works. I was just trying to outline one of the possible mechanisms for valinomycin at certain conditions. A channel could possibly be formed from four valinomycins. This would demand a fourth-power dependency of conductance increase, which does not happen. So I would prefer to say that sandwiching does occur in solution and leave the rest to the others.

Dr. Haynes: About four or five years ago, Dr. Pressman and I looked at enniatin complexation with the resonance techniques. We also looked at the biological activity of enniatin in the mitochondria, and we found only first-power dependence of the transport reaction on enniatin concentration, indicating that there is essentially one enniatin complexing one cation to make one effective complex which transverses a membrane.

In the nuclear magnetic resonance studies at concentrations of about 20 mM we found evidence for sandwich complexation as well, but calculation under relevant biological conditions shows that membrane concentration is much lower than millimolar. Of course with lipid bilayer one can jack this up, but in biological membranes we're working with less than millimolar concentration. I don't see that sandwich complexation or tunnel formation or anything is really necessary for these ionophores.

Dr. Ivanov: As far as the probability or possibility of enniatin sandwiching during the membrane transport is concerned, the bilayer work done at our Institute is published and shows second-power dependence of the potassium transport in bilayers. Some work in mitochondria was also done at the Institute and it shows that K uptake into mitochondria has a very steep dependence on enniatin concentration. One just has to compare the conditions under which this work was done to understand why we have disagreement between the two results. But I can say for sure that on bilayer you can really see the second-order dependency.

DISCUSSION PAPER

Alan Finkelstein

Department of Physiology
Albert Einstein School of Medicine
Bronx, New York 10461

I would like to make some comments on a few of the papers that have been presented today. The first one has to do with a point that Dr. Parsegian raised, and that is the following: in 0.1 M KCl the conductance of a gramicidin A channel, say in a glycerol monooleate bilayer, is $10^{-11}\ \Omega^{-1}$. Now if you make the most naive calculation possible and assume that the channel is filled with 0.1 M K^+ and that the mobility of K^+ within the channel is the same as in free solution, then for a channel 2 Å in radius and 30 Å in length (the presumed dimensions of the gramicidin channel), you predict that its conductance should be $2 \times 10^{-11}\ \Omega^{-1}$. In other words, the observed value is only a factor of two lower than that predicted from this very naive calculation. On the other hand, as Dr. Parsegian has pointed out, the energy required to move an ion into a pore of radius 2 Å surrounded by a medium of dielectric constant of about 2 is something of the order of 14 kilocalories, which means that the partition coefficient of the potassium ion into such a channel should be of the order of 10^{-10}. That is, its concentration in the channel should be one tenbillionth its concentration in bulk solution. If you assume that the mobility of K^+ in the pore is not tremendously high, and I think one would be hard-pressed to justify the assumption of a much higher mobility than in free solution, you would predict from this slightly less simple, but still rather basic calculation, that the conductance of the gramicidin channel should be 10^{-10} that of the previously calculated value; that is, it should be something like $2 \times 10^{-21}\ \Omega^{-1}$.

Incidentally, if you look at the available data for single-channel conductances in bilayers and what's been written about them, you find that there's a general rejoicing at the fact that these conductances roughly follow the predictions of our naive calculation. However, investigators like Adrian Parsegian, who have thought about this, realize that what this means is that there are other energy terms that compensate for the electrostatic energy barrier and fortuitously sum with it to a final result in agreement with the naive calculation.

Offhand there are two possible explanations for the high-conductance of a gramicidin channel, one of which Dr. Parsegian mentioned. It may be that the carbonyl groups lining the channel are much more polar than the OH dipoles of water, so that the effective dielectric constant within the channel is much greater than 80. I'm no expert in this area, but it seems to me that if this is the case, one would expect a much different and greater selectivity by the channel among the alkali cations than that observed (which essentially parallels their relative mobilities in free solution). Another possibility is that the dielectric constant of the gramicidin molecule outside the pore, that is, the outside part of the molecule, is significantly greater than 2. The energetic calculation that I alluded to[1] is very sensitive to the dielectric constant of the bulk phase in which the channel is embedded, particularly the dielectric constant in its immediate vicinity.

I thought I'd raise this issue and also call your attention to the fact that data on single-channel conductances in biological membranes are also in this same gramicidin "ballpark." For example, in the neuromuscular junction the channels are thought to have a conductance of about $3 \times 10^{-11}\ \Omega^{-1}$. Now these channels are, from a priori considerations, not going to have a radius much larger than 2 Å, if that, so this same problem is present in channel data from biological membranes. This at least deserves to be mentioned. (Of course the structure of the biological channels is completely unknown; if they contained negative-charged groups, that would take care of the energetic problem of moving a cation into them.)

The second point I want to make, also about gramicidin A, has to do with the size of the channel. A year or so ago I measured the water permeability of the gramicidin-treated membrane and found that water permeability was indeed enhanced in the presence of gramicidin.[2] That is, water could go through the gramicidin channel. (Again it was rather amusing that the water permeability was within a factor of 8 of the prediction from simple diffusion of bulk water through the channel.) Also, I measured the urea permeability and found essentially no increase above that for unmodified membrane. Thus, the gramicidin channel allows water but not urea to get in, and this is certainly consistent with a pore of radius of about 2 angstroms.

References

1. PARSEGIAN, A. 1969. Energy of an ion crossing a low dielectric membrane: Solutions to four relevant electrostatic problems. Nature **221:** 844.
2. FINKELSTEIN, A. 1974. Aqueous pores created in thin lipid membranes by the antibiotics nystatin, amphotericin B, and gramicidin A: Implications for pores in plasma membranes. *In* Drugs and Transport Processes. B. A. Callingham, Ed.: 241. Macmillan. London, England.

Discussion

DR. M. BENDER: (*Fairleigh Dickenson University, East Rutherford, N.J.*): The way I understand a carrier is that ions per se will not go through the membrane on account of the hydrophobic barrier. A carrier is needed to take them through, but according to the results I've just heard from the previous two speakers, it did appear to me that it's easier for ions to go through than to be carried through.

DR. FINKELSTEIN: That also follows from the calculation that Parsegian made. That is, even bringing in the electrostatic barrier, you conclude that it's easier for an ion to cross the bilayer through a channel than via a carrier. The issue that I raised is that it's too easy.

DR. BENDER: You are saying then that there must be a high dielectric constant in the vicinity of the passage of the ion.

DR. FINKELSTEIN: That's an implication. I don't know how strongly I feel about this; it's certainly something one should consider.

DR. NEBEROR: Dr. Finkelstein, when you make a naive calculation I presume that you use a bulk diffusion constant in the channel?

DR. FINKELSTEIN: Right.

DR. NEBEROR: But that's a sort of theory that is absolutely not valid.

DR. FINKELSTEIN: Well, what would you expect the ion mobility to be? Would you expect it to be faster in the channel than in the bulk solution? Or slower?

DR. NEBEROR: No as a matter of fact; in that case I'll sit down with Dr. Parsegian and we'll give you a new series.

DR. FINKELSTEIN: No, just the sign. I agree with you that mine was an oversimplified calculation, but what do you expect the mobility of the ion in the channel to be? Greater or less than in free solution?

DR. NEBEROR: I would think that the mobility would be considerably less.

DR. FINKELSTEIN: Right; so that makes the problem even worse. This number of $2 \times 10^{-21}\ \Omega^{-1}$ is based on the assumption that the ion concentration within the channel is calculated from the Born energy, and its mobility is the same as in free solution. Most likely that mobility will be even less, which would bring this number down even lower and even more out of line with the observed $10^{-11}\ \Omega^{-1}$.

MEMBRANE EXCITATION THROUGH VOLTAGE-INDUCED AGGREGATION OF CHANNEL PRECURSORS*

Paul Mueller

Department of Molecular Biology
Eastern Pennsylvania Psychiatric Institute
Philadelphia, Pennsylvania 19129

Introduction

Several compounds of fungal or bacterial origin (EIM, alamethicin, monazomycin, DJ400B) readily incorporate into preformed planar lipid bilayers, form molecular channels, and generate voltage-dependent ion conductances.[1-4] When studied by voltage clamp, the kinetic and steady-state characteristics of these conductance changes are in every respect identical to those found in cells, and their major aspects can be quantitatively described by the Hodgkin and Huxley[5] equations. Thus, the steady-state conductance is a steep exponential function of the membrane potential, it saturates at high potentials, and there is rectification. In response to a stepwise displacement of the voltage of the proper sign, the conductance rises with a sigmoid time course. It decays with an exponential time course for potential steps of opposite sign and the time constants of the conductance changes go through a maximum as a function of the potential. Under appropriate conditions the conductances also show inactivation, as seen in the sodium system of nerve and the potassium system of muscle. In addition, there appear for particular pulsing sequences certain kinetic transients that cannot be accounted for by the Hodgkin and Huxley equations, but are also seen in identical form in nerve.[3] Because the kinetics are identical in all excitable cell membranes and also in these bilayers, it is likely that in spite of the diverse chemical nature of the channel-forming molecules in the bilayers and the widely differing ion selectivities in the cellular systems, the mechanism by which the membrane opens and closes for the flow of ions is essentially the same in all cases.

The Gating Mechanism

The kinetic data imply that a cooperative process is involved in the gating action. In principle, two different concepts could account for the kinetics: one involving an intramolecular configurational change within a permanent channel that spans the membrane; the other, the assembly of a channel through the voltage-dependent aggregation of monomeric channel precursors.

In contrast to nerve or muscle where the analysis has primarily relied on voltage clamp data, the bilayer systems allow the modification of the lipid composition and of the chemical nature, concentration, and sidedness of the channel-forming compounds.

Several aspects of these data, in particular the high-order dependence of the steady-state conductance and of the gating time constants on the concentra-

* This work was supported by Grant No. GB 8144-01 from the National Science Foundation.

tion of the channel-formers, suggest an aggregation mechanism, and a detailed model has been proposed[3, 4] in which the gating involves the voltage-induced insertion of all or part of the translocator molecules from the membrane surface into the hydrocarbon region and their subsequent aggregation into open channels by lateral diffusion. The long axis of the translocators is assumed to span the hydrocarbon region forming the walls of the channel like the staves of a barrel (FIGURE 1).

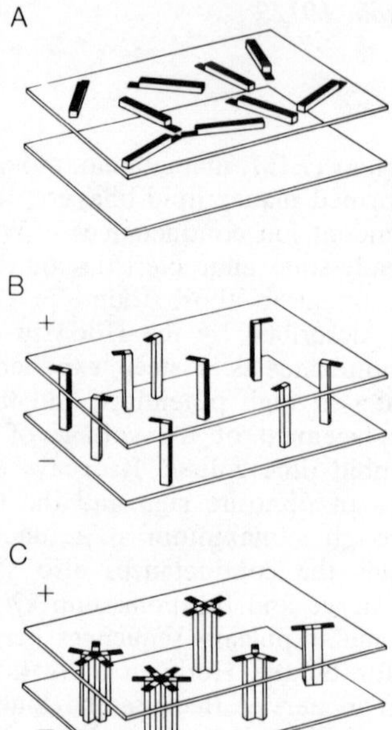

FIGURE 1. A model of the gating process. (A) At rest, the translocator molecules lie flat on one membrane surface. The extension at one end of the molecule represents a water-soluble anchor group. (B) An applied field acting on a gating charge has tilted the molecules into the membrane. (C) Lateral diffusion within the membrane plane leads to aggregation of the monomers into oligomers. Trimers and higher oligomers, but not dimers, form a central opening acting as a channel for the flow of ions. (From Baumann and Mueller.[3] By permission of the publisher.)

This two-step insertion-aggregation mechanism is one of the several possible molecular representations of the Hodgkin-Huxley scheme and its mathematical description accounts quantitatively not only for the conductance-voltage kinetics, including inactivation, but also for a set of complex kinetic features that deviate from the Hodgkin and Huxley kinetics in the sense that the rate constants of the changes are dependent not only on the membrane potential but also on the value of the conductance and on time.[4] The model is also in agreement with single-channel data,[3, 4, 6] and such data can provide additional information about the aggregation process.

In the following I should like to discuss some further details of the aggregation process, the molecular aspects of the channel configuration, and the relevance of the bilayer data to cellular excitability.

Formally the gating process is described by two steps. The first is the insertion or tilting of the monomeric translocators by the applied field into the membrane hydrocarbon region.

Thus:

$$P_o \underset{k_{10}}{\overset{k_{01}}{\rightleftharpoons}} P_1 \qquad (1)$$

where P_o represents the fraction of the translocators at the membrane surface and P_1 that of inserted monomers. The insertion is assumed to be a first-order voltage-dependent process, and the rate constants are of the form:

$$k_{01} = A \exp - [(E_{01} - z0.5\,V\,23.05)/RT] \tag{2}$$
$$k_{10} = A \exp - [(E_{10} + z0.5\,V\,23.05)/RT] \tag{3}$$

k_{01} and k_{10} corresponds to α and β in the Hodgkin-Huxley equations, z is the valence of the gating charge, and the factor 0.5 implies that the gating charge crosses most of the field, i.e., traverses the entire hydrocarbon region.

During the second step, the inserted monomers aggregate according to:

$$P_1 + P_n \underset{k_{n+1,n}}{\overset{k_{n,n+1}}{\rightleftharpoons}} P_{n+1} \tag{4}$$

$$P_1 + P_{m-1} \underset{k_{m,m-1}}{\overset{k_{m-1,m}}{\rightleftharpoons}} P_m \tag{5}$$

there P_n varies from P_1 to P_m and represents the fraction of monomers, dimers, trimers, etc., P_m being the largest oligomer being formed. The total concentration of the translocators

$$C = P_o + \sum_{n=1}^{m} n\,P_n \tag{6}$$

The differential rate equations describing this scheme have the following form:

$$d\,P_o/dt = P_1k_{10} - P_ok_{01} \tag{7}$$

$$d\,P_1/dt = P_ok_{01} - P_1k_{10} + 2P_2k_{21} - \sum_{n=1}^{m-1} P_1P_nk_{n-1,n} + \sum_{n=3}^{m} P_nk_{n,n-1} \tag{8}$$

$$d\,P_n/dt = P_1P_{n-1}k_{n-1,n} + P_{n+1}K_{n+1,n} - P_nk_{n,n-1} - P_1P_nk_{n,n+1};\ n > 1, \tag{9}$$

$$d\,P_m/dt = P_1P_{m-1}k_{m-1,m} - P_mk_{m,m-1} \tag{10}$$

In first approximation, one might assume that the aggregation rate constants are independent of the voltage and of the general form:

$$k_{n,n+1} = A \exp(-E_{n,n+1}/RT) \tag{11}$$

Numerical integrations of the rate equations have shown that the scheme can account for most of the observed kinetics, including inactivation and the state-dependent and multiple time constants.[3, 4] When the aggregation rates are assumed to be much faster than the insertion rate, the scheme reduces to the Hodgkin-Huxley scheme.

The Voltage-Dependence of the Aggregation Rates

In the kinetic scheme described above it was assumed that the aggregation rate constants (Equation 11) do not depend on the membrane potential. However several observations make this assumption untenable.

The first is the fact that both in nerve and in the bilayers the maximal

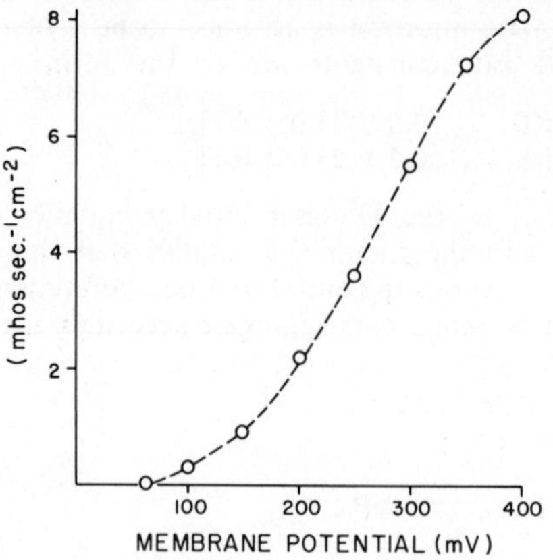

FIGURE 2. The maximal rate of rise of the membrane conductance (ordinate) in response to applied potential steps (abcissa). Voltage clamp. Lecithin bilayer in the presence of 5×10^{-6} monazomycin. The conductance values were corrected for the hyperbolic sine relation between the voltage and the instantaneous conductance of the open channels. The curve represents therefore the true maximal gating rate as a function of the potential displacement. The gating rate continues to rise up to the breakdown potential. Alamethicin and the sodium and potassium channels of nerve show the same phenomenon.

rate of rise of the conductance after a sudden displacement of the membrane potential is approximately a linear function of the potential up to several hundred mV; i.e., for values at which the insertion rates become very fast (FIGURE 2) and since the state-dependent kinetic details can only be accounted for when the aggregation process becomes to some extent rate-limiting, the maximal rate of rise should not increase any further at potentials where the insertion rate is faster than the aggregation rate. Thus, the curve of FIGURE 2 should level off somewhere near 100 mV.

A second observation concerns certain kinetic data that can only be accounted for under the assumption that the aggregation rate constants depend on the voltage in the sense that the ratio of the aggregation to the dissociation rate constants increases with potentials of a sign that increases the conductance. An example is shown in FIGURE 3. When potential steps are applied from

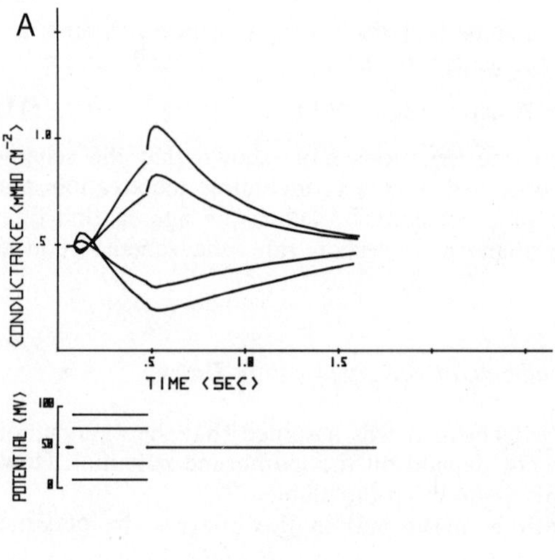

FIGURE 3A. Membrane conductance in response to applied potential steps from a bilayer in 0.01 M NaCl in the presence of 10^{-6} M monazomycin on one side. The membrane solution contained 100 mg oleyl phosphate and 50 mg cholesterol in 1 ml of octane. The conductance was derived from recorded membrane currents. Note that at the beginning of the records the conductance is high due to a holding potential of 50 mV and that it rises transiently for potential steps that lower the steady-state conductance and vice versa. Similar records have been obtained with alamethicin and the potassium system of the squid axon.

a holding potential at which the membrane conductance is high, the corresponding conductance changes show in certain lipids a transient phase of a sign opposite to that expected from the applied potential, i.e., a potential step that leads to a further increase in the steady-state conductance, causes a transient conductance decrease, and vice versa. Quite often, these transients are preceded by a brief transient of the proper sign. The phenomenon depends on the lipid fluidity and temperature, but can be observed with all translocators

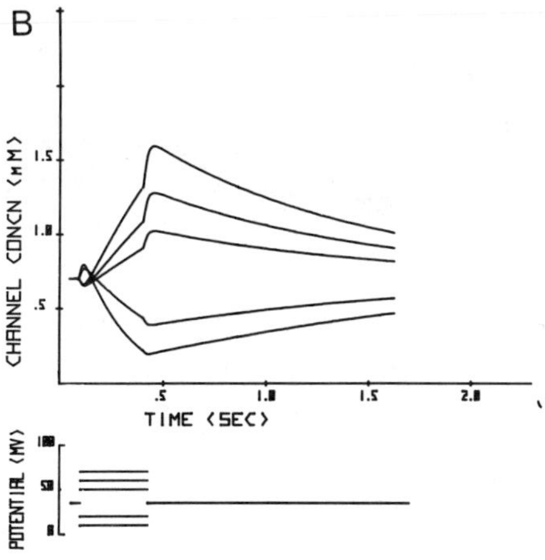

FIGURE 3B. The channel concentration as a function of time and voltage as calculated from the aggregation model. It was assumed that the aggregation rate constants are voltage-dependent. Equations 7 through 10 were integrated by numerical methods and Equations 12 through 14 were used for the rate constants. Only the hexamer was assumed to be conductive. The following constants were used: $C = 1$, $A = 1$, $z = 2$, $E_{01} = 4$, $E_{10} = 2$, E_{12} to $E_{65} = 0$, $\epsilon_1 = 3$, $\epsilon_2 = 2$, $q = 1$, $w_1 = 1$, $w_2 = 0.5$, $w_3 = 0.2$, $w_4 = 0.1$, $w_5 = 0.1$. In the model the transient conductance changes result from transient changes of the monomer concentration, which in turn control the formation of conducting hexamers.

and has also been seen in nerve.[7] A quantitative explanation on the basis of the aggregation scheme (FIGURE 3B) requires that the aggregation rate constants of the lower nonconducting oligomers, such as dimers or trimers, have a stronger voltage-dependence than those of the larger aggregates.

For the kinetic calculations shown in FIGURE 3B, the aggregation rate constants were modified by empirical expressions allowing the adjustment of the voltage-dependence:[3]

$$k_{n,n+1} = A \exp\left[-(E_{n,n+1} - W_n f(V))/RT\right] \tag{12}$$

$$k_{n+1,n} = A \exp\left[-(E_{n+1,n} + W_n f(V))/RT\right] \tag{13}$$

$$f(V) = \frac{\exp\left[-(\epsilon_1 - qV\,23.05)/RT\right]}{\exp\left[-(\epsilon_1 - qV\,23.05)/RT\right] + \exp\left[-(\epsilon_2 - qV\,23.05)/RT\right]} \tag{14}$$

The voltage-dependent term, $W_n f(V)$ modifies the interaction energies between the monomers and the different oligomers as a function of the potential, and ϵ and q control the range and amplitude of the voltage effect. FIGURE 4, which was calculated from equations 12, 13 and 14, shows an example of how the rate constants are assumed to vary as a function of the potential.

Further evidence for the voltage-dependence of the aggregation rate constants comes from the single-channel conductance fluctuations observed with alamethicin.[3, 6, 8, 9] As seen in FIGURE 5, the conductance fluctuates in bursts between several well-defined levels and the spacing of the levels increases from the lower to the higher level in a characteristic manner. The conductance

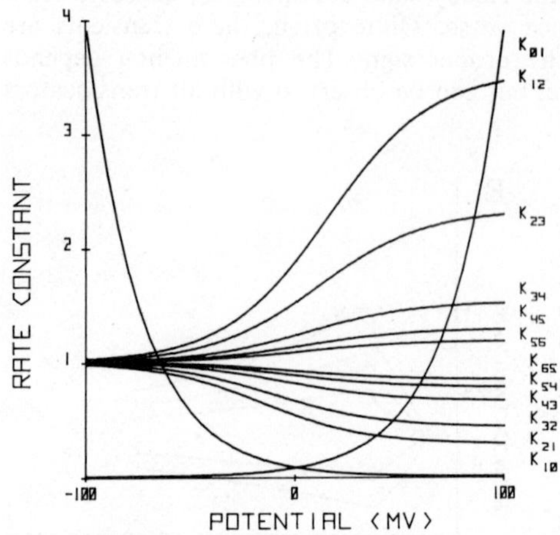

FIGURE 4. An example of the voltage-dependence of the aggregation rate constants. This type of voltage-dependence was used for the calculations in FIGURE 3B. Note that the aggregation constants increase, whereas the dissociation constants decrease with increasing voltage, and that this effect is more pronounced for the lower oligomers. The curves were calculated from Equations 12 through 14 with the following constants: $A = 1$, $z = 2$, $E_{01} = 1.5$, $E_{10} = 1.5$, E_{12} to $E_{65} = 0$, $\epsilon_1 = 1$, $\epsilon_2 = 1$, $q = 0.5$, $w_1 = 1.5$, $w_2 = 1$, $w_3 = .5$, $w_4 = .3$, $w_5 = .2$. The equations permit the adjustment of slopes and range of the voltage-dependence and can be made to fit the voltage-dependent rate constants shown in FIGURE 6B approximately with the following constants: $A = 100$, $z = 2$, $E_{23} = 1.5$, $E_{32} = 2$, $E_{34} = 2$, $E_{43} = 1.5$, $E_{45} = 2.5$, $E_{54} = 1$, $E_{56} = 3$, $E_{65} = 0.5$, $E_{67} = 3.5$, $E_{76} = 0$, $\epsilon_1 = 0$, $\epsilon_2 = 0$, $q = 0.1$, w_1 to $w_5 = 3$.

steps can be interpreted as being caused by the changes of the channel diameter due to the insertion and removal of a single alamethicin molecule into a barrel-stave channel.[3, 4, 6] Accepting this interpretation, the forward and backward rates of formation of the individual oligomers can be determined by dividing the number of transitions, ν, from a given level, σ, to another by the fraction of total observation time the level is occupied ($\tau\sigma$). If the values

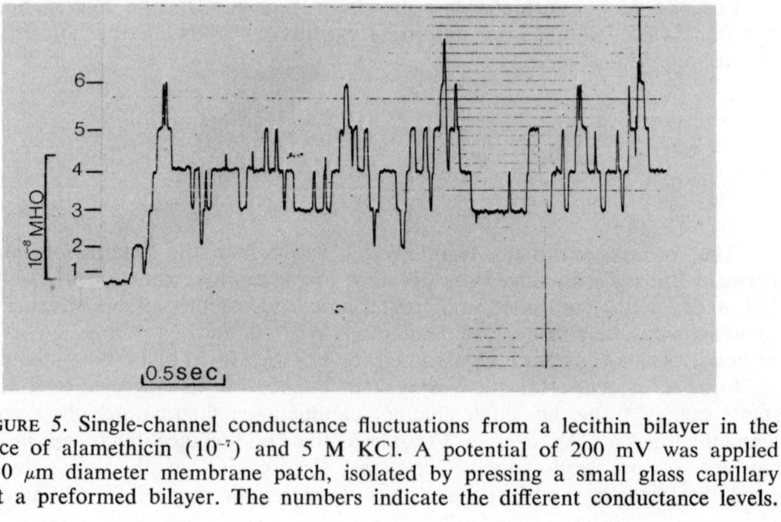

FIGURE 5. Single-channel conductance fluctuations from a lecithin bilayer in the presence of alamethicin (10^{-7}) and 5 M KCl. A potential of 200 mV was applied to a 10 μm diameter membrane patch, isolated by pressing a small glass capillary against a preformed bilayer. The numbers indicate the different conductance levels.

obtained are then divided by the number of monomers, n, assumed to form the channel corresponding to a given conductance level, we obtain the aggregation rates:

$$P_1 P_n k_{n,n+1} = \frac{\nu_{\sigma,\sigma+1}}{\tau_\sigma n}$$

and the dissociation rate constants:

$$k_{n,n-1} = \frac{\nu_{\sigma,\sigma-1}}{\tau_\sigma n}$$

of the conducting oligomers.

Unfortunately, we do not know the concentration, P_1, of inserted monomers and, therefore, $k_{n,n+1}$. Also the number of monomers forming the smallest channel is not known. There is, therefore, no direct way of determining the rate constants of monomer insertion and of formation of the nonconducting smaller oligomers. Nevertheless, several interesting points emerge from a study of the molecular rates as a function of the membrane potential.

As shown in FIGURE 6, the rates of aggregation increase with increasing potential, whereas the rates of disaggregation either decrease or increase; however, the ratio of the aggregation rates to those of dissociation increase e-fold for every 40–60 mV. The voltage-dependence of the individual rate constants

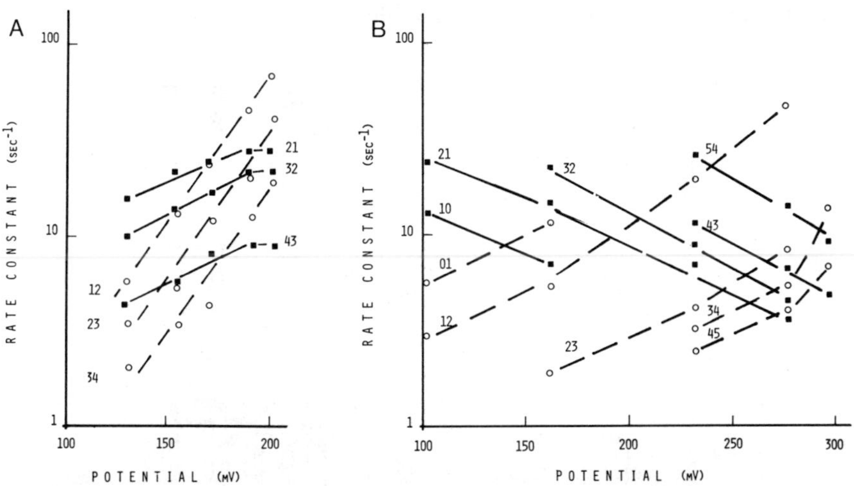

FIGURE 6. The voltage-dependence of the transition rates between different conductance levels of a single alamethicin channel. The solid lines and squares correspond to transitions from a given level to the next lower one. The circles and dashed lines refer to transitions to the next higher level. The numbers indicate the conductance levels between which the transitions take place (see FIGURE 5). The data in (A) were obtained from a membrane made from a 10% solution of purified soybean lecithin in octane. The data in (B) were obtained from a 70% solution of lecithin in octane. Note that the transitions from the higher to the lower levels corresponding to the dissociation rates increase with the voltage in (A) whereas they decrease in (B). However, in both cases the ratio of the upward to downward transition rates, i.e., the ratio of association to dissociation rates increases with increasing voltage.

as far as they can be determined is thus in approximate agreement with the conclusions drawn from the multichannel kinetics.

However, it should be pointed out that the voltage-dependence of the single-channel transitions shown in FIGURE 6 as well as the multichannel kinetic transients in FIGURE 3 are not seen with all membranes and under all conditions. In fluid membranes such as glycerol diolein and at temperatures above 45° C the kinetic transients disappear and the single-channel transition frequencies have also been observed to be independent of the voltage.[10] The phenomenon is thus variable, but sufficiently prominent to play a role in the excitation process.

Possible Causes of the Voltage-Dependence

The cause of the voltage-dependence is not yet clear. Two possibilities can be considered: The part of the translocator inserted into the membrane by the field may be in coiled or helix configuration, and is stretched in proportion to the field as indicated in FIGURE 7A. The stretched configuration favors aggregation and the rate constants change accordingly. However, the large dynamic range of the voltage-dependence (FIGURE 6) speaks against this interpretation.

An alternative mechanism would be a voltage-induced lateral phase separation between the lipids and the translocator molecules. This would squeeze the translocators into clusters, increase their local concentration, and lead

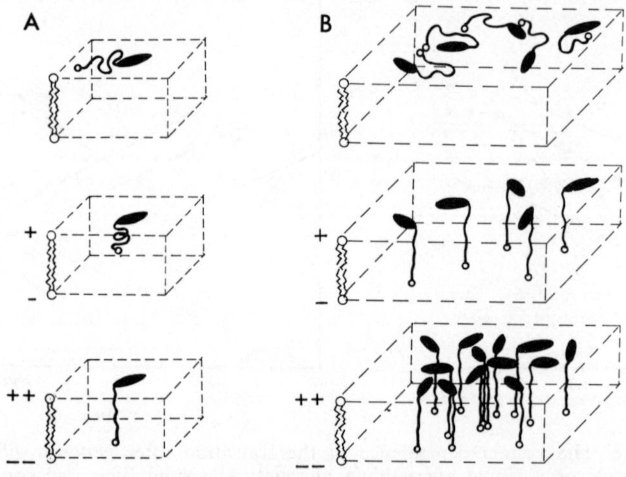

FIGURE 7. Two hypothetical molecular mechanisms causing the voltage-dependence of the aggregation rate constants. The schematic representation of the translocators contains an anchor group (ellipsoid), a flexible channel-forming region long enough to span the bilayer, and a gating charge (circle). In (A) it is assumed that the insertion depth and orientation of the translocator depends on the field strength and that the aggregation rates increase with the insertion depth. In (B) the applied field increases the concentration of the translocators because it induces a lateral phase separation between the lipids and the translocators. The aggregation rates increase accordingly.

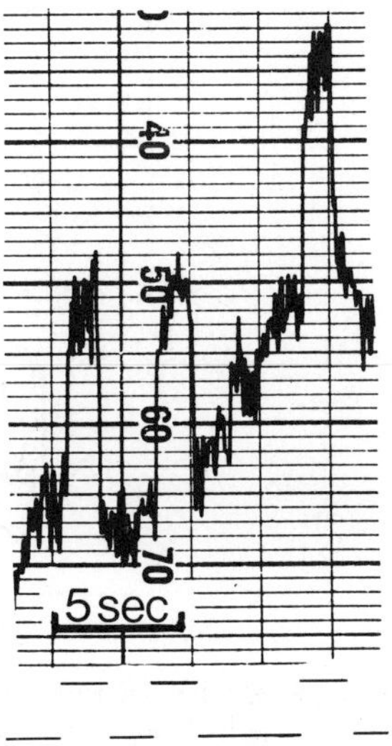

FIGURE 8. Changes of lipid bilayer reflectance (upper record) in response to applied voltage steps of 100 mV (lower trace). The reflectance was measured at an incident angle of 45° from a 20 μm diameter spot of white light. Upward deflection represents a decrease of reflected light intensity. It is caused by the increase in contact angle and contraction of small solvent (octane) lenses in the bilayer.

to an increase of the aggregation rates (FIGURE 7B). The dissociation rates would increase less or decrease, because the lipid concentration near the channel decreases in favor of monomeric translocators, thus raising the free energy of dissociation. Although such a voltage-induced phase separation has not yet been directly demonstrated with any of the translocators, it can be observed to take place for the lipids and the hydrocarbon solvent in a planar bilayer. There, the applied potential squeezes solvent into microscopic "lenses" and the process can be examined by the decrease of light reflectance with an applied potential (FIGURE 8), or if the lenses are large enough, directly through visual observation. A rough estimate based on the voltage-induced rate of growth of a black bilayer region in a grey multilayer membrane indicates that in small (10 μm) patches the concentration of a phasing-out component could double within 5 msec for an applied voltage of 100 mV. The process is therefore fast enough to account for the observed rates of the conductance changes. If the translocators are already preaggregated in clusters (see below), the phase separation would primarily involve the movement of lipids out of the cluster. The molecular mechanics of this voltage-dependent phase separation are not yet fully understood. It is probably not due to a pressure effect of the voltage, which amounts only to a fraction of an atmosphere, but may result from a change of orientation of the lipid head groups and a consequent decrease of the area per lipid molecule. There is additional indirect evidence for the voltage-induced clustering. As can be seen from FIGURE 6, the rate constants of dissociation can either decrease (FIGURE 6B), and this is usually the case, or increase (FIGURE 6A). The increase is more often observed in bilayers with a high solvent content, and since the rate constants of the conductance changes depend very strongly on the lipid fluidity,[4] it seems likely that the concentration of both the solvent and translocator molecules is increased in small, local regions, which then have a higher fluidity than the surrounding areas.

Another piece of evidence for this interpretation is shown in FIGURE 9.

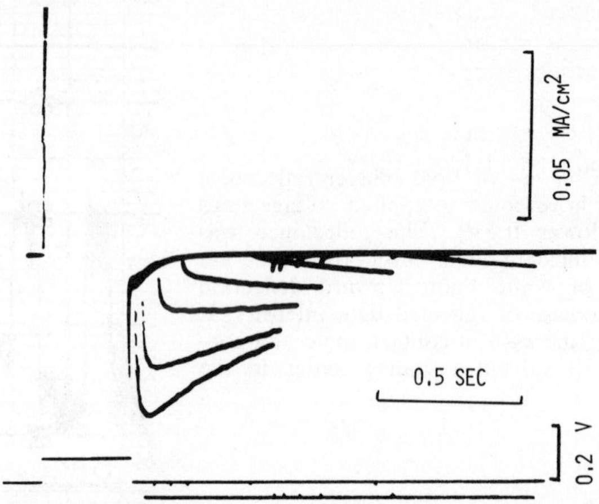

FIGURE 9. Channel inversion in a lipid bilayer. Membrane currents (*top*) in response to applied potential steps (*bottom*) from a lecithin bilayer in the presence of 1M KCl (both sides) and 10^{-6}M alamethicin (one side). Several sweeps are superimposed. During each sweep a positive potential step of 100 mV is followed by negative steps of 70 mV which start at different times after the positive step. The positive steps cause a large and rapidly rising current (going off screen) due to the opening of the alamethicin channels. A brief inverted current tail marks the end of the positive pulse. The negative pulses cause inverted currents whose time constants and amplitudes depend on the delay between the end of the positive pulse and the start of the negative pulse. At very long delays there is no inverted current. At short delays the time constants of the current rise are smaller and the currents decay. At intermediate delays two time constants are visible. During the positive step, alamethicin is transferred to the opposite membrane side where it can form channels in response to negative potentials. The time constants and large currents at short delays are presumably due to the fact that the transfer of alamethicin occurs at localized spots. Lateral diffusion in the membrane plane during the delay time decreases the concentration. As a consequence the currents and the time constants decrease with increasing delay. A quantitative analysis of the effect has shown that the time constants of the inverted currents are much shorter in relation to the conductance than under conditions where the alamethicin is added to the aqueous phase.

When a high potential is applied to a bilayer, some of the channel-forming molecules added to one side are transferred to the opposite side where they now can form channels in response to applied voltages of opposite sign. The number of translocators transferred through the membrane and, therefore, their concentration at the opposite membrane surface is a function of the duration and amplitude of the conditioning potential. As a consequence, the rate constants and the conductances of the inverted gating system increase as a high-order function of the prepulse duration and amplitude just as they vary when the translocator concentration is varied in the aqueous phase. If the inverted gating system is tested with a constant test pulse at different times after the conditioning pulse, the time constants increase and the maximal conductances fall with increasing time after the prepulse. However, the relation between the conductance and the time constants is not the same as when the translocators are added to the aqueous phase. Immediately after the prepulse, the time constants are disproportionately short for the observed conductance, and only at later times is the normal relation between conductance and time constant restored. The behavior of the time constants would follow if the translocators traverse the membrane at local spots so that their concentration is initially high. Subsequent lateral diffusion at the membrane surface lowers the concentration and with it the time constants. The phenomenon is more pronounced at higher prepulse potentials, presumably because during the prepulse the translocators are squeezed into regions of higher concentration.

Obviously the evidence for a lateral phase separation is not yet conclusive, but it does deserve further consideration as a cause of the field-dependent aggregation rates.

Rate-Limiting Steps

As mentioned above, the single-channel conductance fluctuations seen with alamethicin only give information about the open channel, but not about the events leading to the opening. The conductance of the single channel varies only by a factor of 7–8 between the lowest and the highest level, but it changes over several orders of magnitude between the closed state, i.e., the unmodified bilayer conductance and the first open level. Therefore, the events during the silent period between the application of the voltage pulse and the appearance of the first channel opening determine the multichannel conductance kinetics. As seen from FIGURE 10, this period is much longer than the periods between the individual fluctuations from level to level and is therefore the rate-limiting step of the gross conductance changes. This can be also demonstrated by recording the multi- and single-channel kinetics from the same membrane (FIGURE 11). The rate constants of the individual fluctuations are, at the same potential, on the average of 10–20 times faster than the multichannel time constants. Hidden in the silent period are the insertion rate constants and several rate constants of an unknown number of nonconducting oligomers, i.e., dimers, trimers, and perhaps even tetramers. The silent delay period fluctuates statistically and a plot of its average duration as a function of the potential shows that it has the required steepness to account for the steady-state conductance-voltage function (FIGURE 12). The relation between the potential and the average delay time between conductance level 1 and level 4, which gives an indication of the voltage-dependence of increasing

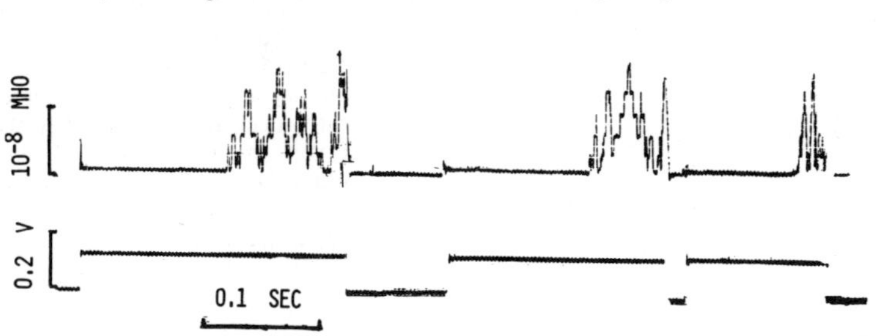

FIGURE 10. Single-channel conductance fluctuations from a lecithin bilayer in the presence of alamethicin (*upper trace*) in response to applied potential steps (*lower trace*). Note that the delay between the beginning of the potential step and the appearance of the conductance fluctuations is much longer than the lifetime of the individual conductance levels.

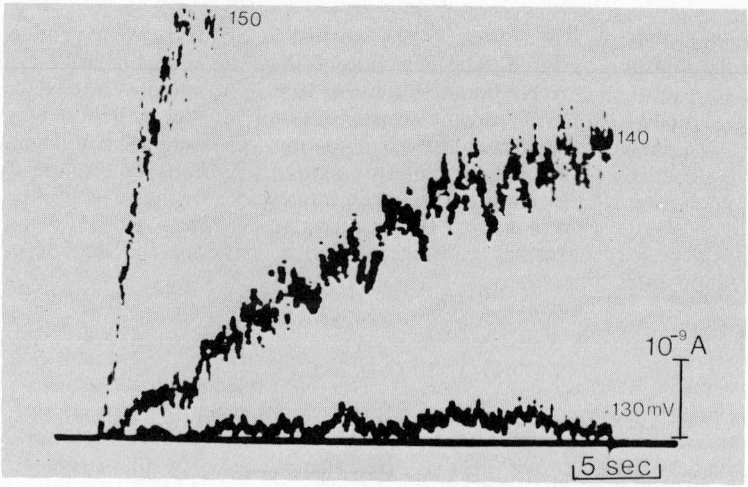

FIGURE 11. Membrane currents in response to voltage steps from a lecithin bilayer in the presence of 10^{-7} M alamethicin and 5 M KCl. At such low concentrations of alamethicin the time constants of the current rise are much slower than the time constants of the single-channel current fluctuations, which are the cause of the noisy current trace.

the oligomer size by three monomers, is less steep, indicating that the minimal channel is composed of more than three monomers or that the insertion is still incomplete within the potential range of the experiment, i.e., 100–150 mV, and therefore adds to the voltage dependence.

The multichannel kinetics require that at high voltages the voltage-depen-

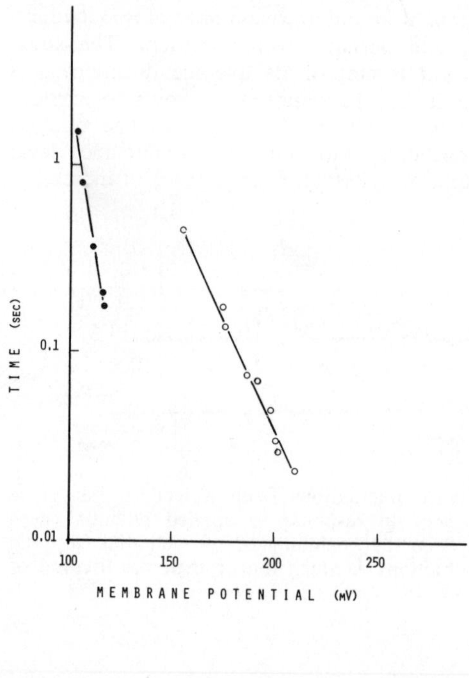

FIGURE 12. *Solid circles:* the average delay between the application of a voltage pulse and the beginning of the single-channel conductance fluctuations. Lecithin bilayers in the presence of alamethicin (FIGURE 10). *Open circles:* the average delay time between conductance level 1 and level 4 as a function of the membrane potential.

dence of the aggregation rate constants is more pronounced for the smaller nonconducting aggregates than that for the higher aggregates. Unfortunately, this assumption cannot be verified directly, but the data of FIGURE 12 are not inconsistent with it. It should also be kept in mind that there are large statistical variations of the local translocator concentrations and of the energies of formation of the individual channels, so that aggregation rate constants

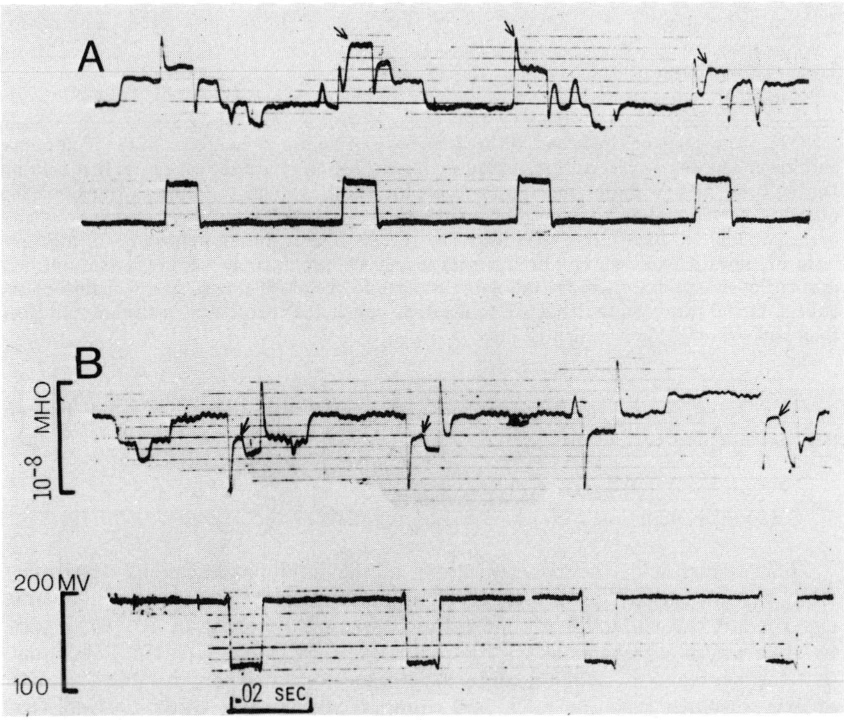

FIGURE 13. The effect of short voltage pulses on the single-channel transition frequencies. Experimental conditions are as in FIGURE 5. In record (A) the potential is increased during the short pulse and the statistical average of the transition frequency from a lower to a higher level (arrows) is larger than the steady-state frequency in the absence of the pulse. In (B) the pulses are of opposite polarity and the transition frequency from a higher to a lower level increases. The short spikes in the current records at the beginning and end of the voltage pulses are capacitative transients. The figure demonstrates that the transition rate constants adjust to their new value after a voltage change within less than 10 msec.

measured from a single channel, which is always the one with the lowest free energy, may not be applicable to the entire channel population.

How fast do the aggregation rate constants change as a function of a potential change? In order to answer this question, a series of brief voltage clamp pulses was superimposed on a high constant voltage, and the statistics of the conductance transitions were evaluated during the brief pulses (FIGURE 13). The results are summarized in TABLE 1. They show that the rate constants adjust to the new value in less than 10 msec and since, in this lipid, the

TABLE 1
RESULTS OF EXPERIMENT SHOWN IN FIGURE 10

Transition Frequency (sec^{-1}) (ν/τ)	At 200 mV	During 10-msec Pulse (+50 mV)	Ratio $\frac{\nu_{\sigma,\sigma+1}}{\nu_{\sigma,\sigma-1}}$	During 10-msec Pulse (−50 mV)	Ratio $\frac{\nu_{\sigma,\sigma+1}}{\nu_{\sigma,\sigma-1}}$
To next higher level ($\nu_{\sigma,\sigma+1}/\tau$)	35	72	3.6	22	0.29
To next lower level ($\nu_{\sigma,\sigma-1}/\tau$)	35	20		75	

NOTE: The sums of the transition frequencies ν/τ from a given level σ to the next higher and lower one ($\nu_{\sigma,\sigma\pm1}/\tau$) during a steady voltage (200 mV) are shown in the first column. During brief 50-mV pulses superimposed on the steady voltage the upward transition frequencies corresponding to aggregation rates increase, whereas the downward frequencies corresponding to dissociation decrease. The reverse is true during pulses of −50 mV. The ratio of upward to downward frequencies changes approximately e-fold for every 40 mV during the short pulse. Because this value is close to the average ratio as a function of the voltage in the steady state (FIGURE 6) it can be concluded that the rate constants adjust to their new value within less than 10 msec.

average fluctuation rates range from 5 to 50 sec^{-1}, the adjustment process is faster and not rate-limiting.

Absolute Reaction Rates and Preaggregation at the Membrane Surface

For alamethicin, the concentrations in the lipid phase can be determined from the measured lipid-water distribution coefficient.[11] In the single-channel experiments the alamethicin concentration was of the order of 10^{-7} M, leading to an average surface concentration of 10^6 to 10^7 molecules cm^{-2}. Assuming that the aggregation rates are diffusion-limited, one may obtain an upper limit of the absolute reaction rates and compare them with those derived from the conductance fluctuations.

The assumption of an encounter-controlled reaction rate implies that only the preexponential factor A of the rate constants (Equation 11) is rate-limiting. This assumption may be justified because the solvent cage effect allows multiple hits between two neighboring molecules before they are separated by the next diffusional jump.

At a surface concentration C, the collision frequency of one molecule is

$$\nu = 2dvC$$

where d is the molecular diameter, and v the diffusional velocity. The diffusion coefficient

$$D = l^2\nu'/2$$

where l is the length of a diffusional jump and ν' the jump frequency. Since the diffusional velocity,

$$v = l\nu',$$

then

$$\nu = 4DdCl.$$

If we neglect steric effects, the preexponential factor

$$A = \nu/C = 4Ddl$$

with a jump length of 8 Å, which is the average distance between two lipid molecules, a diameter of 10 Å for the width of the alamethicin and a diffusion coefficient of 10^{-8} cm^2 sec^{-1}, $A = 5 \times 10^{-8}$ cm^2 sec^{-1} molecule^{-1}.

At a surface concentration of 10^7 molecules cm^{-2} a single channel, depending on size, therefore experiences only 0.5 to 1 collisions per second. The transition frequency between one conductance level and the next higher one is in most lipids 10–100 times larger, and because the estimate neglected steric factors and activation energies, which would slow the reaction even further, it is likely that the alamethicin molecules are already aggregated in local clusters at the membrane surface, interacting through hydrogen bonds between the glutamine and glutamic acid residues. This aggregation may be much more pronounced with EIM or cellular channel formers and, together with the highly unsaturated and fluid lipids of excitable membranes, may account for the high gating rates of these membranes.

Structural Characteristics of Voltage-Gated Channel-Formers

In order to conform to the proposed model a molecule should fulfill several minimal structural requirements: (1) The molecule must either be or contain a structure long enough to span the bilayer hydrocarbon region. (2) At one end this structure must contain a hydrophilic anchor group that by either its size or polar nature keeps that end of the molecule attached to the lipid-water interface. (3) The end that is inserted into the bilayer should either contain a charge, or the inserted part must have a large dipole moment. (4) The inserted part must contain several polar groups facing to one side, which can act as a hydrophilic inner channel wall. The other side should be predominantly hydrophobic in order to interact with the lipid hydrocarbons.

These conditions can be fulfilled by different molecular structures. The cyclic polyene DJ 400B (FIGURE 14) provides an ideal example, with a sugar moiety and a carboxyl group at one end and an amino group at the other. The row of hydroxyl groups faces to one side, the polyene chain to the other, and the long dimension of ~37 Å is sufficient to span the hydrocarbon region. Monazomycin has probably a similar structure, although it contains less double bonds.

Single peptide chains or hairpin loops can also be designed to meet the foregoing requirements and a specific configuration has been postulated for alamethicin.[3, 4] The alamethicin model was based on a cyclic structure. Recent data[12] suggest that the peptide is linear and although this does not exclude the proposed configuration, other possibilities must be considered.

The peptide chain forming the part of the channel wall need not be in β-configuration; a nonpolar α-helix with several polar residues in register, such that their side chains point to one side, may do just as well.

In the aggregated and open-channel configuration the peptide chains need not be straight, like the staves of a barrel, but could be coiled in a multistranded helix, similar to the helix suggested for the gramicidin channel.[13] Finally, several peptide chains may extend from a larger protein which forms the anchor group, and after insertion the channels are formed by association of chains

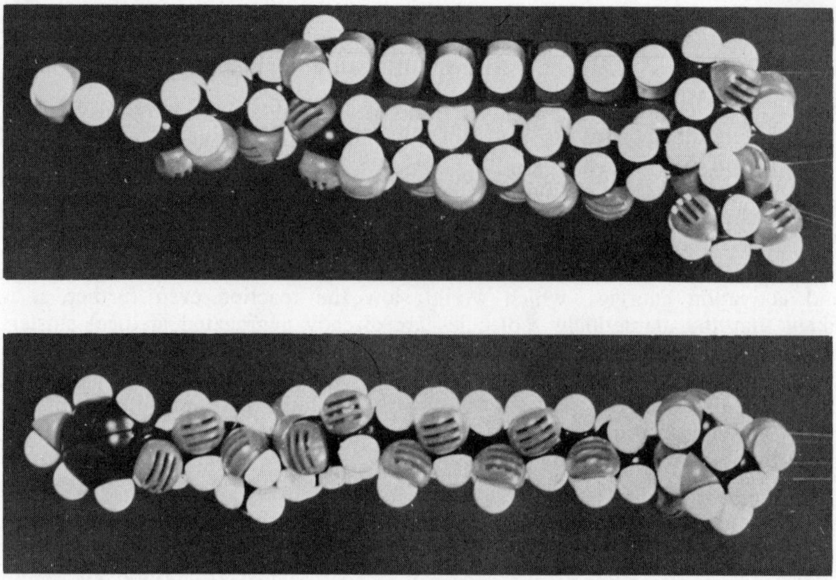

FIGURE 14. Molecular model of the cyclic polyene DJ 400B. The side view (*top*) shows the hydrophobic backbone opposite the row of hydroxyl groups. The sugar group at the right serves as anchor group and the NH_2 at the opposite end as gating charge. The hydroxyl groups, clearly visible in the bottom photograph, provide the hydrophilic channel interior.

belonging to neighboring proteins.[4] If such structures are already aggregated in patches at the membrane surface, only a minimal amount of lateral diffusion or rotation would suffice for the channel formation.

Application to Nerve

In the absence of direct structural information the validity of the aggregation model for nerve rests primarily on the similarities between the steady-state and kinetic data in bilayers and nerve. Recent experiments in squid axon[7] have revealed the same kinetic features that in the bilayers led to the postulation of an aggregation process.

Obviously kinetic arguments alone are not sufficient to establish a mechanism, and in principle the entire kinetics could be derived from a model based on intramolecular configurational changes within a preformed channel. In that case the correspondence between nerve and bilayer would be a fortuitous, albeit a most remarkable coincidence. The high cation selectivity of nerve channels, plus the fact that the sodium channels can be blocked by tetrodotoxin from the outside and the inactivation modified by pronase from the inside,[14] necessitates some modification of the simple model envisioned for the bilayers. One possibility worth considering is that in the case of the Na channels there exist separate ion-selective structures at the outside membrane surface with which the barrel staves link up after insertion and aggregation.

One example for such a series-selectivity filter is provided by the inversion of the selectivity of EIM and alamethicin channels by externally applied protamine.[2] Protamine changes their predominant cation permeability into an anion permeability without affecting the basic gating kinetics.

It may very well turn out that also in nerve the selectivity function can be separated from the gating function, and preliminary observations on the loss of ion selectivity of inverted potassium channels after high depolarization support this notion.[7]

Summary

Electrically excitable lipid bilayers show the same voltage-dependent kinetics as nerve and other excitable cells. In the bilayers the gating process involves the voltage-dependent insertion of channel-forming molecules into the hydrocarbon region and their subsequent aggregation by lateral diffusion into an open "barrel stave channel." This process can account quantitatively for the classical Hodgkin-Huxley kinetics including inactivation as well as for certain kinetic features that lie outside the Hodgkin-Huxley domain. The multi- and single-channel kinetics suggest that both the insertion and the aggregation rate constants are voltage-dependent, and it is argued that a voltage-induced lateral phase separation between the lipids and the channel-forming molecules increases the local concentration of channel precursors and their aggregation rates. Because the observed aggregation rates are faster than those calculated from an upper limit of the diffusion constants and the known average concentration in the lipid phase, it is likely that the channel-formers preaggregate at the membrane surface. The structural characteristics of the channel-formers and the evidence supporting a similar excitation mechanism in nerve are discussed.

Acknowledgments

I should like to thank G. Baumann and D. O. Rudin for many discussions and suggestions, and H. Freifelder, S. Montimore, and C. Williams for help with the preparation of the manuscript.

References

1. Mueller, P. & D. O. Rudin. 1968. Nature **217:** 713.
2. Mueller, P. & D. O. Rudin. 1969. Translocators in bimolecular lipid membranes. *In* Current Topics in Bioenergetics. D. R. Sanadi, Ed. Vol. 3: 157. Academic Press. New York, N.Y.
3. Baumann, G. & P. Mueller. 1974. J. Supramol. Struct. **2:** 538.
4. Mueller, P. 1975. Electrical excitability in bilayers and cell membranes. *In* International Review of Science, Biochemistry Series: Energy Transducing Mechanisms. E. Racker, Ed. Vol. 3:75. Butterworths. London, England.
5. Hodgkin, A. L. & A. F. Huxley. 1952. J. Physiol. (London) **117:** 500.
6. Boheim, G. 1974. J. Membr. Biol. **19:** 277.
7. Adelman, W., R. French & P. Mueller. Unpublished observations.
8. Gordon, L. G. M. & D. A. Haydon. 1972. Biochim. Biophys. Acta **255:** 1014.

9. EISENBERG, M., J. E. HALL & C. A. MEAD. 1973. J. Membr. Biol. **14:** 143.
10. HALL, J. E. Personal communication.
11. CHELACK. W. S. & A. PETKAU. 1973. J. Lipid Res. **14:** 255.
12. WILLIAMS, R. J. P. 1975. Personal communication.
13. URRY, D. W. 1971. Proc. Nat. Acad. Sci. U.S.A. **68:** 672.
14. ARMSTRONG, C. M., F. BEZANILLA & E. ROJAS. 1973. J. Gen. Physiol. **62:** 375.

Discussion

DR. FINKELSTEIN: With regard to the inactivation mechanism that you have talked about, in the case of the individual monomers you have a great excess of monomers over molecules in the channel state. . . .

DR. MUELLER: Not necessarily.

DR. FINKELSTEIN: In the case of biological systems, where presumably things are packaged together, how much of an excess of subunits do you need over the number of subunits that make the conducting channel in order to begin to see this kind of inactivation?

DR. MUELLER: We haven't explored this much, but the best ratio we have obtained is about six to one. That is, there are six times as many monomers as channels at the peak of conductance.

DR. FINKELSTEIN: Do you mean then that if a channel was composed of five monomers you would require thirty subunits in that package in order to have good inactivation?

DR. MUELLER: Yes, something like that.

DR. C. ARMSTRONG (*University of Rochester, Rochester, N.Y.*) Can you make the kinetics of these changes fast enough to be comparable to that of the nerve? The time scale that you illustrated was perhaps two orders of magnitude too slow.

DR. MUELLER: We could make it as fast as the sodium system; 200 microseconds for the turn-on of alamethicin. In this respect I would like to point out that there is evidence that excitable membranes have a very high content of highly unsaturated lipids, which would be ideally suited to provide the liquid environment that we need for rapid transitions.

DR. ARMSTRONG: Does the conductance saturate in this model system?

DR. MUELLER: Yes.

DR. ARMSTRONG: Have yo tried to predict the gating current you obtain when these aggregated channels turn off?

DR. MUELLER: Yes, we have looked at the *on* and the *off* gating currents, and if we make the additional assumption of the voltage-dependence of the aggregation then we can match your gating current.

CURRENTS ASSOCIATED WITH THE IONIC GATING STRUCTURES IN NERVE MEMBRANE*

Clay M. Armstrong

University of Rochester School of Medicine and Dentistry
Rochester, New York 14642

Francisco Bezanilla

Facultad de Ciencias/Universidad de Chile/Viña del Mar, Chile; and
Marine Biological Laboratory/Woods Hole, Massachusetts 02543

INTRODUCTION

The sodium (and potassium) channels of nerve membrane are controlled by charged gating structures that move in the membrane field, producing (at least in the case of the sodium channels) a small but measurable current, the gating current.[1–3] Similar currents have been observed by others and called displacement currents[4, 5] or asymmetry currents.[6] There is strong evidence that at least part of gating current is associated with the sodium channels, and this evidence is briefly reviewed below. Gating current of the potassium channels has not yet been identified.

Gating current is a component of capacity current,[3] and one would predict that membrane capacity should be voltage-dependent and largest in the region where gating charge distribution is most sensitive to voltage.[7] We show here that membrane capacity of squid axons is voltage-dependent in the expected way. A marked frequency-dependence of membrane capacity has been demonstrated by Takashima and Schwann,[8] and we suggest that the frequency-dependent component is due to gating charge movement.

Inactivation of the sodium conductance affects gating current amplitude,[2, 9] and we report further observations on this phenomenon.

METHODS

The essential elements of the technique for measuring gating current are the removal of permeant ions from both sides of the membrane, and subtraction of the linear portion of capacitative current, either by algebraically summing the current from exactly equal pulses of opposite sign, or by the more elaborate procedure described below. Gating current is small, and averaging is necessary to improve the signal-to-noise ratio. Details of the technique involved have been reported previously,[3] and we describe here only several improvements for increasing the speed and accuracy of recording.

Apparatus

FIGURE 1 shows a block diagram of the system currently in use. All digital instruments are in a separate rack, which is connected to the remainder of

* This work was supported by Grant No. NS08951 from the United States Public Health Service.

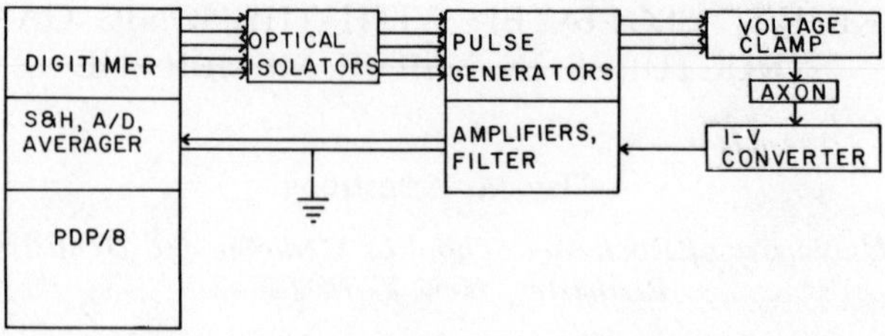

FIGURE 1. Block diagram of apparatus. See text for details.

the electrical equipment by a single signal and ground path. Other signals between the two parts of the apparatus are communicated by way of optical isolators, which are useful in preventing ground loops. The current signal recorded from the axon is digitalized and averaged, and then recorded on the magnetic tape unit of the computer. To increase the duration of the recorded sweep without sacrificing time resolution, the first 128 points in each sweep are taken (usually) at 5 microseconds per point, and the remaining 128 points at a slower rate (usually 50 microseconds per point).

A major limitation on the time resolution of gating current measurements has been the necessity to blank out the first 20 to 50 microseconds following a step, to prevent saturation of the A/D converter. We have overcome this problem by using a technique based on a suggestion by Drs. David Landowne and Joel Brown. The technique (FIGURE 2A) uses a transient generator (RC circuitry and operational amplifiers) to produce a current similar to that coming from the nerve, but opposite in sign. The two currents (from the nerve and from the transient generator) are summed by the I-V converter, resulting in an output voltage that never exceeds the dynamic range of the following amplifiers or the A/D converter. Since the output of the transient generator is a linear function of the input voltage, the sum of its output for a complete cycle of a positive and a negative pulse (or, see below, four small pulses) is zero, so this device in no way alters the gating current signal that we record. Using this technique, blanking is unnecessary, and no time at all is lost following a voltage step: time resolution is limited only by the speed of the clamp, which can complete a voltage step in about 10 microseconds. A second advantage of this technique is that it cancels out a significant amount

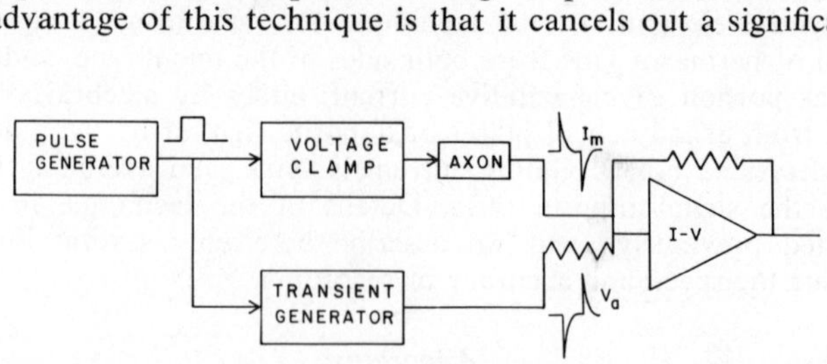

FIGURE 2A. A linear transient generator produces a current of time course similar to that of the axon, but of opposite sign. The two signals are summed by the I-V converter, reducing the amplitude of its output, and preventing saturation of the following amplifiers. See text.

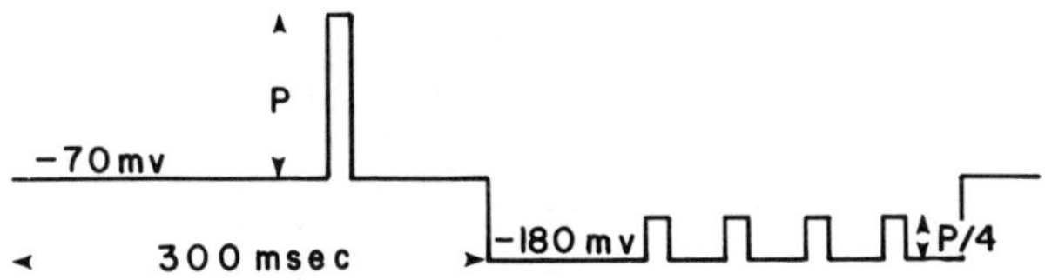

FIGURE 2B. P/4 pulse pattern. Current from the four small steps (amplitude P/4) is digitally subtracted from the current produced by the large step (amplitude P) to eliminate the linear portion of capacitative current.

of the noise introduced on the command signal. A noise transient that produces a positive current from the axon produces a negative current of about the same magnitude and time course from the transient generator. This improves the signal-to-noise ratio of our apparatus by a factor of approximately two.

The pulse sequence which we apply for measuring gating current has been changed, and the new one is diagrammed in FIGURE 2B. The current from four small pulses is digitally subtracted from the positive pulse current. The four pulses cover a negative voltage range where there is reason to believe gating charge movement is slight.[3] Four pulses are now used (instead of a single one with 4× amplification of the current) to improve the signal-to-noise ratio. Digital subtraction of the current signal allows us to use the same pulse throughout the cycle, and the rising phase of this step is thus exactly the same in shape for both the positive step and the four pulses used for subtraction. We have found that using different pulses, with, necessarily, rising phases of slightly different shape, can introduce significant artifact which is not present with the digital subtraction technique. The pattern of FIGURE 2B will be referred to as the P/4 pattern.

Composition of the internal and external solutions used is given in TABLE 1. Plus sign indicates that sucrose was added to give an osmolality of 960 mOs/kg.

TABLE 1

	External Solutions					
	Na(mM)	Tris(mM)	TMA(mM)	Ca(mM)	Mg(mM)	Cl(mM)
5% Na SW	22	418	–	10	50	560
10% Na + 5 Ca + TMA	44	–	476	5	–	530
6 Ca + TMA	–	–	522	6	–	534
Tris SW	–	440	–	10	50	560
20 Ca + 22 Na + Tris	22	498	–	20	–	560

	Internal Solutions						
	Cs(mM)	TMA(mM)	Na(mM)	Tris(mM)	F(mM)	Cl(mM)	Sucrose
125 Cs	125	–	–	–	125	–	+
TMA + Tris	–	100	–	25	100	25	+
133 Cs + 5 Na	133	–	5	–	133	5	+

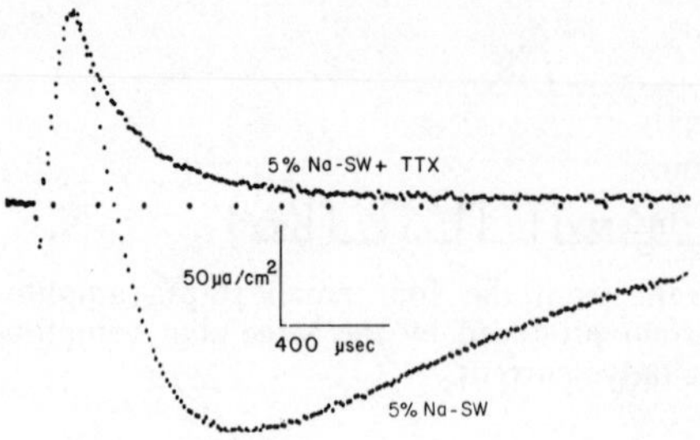

FIGURE 3. Tetrodotoxin eliminates I_{Na} without noticeably affecting the early outward peak of current, the gating current. Positive and negative pulses of 90 mV amplitude were applied from a holding potential of −70 mV (not P/4 pattern). The internal solution was 125 mM CsF + sucrose. 8° C.

RESULTS

When the Na concentration in the external medium is reduced to 5% of its normal value, I_{Na} (sodium current) is much reduced and is preceded by an outward peak of gating current (I_g). I_{Na} can be eliminated by adding tetrodotoxin (TTX) to the external medium (FIGURE 3). The early part of I_g is not detectably affected by TTX, and its later part becomes visible as a current that decays smoothly and has largely subsided by the time of peak I_{Na}. The absence of a TTX effect on I_g suggests that the TTX receptor is far from the gating structures.

Experiments with internal and external application of Zn^{++} provide evidence that I_{Na} and I_g are related (for other evidence, see Reference 3). FIGURE 4 shows I_{Na} and I_g recorded from an axon in 5% Na sea water when internally perfused with a solution containing 10 mM $ZnCl_2$. I_{Na} and I_g decrease approximately in proportion to each other after addition of Zn^{++} to the perfusion solution and both are almost completely abolished after a few minutes. During washout of Zn^{++} from the axon interior, both currents recover at about the same rate, and recovery is essentially complete, except that after Zn^{++}, Na inactivation seems to be slowed or absent.[10]

Externally applied Zn^{++} has quite a different effect, as illustrated in FIGURE 5: both I_{Na} and I_g are slowed by a factor of two or three, but not eliminated. The lower traces show I_g alone with and without external Zn^{++}, and the slowing is quite evident. This is an interesting effect that requires further study, but the parallel slowing of I_{Na} and I_g provides still more evidence that the two currents

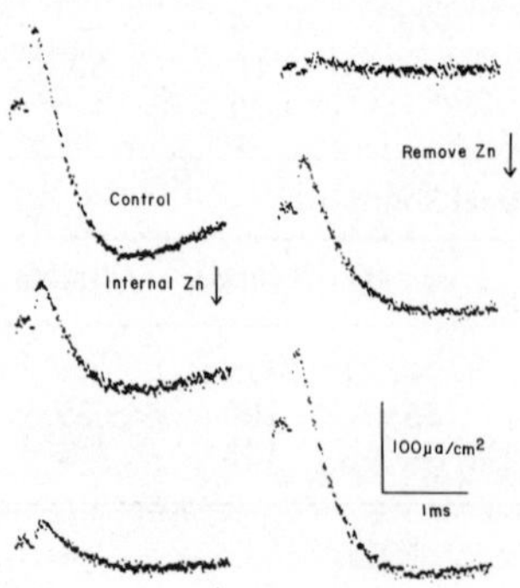

FIGURE 4. Addition of 10 mM $ZnCl_2$ to the internal perfusion medium reversibly eliminates I_{Na} and I_g. Pulses of 80 mV amplitude were applied from a holding potential of −70 mV (not P/4 pattern). 5% Na SW//133 Cs + 5 Na. 8° C.

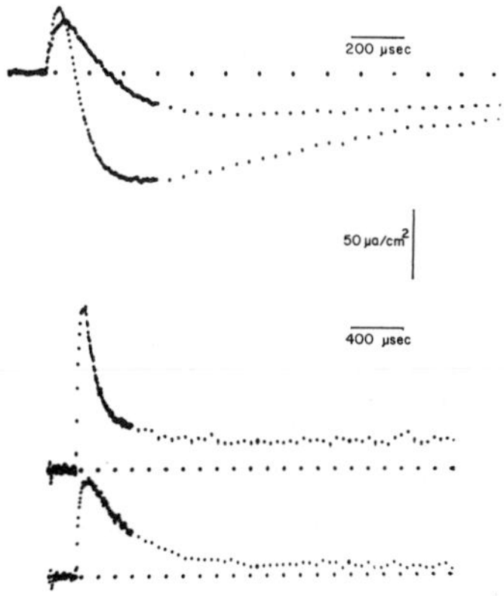

FIGURE 5. The upper traces show the slowing of I_{Na} and I_g by external Zn^{++}. The higher amplitude trace is the control record, in 6 Ca + 30 Zn + TMA//TMA + Tris). For the lower amplitude trace, 27 mM $ZnCl_2$ was added to the external solution. The bottom trace shows I_g (without I_{Na}) in the presence of 30 mM external $ZnCl_2$ (recorded in 6 Ca + 30 Zn + TMA//TMA + Tris). The middle trace is the control record (in 6 Ca^{++} + TMA//TMA + Tris). 8° C. P/4 pattern, with P = 100 mV, from a holding potential of −70 mV.

are associated. (We thank Dr. Bertil Hille for initially informing us of the effects of external Zn^{++} on I_{Na}.)

The Time Course of Gating Current

The rising phase of gating current has not been observed before now because of the large size of the capacity current transient and the necessity to prevent saturation of the A/D converter by blanking, for up to 50 microseconds after the step. The technique described in the *Methods* section allows us to see all of the gating current, beginning at origin of the step. The lower traces in FIGURE 5 were recorded using this technique and the P/4 pattern, which eliminates much of the rising phase seen when positive and negative pulses of equal size are applied from the holding potential.[3] It is interesting that a distinct rising phase remains nonetheless. Possible origins of this rising phase are discussed below.

We have noted before that I_g does not decay with the time course of a single exponential[9] and this is evident in the middle trace of FIGURE 5, which has a clear slow component. Observation of the slower component is always complicated by the fact that the current trace never decays to an absolutely flat baseline: with our present techniques the current goes through a minimum and then slowly increases with time, perhaps as the result of slow alteration of an ionic permeability. Under these circumstances the definition of the baseline, which is, of course, an essential step in any fitting procedure, is rather an arbitrary matter. Our procedure is to fit a least squares straight line to the points between 1.3 and 2.6 msec after the step, and use this fitted line as baseline. The rationale is to select points before the baseline creep becomes serious, but well after the decay of the bulk of I_g. FIGURE 6A is the original trace and FIGURE 6B is the same trace redrawn (by the computer) relative to the fitted baseline.

A slow component is evident in both the original and the redrawn trace of FIGURE 6A and B, and it comes as no surprise that a single exponential fitted by a least squares method to the fall of the gating current (between the two cursors indicated by arrows in FIGURE 6C) is not a good fit. A much better fit is obtained by using two exponentials determined by the proce-

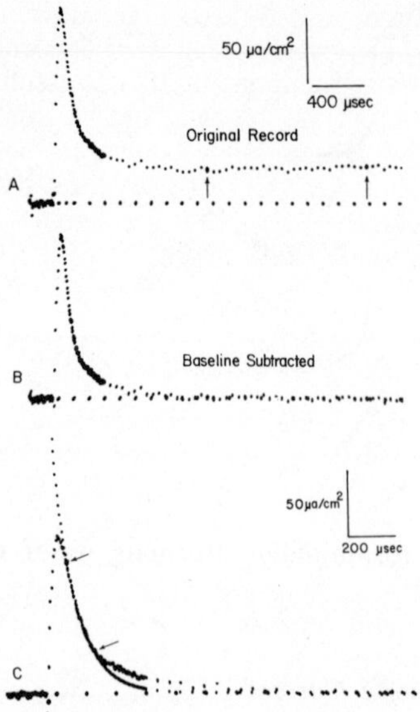

FIGURE 6. Determination of the baseline. (A) A least squares line was fitted to the points between the two cursors (arrows). The trace was then redrawn by the computer relative to this baseline in (B). (C) The same trace, with an exponential curve fitted by a least squares method to the points between the arrows. The falling phase of the trace has a clear slow component, and is not well fitted by a single exponential. P/4 pattern, for a pulse to +50 mV from a holding potential of −70 mV. 6 Ca^{++} + TMA//TMA + Tris. 8° C.

dure illustrated in FIGURE 7. First a single exponential is fitted to the slow component (a least squares fit between the two cursors in FIGURE 7A). This component is then subtracted from the original trace, and an exponential fitted to the remainder (FIGURE 7B). The two component exponentials are then summed to give the curve in FIGURE 7C, which is quite a good fit to the original curve. Reiteration of the process was unnecessary because the time constants of the two component exponentials are quite different, 69 and 382 microseconds. The fast component has a time constant substantially shorter than that of the single exponential fitted in FIGURE 6C.

The slow component shown in FIGURE 7A is surprisingly large, and it

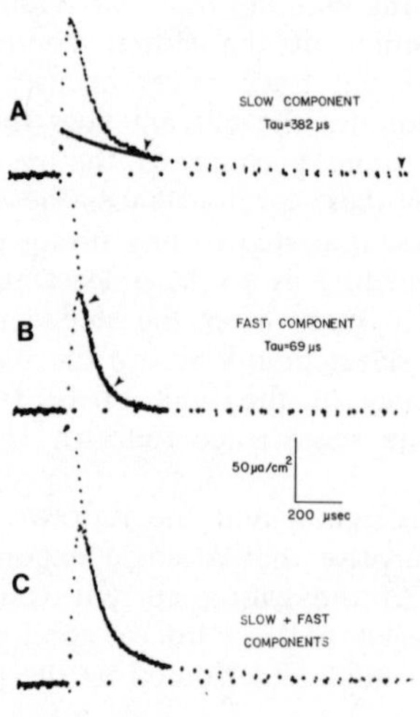

FIGURE 7. Fitting of two exponentials to the falling phase. (A) An exponential was fitted to the trace between the arrows by a least squares method. (B) This exponential was subtracted from the original trace, and the resultant curve was then fitted with an exponential between the two arrows. (C) When summed, the two exponentials are an excellent fit for the falling phase of the current. See the legend of FIGURE 6 for conditions.

would be of interest to know what fraction of the total charge movement it accounts for. Only a crude estimate can be given, because of ignorance regarding the early part of the time course of this component. If it is assumed to have the time course shown in FIGURE 7A, an assumption that we regard as unlikely to be close to the truth, the slow component accounts for more than half of the total charge movement. It seems clear from these observations that a sizeable slow component of gating current exists, but its significance and complete time course are, for the moment, totally obscure.

Membrane Capacity and Gating Charge Movement

Membrane capacity was measured by integrating the current produced by a voltage step. For an ideal capacitor in parallel with a resistance, the integral for an applied step is the sum of a step (the capacitative component) and a ramp (a steady current flowing through the parallel resistance). If the capacitor is imperfect, capacity current is prolonged by the slow polarization of the dielectric. The result is a curve like the experimental trace shown in FIGURE 8: the slope of the integral (of membrane current) curve decreases progressively as capacity current subsides, until the curve becomes (in theory) a straight line. Gating current recorded from the same axon, for a step of the same size, is shown in the lower trace of FIGURE 8. The integral curve

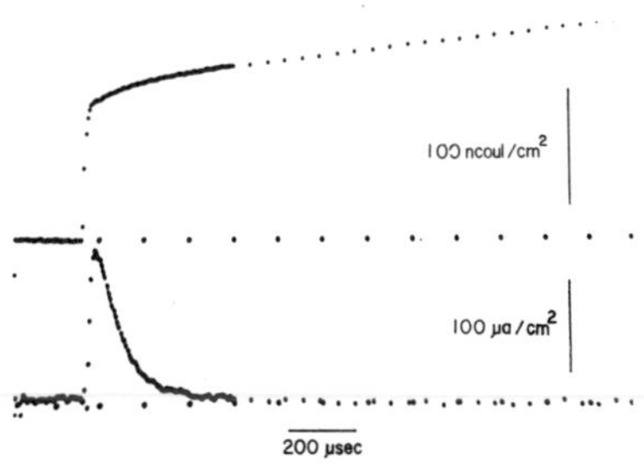

FIGURE 8. The upper trace is the integral of membrane current for a step of Vm from −70 mV (the holding potential) to +50 mV. The lower trace is the gating current produced by the same step (relative to a baseline determined as in FIGURE 6). Tris SW + TTX//125 Cs + 2 mM EDTA. 8° C.

straightens out at about the same time that I_g subsides, which suggests that the slowly polarizing components of the dielectric are the structures that give rise to the gating current.

This suggestion is well supported by a comparison of gating charge movement and membrane capacity. Total capacitative charge movement is obtained from a trace like the one in FIGURE 8 by extrapolating the straight-line part of the trace back to the origin of the step, and measuring the vertical distance (which has units of charge) between the extrapolated line and the integral curve just before the step. As with any method for measuring capacity, this one cannot distinguish between slow polarization of a component of the dielectric and a slow change in a resistance in parallel with a capacity. This problem is exactly analogous to the baseline problem noted above in connection with gating current measurements, and it manifests itself here as a slight curvature of the integral trace at long times, the portion of the trace that is ideally a straight line. We coped with this problem by fitting a least squares straight line to the points between 1.5 and 2.5 msec, the same limits that were used for fitting the baseline of gating current traces. Though somewhat arbitrary, this seems to us the fairest procedure for comparing gating charge movement and capacity measurements.

The upper curve in FIGURE 9 shows the differential capacitance of an axon, $\Delta Q/\Delta V$, for 10 mV intervals as a function V_m. Capacity rises from about 1 $\mu F/cm^2$ at —70 mV to a peak of about 1.35 $\mu F/cm^2$ near —20 mV, and then decreases at more positive voltages. The part of membrane capacitance that is due to gating charge movement is proportional to the slope of the gating charge distribution curve, which is given for the same axon by the lower curve in FIGURE 9. The maximal slope of this curve occurs near —30 mV, not far from the voltage where membrane capacitance has its maximal value. At this voltage $\Delta Q/\Delta V$ has a value of 300 nC cm^{-2} V^{-1}, or 0.3 μF cm^{-2}. This is close to the difference between C_m (membrane capacity) at —70 mV (where there is relatively little charge movement) and C_m at

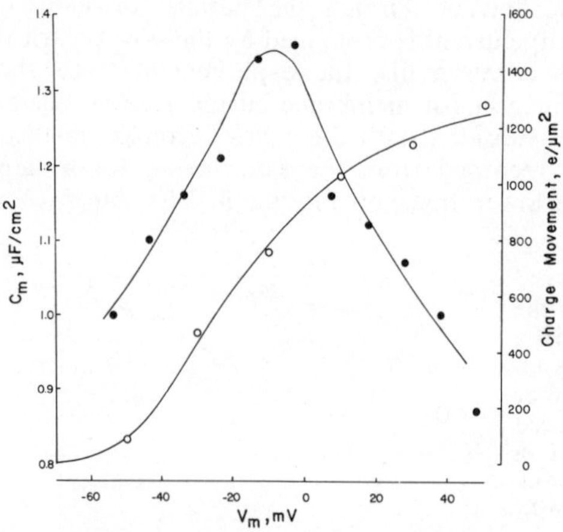

FIGURE 9. The upper curve is the differential membrane capacity, $\Delta Q/\Delta V$, for 10-mV intervals. The lower curve (open circles) represents gating charge movement in units of electronic charges/μm^2 for steps from a holding potential of —70 mV. Conditions are as in FIGURE 7.

—10 mV; i.e., the extra capacity at —10 mV can be largely explained as gating charge movement. Further experiments may improve the quantitation, but it seems clear from the experiment of FIGURE 9 that gating charge movement makes a significant contribution to C_m in the range from —70 to +20 mV.

Inactivation of Gating Charge Movement

In a previous report[2, 9] we showed that a prepulse applied as in FIGURE 10 diminished I_g and I_{Na} in proportion, and we gave evidence that the prepulse acted on I_g by inactivating g_{Na} (the sodium conductance). We have reexamined the effect of a prepulse using the P/4 procedure (see the inset of FIGURE 12 for the pulse pattern), and with this procedure the effect is still present, but somewhat less marked than with the old procedure. (Other experiments indicate that with the pattern of FIGURE 10, a positive prepulse before the negative test pulse enhances the inward current in the test pulse. The significance of this extra inward current is unknown.) An example of the effect

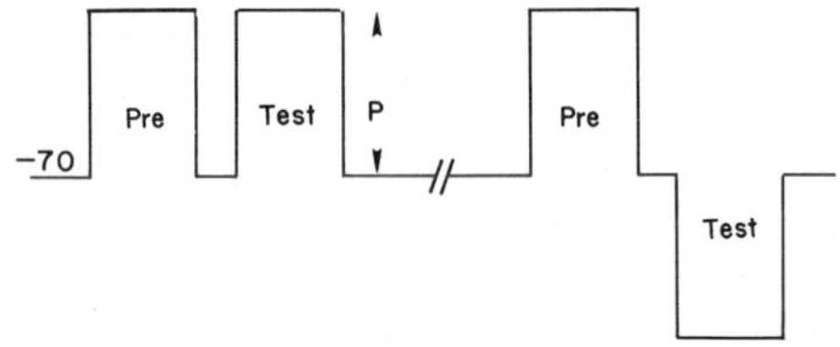

FIGURE 10. Pulse pattern used in previous experiments for demonstrating the effect of inactivation on I_g.

of a prepulse with the P/4 pattern is shown in FIGURE 11. The current trace of larger amplitude is the control, for which there was no prepulse. For the lower amplitude trace, the test pulse (to +20 mV) was preceded by a 10-msec prepulse (to +20 mV) followed by a 1-msec recovery interval at −70 mV. Reduction of gating current by a prepulse was studied quantitatively by integrating I_g traces recorded after recovery intervals of various durations, yielding the results plotted in FIGURE 12. Charge movement in the test pulse

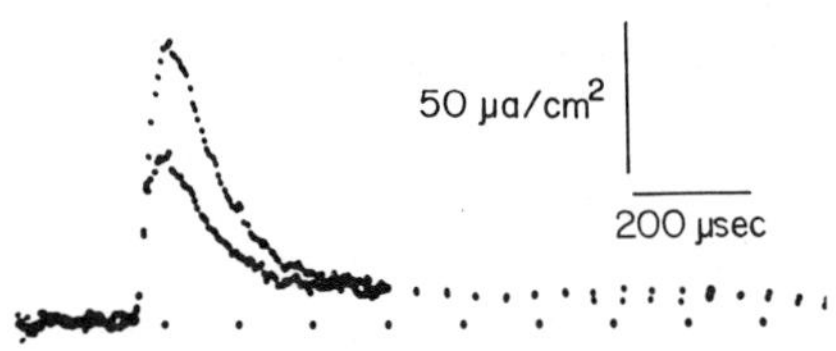

FIGURE 11. The effect of a prepulse on I_g. The lower amplitude trace shows I_g for a step to +20 mV after a prepulse (see the inset of FIGURE 12 for the pulse pattern). The other trace is the control, without prepulse. The prepulse was to +20 mV for 10 msec, and there was a 1-msec recovery interval between pre- and test pulse. Tris SW + TTX//125 Cs, at 8° C.

is strongly dependent on the recovery interval. After recovery for 0.5 msec, charge movement is about 40% of the control movement, but it almost equals control movement after a recovery interval of 12 msec. The prepulse apparently immobilizes more than half of the gating charge, trapping part of it for a period of milliseconds. Ultimately, of course, the gating charge must return to its "closed" position, and be free to move during a test pulse, for the effect of a prepulse is not irreversible. It could be argued that 0.5 msec is

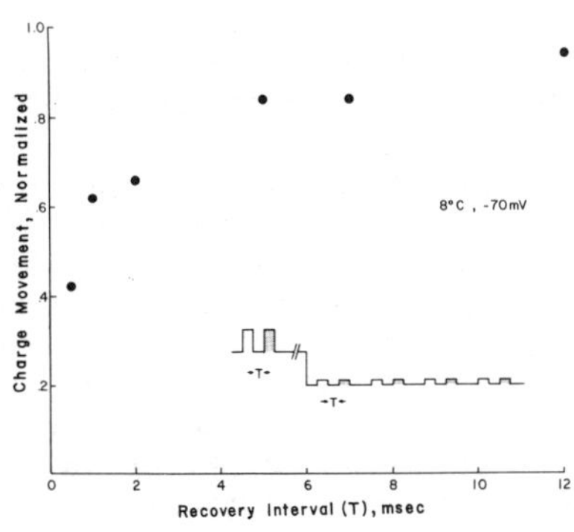

FIGURE 12. Gating charge movement during the test pulse as a function of the interval between conditioning and test pulse. Gating charge movement has been normalized relative to that after a very long recovery interval. The inset shows the pulse pattern. 20 Ca + 22 Na + Tris + TTX//125 CsF, 8° C.

too short an interval to allow even the rapid component of return current to subside. With this in mind we examined the inward gating current tail at the end of the prepulse, and found that it could be fitted fairly well by an exponential with a time constant of 95 μsec. The 0.5-msec recovery interval was thus more than five time constants for this component. The remainder of the gating charge returned over an interval of more than 12 msec (the time required for complete recovery), producing a current that was much too small to detect.

Discussion

There is now a strong case for associating gating current with opening and closing of the Na channels.[2, 3] Briefly, the evidence is that I_g has an appropriate time course; at least a part of it appears to inactivate as does g_{Na}; and both I_{Na} and I_g are blocked reversibly by prolonged depolarization or internal perfusion with Zn^{++}. To this list can now be added another correlation, the slowing of both I_{Na} and I_g by external Zn^{++} ion.

I_g and I_{Na} are clearly related, but regarding the precise manner of their interrelationship there are perhaps more questions than answers at present. A first question is the exact time course of the gating current associated with the opening of the Na channels. The records presented above have a distinct rising phase that lasts for 25–50 μsec. We have shown that much of the rising phase for steps of equal amplitude from the holding potential arises from inward current during the negative step.[3] In the records illustrated here, however, this charge movement has been reduced if not eliminated by using the P/4 pattern: the subtraction pulses cover a very negative voltage range where there is relatively little charge movement. There are three possible origins for the rising phase seen with this procedure. (1) There is enough current even in the negative range near —170 mV to account for it. (2) It is due to a slower-than-expected response of the voltage clamp, which completes a step in 50 μsec rather than 10 μsec. This seems unlikely to us, but not impossible. (3) The rising phase is a feature of the gating current associated with the opening of the Na channels. Physically this would mean that the first steps involved in the opening of the channels involve relatively little charge movement. At present we are unable to decide clearly among these possiblities.

Another question concerns the significance of the slower component of gating current demonstrated in FIGURES 6 and 7. A slow component of current has been noted by us previously,[9] but has not been mentioned by Keynes and Rojas[4, 5] or Meves.[6] This is probably the result of a difference in method. We see this component clearly only when using the P/4 pulse pattern, and in records where the current is followed by a long and relatively flat baseline. It remains to be seen whether the presence of this component will be confirmed by others when they adopt a similar procedure. The origin of the slow component, which accounts for a very significant fraction of total charge movement, is almost totally obscure at present, and it could be associated with any, or none, of the three gating functions, Na activation, Na inactivation, or K activation. At present we are inclined to believe it is associated with Na activation, but the evidence is not strong enough to merit a summary.

Inactivation affects gating current, and appears to immobilize a large fraction of the gating charge. This is suggested by FIGURES 11 and 12, which show that a positive prepulse, which should inactivate I_{Na}, diminishes the gating charge movement during a following test pulse. After a prepulse, only a fraction of gating charge moves back to "closed" position quickly, as can be directly confirmed by integrating the gating current tail at the end of the prepulse. The remainder of the charge apparently moves back over the course of a number of milliseconds, producing a current that is too small to detect with present methods. The evidence that a prepulse affects I_g by inactivating g_{Na} is: (i) recovery of I_g following the prepulse is similar in time course to the recovery of g_{Na}; and (ii) pronase, which destroys Na inactivation, removes the effect of a prepulse on both I_g and g_{Na}.

In spite of this rather clear evidence, an effect of inactivation on I_g has been disputed by others.[6, 12] In our view, confusion has arisen mainly from a failure to distinguish between the inactivation process described by Hodgkin and Huxley[11] and slow inactivation, which is quite a different phenomenon. If pulses are applied to a fiber held at a relatively positive potential, net charge movement is inward, as has been reported by Keynes and Rojas.[12] This current has been interpreted as normal gating current, and taken as evidence that I_g does not inactivate. We confirm the existence of this current, which we observed in our original experiments, but we do not believe that it is gating charge moving in the normal mode, for the following reason. A fiber held at +56 mV for two minutes shows inward current when pulses are applied to it, but when pulsed after a sudden change of the holding potential from +56 to —70 mV, there is initially neither gating current nor I_{Na}. I_g and I_{Na} then recover with parallel time courses, with a half time of about 30 sec.[3] The origin of the inward current seen at positive holding potentials is thus not at all clear, since the fiber can produce I_{Na} neither at +56 mV, nor, for many seconds, at —70 mV after return from +56 mV. There is no evidence to link the inward current at positive holding potentials or the Na activation mechanism.

Information regarding the inactivation process described by Hodgkin and Huxley can be obtained only by means of relatively short depolarizations (not exceeding 50 or 100 msec) applied to an axon held at a potential in the normal range (60 mV or more negative). The only published experiments of this type are our own, and these yield the results described above. As noted by Chandler and Meves,[13] internal perfusion with Cs^+ slows or removes Na inactivation. We find that after prolonged exposure to Cs^+, inactivation of I_{Na} is markedly slowed, and inactivation of I_g becomes more difficult to demonstrate. The removal of inactivation by Cs^+ is fortunate in a sense, for it made simpler the demonstration of equal on and off charge movement (i.e., equality of outward charge movement during opening of the channels and inward movement during closing of the channels), a point important in demonstrating that gating current is capacitative. We stress, however, that in fresh axons, on and off charge movements are equal only for pulses of short duration.

We have shown here that membrane capacity is voltage-dependent, with a maximal value of roughly 1.35 $\mu F/cm^2$ at about -10 mV, and a value of roughly 1 $\mu F/cm^2$ at —60 mV. The difference in capacitance at the two voltages is at least in large part due to gating charge movement. This large contribution of the gating structures to membrane capacitance provides an answer to a frequently asked question: Why do not membrane molecules other than the gating structures contribute to capacitative current? Some contribution from nongating molecules cannot be excluded, but it is clear that a large increase in capacitance from this source would slow action potential propagation and be functionally disadvantageous.

References

1. Armstrong, C. M. & F. Bezanilla. 1973. Currents related to movement of the gating particles of the sodium channel. Nature **242:** 459.
2. Bezanilla, F. & C. M. Armstrong. 1974. Gating currents of the sodium channels: three ways to block them. Science **183:** 753.
3. Armstrong, C. M. & F. Bezanilla. 1974. Charge movement associated with the opening and closing of the activation gates of the Na channels. J. Gen. Physiol. **63:** 533–552.
4. Keynes, R. D. & E. Rojas. 1973. Characteristics of the sodium gating current in squid giant axons. J. Physiol. **233:** 28P.
5. Keynes, R. D. & E. Rojas. 1974. Kinetics and steady state properties of the charged system controlling sodium conductance in the squid giant axon. J. Physiol. **239:** 393.
6. Meves, H. 1974. The effect of holding potential on the asymmetry currents in squid giant axons. J. Physiol. **243:** 847.
7. Chandler, W. K. & M. F. Schneider. Personal communication.
8. Takashima, S. & H. P. Schwann. 1974. Passive electrical properties of squid axon membrane. J. Membr. Biol. **17:** 51.
9. Bezanilla, F. & C. M. Armstrong. 1975. Kinetic properties and inactivation of the gating currents of sodium channels in squid axon. Phil. Trans. R. Soc. Lond., Ser. B. **270:** 449.
10. Fox, J. M., E. Rojas & R. Stampfli. 1974. Blocking of sodium and potassium conductance by internal application of Zn^{++} in the node of Ranvier. Pflugers Arch. **351:** 271.
11. Hodgkin, A. L. & A. F. Huxley. 1952. A quantitative description of membrane current and its application to conduction and excitation in nerve. J. Physiol. **117:** 500.
12. Keynes, R. D., E. Rojas & B. Rudy. 1974. Demonstration of first-order voltage-dependent transition of the sodium activation gates. J. Physiol. **239:** 100P.
13. Chandler, W. K. & H. Meves. 1970. Evidence for two types of sodium conductance in axons perfused with sodium fluoride solution. J. Physiol. **211:** 653.

Discussion

Dr. Racker: Could you describe the pronase preparations you used and tell us whether you have established what is removed from the membrane after pronase treatment? I would also like to ask Dr. Mueller whether this complicates his explanation of inactivation.

Dr. Bezanilla: The pronase preparation used is not highly purified and some components may be inactive in the perfusion solution, which contains fluoride. We don't know what part of the enzyme is really acting, but we do know that the effect is due to the enzymatic activity because the reaction is strongly temperature-dependent. Furthermore, boiling destroys the activity.

DR. MUELLER: Our model is not inconsistent with the pronase data. If the association energy of a nonconducting oligomer is raised by protease activity, this could result in a similar lack of inactivation. The gating currents can be accounted for in similar manner. Of course, we do not claim that the same system is operative in nerve membrane.

DR. URRY: Could you give a value for the gating charge per channel?

DR. BEZANILLA: In order to answer that question I would have to make an assumption about channel density, so instead of giving you that value, I shall give you another and you make a calculation according to the assumption that you want to make. Therefore I am going to tell you that there are twelve hundred electronic charges per μm^2.

DR. LEE: Did you see any effect of pronase treatment on the potassium channel?

DR. BEZANILLA: No.

DR. LEE: Are the gating currents dependent on metabolic energy?

DR. BEZANILLA: No. The gating currents are seen regardless of the ionic electrochemical gradients across the membrane. Also, we generally use fluoride inside and provide no chemical source of metabolic energy.

DR. MILLER: How convinced are you that the rising phase of your gating current is real?

DR. BEZANILLA: In our older procedure we added currents resulting from hyperpolarizing and depolarizing pulses of equal size, from the holding potential. We proposed that the currents for each step, negative and positive, rose instantaneously and then fell, but amplitudes and time constants were different for the two pulses and hence their sum showed a rising phase. In other words, there is some gating charge movement during the negative steps, and this adds onto and distorts the positive step gating current, which is the current we wish to see. In the new procedure we take one pulse from the very negative region of membrane potential, where the charge movement is extremely small, and we still see some rising phase in the averaged current. So the question remains: does the current associated with the opening of channels have a rising phase?

DR. FINKELSTEIN: Can I pursue that point a moment? I thought in your earlier publication you had shown that by proper setting of your pulses you could completely eliminate the rising phase.

DR. BEZANILLA: In those experiments we had a blanking period of 20–50 microseconds and there was therefore some uncertainty with regard to the rising phase. In our present method we have improved the time resolution and we see a rising phase.

DR. MAURO: Dr. Bezanilla would you kindly comment on the temperature-dependence of the gating currents.

DR. BEZANILLA: The temperature-dependence is quite strong with a Q_{10} of the order of 2.5 or 2.7. However, we have to be careful in deciding whether this overall Q_{10} has any meaning because now that we see more than one component we should analyze them separately.

SOME DIELECTRIC PROPERTIES OF MUSCLE MEMBRANE AND THEIR POSSIBLE IMPORTANCE FOR EXCITATION-CONTRACTION COUPLING*

W. Almers

Department of Physiology and Biophysics
University of Washington School of Medicine
Seattle, Washington 98195

R. H. Adrian and S. R. Levinson

Physiological Laboratory
Cambridge CB2 3EG, England

Introduction

Some important physiological functions of excitable cells are regulated by the membrane potential. In nerve, for example, the membrane potential controls the permeabilities to sodium and potassium, and potential-dependent permeability changes to these two ions enable a nerve to conduct impulses. In muscle, the membrane potential in addition regulates contraction. The range of membrane potentials available to excitable cells for such regulation is equal to the amplitude of an action potential, or roughly 120 mV. Within this narrow range, the cell membrane can shut off and turn on fully its ionic permeabilities, as well as the mechanical force output in skeletal muscle. Not unexpectedly, voltage-regulated processes depend extremely steeply on the membrane potential. Sodium and potassium permeabilities change e-fold within only 4 to 6 mV, and the transition from mechanical quiescence in muscle to full tension output occurs over a potential span of only about 20 mV. By comparison, the conductance of a solid-state diode requires a full 25 mV for an e-fold change.

In order to learn about voltage-dependent regulation in excitable cells (called "voltage gating" hereafter), several laboratories have now begun to investigate in some form or another the dielectric properties of cell membranes. Of course, the fact that biological membranes behave like a capacitor need have nothing at all to do with their biological functions; for instance, artificial black membranes formed from pure phospholipids and hydrocarbon solvent have a capacity which is nearly, although not quite, as large as that of excitable membranes.[1, 2] Nevertheless, as pointed out some time ago by Hodgkin and Huxley,[3] voltage gating should involve the movement in the membrane of molecules with a large permanent dipole moment. Such movement would appear as an electric current, a polarization current, which flows in addition to that required to charge the conventional capacity arising in the hydrocarbon portion of the membrane.

There have been extensive studies of the dielectric behavior of chemically

* This work was supported in part by Grant No. AM-17803 from the United States Public Health Service.

well-defined artificial phospholipid bilayers with no additives other than hydrocarbon solvent.[1, 2, 4, 5] Under conditions where the amount of solvent is too small to have significant effect (there is, of course, no solvent in biological membranes), such an artificial bilayer behaves as a slightly leaky, but otherwise perfect, capacitor. Its capacity is said to be independent of potential and frequency at frequencies up to 10 MHz.[4] This fits well with the view that the capacity of artificial black membranes arises in the hydrocarbon layer from field-induced ("electronic") dipoles.

By contrast, any dielectric contribution made by "gating molecules" is likely to have all the characteristics of a dielectric made from permanent dipoles. It should be present over the limited potential range where "gating" occurs—the potential range spanned by an action potential—but should vanish at more extreme potentials to either side of this range. Besides this effect of "dielectric saturation," one might expect that because of the molecular size of their moving parts, gating molecules would constitute a "lossy" delectric in that they may not be able to change their orientation instantly after a step change in electric field. Dielectric saturation and "lossiness" therefore represent features by which one might recognize those dielectric components in biological membranes which may have to do with voltage gating.

Materials and Methods

All experiments were performed in the cold (4° C) on sartorius muscles of *Rana temporaria.* The three-microelectrode voltage-clamp technique of Adrian, Chandler, and Hodgkin[6] was used for most electrical measurements. Experimental solutions were designed to block or minimize the normal ion movements across the muscle membrane; their compositions are given in Table 1. In both of these solutions, the conductance of muscle membrane is between 20 and 50 μmho/cm^2 and therefore ten times less than the resting conductance of a normal muscle fiber. What little ionic conductance remained showed little or no variation with time; the solutions of Table 1 therefore represent favorable conditions for studying the dielectric properties of muscle membrane.

Table 1

Composition of Solutions*

Ingredient	Solution A	Solution B	Purpose
Na^+	–	180	–
Tetraethylammonium$^+$	180	–	Block delayed K^+ channels
Rb^+	10	10	Block inwardly rectifying K^+ channels
$SO_4^=$	95	95	Impermeant anion
Ca_2SO_4	5	5	–
Tetrodotoxin	4×10^{-4}	4×10^{-4}	Block sodium channels
Sucrose	350	–	Block contraction

* All concentrations in mM. Solutions were buffered to pH 7.2–7.4 with 2 mM Tris-maleate buffer. Solution A has 2.5 times the tonicity of Ringer's fluid; solution B is isotonic.

RESULTS AND DISCUSSION

Muscle Membrane as a Dielectric: Polarization Currents, Voltage-Dependent Capacity and Dielectric Saturation

The first indication that the dielectric behavior of muscle membrane differs from that of an undoped artificial lipid bilayer came from the discovery of Schneider and Chandler[7] that the charge that one needs to deposit on the membrane capacity for a given potential displacement depends on the membrane potential. The experiment of FIGURE 1 is similar to theirs and illustrates the effect. First, membrane currents during four small rectangular voltage pulses, "control pulses," were recorded, stored, and summed in a digital computer. These pulses went from —108 to —92 mV and therefore did not include the range of potentials spanned by the muscle action potential. Next, a larger voltage pulse, the "test pulse," was given, which carried the membrane from —100 mV to some less negative potential. The computer then scaled the summed current traces recorded during the control pulse by one quarter times the ratio of test- and control pulse amplitudes, and subtracted this scaled version of the control current trace from that recorded during the test pulse. If the membrane behavior could be described with perfect capacitors and resistors, nothing should be left after the subtraction. Instead, there remains an excess outward current during the pulse, and an excess inward transient after the pulse (FIGURE 1).

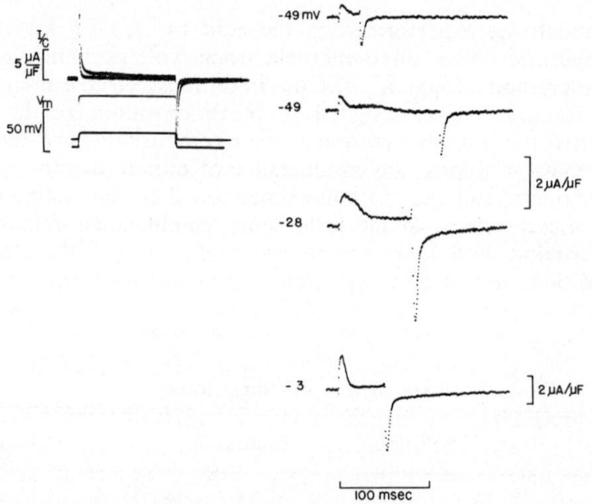

FIGURE 1. Polarization currents in muscle. *Left:* single oscilloscope traces of current (noisy trace) and voltage (bottom) during the small control pulse and a test pulse to —48 mV. I/C, membrane current referred to capacity. *Right:* difference traces compiled from control and test pulses as described in the text; numbers next to each trace give membrane potential during test pulse. Second trace from the top and oscilloscope traces on the left of the figure were from the same measurement. Note presence of a "hump" after the initial transient, which appears on the single current trace (left) as well as on the processed trace (right), which cannot therefore result from the signal-averaging or subtraction procedure and may indicate complex kinetics of polarization currents.[8] Solution A.

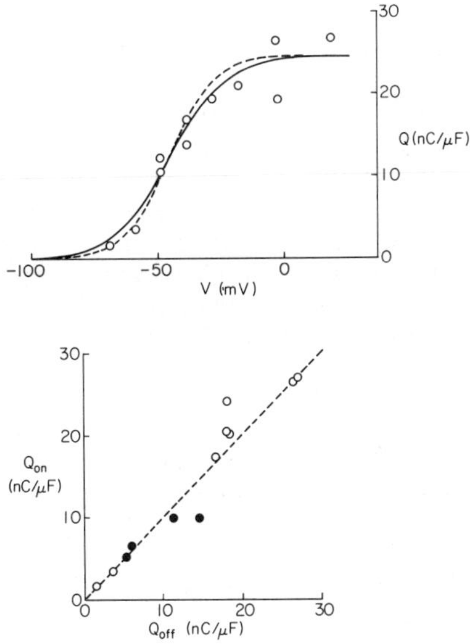

FIGURE 2. Analysis of the experiment in FIGURE 1. *Top:* ordinate, excess charge; abscissa, potential during the test pulse. All measurements from pulses of sufficient duration for the difference current to reach a steady level. The transient part of the current was defined by fitting a straight line to the later part of the trace and extrapolating to the previous voltage step. Continuous line, two-state model (Equation 2) with p/a = 1, V′ = —45 mV, and Q_{max} = 26.2 nC/μF. Dashed line, Langevin-Debye model (Equation 4) with p/a = 3, V′ = —45 mV, and Q_{max} = 36.3 nC/μF. In FIGURE 1, some dipole reorientation must have taken place also during the control pulse; the theoretical curves therefore represent $Q = Q^* - Q^*_{-100\ mV} - A\, dQ^*_{-100\ mV}/dV$ where $Q^*(V)$ is given by Equation 2 or 4, and A is the amplitude of each test pulse.[10] The correction terms were small for the two-state model, but large for the Langevin-Debye model, hence the different values for Q_{max}. *Bottom:* Comparison of "on" and "off" charges. Filled circles, test pulses to —48 mV of varying duration; open circles, test pulses to varying potentials. Dashed line represents equality of on- and off-charge.

Where present, a steady portion of the excess outward current (e.g., trace labeled "—28" in FIGURE 1) undoubtedly represents movement of ions across the membrane, and arises because the small residual ionic membrane conductance is not entirely ohmic. The transient portion, however, does not behave as if it represented transmembrane ion flow. We have observed such transients in a variety of external media, and it is clear that none of the ions present externally in the experiment of FIGURE 1 significantly affects shape or magnitude of the transients. Furthermore, although the peak current amplitudes at the make and break of the test pulse may vary considerably, the amounts of charge displaced by the transients during and after the pulse are closely similar. This "conservation of charge" applies no matter whether the excess transient is varied by changing pulse amplitude or duration (FIGURE 2, bottom). We cannot rigorously disprove that the excess transients result from transmembrane flow of some unknown myoplasmic ion through an unknown ionic channel with unprecedented characteristics. The complete lack of evidence for such a mechanism, however, persuades us to accept the much simpler view that the transients result from slow polarization of the membrane dielectric or some constitutent thereof.

FIGURE 1 (for example, the second trace from the top) suggests that these polarization currents may not follow simple first-order kinetics.[8] However, by far the larger part of a muscle fiber's cell membrane belongs to a system of narrow, tube-like invaginations of the sarcolemma called transverse tubular system. The tubule lumen is only a few hundred angstroms in diameter, and therefore each part acts as an appreciable series resistance. If the excess transients arose in the tubule membrane, this series resistance would greatly complicate a kinetic analysis. Nevertheless, whatever the problems of voltage control in the tubules when the membrane potential changes rapidly, steady-state potential control there should be virtually perfect under the present experimental conditions[8, 9, 10] and analysis of some steady-state properties of polarization currents therefore relatively simple. FIGURE 2 (*top*) plots excess polarization charge against the potential during the test pulse. The Figure suggests that the amount of excess charge that can be displaced by potential changes is finite and saturates at positive potentials. The maximal charge displacement in experiments of this sort is 25 to 35 nC/μF if referred to the low-frequency membrane capacity at —100 mV†; charge displacement is half-maximal at —47 to —30 mV, depending on the composition of the bathing medium.

FIGURE 3 shows the effect of potential on membrane capacity. In measuring the capacity, one integrates with respect to time the membrane current transient during a voltage step of sufficiently small amplitude ($<$10 mV) for the membrane to behave as a linear element; the capacity is then obtained by dividing this integral by the amplitude of the voltage step. Details of method and underlying theory are given elsewhere.[9, 15] In the experiment of FIGURE 3, the membrane was held at —100 mV and periodically voltage-clamped for 200 to 300 msec to various other potentials (abscissa); during this time, the capacity was measured as described. Plotted on the ordinate is the ratio of capacity at the potential on the abscissa (c_T) to the capacity at the holding potential (c_c). FIGURE 3 shows that membrane capacity depends strongly on the membrane potential, showing a sharp peak between —30 and —40 mV and diminishing at either more positive or negative potentials. The average peak capacity in such experiments was 1.43 times that at —90 to —100 mV, and occurred on the average at —40 mV. The extra charge needed in depolarizing the membrane from —100 mV to some potential x can be calculated from FIGURE 3 by

$$Q = c_c \int_{-100\ \mathrm{mV}}^{x} \left(\frac{c_T}{c_c} - 1\right) dV \qquad \textbf{(1)}$$

At 0 mV, the excess charge was 26.9 nC/μF in FIGURE 3, which is in good agreement with the corresponding value from FIGURE 2.

The experiments of FIGURES 2 and 3 strongly indicate the presence in the membrane of permanent molecular dipoles, and we believe that the excess polarization currents as well as the bell-shaped dependence of capacity on voltage represent the dielectric contribution of such polar membrane constituents. At the potential of peak capacity (FIGURE 3) and the potential where excess charge displacement is half-maximal (FIGURE 2), one would expect the dipoles to be randomly oriented (transition potential), whereas at sufficiently extreme potentials, all dipoles become aligned and a further increase in membrane electric field would cause no further polarization of this compo-

† Polarization currents and charges are referred to the membrane capacity measured at —100 mV rather than the membrane area, which in skeletal muscle cannot be measured in any simple manner because of the presence of the transverse tubular system. An approximate figure for the membrane area across which the observed currents flow can be obtained by assuming a specific membrane capacity of 1 μF/cm^2.

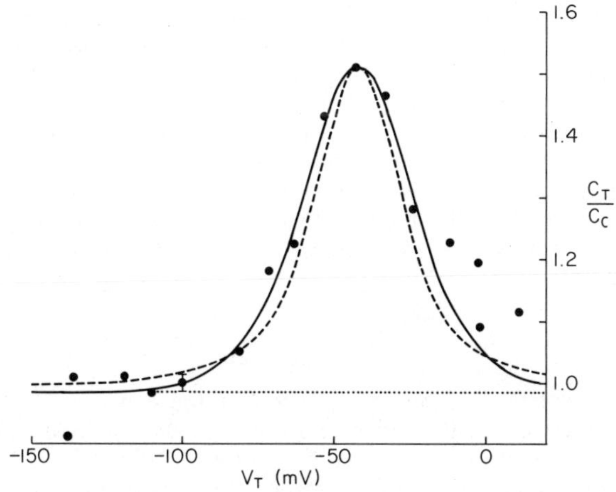

FIGURE 3. Voltage-dependence of capacity. Ordinate, ratio of capacities at test potential (c_T) and holding potential (c_c); abscissa, test potential. Holding potential −100 mV. Error bar at −100 mV represents standard error of mean of five measurements. All points from "on" transients; length of capacity-measuring pulse and integration time, 100 msec. Continuous curve, Equation 3 with p/a = 1, Q_{max} = 25.5 nC/μF, and V′ = −46 mV, c_o/c_c = 0.985 (dotted line). Dashed curve, Equation 5 with the same parameters except p/a = 3. Measurements deviate from theoretical curves at potentials positive to −10 mV; this may be due, at least in part, to voltage- and time-dependent conductance changes during the measurement. Solution A.

nent (dielectric saturation). Below, this model is treated quantitatively. Although unlikely to reveal many molecular details, the treatment will illustrate what properties one may expect in two of the simplest cases.

Estimates of Some Parameters Governing the Dielectric Contribution of Polar Molecules

For simplicity, it will be assumed that the above observations are due to a single population of dipoles, despite suggestions that this may not be strictly correct.[8, 10] The simplest model results if one assumes that dipoles can exist in only two states between which they distribute themselves independently of one another and in accordance with the Boltzmann distribution. To the extent that reorientation during the control pulse in FIGURE 1 can be neglected (as it could be if one took control pulses at sufficiently negative potentials) the excess polarization charge can be described by

$$Q = Q_{max} \frac{1}{1 + \exp\,[2p(V' - V)/akT]} \tag{2}$$

where Q_{max} is the maximal charge that would result if all dipoles switched from one aligned state into the other; it is given by $Q_{max} = 2\ Np/a$ where N is the number of dipoles per unit membrane area. In Equation 2, p is the dipole moment or, more strictly, the product of dipole moment and the component normal to the membrane surface of the change in dipole orientation between the two allowed states. The membrane thickness is given by a; k and T are Boltzmann's constant and absolute temperature; V′ the transition potential; V the membrane potential; and V/a the electric field (assumed constant) acting on the dipole. Instead of dipoles, one can assume the presence of N charged particles per unit membrane area, each of valency z, which can distribute themselves between two positions separated by a distance αa in a direction normal to the membrane. The equation describing that case[7] is obtained by substituting $p = \alpha az/2$ in Equation 2. The capacity defined as $C = dQ/dV$ is given by

$$C = c_0 + \frac{pQ_{max}}{akT} \frac{1}{1 + \cosh[2p(V' - V)/akT]} \tag{3}$$

where c_0 is the voltage-independent portion of the capacity. At the transition potential, V′, the voltage-dependent portion has its maximal value given by $pQ_{max}/2akT$. The continuous curves in FIGURES 2(top) and 3 were drawn on the basis of Equations 2 and 3, respectively, assuming p/a equal to one electronic charge, and $Q_{max} = 25.5$ nC/μF. The curves reproduce the main features of the observations and suggest the presence of about 8×10^{10} dipoles for every μF of membrane capacity measured at —100 mV. The dipole moment is equal to that of a rod as long as the membrane is thick, bearing one positive or negative electronic charge at the ends. If the membrane thickness is 39 Å, the dipole moment is 187 Debye units or 100 times that of a water molecule. If the specific capacity of tubule membrane and sarcolemma is 1 $\mu F/cm^2$, there are about 800 dipoles/μm^2 of membrane.

Both number and moment of dipoles are strongly model-dependent. If the dipole could occupy in the absence of an electric field all angular orientations with equal probability, the excess polarization charge would be described by a modification of the well-known Langevin-Debye relation (see, e.g., Reference 11)

$$Q = Q_{max}\left[\frac{akT}{2p(V' - V)} + \frac{1}{1 - \exp[2p(V' - V)/akT]}\right] \tag{4}$$

and the capacity would be given by

$$C = c_0 + \frac{pQ_{max}}{akT}\left[\frac{(akT)^2}{2p^2(V' - V)^2} - \frac{1}{\cosh[2p(V' - V)/akT] - 1}\right] \tag{5}$$

Here, the maximal value reached by the voltage-dependent portion of the capacity at $V = V'$ is $pQ_{max}/6\,akT$, which is three times less than in the model of Equation 3. In fitting Equations 4 and 5 to the data, $p/a = 3$ was therefore assumed, which corresponds to a dipole moment of 561 Debye units at a membrane thickness of 39 Å. Retaining the experimentally given value of 25.5 nC/μF in FIGURE 3, the number of dipoles must be reduced threefold to about 265/μm^2 in this model. The dipole moment could be yet higher and the number of dipoles yet lower if, for example, there existed an energetically favored dipole orientation in the plane of the membrane.

The data are not sufficiently accurate to distinguish between these two or other, more complicated models. It is nevertheless probable that one is dealing with polar molecules inside the membrane which could serve as "gating molecues"; their dipole moment is sufficiently high and their transition potential appropriate for them to respond strongly to the relatively small changes in membrane potential found in living cells. Perhaps the most important question at this stage concerns the physiological role of these polar molecules.

Excess Charge and the Regulation of Sodium and Potassium Permeabilities

The results described here are similar in some ways to those obtained by Armstrong and Bezanilla[12, 13] and then by others[14, 15] in the squid giant axon; the maximal excess polarization charge in squid giant axons is about 30 nC/μF[14] and almost identical to that in muscle. Polarization currents in the squid axon are thought to accompany the gating of sodium channels during the action potential, and since skeletal muscle also has sodium channels, it is natural to ask to what extent polarization currents in muscle are concerned with sodium channel gating.

From the start, the dielectric events in muscle are too slow to be concerned solely with sodium channel gating. The most direct way towards more quantitative information about the contribution of sodium channels is to study the binding to muscle of the drug tetrodotoxin. Tetrodotoxin blocks sodium channels at nanomolar concentrations and has no other known effects on excitable cells. The accepted view is therefore that at sufficiently low concentrations, tetrodotoxin binds predominantly to sodium channels, and many authors have measured uptake of this drug by excitable tissues in order to determine the sodium channel density. FIGURE 4 shows uptake by frog sartorius muscle

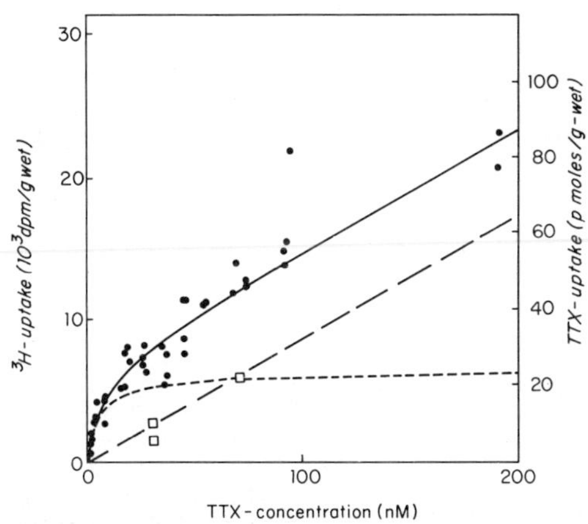

FIGURE 4. Uptake of tritium-labeled tetrodotoxin by frog sartorius muscle. Incubation for 6 hours in Ringer's fluid at 4° C. Continuous line, least-squares fit of U = M[TTX]/(K+[TTX]) + b[TTX] to the filled circles; U = uptake, M = capacity of saturable component, K = equilibrium constant, [TTX] = toxin concentration, b = a constant. Saturable (dotted line) and linear binding components (dashed line) as obtained from the curve-fitting procedure are also drawn separately. Parameters for the saturable component given in the text. Open squares represent measurements to test the two-component model. They represent uptake in the presence of 1–2 μM unlabeled toxin, which should prevent saturable binding of tritiated toxin by saturating the receptors. As predicted only the linear binding component remains (dashed line). (From Almers and Levinson.[17] By permission of the *Journal of Physiology*.)

of tritium-labeled tetrodotoxin. The method of tritium labeling, assay of specific activity, and other experimental details are given elsewhere.[17] As is usual in this type of experiment,[16] the uptake of radioactivity consists of two components, one showing no indication of saturation at concentrations up to 200 mM, the other saturating and behaving as if the drug were bound to a finite number of high-affinity receptors. The solid curve is the least squares fit of this two-component model to the data; saturable (dotted line) and unsaturable (dashed line) components are also drawn separately. The saturable component has a maximal value of 23 pmol/g wet tissue weight, and an apparent dissociation constant of 5.5 nM, which is in reasonable agreement with the value derived from the electrical effects of the drug on sodium channels. Other experiments, some of them with unlabeled tetrodotoxin, confirm that the saturable uptake of radioactivity in FIGURE 4 represents binding of tetrodotoxin to muscle. If each tetrodotoxin receptor represents one sodium channel, it can be concluded that an average cylindrical fiber of 80 μm diameter has 350 to 400 sodium channels per square micrometer of cylinder surface. The squid giant axon has 550 binding sites per square micrometer,[18] a very similar number. Due to the transverse tubular system, however, the membrane capacity associated with unit surface area of a muscle fiber is about 7 μF/cm in a fiber of 80 μm diameter, whereas the membrane capacity of squid axons is 0.9 $\mu F/cm^2$.[19] When referred to the membrane capacity, muscle has therefore ten times fewer sodium channels than the squid axon. Even if the maximal polarization charge of 33 nC/μF in squid nerve[14] is associated entirely with sodium channels, their contribution to excess charge movements in muscle could be no more than about 3 nC/μF. In conclusion, muscle has far too few sodium channels to account for the observed excess charge.

Superficially, it seems more likely that the excess charge movements are associated with the "gating" of potassium channels, since polarization and delayed potassium currents occur on roughly similar time scales, and the former precede the latter. However, evidence against this possibility can be obtained by comparing the effects of the local anesthetic tetracaine on the two currents (Almers; in preparation). FIGURE 5 shows polarization currents and delayed

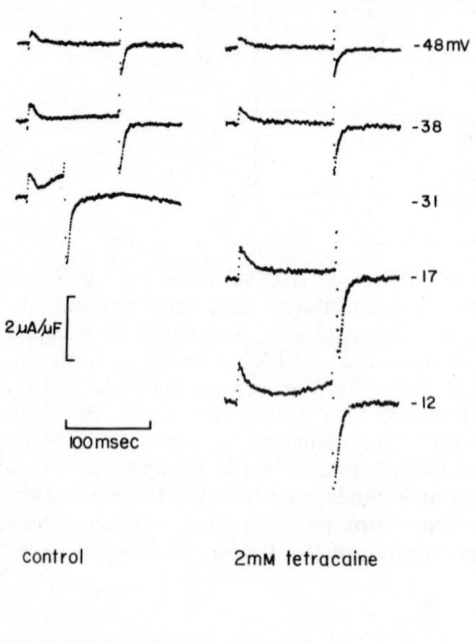

FIGURE 5. Polarization currents in the absence (*left*) and presence (*right*) of 2 mM tetracaine. Both fibers from the same animal; experimental details as in FIGURE 1. Numbers to the right give potential during the test pulse and refer to both columns. In the control fiber the pulse to —31 mV resulted in a visible contracture; the others did not. Curvature of trace at —31 mV is a movement artifact. Solution B.

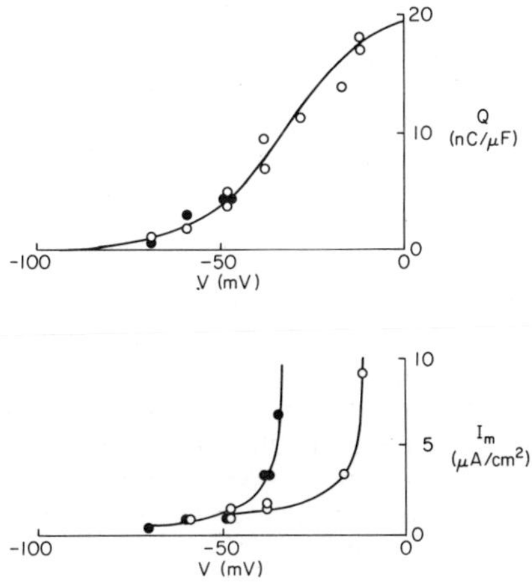

FIGURE 6. Analysis of the experiment in FIGURE 5. Filled circles, control; open circles, tetracaine-poisoned fiber. *Top:* excess polarization charge; the continuous curve is Equation 2 with p/a = 1, V′ = −32 mV, and Q_{max} = 20.3 nC/μF. *Bottom:* final values of membrane current during the test pulses, corrected for leakage. Tetracaine shifts the threshold for appearance of delayed currents, but not the amount of polarization charge transferred at a given potential.

currents from a muscle in tetraethylammonium-free isotonic medium (solution B), the right column with, the left without tetracaine. It is seen that the drug has little or no effect on either size or kinetics of excess polarization currents, at least over the potential range where a comparison can be made. On the other hand, while delayed currents are prominent at —27 mV in the control, the presence of 2 mM tetracaine makes a depolarization of —12 mV necessary to elicit the first detectable sign of delayed currents. FIGURE 6 analyzes the experiment further; the upper portion shows the potential dependence of excess charge, and the lower, the membrane current corrected for leakage. The potassium conductance threshold in tetracaine (open circles) is about 20 mV more positive than in the control fiber (filled circles) whereas no such shift appears on charge movements. Fitting the delayed K^+-currents with Hodgkin-Huxley kinetics[3] shows that 2 mM tetracaine, besides reducing the maximal K^+ conductance two- or threefold, slows the rate constants ($1/\tau_n$) about fourfold and shifts the steady-state relationship between potassium conductance and potential from 20 to 25 mV more positive potentials. Tetracaine has therefore large effects on the gating of K^+ channels, yet little or no effect on polarization currents.

In conclusion, the major portion of the observed polarization currents is probably concerned neither with sodium nor potassium channels, although, of course, both sodium and potassium channel gating currents may make some small contribution.

Possible Role of Polarization Currents in Excitation-Contraction Coupling

A third strongly voltage-dependent physiological function of muscle is its mechanical activity. From the work of Hodgkin and Horowicz[21] and Heistracher and Hunt[32] it is known in some detail how contractile activity depends on the membrane potential. It is also clear that electrical signals traveling along a fiber are conducted radially into the interior[23, 24] via the transverse tubular system (for the anatomy of transverse tubules see Reference 25). Ultimately, electrical excitation in the transverse tubules leads to calcium release into the myoplasm from an intracellular store called sarcoplasmic reticulum (SR). Inside a muscle fiber, transverse tubules and SR must therefore interact in such a way that the appropriate membrane-potential change in the former will cause calcium release from the latter. Electronmicroscopy has revealed the probable site of this interaction by showing that SR and transverse tubules meet in close apposition and form a structure called "triad." At the triad, the membranes of the SR- and T system are within 200 Å of each other; the gap is periodically bridged by electron-dense structures which are sometimes referred to as "feet."[26]

In conjunction with the observed polarization currents, these anatomical data suggest a possible mechanism for excitation-contraction coupling.[7] The transverse tubule membrane at the triad might contain molecules that extend across the junctional gap into the SR membrane; these molecules might possess a dipole moment and become sufficiently distorted by changes in the electric field across the tubule membrane to serve as gating molecules for a nearby calcium pore in the SR. Unlike the case in nerve, therefore, where gating and ion transport happen in the same membrane, this hypothesis for excitation-contraction coupling in muscle postulates the gating molecule to sense the potential across the tubule membrane and control a calcium-permeable channel in the SR.

A number of findings fit well with this hypothesis: (i) The voltage range over which the observed molecular dipoles change orientation spans that over which regulation of contraction occurs. (ii) The muscle's contractile response to changes in membrane potential disappears under maintained depolarization[21] and so do the observed polarization currents;[20, 37] contractile activity and polarization currents recover with approximately the same time course after repolarization.[20] (iii) The number of electron-dense "feet" at the triadic junction may be similar to the number of dipoles that give rise to the observed polarization currents. As discussed above, the dielectric data can be fitted by assuming the presence of about 800 dipoles/μm^2 of membrane; the number of electron-dense "feet" is 500–1000/μm^2 of tubule membrane.[26] Of course, this similarity as well as the considerations about to follow depend on the model used to estimate the number of gating molecules in the membrane.

It is of interest to ask how many calcium ions per unit of time the SR would have to release for each postulated gating dipole in the activating position. In an average cylindrical fiber of 80 μm diameter, with 7 μF capacity per square centimeter of cylinder surface and an average 10^{11} gating dipoles per microfarad, the "myoplasmic concentration" of gating molecules is around 0.5 μM. By comparison, the concentration of sites where calcium can bind to troponin[27] is roughly 200 times higher, or about 100 μM. In the experiment represented in FIGURE 5 (control fiber) the pulse from —100 to —38 mV transferred 6.8 nC/μF of charge or, in the present hypothesis, about 25% of all gating molecules into the activating position. No sign of contraction was visible by microscopic observation during and after the pulse. Ford and Podolsky[28] have measured the rate of calcium uptake by the SR in skinned muscle fibers at room temperature and have described it by

$$v = \xi[Ca]^{++} \qquad \textbf{(6)}$$

where v is the uptake rate (mmol/l fiber volume and second), $[Ca]^{++}$ the myoplasmic calcium concentration, and $\xi = 1.4 \times 10^3$/sec by their measurement. In their experiments, the myoplasmic bulk calcium concentration at contraction threshold was 0.2 μM, and the calcium uptake rate under this condition therefore 280 $\mu M sec^{-1}$. Since the threshold potential for contraction was very close to —38 mV in the fiber represented in FIGURE 5, it is reasonable to suppose that calcium uptake and release must have been equal at that potential and of the order of 280 μMsec^{-1}. Therefore, the rate of calcium release per gating molecule in the activating position was (280 μM Ca^{++} sec^{-1})/0.125 μM sites or 2200 calcium ions per second. Another argument,[8] based on the latency period between stimulus and onset of the active state, leads to the similar number of 10^4 ions per second.

These estimates could be in error by as much as an order of magnitude. For instance, calcium uptake by the SR is likely to be slower in the cold than at room temperature; on the other hand, Ford and Podolsky's measurement of ξ may be a lower limit. Nevertheless, an ion flux of 10^3 to 10^5 ions per second could easily be handled by a single ion-conducting pore; in artificial membranes, transit rates through pores of 10^7 ions/sec (gramicidin A channel[29]) and even 10^9 ions/sec (alamethicin channel[30]) have been measured. However crude and speculative, these considerations show that one would need only a single calcium-permeable pore next to each gating particle in order to satisfy the calcium requirements for contractile activation in muscle.

Pharmacological Studies of Polarization Currents

In searching for additional evidence that might link excess polarization charge to contraction, we have investigated pharmacological agents that block the contractile response to depolarization. The local anesthetic tetracaine has already been mentioned; at 2 mM concentration, it blocks contraction in intact fibers[36] but not in skinned fibers when calcium ions are applied directly (Almers and Best; in preparation). One may therefore assume that tetracaine leaves the contractile proteins intact and instead interferes with some process in excitation-contraction coupling which precedes the binding of calcium to troponin, such as calcium release. Perhaps unexpectedly, the drug shows no noticeable effects on polarzation currents. Another interesting agent is formaldehyde; a 30-minute exposure to media containing 10 mM formaldehyde (about 0.05%) abolishes depolarization-contraction coupling although more than 50% of the excess charge remains (Almers; unpublished material).

Besides maintained depolarization, the only procedure known so far that attenuates contraction as well as polarization currents in glycerol withdrawal,[20, 35] a technique that is known to disrupt the electrical[32] and anatomical[33, 34] continuity between transverse tubules and sarcolemma. However, the failure of formaldehyde treatment and tetracaine to abolish excess polarization currents is no evidence against a role of these currents in excitation-contraction coupling. In nerve, for instance, tetrodotoxin blocks ion transport through the sodium channel, but does not change membrane currents attributed to the gating of that channel.[13, 14] Similarly, tetracaine might block calcium channels in the SR without interfering with the movement of the gating molecule.

CONCLUSION

Voltage-clamp studies of polarization currents have shown the presence of polar molecules in the membrane of skeletal muscle. Their high dipole moment and favorable transition potential suggest that they may be "gating molecules" which serve to regulate some membrane potential-dependent physiological function. Experiments with the drugs tetrodotoxin and tetracaine show that these polar molecules are unlikely to play a role in regulating the sodium and potassium channels responsible for electrical excitation. Instead, they may regulate calcium release from the sarcoplasmic reticulum, which also depends steeply on the cell membrane potential. Schneider and Chandler's hypothesis whereby calcium release from the sarcoplasmic reticulum is regulated by voltage-sensitive "gating molecules" in the tubule membrane is still unproven, but remains attractive and plausible.

References

1. Hanai, T., D. A. Haydon & J. Taylor. 1965. The influence of lipid composition and of some adsorbed proteins on the capitance of black hydrocarbon membranes. J. Theoret. Biol. **9:** 422–432.
2. Fettiplace, R., D. M. Andrews & D. A. Haydon. 1971. The thickness, composition and structure of some lipid bilayers and natural membranes. J. Membr. Biol. **5:** 277–296.
3. Hodgkin, A. L. & A. F. Huxley. 1952. A quantitative description of membrane current and its application to conduction and excitation in nerve. J. Physiol. **117:** 500–544.
4. Hanai, T., D. A. Haydon & J. Taylor. 1964. An investigation by electrical methods of lecithin-in-hydrocarbon films in aqueous solutions. Proc. Roy. Soc. London, Ser. A **281:** 377–391.
5. Andrews, D. M., E. D. Manev & D. A. Haydon. 1970. Composition and energy relationships for some thin lipid films, and the chain conformation in monolayers at liquid-liquid interfaces. Spec. Discuss. Faraday Soc. **1:** 46–56.
6. Adrian, R. H., W. K. Chandler & A. L. Hodgkin. 1970. Voltage-clamp experiments in striated muscle fibres. J. Physiol. **208:** 607–644.
7. Schneider, M. F. & W. K. Chandler. 1973. Voltage-dependent charge movement in skeletal muscle: A possible step in excitation-contraction coupling. Nature **242:** 244–246.
8. Almers, W. 1975. Observations on intramembrane charge movements in skeletal muscle. Phil. Trans. Roy. Soc. London **270:** 506–511.
9. Adrian, R. H. & W. Almers. 1974. Membrane capacity measurements on frog skeletal muscle in media of low ion content. J. Physiol. **237:** 573–605.
10. Adrian, R. H. & W. Almers. 1975. The voltage dependence of membrane capacity. J. Physiol. In press.
11. Debye, P. 1929. Polar Molecules. Dover. New York, N.Y.
12. Armstrong, C. M. & F. Bezanilla. 1973. Currents related to movement of the gating particles of the sodium channels. Nature **242:** 459–461.
13. Armstrong, C. M. & F. Bezanilla. 1974. Charge movement associated with the opening and closing of the activation gates of the Na channels. J. Gen. Physiol. **63:** 533–552.
14. Keynes, R. D. & E. Rojas. 1974. Kinetics and steady-state properties of the charged system controlling sodium conductance in the squid giant axon. J. Physiol. **239:** 393–434.
15. Meves, H. 1975. The effect of holding potential on the asymmetry currents in squid giant axons. J. Physiol. **243:** 847–867.
16. Colquhoun, D., R. Henderson & J. M. Ritchie. 1972. The binding of labelled tetrodotoxin to non-myelinated nerve fibres. J. Physiol. **227:** 95–126.
17. Almers, W. & S. R. Levinson. 1975. Tetrodotoxin binding to normal and depolarized frog muscle and the conductance of a single sodium channel. J. Physiol. **247:** 483–509.

18. LEVINSON, R. S. & H. MEVES. 1975. The binding of tritiated tetrodotoxin to squid giant axons. Phil. Trans. Roy. Soc. London **270:** 349–352.
19. HODGKIN, A. L., A. F. HUXLEY & B. KATZ. 1952. Measurement of current voltage relations in the membrane of the giant axon of *Loligo*. J. Physiol. **116:** 424–448.
20. CHANDLER, W. K., M. F. SCHNEIDER, R. F. RAKOWSKI & R. H. ADRIAN. 1975. Charge movements in skeletal muscle. Phil. Trans. Roy. Soc. London **270:** 501–505.
21. HODGKIN, A. L. & P. HOROWICZ. 1960. Potassium contractures in single muscle fibres. J. Physiol. **153:** 386–403.
22. HEISTRACHER, P. & C. C. HUNT. 1969. The relation of membrane changes to contraction in twitch muscle fibres. J. Physiol. **201:** 589–611.
23. COSTANTIN, L. L. 1970. The role of sodium current in the radial spread of contraction in frog muscle fibers. J. Gen. Physiol. **55:** 703–715.
24. HUXLEY, A. F. & R. E. TAYLOR. 1958. Local activation of striated muscle fibres. J. Physiol. **144:** 426–441.
25. PEACHEY, L. D. 1965. The sarcoplasmic reticulum and transverse tubules of the frog's sartorius. J. Cell Biol. **25:**209–231.
26. FRANZINI-ARMSTRONG, C. 1970. Studies of the triad. I. Structure of the junction in frog twitch fibers. J. Cell Biol. **47:** 488–499.
27. EBASHI, S. & M. ENDO. 1968. Calcium ion and muscle contraction. Progr. Biophys. Mol. Biol. **18:** 123–183.
28. FORD, L. E. & R. J. PODOLSKY. 1972. Intracellular calcium movements in skinned muscle fibres. J. Physiol. **223:** 21–33.
29. HLADKY, S. B. & D. A. HAYDON. 1972. Ion transfer across lipid membranes in the presence of gramicidin A. I. Studies of the unit conductance channel. Biochim. Biophys. Acta **274:** 294–312.
30. GORDON, L. G. M. & D. A. HAYDON. 1972. The unit conductance channel of alamethicin. Biochim. Biophys. Acta **255:** 1014–1018.
31. HUTTER, O. F. 1969. Potassium conductance of skeletal muscle treated with formaldehyde. Nature **224:** 1215–1216.
32. GAGE, P. W. & R. S. EISENBERG. 1969. Capacitance of the surface and transverse tubular membrane of frog sartorius muscle fibers. J. Gen. Physiol. **53:** 265–278.
33. EISENBERG, G. & R. S. EISENBERG. 1968. Selective disruption of the sarcotubular system in frog sartorius muscle. J. Cell Biol. **39:** 451.
34. FRANZINI-ARMSTRONG, C., R. A. VENOSA & P. HOROWICZ. 1973. Morphology and accessibility of the transverse tubular system in frog sartorius muscle after glycerol treatment. J. Membr. Biol. **14:** 197–212.
35. DULHUNTY, A. F. & P. W. GAGE. 1973. Differential effects of glycerol treatment on membrane capacity and excitation-contraction coupling in toad sartorius fibres. J. Physiol. **234:** 373–408.
36. LÜTTGAU, H. C. & H. OETLIKER. 1968. The action of caffeine on the activation of the contractile mechanism in striated muscle fibres. J. Physiol. **194:** 51–74.
37. ADRIAN, R. H. & W. ALMERS. 1975. Charge movement in the membrane of striated muscle. J. Physiol. In press.

DISCUSSION

DR. BENDER: You referred to dipoles of large values. Do you have a hypothesis concerning possible atomic arrangements resulting in a dipole moment of this magnitude?

DR. ALMERS: A dipole moment of 180 Debye units, which is the lower limit that would fit our data, is large for a small molecule such as water, but not very large for proteins. For instance, β-lactoglobulins in aqueous solution are said to have dipole moments of 700 Debye units (Ferry and Oncley. 1941. J. Am. Chem. Soc. **63:** 272). Even this large number is lower than that calculated if one added up the dipole moments of all amino acids that make up such a protein. So for large molecules such as proteins, a dipole moment of 180 Debye units is not very large.

DR. MCLAUGHLIN: If I understand you and Dr. Bezanilla correctly, the capacitance should be maximal when the charges that carry the gating current are more randomly distributed. If this is so, why does this occur not at zero transmembrane potential but at —50 mV in your case and at —20 mV in Dr. Bezanilla's case? Could this be a manifestation of an asymmetric surface charge or do you think it is due to something else?

DR. ALMERS: It could be due to a number of things, including an asymmetric surface charge. Another possibility would be that one position of the molecules in the membrane may be energetically favored over another, which would also introduce some kind of bias. We cannot yet distinguish between these two possibilities.

DR. URRY: Each element within an alpha-helical structure contributes a dipole moment of 70 Debye units, so that one would require only three of these aligned in parallel to give 210 Debye, and of course one would probably want more than that to construct a channel.

DR. MEISLER: What was the resting potential in these fibers, and was there any electrical response from them? Is it possible that part of the response may be due to a regenerative calcium release from the transverse tubules and sarcoplasmic reticulum?

DR. ALMERS: The zero-current-potential was about —60 mV, and we have to hold the fiber at —100 mV for about 5 minutes in order to bring the membrane back into a state resembling normality. Even then, we find little or no electrical response other than the observed transients.

Regarding the role of calcium ions, the observed polarization currents do not change noticeably if one diminishes the external Ca^{++}-concentration by a factor of four, suggesting that we are not dealing with calcium movements across the transverse tubule membrane. The capacitive behavior of the currents makes us disbelieve, anyway, that the observed currents result from ion flow across the cell membrane. As far as regenerative calcium release from the SR, it is not clear whether it ever occurs in intact muscle fibers. Even if it does we have no reason to expect it to have an effect on the currents measured by us. These currents flow across sarcolemma and tubule membranes, and, we believe, not across the membrane of the SR. All evidence gathered during the last 20 years strongly suggests that the SR is electrically invisible when one measures currents flowing between inside and outside of muscle cells.

SPECIFIC CHEMICAL GROUPS INVOLVED IN THE CONTROL OF IONIC CONDUCTANCE IN NERVE*

Peter Shrager

Department of Physiology
University of Rochester Medical Center
Rochester, New York 14642

Introduction

The electrophysiological characteristics of ion-permeability control in excitable membranes are becoming increasingly clear. However, the identification of the biochemical constituents involved remains relatively unexplored. Experiments designed to detect chemical changes in ionic channels and gates (the words are not intended to imply any particular mechanism) during excitation are difficult because the events are rapid and the number of sites involved, relative to total surface area, is very small. Electrophysiological studies have been greatly aided by the discovery of pharmacological agents, most notably tetrodotoxin and tetraethylammonium ion, that block either sodium or potassium channels with a very high degree of selectivity. The chemical interaction of these compounds with excitable membranes is, however, not well defined. Reagents that have equally good physiological selectivity, but that in addition are known to bind to membrane components in predictable ways, would therefore be of value in further analysis. We have initiated studies of nerve fibers modified by chemical reagents of known specificities for various reactive groups normally present in proteins and lipids. We have sought to correlate structural perturbations with changes in conductance parameters by means of the voltage-clamp technique. Our experiments suggest that it is possible to identify and covalently modify particular groups at sites involved with either sodium or potassium permeation.

Methods

Medial giant axons from the crayfish *Procambarus clarkii* were voltage-clamped by an axial wire technique as previously described.[1] The axon was cannulated with a combined current wire and potential pipette electrode. Currents were collected by a guarded platinum electrode and fed into a current-to-voltage transducer. Temperatures used are noted in results.

Voltage-clamped axons were internally perfused by a modification of the procedure of Shrager *et al.*[2] The voltage-clamp electrode was inserted as for intact (nonperfused) axons, although the lacquer insulation that normally served to seal the entry hole was kept thin for perfusion experiments. A short perfusion pipette was inserted near ganglion 7 (counting the brain as ganglion 1). Flow was driven by gravity (3–4 cm of water) except during solution changes, when a syringe pump provided the additional pressure re-

* This work has been supported by Grants 5-RO1-NS10500 and 5-PO1-NS10981 from the National Institutes of Health.

quired to prevent collapse of the axon. The axoplasm of crayfish giant fibers is relatively fluid and is readily washed away.

The normal external saline (NVH) was that of Van Harreveld,[3] and contained (mM): NaCl 205, KCl 5.4, $MgCl_2$ 2.6, $CaCl_2$ 13.5, $NaHCO_3$ 2.3, pH adjusted to 7.5–7.6 with 1N HCl. The standard internal perfusate[2] contained (mM): KF 109, K_3 citrate 37, NaCl 15, mannitol 96, pH adjusted to 7.5 with a small amount of citric acid. Modifying reagents were added to solutions immediately before use. Vigorous stirring was employed with relatively insoluble compounds rather than using "carrier" organic solvents.

The notation used for conductances will be that of the formulation of Hodgkin and Huxley.[5]

Evidence for the Interaction of Histidine Side Chains with Potassium Gates

When the external pH is lowered, potassium currents in crayfish axons are markedly slowed relative to sodium currents.[1] This is illustrated in FIGURE 1, which shows data from an intact axon. At pH 5.8 the time to peak inward

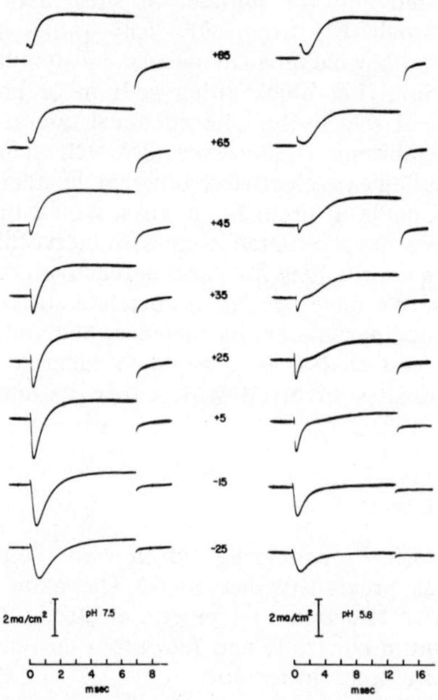

FIGURE 1. Voltage clamp currents at normal and low external pH.[1] *Left:* pH 7.5. *Right:* pH 5.8. Depolarizations were from the holding potential of −75 mv to the levels indicated. Note difference in time scales. The external solution contained (mM): NaCl 197.3, KCl 5.4, $MgCl_2$ 2.6, $CaCl_2$ 13.5, bis-tris 10, pH adjusted with HCl. T = 8.5°C. (From Shrager.[1] By permission of the *Journal of General Physiology*.)

current is about 50% longer than its corresponding value at pH 7.5. The rise of outward K current, however, is slowed on the order of fourfold (measured more clearly in the presence of tetrodotoxin, which specifically blocks Na currents[4]), and Na and K currents become temporally separated. The dependence of potassium currents on external pH is illustrated in FIGURE 2. Increasing the pH from the normal value of 7.5 has almost no effect. As the pH is lowered below 7.5, K currents are slowed and steady-state levels are reduced. Effects of low pH are almost completely reversible.

FIGURE 2. Dependence of potassium currents on external pH. Currents resulting from depolarizations to +44 mv from the holding potential of −76 mv. The time bar in each record indicates 5 msec. T = 8.5° C, External solution (mM): NaCl 197.3, KCl 5.4, $MgCl_2$ 2.6, $CaCl_2$ 13.5, tris-maleate 10, TTX 100 nM.

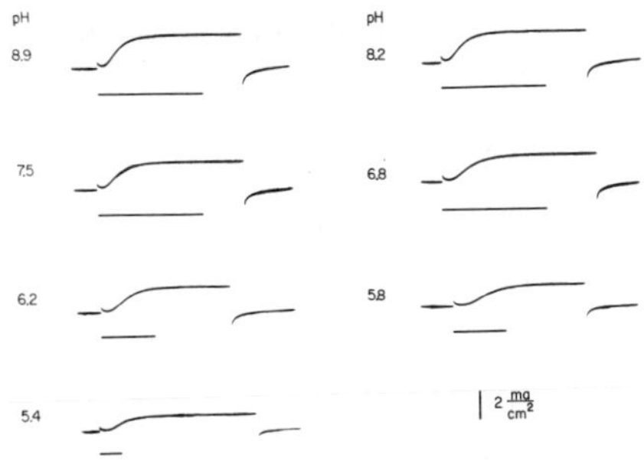

Steady-state currents are significantly depressed at low depolarizations, but are unaffected by the pH change at strong depolarizations (FIGURE 1). This suggested that one effect of the lowered pH might be a change in fixed surface charge resulting from protonation of a membrane group.[14] Steady-state K conductance, $g_{K} \infty$, is shifted along the voltage axis by a constant amount in a depolarizing direction at low pH. However, it is not yet clear whether this voltage shift is sufficient to account for the observed changes in K current kinetics as well.[1] If protonation of a membrane group at the external surface is involved, then effects of low pH might be quite different if applied internally. FIGURE 3 illustrates results of pH changes on a perfused axon in the presence of tetrodotoxin (TTX). Lowering the pH of the standard perfusate to 5.8 by addition of small amounts of citric acid and KH_2PO_4 depresses steady-state levels slightly, but results in no change in kinetics (FIGURE 3b). The change in amplitude is reversed on returning the internal pH to 7.5 (FIGURE 3c). Lowering the external pH resulted in a slowing of K currents as in intact axons (FIGURE 3d). However, unlike results in intact fibers, steady-state currents increased at low external pH. This point has not as yet been investigated in detail. Perfused axons often had lower resting potentials than intact axons. It is possible that a hyperpolarizing surface charge introduced at low external pH removed long-term K inactivation that was not removed by the prepulse

FIGURE 3. Variation of the internal and external pH of a perfused axon. Records were taken in order: a-e. Holding potential, −70 mv. In each family of currents a hyperpolarizing prepulse to −80 mv (150 msec) preceded depolarizations to: −30, −10, +10, +30, +50, +70 mv. (a), (c), (e): control currents at external pH 7.5, internal pH 7.5. (b): internal pH lowered to 5.8 with citric acid and KH_2PO_4. (d): external pH lowered to 5.8 with HCl. Note different time scale in (d). Solutions: External; (mM) NaCl 202.7, KCl 0, $MgCl_2$ 2.6, $CaCl_2$ 13.5, bis-tris 10, pH adjusted with HCl, TTX 100 nM. Internal; standard perfusate (see *Methods*). T = 8° C.

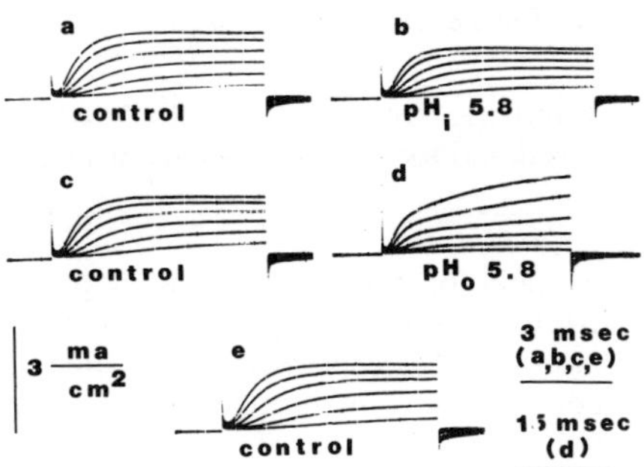

used. Cahalan (unpublished data) has observed removal of long-term K inactivation in nodes of Ranvier bathed in high K media when the external calcium concentration was increased.

An attempt has been made to identify the membrane group involved in the alteration of K-current kinetics at low external pH. We consider as a model the possibility that protonation of the group, either by changes in surface charge, activation energy, or both, renders the probability of a K gating compo-

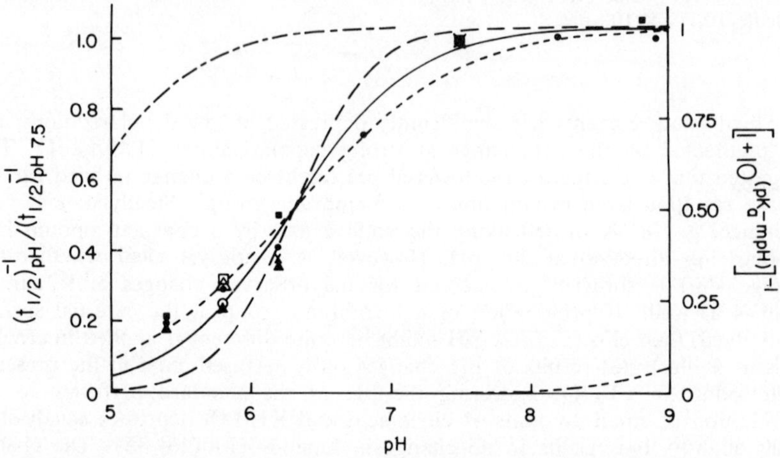

FIGURE 4. The dependence of potassium kinetics on external pH. Points represent experimental data, taken from several intact axons, measured as the ratio of the reciprocal of the time required for I_K to reach one-half its steady-state level at the pH value on the abscissa, relative to that at pH 7.5. Depolarizations were from the holding potential (−75 to −78 mv) to +65 mv. For the initial K conductance (g_{Ko}) low, and strong depolarizations, the left ordinate is close to the ratio of rate constants for gating components to open: $\alpha_{n,pH}/\alpha_{n,pH\ 7.5}$. The curves are calculated from the function on the right ordinate, which represents the fraction of time that residue A is unprotonated. ——, $pK_a = 6.3$, $m = 1$; — - —, $pK_a = 4.7$, $m = 1$; — — — —, $pK_a = 10.1$, $m = 1$; -------- $pK_a = 4.7$, $m = 0.75$; — — $pK_a = 10.1$, $m = 1.6$. Buffers: ▲, phosphate; ■, tris-maleate; ●, tris-maleate (+TTX); ○, bis tris Cl; □, Δ, bis tris Cl(+TTX); X, maleate. T = 8.5° C. (From Shrager.[1] By permission of the *Journal of General Physiology*.)

nent moving from "closed" to "open" state effectively zero. The group (A) is considered to bind protons according to the reaction:

$$A + mH^+ \rightleftarrows A\,(H^+)_m$$

Since rates of protonation and deprotonation are fast compared to rate constants for gates, the equilibrium fraction of A that is unprotonated may also be considered the percentage of time that any particular group A is unprotonated. We can then compare titration curves for the above reaction with experimental data for I_K kinetics (FIGURE 4).[1] The solid line represents a first-order reaction with histidine residues ($m = 1$; $pK_a = 6.3$), and provides a reasonable fit to the experimental points. First-order curves for carboxyl ($pK_a = 4.7$) and amino ($pK_a = 10.1$) groups do not fit. A non-first-order

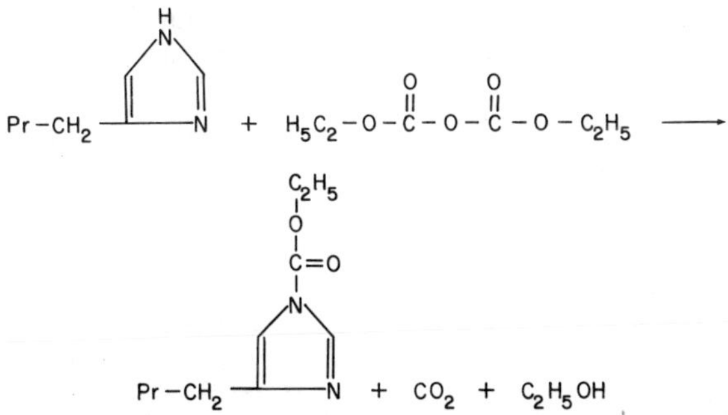

FIGURE 5. The reaction of diethylpyrocarbonate with histidine imidazole groups of a protein (Pr).[1, 20, 21] (From Shrager.[1] By permission of the *Journal of General Physiology.*)

(m = .75) titration of carboxyl groups provides a fit about equally as good as a first-order reaction with histidine, but would require that an individual gating component have four binding sites, with the protonation of any three sufficient to produce block.

The simplest scheme thus suggests titration of histidine imidazole groups, and in order to provide an independent test we have examined results of chemical modification of this side chain. Diethylpyrocarbonate (DEP) has been reported to be a specific modifier of histidine when used at pH 6.0, although block of other groups is possible at higher pH levels or at high concentrations of reagent.[6–9] The probable reaction with histidine is shown in FIGURE 5. When added to the external bath at pH 6.0, diethylpyrocarbonate slowed and suppressed potassium currents, as illustrated in FIGURE 6.[1] Sodium currents were unaffected, and effects on delayed currents were unchanged by TTX, and were irreversible. We have tested DEP internally in perfused axons, and found it to be without effect, as shown in FIGURE 7. However, it should be noted that if DEP is highly permeable, its effective concentration at the region of current measurement could be less than 0.75 mM. Results of experiments on chemical modification by DEP are thus consistent with those at low pH, both in suggesting that a histidine side chain is closely associated with

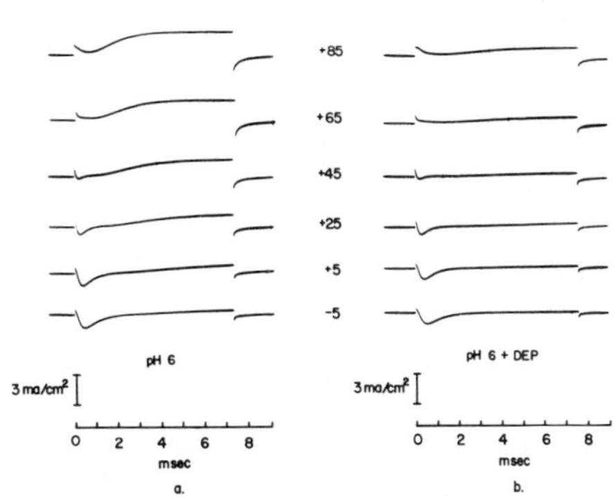

FIGURE 6. Voltage-clamp results of chemical modification by diethylpyrocarbonate (DEP).[1] (a) Control records at pH 6.0. The external solution contained (mM): NaCl 185, KCl 5.4, $MgCl_2$ 2.6, $CaCl_2$ 13.5, $NaHCO_3$ 2.3, Na maleate-maleic acid 20. Depolarizations from the holding potential of −75 mv to the potentials indicated. (b) Results 20 minutes after addition of 0.75 mM DEP to the solution of (a). Potentials as in (a). T = 8.5° C. (From Shrager.[1] By permission of the *Journal of General Physiology.*)

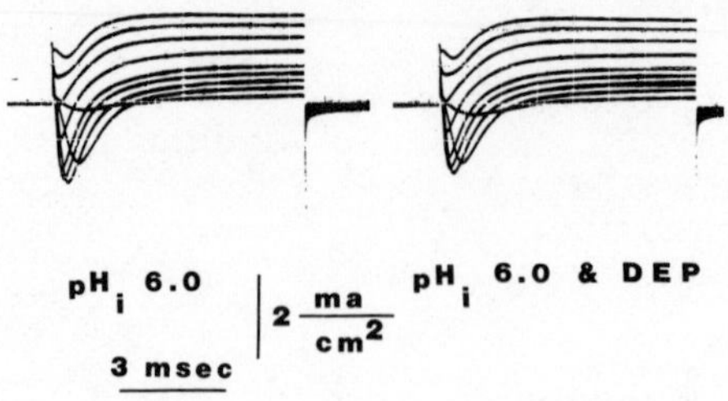

FIGURE 7. Internal perfusion of DEP. *Left:* control records at internal pH 6.0. Depolarizations from the holding potential of −78 mv to −38, −18, −8, +2, +22, +42, +62, +82 mv. Solutions: External (mM); NaCl 203, $MgCl_2$ 2.6, $CaCl_2$ 13.5, bis Tris 10. Internal; standard perfusate with pH adjusted to 6.0 with citric acid and KH_2PO_4. *Right:* 20 minutes after adding 0.75 mM DEP to the internal perfusate. T = 8° C.

potassium gating components, and in showing that the group may be accessible only at the external membrane surface.

Chemical Modification of Sulfhydryl Groups

The consequences of modification of sulfhydryl groups have been studied with several reagents. 5,5′-Dithiobis(2-nitrobenzoic acid) (DTNB) reacts with sulfhydryl groups with high specificity according to the reaction shown in FIGURE 8 (top).[10] 1.5 mM DTNB added to the external saline at pH 8.0 was without effect on ionic currents (FIGURE 9 a and b). The fiber became noticeably yellow, suggesting that reaction had occurred at sites not involved in ion permeation. *N*-ethymaleimide (NEM) also shows a high specificity for sulfhydryl groups when used at a low concentration and at neutral pH[11] (FIGURE 8, bottom). After the unreacted DTNB was washed away, in the experiment of FIGURE 9, the external pH was lowered to 7.0. After 20 minutes at pH 7.0 the records of FIGURE 9c were taken, and 1 mM NEM was then added to the bath. Sodium currents decreased progressively with little change in kinetics. FIGURE 9d shows results 30 minutes after addition of NEM. Potas-

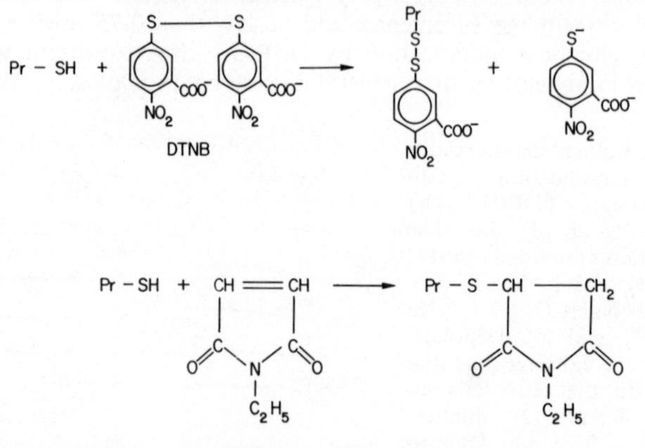

FIGURE 8. Sulfhydryl group modification. *Top:* The reaction of 5,5′-dithiobis (2-nitrobenzoic acid) (DTNB) with protein sulfhydryl groups.[10] *Bottom:* The reaction of *N*-ethylmaleimide (NEM) with sulfhydryl residues.[22]

FIGURE 9. Voltage-clamp records after modification of sulfhydryl residues in an intact axon. Depolarizations from the holding potential (—76 mv) to: —26, —16, —6, +4, +24, +44, +64, +84 mv. (a) Control records at pH 8.0. External solution: NaCl 197.3, KCl 5.4, $MgCl_2$ 2.6, $CaCl_2$ 13.5, Tris Cl 10. (b) Currents recorded 1 hr after adding 1.5 mM DTNB to the bath. (c) Records 20 min after washing away unreacted DTNB at pH 8.0, and changing to (mM): NaCl 197.3, KCl 5.4, $MgCl_2$ 2.6, $CaCl_2$ 13.5, Tris maleate 10, pH 7.0. (d) Currents 30 min after adding 1 mM NEM to the bath at pH 7.0. (e) Currents after washing the fiber for 45 min in normal Van Harreveld's solution. T = 8.0° C.

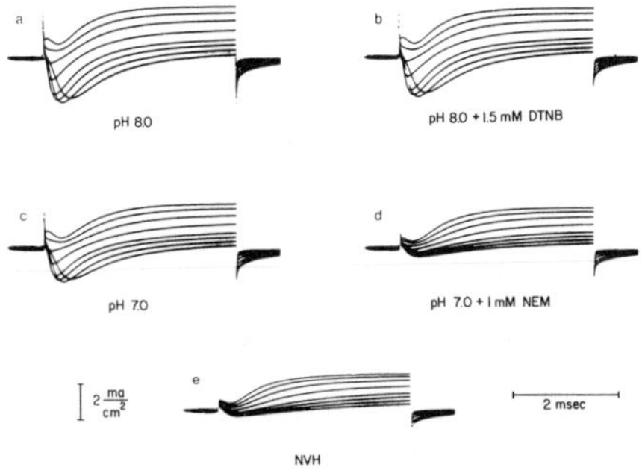

sium currents were only slightly reduced, while sodium currents were almost completely inhibited. Effects of NEM were irreversible on washing (FIGURE 9e). Results were identical if NEM was used without previous exposure of the fiber to DTNB. Preliminary evidence suggests that the inhibition of Na currents may result at least partially from a voltage shift in steady-state inactivation. After modification by NEM, sodium currents were almost completely restored if test depolarizations were preceded by a 350-msec hyperpolarization to —112 mv. Further experiments aimed at quantitating this result are in progress. Other preliminary work has shown that $HgCl_2$ and *p*-chloromercuriphenylsulfonic acid (PCMBS), which also react with sulfhydryl groups, inhibit sodium currents.

Modification of Amino Groups

Tamaoki *et al.*[12] have reported that 2-methoxy-5-nitrotropone (MNT) reacts with amino groups with high specificity. Added to the external saline at 0.5–1 mM, MNT inhibited sodium currents incompletely and without significant changes in kinetics (FIGURE 10). Although potassium currents were unchanged

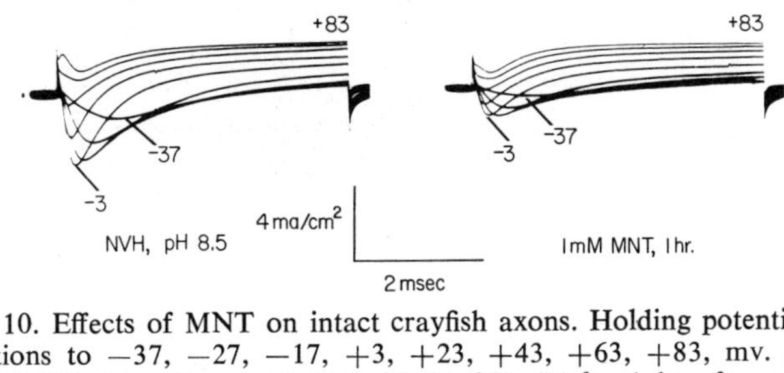

FIGURE 10. Effects of MNT on intact crayfish axons. Holding potential: —77 mv. Depolarizations to —37, —27, —17, +3, +23, +43, +63, +83, mv. *Left:* control currents in NVH, brought to pH 8.3 with NaOH. *Right:* 1 hr after adding 1 mM MNT to the bath (final pH 8.1). T = 9.5° C.

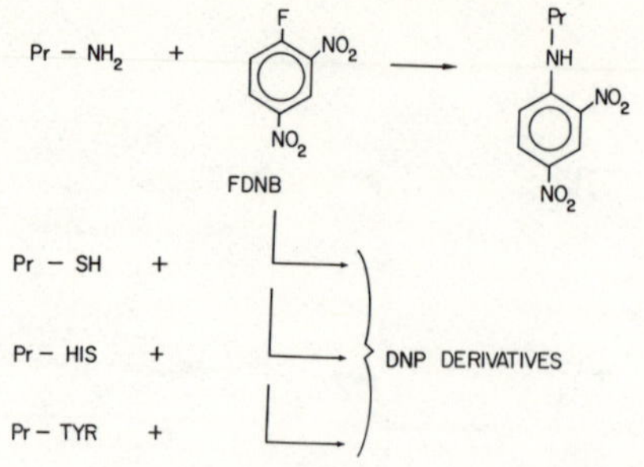

FIGURE 11. The reactions of FDNB with amino, sulfhydryl, histidyl and tyrosyl groups in proteins.[13]

in the experiment illustrated, in other fibers the reagent appeared to be less specific, and reduced steady-state currents typically by about 20%. Another difficulty with MNT is its insolubility in aqueous media. It proved difficult to place in solution, even at 1 mM, and the solution became highly colored as it went in, suggesting some hydrolysis. The final concentration is therefore somewhat uncertain.

1-fluoro-2,4-dinitrobenzene (FDNB) reacts with amino, histidyl, sulfhydryl, and tyrosyl groups to form dinitrophenyl derivatives (FIGURE 11).[13] It was considered possible, therefore, that this compound might combine the effects of MNT, DEP, and NEM in inhibiting Na currents with little change in kinetics, and in inhibiting K currents with significant slowing. This could be the case if the result of blocking a particular group were relatively independent of the nature of the ligand. Effects of adding 2 mM FDNB to the external medium are illustrated in FIGURE 12. Results are consistent with the notion that this rather nonspecific reagent blocks histidyl and either amino or sulfhydryl (or both) groups. Further experiments are in progress that are designed to test whether the effects of FDNB on different currents are in fact due to reaction with different membrane residues.

DEP, NEM and FDNB, at the concentrations used, all produce the reported results within 20–40 minutes, and with no change in resting potential. If exposure to these reagents is extended beyond this time, or if high concentrations of compounds are used, in all cases axons slowly become leaky and gradually depolarize. It is not possible to judge from the experiments performed thus far whether these later effects are due to reaction with other groups less accessible or less reactive than the residues involved in excitation, or whether the ligand slowly induces secondary effects on membrane structure.

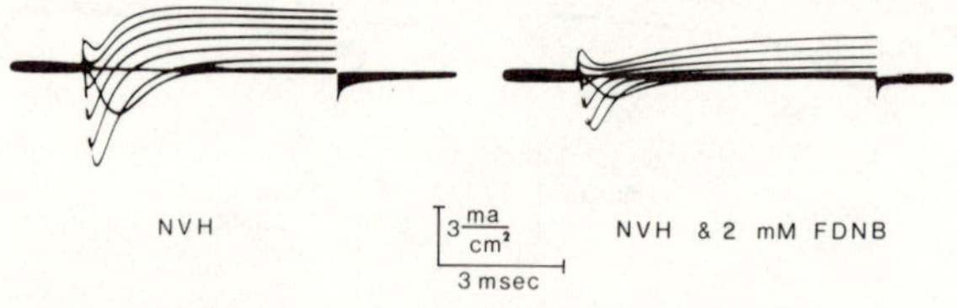

FIGURE 12. Effects of FDNB on ionic currents of an intact axon. Holding potential: −75 mv. Depolarizations to −55, −35, −15, +5, +25, +45, +65, +85 mv. *Left:* control records in NVH. *Right:* results 20 min after adding 2 mM FDNB to the bath. T = 8.0° C.

Discussion

These experiments have sought to identify particular biochemical groups present in close association with the nerve membrane components responsible for the control of ion permeation. Reagents used for chemical modification of proteins and lipids are rarely truly specific for a single reactive group. It is therefore valuable, when possible, to utilize a series of compounds of overlapping specificities, or to attempt independent verification of the suspected reaction. Results at low external pH suggest that a group with a pK_a on the order of 6.3 is closely associated with potassium gating components. An uncertainty in these experiments derives from the possible accumulation or depletion of protons in the immediate vicinity of the membrane due to a non-zero net fixed surface charge.[15] Future experiments, with variation of divalent cation concentrations, are planned in order to estimate the degree of correction required. However, experiments on other nerve fibers suggest that the net surface charge in the neighborhood of potassium gates is low.[16, 17] Among protein side chains, histidine imidazole groups have values of pK_a closest to that measured at potassium gates. The slowing and inhibition of potassium current by DEP, a reagent with a reportedly high selectivity for histidine residues, provides further evidence for the participation of this amino acid. Moreover, both DEP and low pH were effective in slowing potassium currents only when applied at the external membrane surface, a result supportive of a common site of action.

Of the two sulfhydryl reagents tested extensively thus far, DTNB had no effect on excitation, while NEM acted specifically on sodium currents. DTNB is a rather large compound, and bears a net charge of —2. Its lack of activity could be due to inaccessibility to a critical membrane site, either because of size or electrostatic considerations, or to impenetrability if the site is at the internal membrane surface. NEM is smaller and neutral, and is known to be permeable in red blood cells.[19] Future experiments with perfused axons may help to resolve these questions. In experiments on unclamped fibers, NEM has been shown to block conduction irreversibly.[24–26] Marquis and Mautner[26] have reported a potentiation of this conduction block by stimulation. It is possible that this effect may be related to the voltage shift in inactivation that we have observed. Reports of a positive effect of DTNB on conduction in unclamped squid axons,[26] and of a negative effect of NEM in reducing sodium currents irreversibly in the node,[23] are at variance with the results reported here.

Data with FDNB are consistent with its known spectrum of reactivity and with the effects of other, more specific reagents. Cooke *et al.*[18] have shown that FDNB suppresses both Na and K currents in squid axons, but did not examine kinetics.

By choosing proper conditions of concentration, pH, etc., we have been able to block sodium or potassium currents independently, and with well-defined chemical procedures. In each case the reactive group involved suggests that a membrane protein is closely associated with gating. It is hoped that similar procedures will provide further information with regard to the biochemical nature of these important membrane sites. The irreversibility of many chemical modification procedures, while a disadvantage in electrophysiological experiments, may be of value in future biochemical analysis.

References

1. Shrager, P. 1974. Ionic conductance changes in voltage clamped crayfish axons at low pH. J. Gen. Physiol. **64:** 666–690.
2. Shrager, P., R. I. Macey & A. Strickholm. 1969. Internal perfusion of crayfish giant axons: Action of tannic acid, DDT and TEA. J. Cell. Physiol. **74:** 77–90.
4. Narahashi, T., J. W. Moore & W. R. Scott. 1964. Tetrodotoxin blockage
3. Van Harreveld, A. 1936. Physiological saline for crayfish. Proc. Soc. Exp. Biol. Med. **34:** 428.
of sodium conductance increase in lobster giant axons. J. Gen. Physiol. **47:** 965.
5. Hodgkin, A. L. & A. F. Huxley. 1952. A quantitative description of membrane current and its application to conduction and excitation in nerve. J. Physiol. **117:** 500.
6. Muhlrad, A., G. Hegyi & M. Horányi. 1969. Studies on the properties of chemically modified actin. III. Carbethoxylation. Biochim. Biophys. Acta **181:** 184.
7. Ovádi, J., S. Libor & P. Elödi. 1967. Spectrophotometric determination of histidine in proteins with diethylpyrocarbonate. Acta Biochim. Biophys. Acad. Sci. Hung. **2:** 455.
8. Rosén, C.-G., T. Gejvall & L.-O. Andersson. 1970. Reaction of diethylpyrocarbonate with indole derivatives with special reference to the reaction with tryptophan residues in a protein. Biochim. Biophys. Acta **221:** 207.
9. Wolf, B., J. A. Lesnaw & M. E. Reichmann. 1970. A mechanism of the irreversible inactivation of bovine pancreatic ribonuclease by diethylpyrocarbonate. Eur. J. Biochem. **13:** 519.
10. Ellman, G. L. 1959. Tissue sulfhydryl groups. Arch. Biochem. Biophys. **82:** 70.
11. Riordan, J. F. & B. L. Vallee. 1972. Reactions with N-ethylmaleimide and p-mercuribenzoate. *In* Methods in Enzymology. C. H. W. Hirs and S. N. Timasheff, Eds. Vol. XXV: 449. Academic Press. New York, N.Y.
12. Tamaoki, H., Y. Murase, S. Minato & K. Nakanishi. 1967. Reversible chemical modification of protein by 2-methoxy-5-nitrotropone. J. Biochem. (Tokyo) **62:** 7.
13. Sanger, F. 1945. The free amino groups of insulin. Biochemistry **39:** 507–515.
14. Frankenhaeuser, B., & A. L. Hodgkin. 1957. The action of calcium on the electrical properties of squid axons. J. Physiol. (London) **137:** 218.
15. Gilbert, D. L., & G. Ehrenstein. 1970. Use of a fixed charge model to determine the pK of the negative sites on the external membrane surface. J. Gen. Physiol. **55:** 822.
16. Hille, B. 1968. Charges and potentials at the nerve surface: divalent ions and pH. J. Gen. Physiol. **51:** 221.
17. Begenisich, T. Magnitude & location of fixed charges on *Myxicola* giant axons. In preparation.
18. Cooke, I. M., J. M. Diamond, A. D. Grinnell, S. Hagiwara, & H. Sakata. 1968. Suppression of the action potential in nerve by nitrobenzene derivatives. Proc. Nat. Acad. Sci. U.S.A. **60:** 470.
19. Jacob, H. S. & J. H. Jandl. 1962. Effects of sulfhydryl inhibition on red blood cells. I. Mechanism of hemolysis. J. Clin. Invest. **41:** 779.
20. Muhlrad, A., G. Hegyi & G. Toth. 1967. Effect of diethylpyrocarbonate on proteins. I. Reaction of diethylpyrocarbonate with amino acids. Acta Biochim. Biophys. Acad. Sci. Hung. **2:** 19.
21. Hill, T. L. & G. E. K. Branch. 1940. Resonance and the chemistry of histidine. Science **91:** 145.

22. LEACH, S. J. 1966. The estimation of thiol and disulfide groups. *In* A Laboratory Manual of Analytical Methods of Protein Chemistry. P. Alexander and H. P. Lundgren, Eds. Pergamon Press. Oxford, England.
23. KEANA, J. F. W. & R. STÄMPFLI. 1974. Effect of several "specific" chemical reagents on the Na^+, K^+ and leakage currents n voltage-clamped single nodes of Ranvier. Biochim. Biophys. Acta **373:** 18–33.
24. SMITH, H. M. 1958. Effects of sulfhydryl blockade on axonal function. J. Cell Comp. Physiol. **51:** 161.
25. HUNEEUS-COX, F., H. L. FERNANDEZ, & B. H. SMITH. 1966. Effects of redox and sulfhydryl reagents on the bioelectric properties of the giant axon of the squid. Biophys. J. **6:** 675.
26. MARQUIS, J. K. & H. G. MAUTNER. 1974. The effect of electrical stimulation on the action of sulfhydryl reagents in the giant axon of squid: Suggested mechanisms for the role of thiol and disulfide groups in electrically-induced conformational changes. J. Membr. Biol. **15:** 249.

Discussion

DR. SHAW: Do any of these reagents have any effect on the binding of tetrodotoxin?

DR. SHRAGER: These experiments are being done in collaboration with Dr. Enrique Jaimovich. We have only very preliminary results thus far. Neither diethylpyrocarbonate nor NEM seems to reduce TTX binding under conditions similar to those used here.

DR. D'ARRIGO: Does the fact that a hyperpolarizing prepulse could reverse the sodium inactivation caused by NEM imply that this modification changes the local field near the gate without blocking a channel?

DR. SHRAGER: The reaction of NEM with sulfhydryl groups would not be expected to alter membrane fixed charge at pH 7. This modification procedure might therefore alter the activation energy for the transition of the gate from one state to another, for example, by steric hindrance, which can be overcome by strong hyperpolarization.

DR. D'ARRIGO: Do you find any complication with the fact that kinetics were not changed although according to that hypothesis the activation energy was changed?

DR. SHRAGER: The inactivation produced by NEM is a "slow" inactivation: 350 msec hyperpolarizing prepulses are required to reverse it. After partial reaction, remaining sodium currents without prepulses therefore would not show altered kinetics. However, we have not yet fully analyzed the situation with strong prepulses.

MECHANISM FOR CHANNEL GATING IN EXCITABLE BILAYERS

Harold Lecar, Gerald Ehrenstein, and Ramon Latorre*

Laboratory of Biophysics, IR
National Institute of Neurological and Communicative Disorders and Stroke
National Institutes of Health
Bethesda, Maryland 20014

Excitation and Gating

Electrical excitation in the nervous system is produced by the transient increase of the permeability of a particular ion through the membrane. Action potential generation, for example, is often described in terms of a sodium conduction pathway having a negative-resistance current-voltage characteristic.[1, 2] The negative resistance and its attendant instability are the result of a steeply voltage-sensitive ionic conductance which approaches to new values with a voltage-dependent relaxation time.[3]

As Mueller and Rudin[4, 5] first showed, certain additives, such as EIM (excitability inducing material, a protein extracted from aerobacter bacteria), alamethicin and monazomycin, endow synthetic lipid bilayer membranes with the type of voltage-dependent conductance necessary for excitability. FIGURE 1 shows the voltage-dependent conductance and relaxation time for EIM.[6]

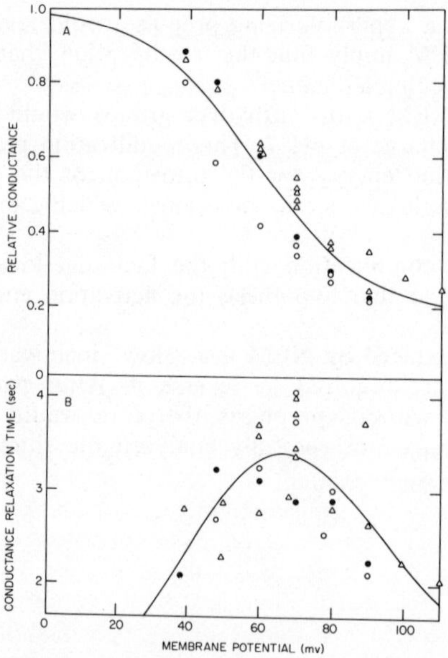

FIGURE 1. Normalized voltage-dependent conductance and conductance relaxation time for an EIM-doped lipid bilayer membrane (From Ehrenstein *et al.*[6] By permission of the *Journal of General Physiology.*)

* Present address: Department of Physiology and Pharmacology, Duke University, Durham, North Carolina 27710.

The curves are reminiscent of the Hodgkin-Huxley parameters which describe axonal excitation.[1, 2]

Observations on Single Channels

Much of what we know about the excitable bilayers comes from the fortunate circumstance that it is often possible to measure conductance changes at extremely low concentrations of additive. Bean and his coworkers[7] observed that for low levels of EIM added to the solution bathing a bilayer the conduc-

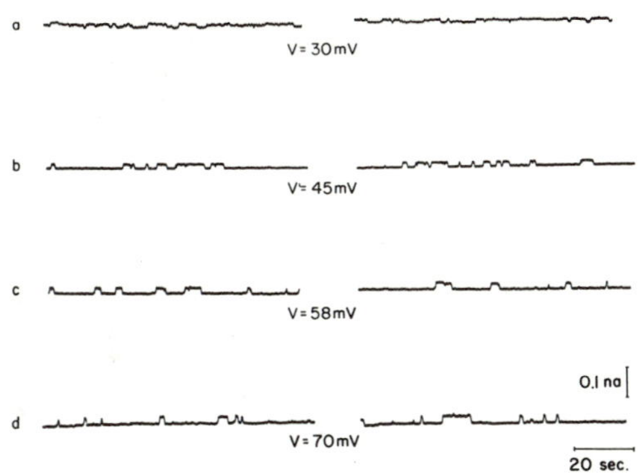

FIGURE 2. Conductance jumps for a single EIM channel in an oxidized cholesterol membrane. (From Ehrenstein *et al.*[6] By permission of the *Journal of General Physiology.*)

tance changes appeared as discrete steps. These steps represent the change in current through individual ion-conducting units. All the units seemed to be identical[8] and to conduct a high flux of ions (4×10^{-10} mho or about 10^8 ions per sec with 100 mV across the membrane). The conductance steps thus represent discrete channels through the membrane; channels, not domains, because the conductances are very uniform[8]; channels, not carriers, because of the high flux per unit.[3]

By diluting the EIM solution still further, we eventually were able to observe single channels which remained in isolation for long times, often an hour or more.[9] With such isolated conducting channels, one can vary the membrane potential, observe a pulse train, and learn whether the amplitudes of the jumps or their frequency vary in a significant way over the negative-resistance region. One can answer the question of whether the negative resistance is a transport property of each individual channel or whether it is a statistical result of the random channel jumps. EIM jumps recorded with various membrane potentials on a single channel are shown in FIGURE 2.

These jumps in current are caused by transitions of the channel from a low-conductance state to a high-conductance state. We call these states "closed" and "open." The conductance in the open state is a property of the EIM channel independent of the lipid used. The closed-state conductance can vary depending upon the lipid (up to one-fourth the open conductance for EIM in oxidized cholesterol). Thus, when a channel is closed by application of membrane potential, the channel structure is altered, but the channel does not break up completely.

Transport through a Single Channel

The jump amplitudes plotted as a function of voltage are shown in FIGURE 3 for various alkali cations. These curves are current-voltage curves for the ion transport through a single channel. The curves are approximately linear, but it is more important to note that they are monotonically increasing through the region of voltage corresponding to negative resistance. Thus, there is nothing remarkable about the transport process, and the origin of excitability resides in the conductance transitions of the EIM channels.

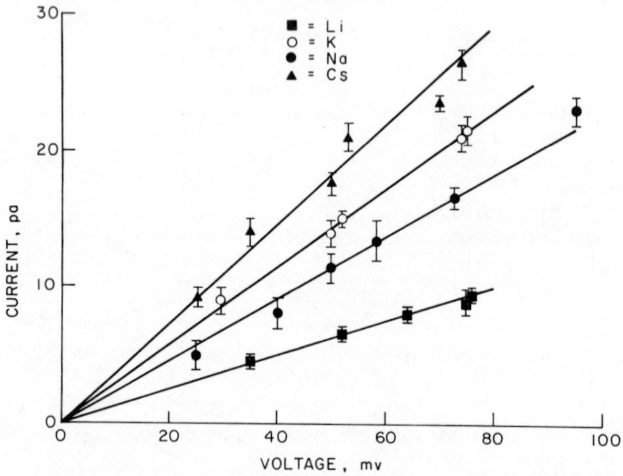

FIGURE 3. Current-voltage curves for a single EIM channel. Measurement is current-jump amplitude as a function of steady potential. All measurements in 100 mM chloride solutions of ions indicated. (From Latorre *et al.* By permission of the *Journal of General Physiology*.)

The single-channel current-voltage curves can also be used to measure ionic selectivity. The slopes of the curves in FIGURE 3 give the relative permeabilities of the channel to the various alkali cations. The permeabilities determined from these slope measurements (and confirmed by bi-ionic potential measurements on many-channel membranes) are in proportion to the free-solution mobilities of the ions.[8] Thus the EIM channels do not select among cations. Because of this and because EIM channels have such a large conductance per site, one may reasonably conclude that the channels are aqueous, possibly negative-charge-lined aqueous pores, since they are cation-selective. This conclusion has been confirmed by Latorre, Alvarez, and Verdugo,[10] who found the temperature-dependence of open EIM channel conductance to be the same as that of aqueous electrolyte. Open alamethicin channels also appear to behave as aqueous pores.[11] Hence, although these gated channels possess the major property of excitation—gated ion fluxes—they do not have the high selectivity of the sodium or potassium channels found in natural excitable membranes.

Voltage-Dependent Transition Probabilities

From a long train of conductance jumps, one can measure the fraction of time that a channel stays open at a given potential.[9] If all the channels are identical, we might expect the time average of the probability of a single channel being open to be equal to the steady-state average of the fraction of channels open for a many-channel membrane at the same voltage. FIGURE 4 shows this comparison. The points are time-averages for single- or few-channel membranes. The curve is the normalized voltage-dependent conductance of a many-channel membrane. This experiment shows that the permeability is voltage-dependent because the channel has two conducting states and the relative probability of being in either state is a function of voltage.

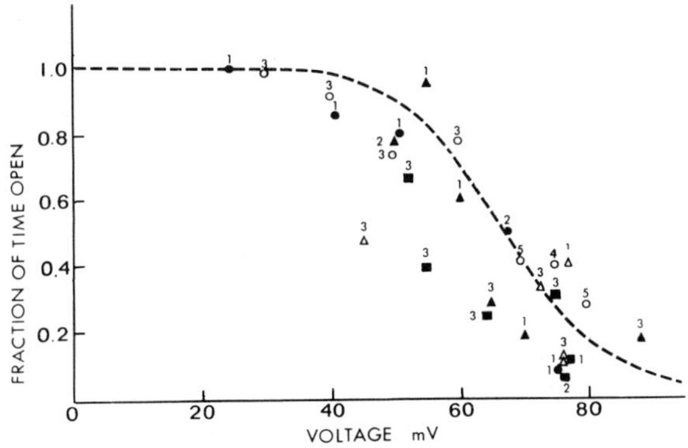

FIGURE 4. Fraction of time spent in open-conducting state as a function of membrane potential. Data taken on few-channel membranes with number of channels indicated next to experimental point. Different symbols indicate different cations. Dashed curve shows fraction of open channels as predicted from Equation 5.

Statistical Mechanics of the Channels

We can show that both the sigmoid voltage-dependent conductance and the bell-shaped relaxation time have their origin in the simplest possible picture of the channel transitions. Imagine a channel structure that has two states. Let us say that the energy difference between the two states is given by the sum of some intrinsic configuration energy and the work needed for the electric field to push charges around or twist dipoles in going from one state to the other. That is

$$\Delta W = \Delta W_{conf} + \Delta W_{elec} \tag{1}$$

and

$$\Delta W_{elec} = aV \tag{2}$$

where V is membrane potential and a is a constant. Since at the potential V_0, for which half the channels are in each state, the ΔW_{elec} just cancels ΔW_{conf}, we can write the energy as

$$\Delta W(V) = a(V - V_0) \tag{3}$$

If we denote the total number of channels by N, the number open by n_0 and the number closed by n_c, we can say that for thermal equilibrium at any potential

$$n_0/n_c = \exp(-\Delta W/kT) \quad (4)$$

where T is absolute temperature and k is Boltzmann's constant. Since $n_0 + n_c = N$, Equation 4 leads to the following expression for the fraction of channels open at any voltage

$$f(v) = n_0/N = \left(1 + \exp\left[\frac{a(V - V_0)}{kT}\right]\right)^{-1} \quad (5)$$

Substitution for ΔW from Equation 3 immediately gives the sigmoid conductance function of FIGURES 1 and 4.

One further assumption is needed to get the bell-shaped relaxation time. If we denote the rate for a closed channel to open as α and that for an open channel to close as β, we know from the principle of detailed balance that

$$\alpha/\beta = n_0/n_c = \exp\left[\frac{-a(V - V_0)}{kT}\right] \quad (6)$$

We now assume that α and β have opposite voltage-dependence, that is

$$\alpha = \lambda \exp(-a(V - V_0)/2kT) \qquad \beta = \lambda \exp(+a(V - V_0)/2kT) \quad (7)$$

where λ is an arbitrary rate.

In this case the relaxation time, τ, can be written as

$$\tau(V) = (\alpha + \beta)^{-1} = (\tfrac{1}{2}\lambda) \operatorname{sech}[a(V - V_0)/2kT] \quad (8)$$

This shows that the bell-shaped time constants, analogous to the well-known Hodgkin-Huxley time constants for axonal excitation,[3] are inherent in this type of voltage-dependent transition.

Equation 7 stated the only assumption, a specific form for the rate constants. This assumption can be tested experimentally. One of the nice features of single-channel jump measurements is that one can measure forward and backward transition rates independently.[6] The measurements are done by statistical analysis of the distribution of jump durations. The distribution of jump widths (open channel times) is exponential, indicating that the channels obey Poisson statistics; i.e., the chance of an open channel closing or a closed channel opening is uniform in time. For a Poisson distribution it can be shown that the mean open (closed) time is equal to the reciprocal of the rate constant for closing (opening).

FIGURE 5 shows the results of rate constant measurements on three different membranes, each having one or a few channels. The log plot shows that the rates are exponential functions of potential in rough agreement with Equation 7. These experiments demonstrate that the energy of each conductance state is linear in the applied electric field.

Complications; Other Channels

EIM channels in oxidized cholesterol bilayers exhibit simple gating behavior. The channel jumps between two states. The kinetics indicate a simple first-order Poisson process, with transition rates varying as an exponential function of potential. Latorre, Alvarez, and Verdugo[10] have shown that the

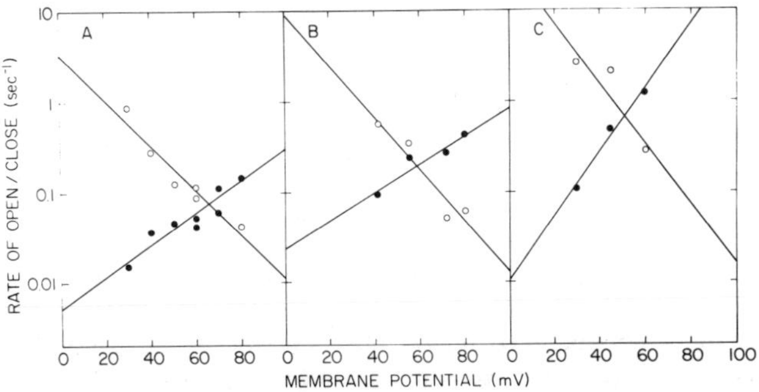

FIGURE 5. Opening and closing transition rates as a function of voltage for three different membranes: (A) one, three, and five channels; (B) two, four, and six channels; (C) one channel. Varying number of channels was necessitated by birth or death of channels during the course of a long experiment. (From Latorre *et al.* By permission of the *Journal of General Physiology*.)

two conductance states of EIM have very different temperature coefficients for ionic transport. This suggests that the open and closed (not completely closed) states have physically different modes of transport.

Bean[12] has shown that EIM in different lipids gives different closed-state conductances and multiple step-height jumps. Thus, although EIM behavior in other lipids resembles that in oxidized cholesterol, there are some differences. This provides interesting possibilities for the study of the interaction between the protein channel structure and the underlying lipid matrix. FIGURE 5 shows one observation having to do with a kind of local viscosity; the different channels, all in oxidized cholesterol, have the same voltage-dependence, but do not have the same absolute rate scale. The energy difference between open and closed state is thus uniform throughout the membrane, and probably a property of the channel, but the activation energy for the transition is not uniform, possibly because of inhomogeneities in the interaction between channel and lipid.

There are other channels for which single-channel jumps have been observed. Alamethicin, described by Eisenberg, Hall, and Mead,[11] produces a channel capable of transitions between half a dozen sublevels of conductance. These sublevels represent the stable states of the channel structure, probably various stable states of agglomeration of alamethicin subunits. Channel jumps have also been studied for bilayers doped with keyhole limpet hemocyanin. These jumps lead to a picture of discrete conductance states, but with a nonlinear current-voltage relation prevailing in each state, possibly due to a set of finer, more rapid structural changes.[14] Despite the complexity caused by the large number of conductance states, the alamethicin and hemocyanin preparations are valuable because they contain gating channels made of molecules for which there is structural information.

Inactivation and Molecular Pictures of Gating

Baumann and Mueller[13] have described the kinetics of membrane channels made of subunits that are tipped into the membrane by the electric field, and are then free to aggregate into polar-lined pore structures. One of the attractive features of this model is that it provides a natural description of inactivation—the process by which a channel, such as the sodium channel of a nerve axon, opens and then spontaneously closes again during excitation. In the subunit (or oligomer) model, inactivation is a result of the relaxation of the population of channels to nonconducting states of aggregation.

EIM-doped bilayers have an electric-field behavior opposite to that of alamethicin. As can be seen from FIGURE 1, the channels are open when the membrane potential is zero and closed for high potential. The Figure shows only positive potentials but there is a similar, although not identical, voltage-dependent conductance region observed for negative potentials. Recent experiments[15] show that this two-sided gating is a property of individual channels and not a superposition of the responses of oppositely oriented channels.

If an EIM channel is made of subunits, like an alamethicin channel, then we might envision a situation in which a large electric field induces a closed state consisting of monomers assembled in some channel-precursor structure (possibly at the membrane-solution interface). Removal of the field allows the entire assembly to turn into the membrane to form a stable pore. One can think of the closed state as a group of loosely connected barrel staves floating at the surface of the membrane and the open state as the reconstituted barrel.

One consequence of this "barrel-stave" picture, as illustrated in FIGURE 6, is that it gives a natural explanation of why EIM has two distinct closed states for high fields in either direction across the membrane. In the picture, high fields in either direction orient the membrane dipoles to push the EIM subunits out towards a membrane surface. This two-sided gating leads to

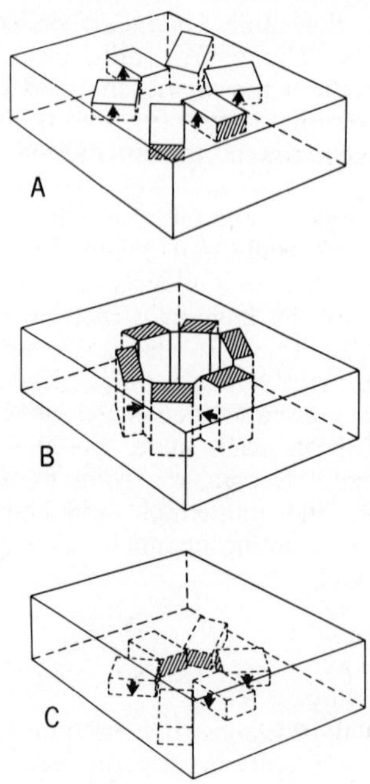

FIGURE 6. Highly schematic picture of hypothetical EIM channel structure made of subunits. Arrows indicate direction of dipole moment. State B is open state in absence of field, with subunits forming a pore. State A is closed state corresponding to one direction of transmembrane field (upward in figure), and state C is closed state corresponding to opposite direction of transmembrane field.

conductance inactivation when the voltage is switched from one polarity to the other. The channels are open transiently because they must pass through the open state in going between the two closed configurations.[15] This mechanism for inactivation can be regarded as a simple three-state model for inactivating channels such as the axon sodium channel. The need for a three-state model in the description of axon kinetics has recently been discussed by Goldman and Schauf.[16]

The barrel-stave model can also explain the finding of Latorre *et al.*[10] that the temperature-dependence of the conductance of the open channel is the same as the temperature-dependence of the mobility of the bathing electrolyte, but the conductance of the closed channel decreases with increasing temperature. In the model, the array of EIM subunits at the membrane interface might affect the membrane structure, resulting in a partially-open channel. As temperature is increased, the lipids become more mobile, and can more effectively disrupt the ionic pathway, thus reducing the conductance. The channel-precursor states of the model are necessary to explain the observation that a membrane may contain one, and only one, channel over a period of many minutes. If an oligomer always broke up into disconnected monomers, the number of channels would fluctuate considerably.

Channels in Natural Membranes

There is considerable evidence for discrete conducting channels in natural excitable membranes. Experiments with blocking agents, such as tetrodotoxin, provide an estimate of the minimal conductance blocked by a single drug molecule.[17] The recently discovered gating currents[18] represent the actual charge movement or dipole reorientation associated with the gating process. There is in addition a body of indirect evidence for the existence of conducting channels.[3]

Direct observation of single-channel fluctuations is difficult for natural membranes. It was possible to observe the channel jumps in bilayers because the conductance of an undoped lipid bilayer is so low that current fluctuations associated with leakage pathways cause relatively small background noise. For natural membranes, the most promising approach at present is the measurement of the electrical noise produced when many channels undergo transitions. For a conductance relaxation process, such as EIM gating, the electrical noise produced by the random opening and closing of the channels has a characteristic frequency spectrum that can readily be identified. Electrical noise spectra have been studied for the acetylcholine-induced gating of the postsynaptic membrane,[19–20] for the electrical gating of the axon membrane,[22, 23] and for photoreceptors.[24] The study of channel noise promises to yield much of the same kind of detailed information that we now have about the gating process in doped bilayers.

References

1. Hodgkin, A. L. & A. F. Huxley. 1952. J. Physiol. **117:** 500.
2. Cole, K. S. 1968. Membranes, Ions and Impulses. University of California Press. Berkeley, California.

3. Ehrenstein, G. & H. Lecar. 1972. Ann. Rev. Biophys. Bioeng. **1:** 347.
4. Mueller, P. & D. O. Rudin. 1963. J. Theor. Biol. **4:** 268.
5. Mueller, P. & D. O. Rudin. 1969. Curr. Top. Bioenerg. **3:** 157.
6. Ehrenstein, G., R. Blumenthal, R. Latorre & H. Lecar. 1974. J. Gen. Physiol. **63:** 707.
7. Bean, R., W. C. Shepherd, H. Chan & J. T. Eichner. 1969. J. Gen. Physiol. **53:** 741.
8. Latorre, R., G. Ehrenstein & H. Lecar. 1972. J. Gen. Physiol. **60:** 72.
9. Ehrenstein, G., H. Lecar & R. Nossal. 1970. J. Gen. Physiol. **55:** 119.
10. Latorre, R., O. Alvarez & P. Verdugo. 1974. Biochim. Biophys. Acta **367:** 361.
11. Eisenberg, M., J. E. Hall & C. A. Mead. 1973. J. Membr. Biol. **14:** 143.
12. Bean, R. C. 1972. J. Membr. Biol. **7:** 15.
13. Baumann, G. & P. Mueller. 1974. J. Supramol. Biol. In press.
14. Alvarez, O., E. Diaz & R. Latorre. 1975. Biochim. Biophys. Acta. In press.
15. Ehrenstein, G. & H. Lecar. 1975. Biophys. J. **15:** 167a.
16. Goldman, L. & C. L. Schauf. 1973. J. Gen. Physiol. **61:** 361.
17. Keynes, R. D., J. M. Ritchie & E. Rojas. 1971. J. Physiol. **213:** 235.
18. Armstrong, C. M. & F. Bezanilla. 1974. J. Gen. Physiol. **63:** 533.
19. Katz, B. & R. Miledi. 1971. Nature New Biol. **232:** 124.
20. Anderson, C. R. & C. F. Stevens. 1973. J. Physiol. **235:** 655.
21. Sachs, F. & H. Lecar. 1973. Nature New Biol. **246:** 214.
22. Verveen, A. A. & L. J. De Felice. 1974. Progr. Biophys. Mol. Biol. **28:** 191.
23. Fishman, H. 1973. Proc. Natl. Acad. Sci. U.S.A. **70:** 876.
24. Hagins, W. A. 1965. Cold Spring Harbor Symp. Quant. Biol. **30:** 403.

Discussion

Dr. Mueller: I would like to emphasize that the conductance of these membranes is controlled by the fluctuations between the zero conductance level and the first step because that is where the very large change of the conductance occurs. The later step is only a three- to fivefold change in the conductance, whereas during this first part, if you study many channel kinetics, the conductance varies over three, four, or five orders of magnitude. In the sphingomyelin membranes one finds high-order kinetics, whereas in the oxidized cholesterol system, where the conductance varies only between three- or fourfold one has first-order kinetics. After the channel has assembled in the field, it can move back and forth, with the possibility that the lipid heads may tend to close up the opening, resulting in a partially closed channel; and yet there are other high-order mechanisms superimposed on all that.

Dr. Lecar: There are a number of interesting experiments involving changes in the lipids. EIM can be made to look like alamethicin if you take it out of oxidized cholesterol; in fact alamethicin can be made to look more like EIM in some lipids. Further, there is the possibility of having a time delay in activating the channels in a membrane without the use of a first-order variable raised to a power as in the Hodgkin-Huxley equations. In that type of kinetic description you have to go to a very high power to explain long time delays, and that seems unreasonable especially in the models where the power is associated with the number of particles involved. Furthermore, the power is different in different species. The Frankenhaeuser-Huxley equations that describe the frog node have the sodium conductance proportional to m^2 instead of to m^3, and potassium conductance proportional to n^2 instead of n^4. Perhaps the gating-current or noise-spectrum experiments will provide new information on the physical origin of the time delays.

DR. PRESSMAN: I would like to suggest another possiblity with respect to the barrel stave concept developed by Mueller and others from alamethicin properties. If these staves are attached to an ionophore in the shape of a doughnut which also serves as an anchor in one side of the membrane, then when the staves move across in the field they would arrive oriented to form the channel, eliminating the time-consuming aggregation process. The doughnut would be able to recognize the ions and confer on this channel a degree of ion-selectivity that most of the channels described up to now at this conference do not. Finally, within that doughnut there is a large amount of material that could serve as the mechanism for the chemistry involved in the inactivation of those channels which do inactivate.

DR. FINKELSTEIN: In the case of alamethicin you can change the delay times, the lag in the S shape behavior, etc., by the amount of material you add to the membrane because the absolute rates are going to be determined by the number of monomers that you have in the system and hence, the more you have the faster things go. In the EIM case, where you have a prefabricated channel, the kinetics for the macroscopic system are reflected in the kinetics of the single channel and you don't have that degree of play.

PART VII. PERMEATION MECHANISM ACROSS BIOLOGICAL MEMBRANES: I

IONIC MECHANISMS IN EXCITATION OF PHOTORECEPTORS

W. A. Hagins* and S. Yoshikami†

**Laboratory of Chemical Physics*
National Institute of Arthritis, Metabolism, and Digestive Diseases; and
† Laboratory of Vision Research
National Eye Institute
National Institutes of Health
Bethesda, Maryland 20014

The response of a photoreceptor to a single absorbed photon is one of the clearest examples of a quantized response to a quantized stimulus. Not only are photoreceptors good photon counters with high quantum efficiencies, but they are also relatively free of spontaneous responses in darkness.[1, 2] In all photoreceptor cells that have been studied in sufficient detail, it seems that the photochemical reactions produced by light make themselves felt as sensory responses through a photoelectric conversion process. That is, the light acts on the cell by controlling the flow of an electric current through its plasma membrane, thus changing the membrane potential. The membrane potential change, in turn, produces an action potential in the photoreceptor cell's axon, or, if the cell is short, simply controls a voltage-sensitive leak of a chemical transmitter substance at the cell's synapse with a distal neuron of the visual system.

Although there are many types of photoreceptor cells within the animal kingdom, all can be divided into two types: those that respond to light by increasing their membrane permeability to some ion or ions, and those that respond by a decreased permeability. Examples of the first type are the lateral and ventral photoreceptors of *Limulus*[3, 4, 7] and the photoreceptors of cephalopod eyes.[5, 6] The photoelectric conversion process in the squid eye has been studied in somewhat greater detail than most such receptors. It has been shown that the single photon response consists of a local and relatively ion-specific increase in Na^+ permeability of perhaps 20 pmho that lasts for about 50 msec at 10° C, during which about 10^6 Na^+ ions enter the photoreceptor cell.[6] The responses of *Limulus* photoreceptors are obviously far more complex. Single photon responses occur as very large discrete pulses of inward membrane current that would require that the conductance of the photosensitive membrane transiently increase by more than 10 nmho.[7, 8] While it is conceivable that the squid photon responses arise from opening of single molecular gates in the plasma membrane, the quantal responses of *Limulus* require some additional gain mechanism. For, if the conductance of an elementary photon-induced hole of diameter 2a in the rhabdomal membrane of a *Limulus* retinula cell were as much as 10 nmho, then the combined convergence resistance of the medium around the hole and the resistance of the hole itself within the 70 Å membrane thickness would require that 2a be greater than 30 Å (Refs. 6 and 18, Eq. 19). Such a large transient disturbance in molecular structure is hard to visualize in terms of the reversible transformations that membrane proteins are known to undergo. In 1959 Fuortes[3] suggested a possible way out of this difficulty: the release of a chemical transmitter substance by the light. Although his suggestion was originally devised to explain transfer of excitation from *Limulus* retinula cells to closely adjacent dendrites of excentric cells, much evidence has recently accumulated in support of the existence of transmitter substances in vertebrate vision. In the remainder of this paper, we shall consider evidence bearing on the existence and nature of such an agent in excitation in vertebrate rods and cones and on the membrane conductance mechanism under its control.

Vertebrate retinal receptors belong to the second category of photoreceptors mentioned above: those whose membrane permeability is reduced by light. The first evidence of this was provided by Tomita and his colleagues[9, 10] and by Bortoff and Norton,[11, 12] who found by intracellular recording methods that cones have resting membrane potentials of only 30–40 mV in darkness, but that bright light causes prompt and reversible hyperpolarization of 10–20 mV. At the same time, the resistance of the recording electrode increases by several megohms, presumably because the membrane resistance of some part of the cell increases. By measuring the voltage gradients in the interstitial space of the receptor layer of rat and frog retinas and computing the divergence of the extracellular currents, it was possible to show that the relatively low membrane potential of vertebrate rods was accompanied by the flow of a steady "dark current" into the plasma membrane covering the outer segments and that light transiently reduced the current. If the light was bright enough to cause about 30 photons to be absorbed in each outer segment, the dark current was halved. Brighter flashes produced greater and more long-lasting reductions. If the outer segments were locally illuminated, the dark current was locally suppressed.[13, 14] The relation between the energy of a stimulus flash and the degree of suppression of the current was hyperbolic,[15] as is the hyperpolarization of cones observed by the intracellular recording technique.[16] The electrical circuits thought to underly the flow of the dark current in a generalized rod are shown in FIGURE 1.

There are two compelling arguments supporting the notion that an internal chemical transmitter links photochemistry in the photopigments of outer segments with the electrical responses just described. The first is based on the topology of outer-segment membranes, the second on the need for numerical

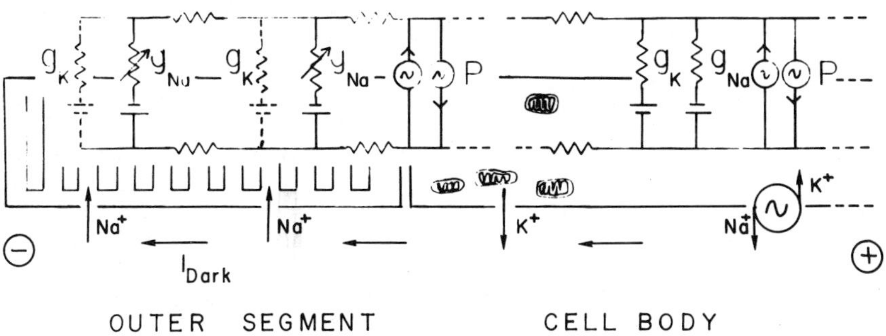

FIGURE 1. Schematic drawing of the electrical circuits and ionic fluxes thought to underlie the dark current of vertebrate rods and cones. From Hagins *et al.*[21]

gain in the light response. The topological evidence is that ultrastructural,[17] electrical,[15, 19] and dye-staining properties[20] of rod outer segments indicate that the photopigment that absorbs the light is located in internal disk membranes that are not continuous with the plasma membrane. Communication between the two structures thus seems to require a diffusion step. The significance of numerical gain stems from the observation that a single photon absorbed in one of the 30 million photopigment molecules in a rat rod suppresses about 3% of the dark current. This effect could be achieved if a photon were to open a K^+ channel of large conductance in the plasma membrane,[21] but this would conflict with the observed light-induced increase in its resistance. If a 3% reduction in the conductance of the plasma membrane is to be caused by a photon, it can be argued that the observed speed of the response requires that many particles released by light must react with the sites admitting the dark current.[18] This point will be considered again after the ionic nature of the dark current is taken up.

The Driving Force for the Dark Current

Estimation of the ionic batteries available to drive dark current around the path shown in FIGURE 1 requires knowledge of the ionic composition of retinal photoreceptors. This is an extremely difficult task, because the retina consists of many cell types and there is no known way to separate them so as to permit analyses by the usual wet chemical methods. We have therefore used the technique of electron-probe microanalysis to study diffusible ions in frozen frog retinas. Dark-adapted frogs were decapitated and their retinas dissected out in a dish of oxygenated physiological solution whose composition appears in TABLE 1. Infrared illumination and a dissecting microscope fitted with image converters were used to assure that the rods were fully dark-adapted. Each retina was attached to a Millipore HA filter as previ-

TABLE 1

ELEMENTAL COMPOSITION OF *R. pipiens* ROD OUTER SEGMENTS BY ELECTRON MICROPROBE ANALYSIS OF QUICK-FROZEN AND CLEAVED DARK-ADAPTED RETINAS*

	Concentration ($mmol \cdot liter^{-1} \pm$ S.E.M.)				
Element	Perfusion Fluid	Gross Outer-Segment Layer	Outer Segments Corrected for Interstitial Fluid	Outer-Segment Water	Nernstian Potential E_n (mV)
K	2.5	15 ± 4	22 ± 5	37	−68
Ca	0.1	2.5 ± 0.8	3.8 ± 1		
Cl	125	62 ± 5†	35 ± 8†	58	−18
S	0	25 ± 5	38 ± 8		
P	0.7	190 ± 20	290 ± 30		
Na	115			[68] ‡	[+20] ‡

* Averages of analyses at −150° C of four 40 × 50 μm areas on two retinas. See FIGURE 2 for a typical specimen.

† See text for cautions about interpretation of Cl analyses.

‡ Calculated by difference from cationic content of perfusion fluid and K content of outer segments. Direct Na determinations were of doubtful accuracy for technical reasons.

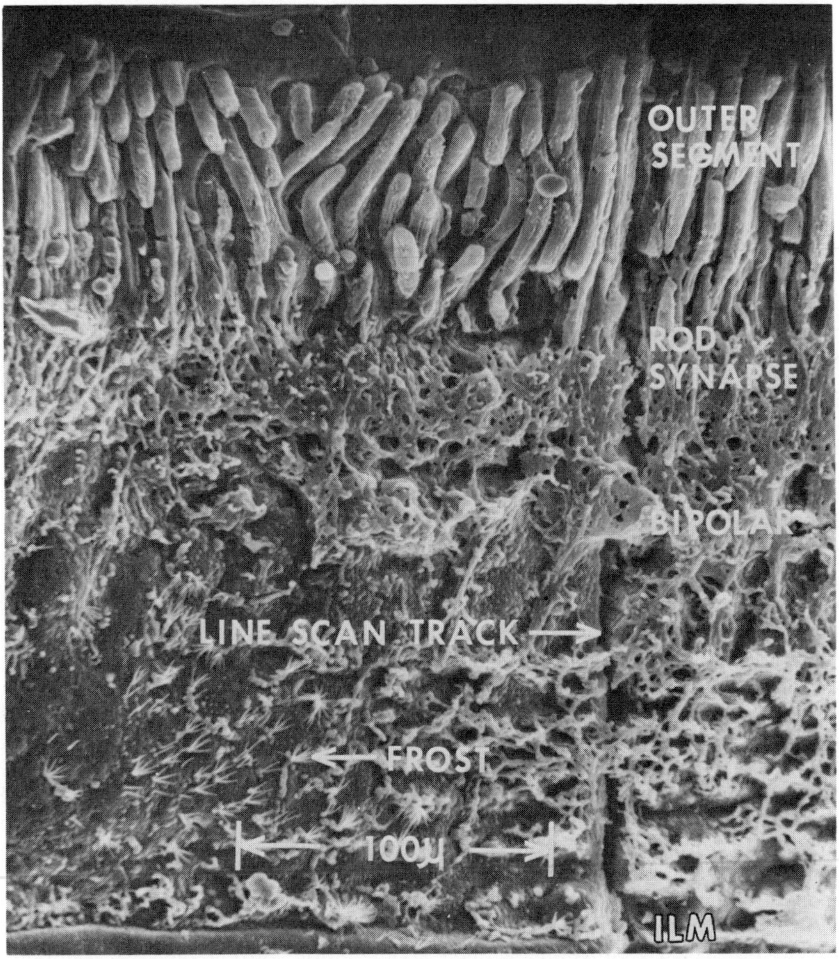

FIGURE 2. Secondary electron scan of a quick-frozen frog (*R. pipiens*) retina cleaved along a plane passing radially through the retinal layers. Etched 18 hr in vacuum at —150° C, coated with ~200 Å of Al and scanned in the secondary electron mode. Small frost crystals and a groove etched by a 5-min line profile element scan of the specimen with a ~7.5 nA electron beam are visible. Inner limiting membrane (ILM). The rod outer segments are about 60 μm long. The characteristic $K\alpha$ x-ray emission lines of the elements shown in FIGURE 3 were obtained from 40 × 50 μm rectangular areas in the outer segment layer, raster-scanned without coating.

ously described.[14] The filter and retina were quickly placed between two polished copper blocks held at —190° C and compressed so that freezing was complete in less than 40 msec. Frozen samples were stored in liquid N_2 until they could be analyzed two or three days later.

FIGURE 2 shows the cleaved edge of such a preparation held at —150° C

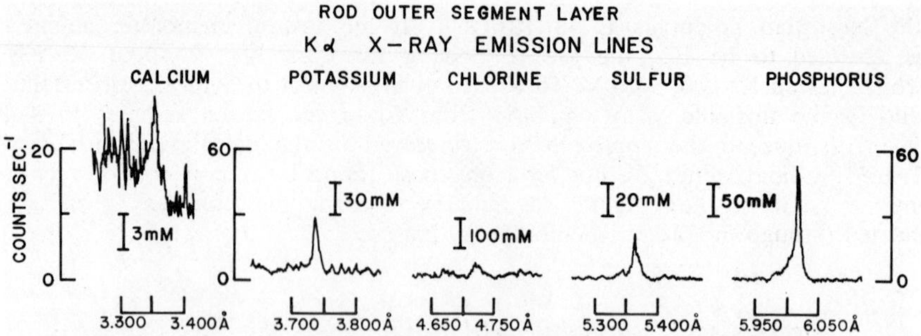

FIGURE 3. Characteristic Kα emission lines of several elements excited in the rod outer-segment layer of a preparation like that described in FIGURE 2. Each element was detected by scanning the wavelength-dispersive x-ray spectrometer across its characteristic emission wavelength. The spectrometer was then set on each peak and a 5-min count taken. This was then corrected for the x-ray continuum by detuning the spectrometer and counting the background about 0.01 Å from the line for 5 minutes in each case. Emission intensities were calibrated in terms of element concentrations in the frozen tissue by comparison with quick-frozen standards of known composition. Temperature −150° C. Electron beam acceleration, 10 kV; specimen current, 7.5 nA.

on the cold stage of a scanning electron microscope. It was 18-hours-old at the time of the scan and had lost about 5 μm of ice by sublimation, mainly because of slight heating during many repeated electron scans with a beam yielding a specimen current of 7.5 nA. It is coated with a thin film of Al. FIGURE 3 shows the Kα x-ray emission lines of several elements excited by scanning an area 40 × 50 μm in the outer segment layer about ten times per second with a 10-kV electron beam. Plots of emission intensity vs. wavelength measured with a crystal spectrometer fitted with a focusing pentaerythrytol (PET) crystal are given. The emission intensities are calibrated in element concentration by comparison with cleaved layers of similarly frozen electrolyte solutions of known composition. Area maps of entire retinal specimens are shown elsewhere.[21] The x-ray emission signals were quite stable with 7.5-nA specimen currents and scanning periods of 10 minutes for all elements except Cl. The Cl emission declined about tenfold during the first 2 minutes of the scanning period in both the retina and in the ice standards, presumably because of radiochemical reactions with free radicals produced by the electron beam. The estimates of Cl are thus less reliable than those of the other elements. The Ca signals were taken from outer segment regions that had been checked for freedom from Ca-bearing dust that often adhered to the frozen specimens. The results of some analyses on two specimens are shown in TABLE 1.

It is clear from the analyses that frog rod outer segments have elemental compositions like those of other nerve cells and thus have similar electromotive forces available for driving ionic currents. The last column in TABLE 1 gives estimates for the Nernst potentials for Na, Cl, and K on the assumption that these ions are all unbound and are entirely in the rod cytoplasm. No allowance was made for the intradisk water. The size of intradisk space is still not known exactly, though recent small-angle x-ray diffraction data suggest a value of 10–20 Å, corresponding to perhaps 10% of the total rod water.[22]

The driving forces on the principal ions in rods can be estimated from the Nernstian potentials E_n in TABLE 1. If the resting membrane potential is assumed to be E_{Cl}, the inward driving force on Na^+ is about 40 mV, while that on K^+ is —50 mV. These figures are subject to a 10-mV uncertainty due to the unstable x-ray emission from Cl in the frozen samples as well as errors due to the approximations involved in the calculations in TABLE 1, but the final values are not far from those found by microelectrode recording.[16, 23, 24] Given these figures, it remains to show that the dark current is carried through the plasma membrane by Na^+.

Effects of Ionic Substitution and the Carriers of the Dark Current

Tomita and his coworkers found that the transretinal PIII wave of retinas treated with synaptic-blocking aspartate ions disappeared reversibly when Li^+ was substituted for Na^+ in the perfusing fluid.[10] The same result is obtained when the photocurrent responses of the rod layer of an isolated rat retina are observed by the extracellular voltage-gradient recording method[13, 14] and Na^+ is replaced by $Tris^+$ or Li^+ (TABLE 2). The result of such an experiment is shown in FIGURE 4. A small fragment of rat retina was mounted in a flow chamber with a volume of ca. 20 microliters. Prewarmed perfusion fluid at 37° C flowed rapidly tangentially to the receptor layer, which lay uppermost. Using infrared microscopy, a pair of microelectrodes were placed in a radial

TABLE 2

IMMEDIATE EFFECTS (<10 SEC) OF IONIC SUBSTITUTIONS ON THE PHOTOVOLTAGE RESPONSE (V_L) AND RADIAL DARK VOLTAGE GRADIENT (V_D) IN THE RECEPTOR LAYER OF ISOLATED RAT RETINAS AT 37° C*

	Perfusion Solution (mM)								
	[Na+]	[K+]	[Tris+]	[Li+]	[Ca++]	Major Anion	V_D	V_L	
1.	136	2.7			1.36	Cl^- 145 mM	1.0	−1.0	Ringer II
2.	60	2.7	76		1.36		U†	−0.4	
3.	20		116		1.36		U	0.0	
4.	<0.10		136		1.36		U	0.0	
5.	1.0		135		10^{-2}		U	−0.1	
6.	60		76		1.36		U	−0.4	
7.	136	0			1.36		1.0	−1.0	
8.	126	13			1.36		1.0	−1.0	
9.	136	2.7			1.36	$Isethionate^-$ 145 mM	U	−0.9	
10.	136	2.7			1.36	$SO_4^=$ 68 mM + sucrose 68 mM	U	−1.0	
11.	0	2.7		136	1.36	Cl^-	U	0.0	

* Extracellular recording by the method of reference 15. V_L is the peak change in V_D in response to a 560-nm test flash causing about 30 photons absorbed · rod^{-1}. V_L and V_D are normalized to −1.0 and 1.0, respectively. Flow chamber experiments as in FIGURE 4.

† Not measurable because of diffusion potentials at the electrode tips. See FIGURE 4 and text.

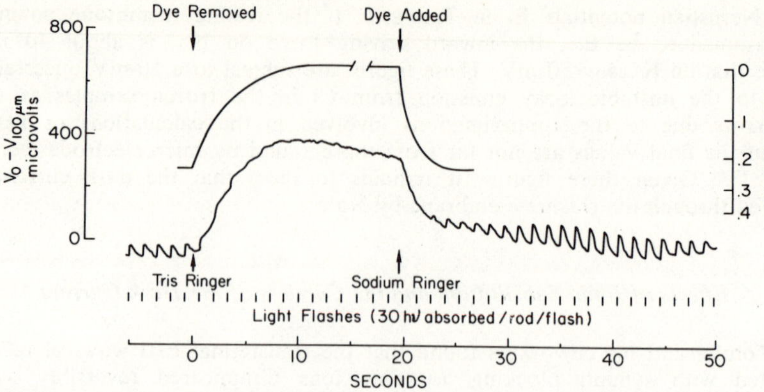

FIGURE 4. Effect of substitution of $Tris^+$ for Na^+ in perfusion bath on radial dark voltage gradient V_D and 30 photon rod^{-1} light responses (V_L) in the rod layer of the rat retina at 37° C. *Upper curve:* dye wash-out and wash-in curve measured spectrophotometrically with a light beam passing across the perfusion chamber just above the retina. Dye: Methylene blue. *Lower curve:* Voltage difference between a 150-mM NaCl-filled pipette at the rod tips and a similar one about 100 μm deep in the retina (at the rod synapses). The large offset voltage produced by $Tris^+$ substitution is a diffusion potential at the electrode tips. The light responses disappear and reappear in less than 10 sec when Na^+ is removed and restored.

line, one at the distal ends of the rod outer segments and the other near the rod-bipolar synapses 100 μm deeper in the retina. The radial voltage gradient in the receptor layer due to flow of the dark current and its changes in light were thus recorded directly as previously described.[15] The fluid flow was sufficient to wash out the small chamber with a half-time of 2 sec, as shown by the dye wash-in and wash-out curves in the upper half of the Figure. As the $Tris^+$ replaced the Na^+, a d.c. offset potential between the two electrodes developed with a time course virtually identical with the dye flux curve. This was a diffusion potential between the electrode tips produced by the gross change in ionic composition of the bathing fluid, not a retinal response. However, the rod responses to an ongoing train of 1-μsec stimulus flashes were immediately and reversibly abolished as the Na^+ was replaced by $Tris^+$.

A summary of similar ionic substitution experiments is shown in TABLE 2. Of the substitutions listed, only those in which (Na^+) was reduced had any immediate effect on the size of the light responses. Partial replacement of Na^+ reduced them roughly in proportion to the amount of the ion remaining, in accordance with the prediction of constant field theory. However, when (Na^+) was reduced below 20 mM in the presence of 1.4 mM external Ca^{++}, V_L vanished entirely (line 3, TABLE 2), but did not reverse in sign as constant field theory would predict for the model of FIGURE 1. This observation agrees with those of Tomita *et al.*[10] and suggests that when the driving force on the dark current is reversed by drastic Na^+ replacement, a rectification prevents reversed light responses. The reason for this is suggested by line 5 in TABLE 2. If (Ca^{++}) is reduced to 10^{-5} M, lowering the external (Na^+) to as little as 1 mM still permits light responses of substantial size. The conductance mechanism for the dark current is suppressed by the combination of low

(Na^+) and relatively high external (Ca^{++}), as if a Na:Ca exchange pump were operating in the plasma membrane.

Since the presence of Na^+ in the perfusate seems essential to the photocurrent of rat rods, the dark current evidently enters the plasma membrane of the outer segments as a sodium ion influx. In Ringer II, this flux is very large in relation to the small volume of a rat rod. Estimates based on differential electrode recordings in the receptor layer[14] yield values of not less than 20 and perhaps as much as 70 pA for the dark current, a value sufficient to turn over all of the cell's cations in ca. 1 minute. Thus it is understandable that ouabain causes V_D and V_L to disappear with a half-time of about 30 sec.[25] Given the driving force of 40 mV on Na^+ (TABLE 1), This implies that the total Na^+ conductance of the plasma membrane of a single outer segment of area 1.3×10^{-6} cm^2 is 500–1700 pmho, or about 400–1200 μmho cm^{-2}. The question of the molecular basis of this conductance will now be considered.

Estimation of the Specific Conductance of the Dark Current Mechanism

The model upon which the estimate is based is shown in FIGURE 5. Evidence for the model is discussed elsewhere.[25–27] Its essence is that there are discrete sites on the plasma membrane, "D sites," that admit Na^+ as a flux driven inward by passive electrochemical forces of potential and concentration gradients. These sites combine reversibly with Ca^{++} ions in the cytoplasm. While in the Ca^{++} form, they do not admit Na^+. Light releases additional Ca^{++} from the internal disks of the outer segment, thus raising the free Ca^{++} activity in the cytoplasm and hence the fraction of the D sites that are blocked. At the same time, leakage of Ca^{++} through the plasma membrane from the extracellular fluid, countered by active Ca^{++} extrusion, serves to give a small but finite steady-state free Ca^{++} activity in the cytoplasm. Additional Ca^{++}-binding buffer sites Y in the cytoplasm are probably present as well.[31] The binding kinetics of the D sites are presumed to be first-order and thus to give a hyperbolic relation between cytoplasmic Ca^{++} activity and dark current amplitude. If the quantity of Ca^{++} released from the disks by light is proportional to stimulus amplitude, and the Y sites are already saturated with Ca^{++}, the experimentally observed hyperbolic relation between stimulus and rod response is obtained.[14–16, 28]

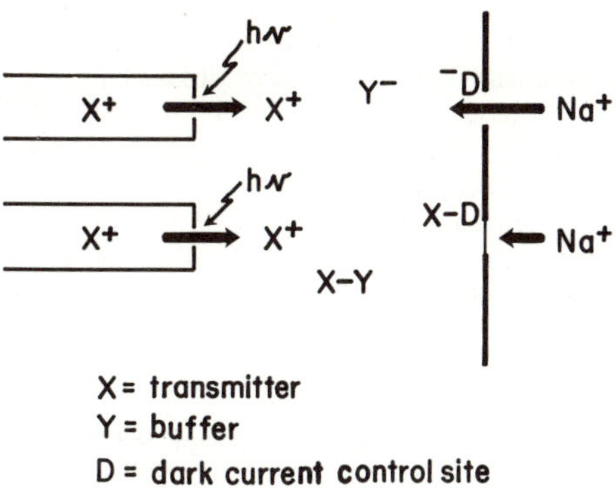

FIGURE 5. Model of the ionic excitatory mechanism for retinal rods. Transmitter ions X^+ are stored in disks at left and react with plasma membrane (vertical line at right) when released by photons. The buffer sites Y are Ca^{++} binding sites not associated with control of the Na^+ current through the D sites on the rod plasma membrane. The Y sites are only apparent under severe conditions of Ca^{++} depletion.[21, 31]

Since the model postulates a steady state in darkness in which a mean fraction $1 - \alpha$ of the channels in an outer segment are blocked by bound Ca^{++}, the exact number of unblocked sites will fluctuate, and so will the size of the dark current. If these fluctuations are to be small compared to the 3% perturbation in the dark current produced by the absorption of a single photon, then the relative standard deviation σ of the instantaneous number of open channels, given by

$$\sigma = (N)^{-1/2}[(1 - \alpha)/\alpha]^{1/2} \tag{1}$$

must be less than .03. Now, in Ringer II, the fraction of D sites blocked at any time is about 0.8, since the dark current can be increased at least fivefold by reducing the (Ca^{++}) mm to $<10^{-5}$ M. Thus $\alpha \simeq 0.2$ and Equation 1 therefore requires that the total number N of D sites must be greater than 3600 if the rod is to be a good photon-counter.[18, 25] Of these sites, at least 600 must contribute to the Na^+ conductance and so each site need contribute no more than about 0.7–2. pmho. The Na^+ current through each unblocked site would therefore be $<5 \times 10^5$ electronic charges sec^{-1}. Such a turnover number would seem consistent with conventional carrier fluxes, a conclusion supported by the high ionic specificity of the dark conductance. This conductance thus is quite unlike that underlying the Na mechanism of nerve action potentials, and it is also insensitive to tetrodotoxin (Yoshikami and Hagins; unpublished material).

The Calcium Permeability of the Plasma Membrane

The prompt suppression of the dark current by added external Ca^{++}[25] is explained in the model of FIGURE 5 by supposing that the plasma membrane has a finite resting permeability to Ca^{++}. A simple model to allow quantitative treatment of the Ca^{++} fluxes, given previously,[26] expresses the activity C of Ca^{++} in the outer segment cytoplasm by the differential equation

$$v\frac{dC}{dt} = S(Pe^{\beta}C_0 - (P + R)C) \tag{2}$$

where v is the cytoplasmic volume of the outer segment, C_0 is the Ca^{++} activity in the extracellular fluid, P is the permeability coefficient of the plasma membrane for Ca^{++}, R is the coefficient of Ca^{++} extrusion by pumping mechanisms in the plasma membrane, and $\exp(\beta)$ is the biasing effect of the membrane potential on the passive fluxes. For a rod membrane potential near E_{Cl} in TABLE 1, this factor will be about 10. S is the surface area of the Ca^{++}-permeable part of the rod, here assumed to be the outer segment envelope and v is the cytoplasmic volume.

Direct measurements of P and R are not yet available for rods, but crude estimates of their magnitudes can be made from rods treated with the Ca^{++} ionophores, X537A[26] and A23187 (Yoshikami and Hagins; unpublished material). When rods are treated with these substances, their dark current can be suppressed by external Ca^{++} activities at least 2000 times less than is required in untreated rods (FIGURE 6). On the assumption that $P \gg R$ in X537A, the internal calcium necessary for half-saturation of the D sites can be estimated to be not more than 10^{-6} M. Then, if untreated rods show half-saturated D sites

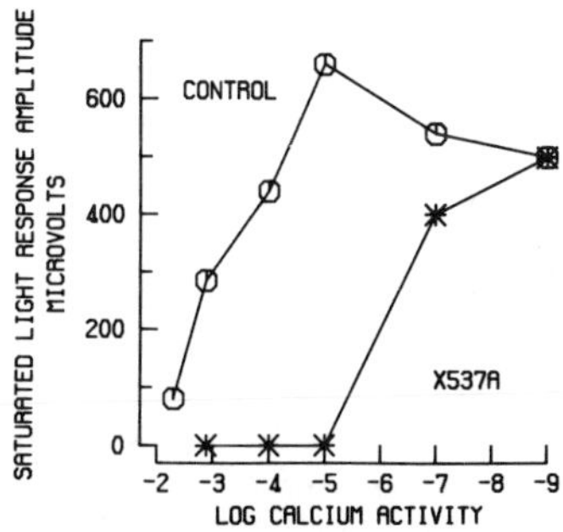

FIGURE 6. Effect of the divalent cation ionophore X537A on the Ca^{++}-sensitivity of rat rods. The maximal amplitude light response V_L (max) (which is almost equal to the dark voltage gradient V_D) is plotted vs. EGTA-buffered Ca^{++} ion activity in the perfusion fluid. Circles: without X537A; asterisks: in presence of 10 μm X537A added to the perfusate in ethanol. (From Hagins and Yoshikami.[26])

at 5×10^{-4} M Ca^{++}, R $\gg$ P in the absence of added ionophores. In the steady state, Equation 2 in fact requires that

$$R \geqq 5000\,P \tag{3}$$

Now, adding the observation that quick addition of extra Ca^{++} to the perfusate will suppress the dark current of normal rods with a time constant τ of about 1 sec (Reference 25, Figure 4), Equation 2 predicts that

$$\tau = v/[S(P + R)] \tag{4}$$

Combining Equations 3 and 4 and substituting $v = 28 \times 10^{-12}$ cm^3 and $S = 1.3 \times 10^{-6}$ cm^2 for rat rod outer segments we obtain

$$P \leqq 4 \times 10^{-9}\ cm\ sec^{-1}.$$

For an internal (Ca^{++}) of 5×10^{-4} M and a membrane potential of -30 mV, the Ca^{++} influx in a rod would be about 0.02 pmol cm^{-2} sec^{-1} or 16,000 Ca^{++} ions sec^{-1}. This small flux is quite consistent with a Na:Ca exchange mechanism like that of other nerve cells.

Light-Induced Ca^{++} Fluxes from the Disks

Although the foregoing calculation is obviously crude, it yields figures for Ca^{++} fluxes quite comparable to those found in squid axons.[29] So it is interesting to estimate the light-induced Ca^{++} fluxes from the disks that would be required to operate the model of FIGURE 5. From the data of Penn and Hagins[15] it can be seen that a steady light causing half-suppression of the dark current in Ringer II (or in any Ringer having more than 10^{-6} M Ca^{++}) causes absorption of about 400 photons $rod^{-1} \cdot sec^{-1}$. This light is thus about equivalent to a doubling of the external Ca^{++} activity, which itself will double the passive Ca^{++} influx. Thus 400 photons absorbed per second might release as few as 16,000 Ca^{++} ions per second.

The numerical gain of the transmitter release mechanism might thus be as small as 40. This number should be compared with two others: (a) the shot noise fluctuations in the transmitter concentration in the cytoplasm of an outer segment, and (b) the minimal number of D sites that must be blocked in a single photon response.

Because the fluid volume of a rod outer segment is so small and the concentration of the presumed transmitter, Ca^{++}, is so low, fluctuations in $[Ca^{++}]$ should be apparent. At 10^{-6} M, there would be only about 15,000 free Ca^{++} ions present. Thermodynamic considerations[32] predict total r.m.s. fluctuations of at least ± 120 in this number of free ions, while if the influx of Ca^{++} by leakage and outflux by extrusion are uncorrelated, additional fluctuations in the internal Ca^{++} activity would exist. Thus the transmitter release mechanism must exceed these fluctuations if the rod is to be a reliable quantum detector in the presence of 10^{-6} M free internal Ca^{++}.

The minimal number of D sites obstructed in a photon response yields a third figure for the minimal numerical gain in transmitter release. If at least 3% of the more than 600 open D sites in Ringer II must react with transmitter molecules to yield the observed change in the dark current, about 20 would be needed.,

These three estimates of >40, >120 and >20 for the numerical gain of transmitter release are based on essentially different considerations, but they all suggest that no excitatory mechanism that produces just one transmitter molecule per photon can perform satisfactorily. Thus a limit is set on the minimal calcium content of rod disks. If each disk is to provide at least 100 transmitter ions in a photon response and is capable of the equivalent of at least ten quantized levels of response to multiple photon absorptions (i.e., if it contains at least 100 photoreceptive units[15]), a rat rod must contain a store of at least 10^6 Ca^{++} ions. Thus the outer segment must have a Ca concentration of at least 5×10^{-5} M in its bulk. How does this compare with the microprobe analyses of TABLE 1?

Calcium in the Rod Disks

The electron microprobe results in TABLE 1 suggest that the total Ca in rod outer segments is not more than 3.8 mmol/liter of rod volume. This figure should be taken to be an upper limit, because the method of cleavage of the specimens in a bath of liquid N_2 usually resulted in calcium-rich particles of dust adhering to their surfaces (see Figure 2 of Reference 21). This quantity of Ca would be much more than sufficient to provide a reservoir of transmitter for rod excitation even in bright light. A single disk with a 20-Å aqueous core containing only 1 mM Ca^{++} would hold more than 10,000 Ca^{++} ions. Released in quantities of 40 with each absorbed photon, a disk could supply signals quantized in at least 200 levels. Of course the kinetics of transmitter release and reaccumulation are still not known, particularly as functions of regeneration of rhodopsin, and further speculation on the point seems unwarranted now. It should be noted, however, that the Ca content of rod outer segments of quick-frozen retinas is much less than that reported for standard biochemical preparations.[30] Study of Ca metabolism in isolated rods thus seems to promise a challenge for experimenters.

Conclusion

Both measurements and rough theoretical calculations suggest that both sites in the plasma membrane where the dark current enters and the mechanism that produces the intracellular excitatory transmitter of rod vision operate via ionophoric mechanisms with very low specific conductivities and high ionic specificities properties similar to those of carriers in artificial and natural membranes.

Acknowledgments

We thank W. C. Lane, Tung Hsu, and Nelson Yew of Etec Corporation, Hayward, California for their great efforts in behalf of the electron probe microanalyses.

References

1. HECHT, S., S. SHIAER & M. H. PIRENNE. 1942. J. Gen. Physiol. **25:** 819–840.
2. BARLOW, H. B. 1957. J. Physiol. (London) **136:** 469.
3. FUORTES, M. G. F. 1959. J. Physiol. (London) **148:** 14–28.
4. BORSELLINO, A., M. G. F. FUORTES & T. G. SMITH. 1965. Cold Spring Harbor Symp. Quant. Biol **30:** 429–443.
5. HAGINS, W. A., R. G. ADAMS & H. V. ZONANA 1962. Nature **194:** 844–847.
6. HAGINS, W. A. 1965. Cold Spring Harbor Symp. Quant Biol. **30:** 403–418.
7. MILLECCHIA, R. & A. MAURO. 1969. J. Gen. Physiol. **54:** 310–51.
8. SREBRO, R. & S. YEANDLE. 1970. J. Gen. Physiol. **56:** 751–67.
9. TOMITA, T. 1965. Cold Spring Harbor Symp. Quant. Biol. **30:** 569–86.
10. TOMITA, T. 1970. Quart. Rev. Biophys. **3:** 179–222.
11. BORTOFF, A. & A. L. NORTON. 1965. Vision Res. **5:** 527–33.
12. BORTOFF, A. & A. L. NORTON. 1967. Vision Res. **7:** 253–63.
13. PENN, R. D. & W. A. HAGINS. Nature **223:** 201–205.
14. HAGINS, W. A., R. D. PENN & S. YOSHIKAMI. 1970. Biophys. J. **10:** 380–412.
15. PENN, R. D. & W. A. HAGINS. 1972. Biophys. J. **12:** 1073–1094.
16. BAYLOR, D. & M. G. F. FUORTES. J. Physiol. **207:** 77–92.
17. COHEN, A. I. 1970. Vision Res. **10:** 445–453.
18. CONE, R. A. 1973. *In* Biochemistry and Physiology of Visual Pigments. H. Langer, Ed.: 275–282. Springer-Verlag. New York, N.Y.
19. RÜPPEL, H. & W. A. HAGINS. Ibid: 257–261.
20. YOSHIKAMI, S., W. E. ROBINSON & W. A. HAGINS. 1974. Science **185:** 1176–1179.
21. HAGINS, W. A., W. E. ROBINSON & S. YOSHIKAMI. 1975. *In* Energy Transformation in Biological Sytems: Ciba Foundation Symposium. Vol. 31: 169–189.
22. WORTHINGTON, C. R. 1974. Ann. Rev. Biophys. Bioeng. **3:** 53–80.
23. BROWN, J. E. & L. H. PINTO. 1974. J. Physiol. (London) **236:** 575–92.
24. BAYLOR, D. & A. L. HODGKIN. 1973. J. Physiol. (London) **234:** 163–198.
25. YOSHIKAMI, S. & W. A. HAGINS. 1973. *In* Biochemistry and Physiology of Visual Pigments. H. Langer, Ed.: 245–255. Springer-Verlag. New York, N.Y.
26. HAGINS, W. A. & S. YOSHIKAMI. 1974. Exp. Eye Res. **18:** 299–305.
27. HAGINS, W. A. 1972. Ann. Rev. Biophys. Bioeng. **1:** 131–158.
28. BAYLOR, D. A., A. L. HODGKIN, & T. D. LAMB. 1974. J. Physiol. (London) **242:** 685–727.
29. MULLINS, L. J. & F. J. BRINLEY. 1975. J. Gen. Physiol. **65:** 135–152.
30. HENDRIKS, TH., DAEMEN, F. J. M. & S. L. BONTING. 1974. Biochem. Biophys. Acta **345:** 468–473.
31. YOSHIKAMI, S. & W. A. HAGINS. 1975. Biophys. J. **15:** 169a.
32. TOLMAN, K. C. 1938. Statistical Thermodynamics, 1st Ed.: 641–645. Oxford University Press. London, England.

ISOLATION AND CHARACTERIZATION OF TRYPTIC FRAGMENTS OF THE SARCOPLASMIC RETICULUM ADENOSINE TRIPHOSPHATASE*

Peter S. Stewart and David H. MacLennan

Banting and Best Department of Medical Research
Charles H. Best Institute
University of Toronto
Toronto, Ontario, Canada, M5G 1L6

The classical approach to biochemical problems is one of resolution to the simplest forms followed by reconstruction of activity from isolated components. This approach has been utilized in our laboratory for the understanding of Ca^{2+} transport in sarcoplasmic reticulum. We initiated the program by isolating and characterizing the $Ca^{2+} + Mg^{2+}$-dependent adenosine triphosphatase.[1-4] A highly active preparation was found to contain three components. The major component, a protein of mol. wt. 102,000, was phosphorylated by ATP-γ-^{32}P and was therefore believed to contain the active site of ATP hydrolysis. A proteolipid[3] of mol. wt. 6,000–12,000 was a minor component. If it constituted 5–6% of the total protein and if its molecular weight were 6,000, its concentration would be stoichiometric with that of the major component. The role of the proteolipid in ATPase activity, in Ca^{2+} transport activity and in structure of the membrane is still undetermined. The third component is phospholipid, which is essential for ATPase activity and for membrane formation.[1, 2]

The ATPase protein, proteolipid, and phospholipid formed membranes that retained many of the structural features of intact sarcoplasmic reticulum.[2, 4] The hydrophobic interior of the membranes contained 9-nm globular particles which were exposed by freeze-fracture. In contrast to intact sarcoplasmic reticulum, in which particles were arranged asymmetrically in the membrane,[5] these particles were symmetrically arranged, being of equal content in either leaflet. Since the ATPase freed of phospholipid and proteolipid appeared by negative staining techniques as globular particles about 9 nm in diameter and since fractures of vesicles containing only the ATPase and phospholipid were free of hydrophobic globules,[3] it is probable that the globules represent penetration of the ATPase protein into the lipid bilayer of the sarcoplasmic reticulum.

The ATPase membranes also contain surface particles first observed by Ikemoto *et al.*[6] in intact sarcoplasmic reticulum. They are about 4 nm in diameter and extend about 6 nm out of the membranes. These particles are observed only by negative staining and do not appear in freeze-etch pictures although lumps of low profile are observed on the external etch surface of sarcoplasmic reticulum membranes.[3] We cannot be sure that these particles protrude into the aqueous environment at all times. Nevertheless, they project from the lipid bilayer in the presence of phosphotungstate and therefore are probably external in location and rather polar in nature.

The surface particles also appear to be part of the ATPase enzyme since

* This research was supported by grants from the Medical Research Council of Canada and the Muscular Dystrophy Association of Canada.

their presence in reformed membranes can only be accounted for by the ATPase protein.[4] Moreover, in spite of contrary reports,[7-9] we have found that loss of surface particles is coincident with the loss of ATPase activity. It is our contention, therefore, that the surface particles represent a functional part of the ATPase protein.

Thin sections of membranes reformed from the ATPase reveal hollow vesicles with occasional onion-skin formations composed of membranes reformed within other membranes. The vesicles are free of any matrix structure observed in intact vesicles and it has been suggested[4, 10] that this matrix is composed of calsequestrin[11] and the high-affinity Ca^{2+} binding protein.[12]

The isolated ATPase enzyme carries out ATP-ADP exchange[1] and ATP-^{32}Pi exchange,[13] and it is phosphorylated by ATP-γ-^{32}P in a Ca^{2+}-dependent reaction[1, 2] and dephosphorylated in a Mg^{2+}-dependent reaction.[14] These are the essential steps in an enzymatic model of ATP-dependent Ca^{2+} transport[15, 16] which has strong experimental support. The enzyme will also carry out ATP-dependent Ca^{2+} transport when incorporated into excess phospholipid,[13, 17, 18] thus demonstrating that the ATPase, consisting of the three components, is a transporting unit.

The role of specific phospholipids and the proteolipid in various reactions has not been completely defined. Phosphorylation of the enzyme does not require phospholipid, but there is a phospholipid requirement for dephosphorylation and for Ca^{2+} transport.[19] Warren *et al.*[18] replaced 99% of the endogenous lipid of the ATPase with dioleyl lecithin and found that the enzyme was active in ATP hydrolysis. When the enzyme was incorporated into excess dioleyl lecithin, it was partially active in Ca^{2+} transport. If it were incorporated into excess sarcoplasmic reticulum lipid it was much more active in Ca^{2+} transport. These experiments imply that lecithin is the only essential cofactor for the ATPase protein in ATP-dependent Ca^{2+} transport. It was not clear in the experiments of Warren *et al.*[18] whether the proteolipid was present or not, and so the contribution of this protein to transport was not evaluated. Racker and Eytan[17] have found that phosphatidyl ethanolamine is required together with lecithin for transport activity, and Racker *et al.*[20] have found that a fraction of lipid, probably containing the proteolipid, stimulates ATP-dependent Ca^{2+} transport in their reconstituted system.

The ATPase enzyme freed of lipid and proteolipid and lightly succinylated has been shown to act as a Ca^{2+}-dependent and-selective ionophore.[21] When the protein was incorporated into a bilayer of oxidized cholesterol in the presence of 3 mM $CaCl_2$ and an applied electrical current, the resistance of the bilayer to Ca^{2+} movement was greatly reduced. Activity was also found in tryptic digests of the purified ATPase protein. Similar experiments carried out with the proteolipid were essentially negative.

These data indicate the possibility that specific phospholipids and the proteolipid may play a role in ATP-dependent Ca^{2+} transport and further investigation of these roles will be informative. The data also show that individual sites within the ATPase protein can be identified and investigation of these sites should also provide information critical to understanding of ATP-dependent Ca^{2+} transport. To achieve this understanding resolution must be carried beyond the point of isolation of a single protein into the area of isolation, from the degraded protein, of the specific sites of ATP hydrolysis and ion carrier activity. This is our experimental approach.

One can look for ATPase and ionophoric sites in proteolytic or chemical digests of the isolated ATPase protein. In this case relationships are lost and reconstruction of a molecule of 100,000 mol. wt. from fragments averaging 1% of this size is extremely difficult. An alternative method, permitting fragmentation while retaining relationships, is available. When the sarcoplasmic reticulum is subjected to tryptic digestion for a few seconds the ATPase is cleaved nearly quantitatively to two fragments of mol. wt. 45,000 and 55,000 while retaining ATPase and Ca^{2+} transport activity.[4, 22, 23] This indicates that one site in the ATPase is open to tryptic digestion and that all of these sites are oriented on the exterior surface. A close analogy is the effect of proteolytic enzymes on myosin, splitting it to LMM, S_2 and S_1 at flexible regions.[24] If the digestion is continued for 60 minutes the ATPase is fragmented to less than 5,000 molecular weight. On the other hand, if the digestion is carried out in the presence of 1 M sucrose, digestion is largely limited to the further formation of fragments of 30,000 and 20,000 mol. wt. and after about 15 minutes a mixture of components of 55,000, 45,000, 30,000, and 20,000 mol. wt. is present with only small amounts of low molecular weight material (FIGURE 1).

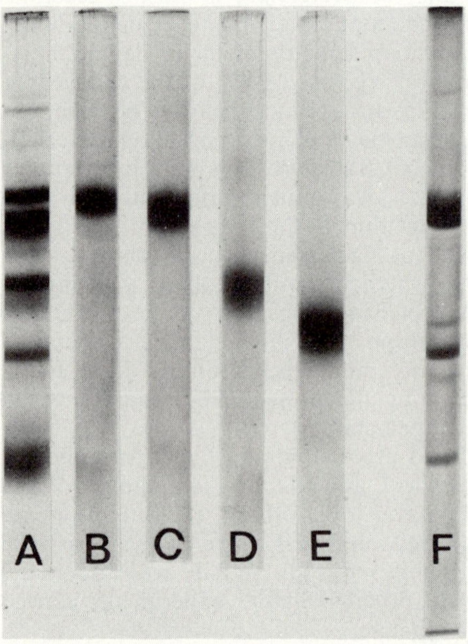

FIGURE 1. SDS gel electrophoresis of fragments purified from trypsin-treated sarcoplasmic reticulum. A, sarcoplasmic reticulum digested for 15 minutes in the presence of 1 M sucrose; B, purified 55,000-mol. wt. fragment; C, purified 45,000-mol. wt. fragment; D, purified 30,000-mol. wt. fragments; E, purified 20,000-mol. wt. fragment; F, sarcoplasmic reticulum digested for 15 minutes in the presence of 1 M sucrose. The system of Weber and Osborn[27] was used for gels A to E. The system of Laemmli[25] was used for gel F. In this system the 45,000- and 55,000-mol. wt. fragments were not separated but the 30,000-mol. wt. material was resolved into three components.

If these digestions are carried out with membranes reformed from the purified ATPase, the protein is not quantitatively cleaved in the first minute to two large fragments. Only about 70% of the protein is cleaved.[14] Prolonged digestion in the presence of 1 M sucrose also fails to cleave more than about 70% of the ATPase protein. These observations indicate that in the reformed membranes the ATPase is not oriented asymmetrically (as in the intact sarcoplasmic reticulum), but rather is oriented symmetrically with a large proportion of molecules pointing inward. This observation complements the earlier results from freeze-fracture of these membranes.[2]

Digestion to large fragments does not significantly lower the ATPase activity, and Ca^{2+} transport is retained long after the first cleavage is completed.[4] Therefore, cleavage within the peptide can occur without loss of either of these functions. Prolonged digestion leading to formation of more than two fragments results in loss of Ca^{2+} transport but not of ATPase activity. In order to associate function with these peptides it is essential to isolate them and to characterize them with respect to both structure and function; a continuation of the process of resolution to simpler forms.

Resolution can be achieved by polyacrylamide gel electrophoresis in the presence of sodium dodecyl sulfate. The purified ATPase was digested with trypsin for 15 minutes in the presence of 1 M sucrose so that it retained activity. It was then phosphorylated with ATP-γ-^{32}P, and fragments were separated in acidic gels. Bands of 100,000, 55,000 and 30,000 mol. wt. retained phosphoryl groups showing that the site of ATP hydrolysis had been reduced in size from 100,000 to 30,000.[14] These results are identical to those reported by Thorley-Lawson and Green.[22] The site of ionophoric activity cannot be so readily followed, but must be measured following isolation of peptides.

We have isolated the various peptides in a series of steps involving chromatography in sodium dodecyl sulfate on Biogel A 1.5 M, Sephadex G-150 and Biogel HTP. FIGURE 1 shows the purity of the fractions obtained. The 55,000-, 45,000- and 20,000-mol. wt. fractions are pure. The 30,000-mol. wt. material was not homogeneous and could be separated in the gel electrophoretic system described by Laemmli[25] into one major and two minor components if the source of the tryptic digest was the intact sarcoplasmic reticulum. It was separated into two components if the source was the purified ATPase previously freed of an intrinsic 30,000-mol. wt. protein found in the intact sarcoplasmic reticulum. A further purification step will be required to obtain the 30,000-mol. wt. fragment containing the site of ATP hydrolysis in homogeneous form. The purified 20,000-mol. wt. fragment is active in ionophoric assays but, to some extent, so is the combined 30,000-mol. wt. material.[26] This result can be explained if the minor 30,000-mol. wt. fragment contains a peptide sequence in common with the 20,000-mol. wt. material.

The amino acid compositions of the various fragments is listed in TABLE 1. Nonpolar amino acids make up 60% and polar amino acids make up 40% of the 45,000-mol. wt. fragment. By contrast, these two classes are equally represented in the 55,000-mol. wt. fragment. Thorley-Lawson and Green[22] have suggested that the 45,000-mol. wt. fragment is buried in the hydrophobic interior of the membrane from their observation of limited iodination of this peptide upon exposure to ^{131}I and lactoperoxidase. Our observations of greater hydrophobicity of the 45,000-mol. wt. fragment would support this suggestion. The 20,000-mol. wt. fragment has 53% hydrophilic and 47% hydrophobic amino acid residues. There are four methionine residues in the fragment indicating that treatment with CNBr should yield five peptides.

We have raised antibodies in chickens against each of these three fragments (FIGURE 2). Antibodies against the 45,000- and 55,000-mol. wt. fragments do not crossreact with each other, but they both react with the ATPase protein. This would indicate that they are completely different fragments derived from the cleavage of one linear chain. Antibodies against the 55,000-mol. wt. fragment but not antibodies against the 45,000-mol. wt. fragment crossreact with the 20,000-mol. wt. fragment, indicating an origin of the 20,000-mol. wt. fragment exclusively in the 55,000-mol. wt. fragment. These observations, together with

TABLE 1

AMINO ACID COMPOSITION (MOL/100 MOL) OF TRYPTIC FRAGMENTS OF THE ATPASE*

Amino Acid	55,000	45,000	20,000
Lysine	6.71	4.18	6.97
Histidine	1.21	1.23	2.01
Arginine	4.40	4.92	5.42
Aspartic	9.59	9.10	7.93
Threonine	7.70	5.90	5.65
Serine	7.39	4.92	6.46
Glutamic	12.26	9.34	12.71
	49.26	39.59	46.61
Proline	4.09	5.90	4.93
Glycine	7.29	8.11	6.87
Alanine	8.65	9.10	8.67
Cysteic acid	2.41	2.95	0
Valine	7.15	6.63	6.85
Methionine†	1.34	3.64	1.99
Isoleucine	5.74	7.13	5.34
Leucine	8.81	10.32	10.78
Tyrosine	1.57	2.46	2.63
Phenylalanine	3.69	4.18	4.41
	50.74	60.41	52.47

* Tryptophan not determined.

† Methionine determined as methionine sulfone.

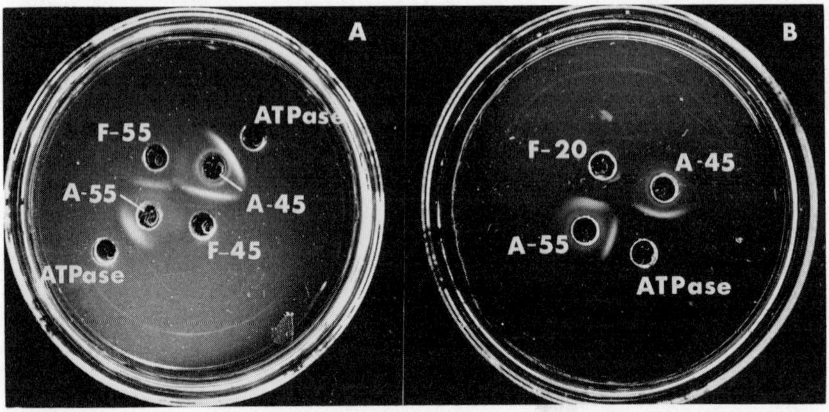

FIGURE 2. Double diffusion patterns in agarose of antibodies against the ATPase and ATPase fragments. A, lack of crossreaction between anti-55,000 antibodies and the 45,000-mol. wt. fragment; lack of crossreaction between anti-45,000 antibodies and the 55,000-mol. wt. fragment. B, crossreaction of anti-55,000 antibody with the 20,000-mol. wt. fragment; lack of crossreaction of anti-45,000 antibody against the 20,000-mol. wt. fragment. (F-20, F-45, F-55, are the 20,000-, 45,000- and 55,000-mol. wt. fragments, respectively; A-45 and A-55 are the antibodies raised against the 45,000- and 55,000-mol. wt. fragments.)

phosphorylation data, show that both the 30,000- and 20,000-mol. wt. fragments originate in the 55,000-mol. wt. fragment.

Our interpretation of these data is as follows. The first cleavage creates two separate fragments of 45,000 and 55,000 mol. wt. The 45,000-mol. wt. fragment is probably then protected from further degradation. The second linear cleavage occurs in the 55,000-mol. wt. fragment, creating two more fragments of about 20,000- and 30,000-mol. wt. One contains the site of ATP hydrolysis, the other, ionophoric activity.

We have used antibodies prepared against the various fragments to learn something of their localization. If the antigenic sites on the fragments are located on the external membrane surface they will react with antibodies applied externally. We have found that such reactions can be detected by aggregation of the membranes if a second antibody, raised against the first, is added to carry out a "sandwich" reaction. Controls for this reaction are invariably negative. We have used the reaction to show the absence of calsequestrin and the high-affinity Ca^{2+} binding protein on the exterior of sarcoplasmic reticulum vesicles and the presence of the ATPase on the external surface.[14]

FIGURE 3 is an example of the effect on membrane aggregation of reacting vesicles, first with purified chicken globulin raised against the intact ATPase

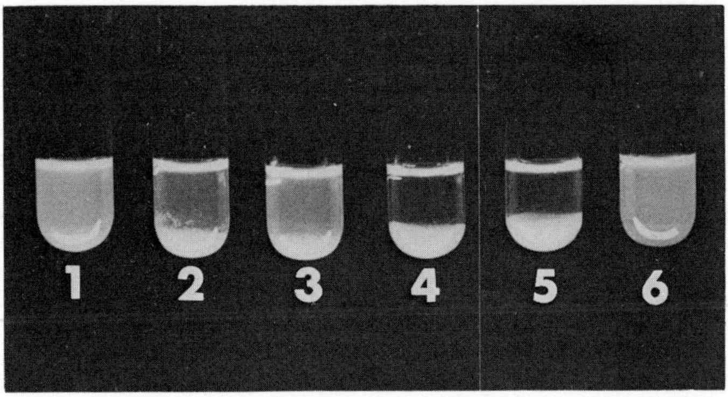

FIGURE 3. The aggregation of intact sarcoplasmic reticulum vesicles with antibodies raised against the ATPase and tryptic fragments of the ATPase. Sarcoplasmic reticulum was treated with purified chicken globulin raised against the purified ATPase or its tryptic fragments. After washing, the vesicles were incubated with rabbit anti-chicken serum. (1) Control chicken globulin; (2) chicken anti-20,000-mol. wt. fragment; (3) chicken anti-45,000-mol. wt. fragment; (4) chicken anti-55,000 mol. wt. fragment; (5) chicken anti-ATPase; (6) no chicken antibody added.

or the various purified fragments, followed by rabbit anti-chicken serum. Aggregation of the sarcoplasmic reticulum vesicles occurred with antibodies against the ATPase and the 55,000-mol. wt. fragment. Aggregation with the antibodies raised against the 45,000- and 20,000-mol. wt. fragments was minimal. We can deduce from these experiments that the 55,000-mol. wt. fragment is exposed on the exterior surface of the membrane. The 45,000-

mol. wt. fragment, however, is largely buried in the membrane. The 20,000-mol. wt. fragment containing the ionophoric site is partly exposed. We do not know whether the site of ATP hydrolysis will be exposed to externally applied antibodies. Antibodies raised against the ATPase and the 45,000- and 55,000-mol. wt. fragments did not inhibit the ATPase enzyme nor Ca^{2+} transport.

Data from iodination[22] and from amino acid composition would indicate that more of the 55,000-mol. wt. fragment than of the 45,000-mol. wt. fragment is located externally. This would indicate that the 55,000-mol. wt. fragment makes up a large portion of the surface particles while the 45,000-mol. wt. fragment comprises a large portion of the hydrophobic globules. However, while much of the 45,000-mol. wt. fragment may be buried within the lipid it is obvious from the antibody aggregation experiments that not all of the molecule is buried in the membrane. It will be important to discover whether any of these peptides extend through the membrane and react with antibodies at the interior surface. Experiments to test this point, using ferritin conjugates, are in progress.

These experiments, although far from complete, show the potential of degrading the Ca^{2+} transport ATPase to fragments and of analyzing the fragments for activity and localization. It is probable that several questions can be answered. It is possible, since the peptides are active after cleavage and prior to separation, that they can be separated in such a way that reconstruction can be carried out and that direct tests of function in each fragment can be measured. It is also possible that the fragments can be mapped for insertion into and protrusion from the membranes. Therefore, it should be possible to build a topographical model of the ATPase relative to its active sites. It should be possible to continue the degradation to isolate the smallest sites of ionophoric activity and of ATP hydrolysis so that their three-dimensional structures can be understood. Finally, it should be possible to map the sites relative to each other and to the membrane topology and, by reconstruction, to understand how the two sites interact and control their interrelated activities.

Acknowledgments

The expert technical assistance of Mr. Vijay Khanna is gratefully acknowledged.

References

1. MacLennan, D. H. 1970. Purification and properties of an adenosine triphosphatase from sarcoplasmic reticulum, J. Biol. Chem. **245:** 4508–4518.
2. MacLennan, D. H., P. Seeman, G. H. Iles & C. C. Yip. 1971. Membrane formation by the adenosine triphosphatase of sarcoplasmic reticulum. J. Biol. Chem. **246:** 2702–2710.
3. MacLennan, D. H., C. C. Yip, G. H. Iles & P. Seeman. 1972. Isolation of sarcoplasmic reticulum proteins. Cold Spring Harbor Symp. Quant. Biol. **37:** 460–478.
4. Stewart, P. S. & D. H. MacLennan. 1974. Surface particles of sarcoplasmic reticulum membranes: Structural features of the adenosine triphosphatase, J. Biol. Chem. **249:** 985–993.
5. Deamer, D. W. & R. J. Baskin. 1969. Ultrastructure of sarcoplasmic reticulum preparations. J. Cell Biol. **42:** 296–307.

6. IKEMOTO, N., F. A. SRETER, & J. GERGELY. 1966. Localization of Ca-uptake and ATPase activity in fragments of sarcoplasmic reticulum. Fed. Proc. **25:** 465.
7. IKEMOTO, N., F. A. SRETER, A. NAKAMURA & J. GERGELY. 1968. Tryptic digestion and localization of calcium uptake and ATPase activity in fragments of sarcoplasmic reticulum. J. Ultrastruct. Res. **23:** 216–232.
8. IKEMOTO, N., F. A. SRETER & J. GERGELY. 1971. Structural features of the surface of the vesicles of FSR—lack of functional role in Ca^{2+} uptake and ATPase activity. Arch. Biochem. Biophys. **147:** 571–582.
9. INESI, G. & H. ASAI. 1968. Trypsin digestion of fragmented sarcoplasmic reticulum. Arch. Biochem. **126:** 469–477.
10. MEISSNER, G. 1974. Heterogeneity of sarcoplasmic reticulum (SR) vesicles. Fed. Proc. **33:** 1283.
11. MACLENNAN, D. H. & P. T. S. WONG. 1971. Isolation of a calcium-sequestering protein from sarcoplasmic reticulum. Proc. Nat. Acad. Sci. U.S.A. **68:** 1231–1235.
12. OSTWALD, T. J. & D. H. MACLENNAN. 1974. Isolation of a high affinity calcium-binding protein from sarcoplasmic reticulum. J. Biol. Chem. **249:** 974–979.
13. RACKER, E. 1972. Reconstitution of a calcium pump with phospholipids and a purified Ca^{++}-adenosine triphosphatase from sarcoplasmic reticulum. J. Biol. Chem. **247:** 8198–8200.
14. MACLENNAN, D. H. 1975. Resolution of the calcium transport system of sarcoplasmic reticulum. Can. J. Biochem. **53:** 251–261.
15. KANAZAWA, T. & P. D. BOYER. 1973. Occurrence and characteristics of a rapid exchange of phosphate oxygens catalyzed by sarcoplasmic reticulum vesicles. J. Biol. Chem. **248:** 3163–3172.
16. YAMAMOTO, T. 1972. The Ca^{2+}-Mg^{2+}-dependent ATPase and the uptake of Ca^{2+} by the fragmented sarcoplasmic reticulum. *In* Muscle Proteins, Muscle Contraction and Cation Transport. Y. Tonomura, Ed.: 305–356. University of Tokyo Press. Tokyo, Japan.
17. RACKER, E. & E. EYTAN. 1973. Reconstitution of an efficient calcium pump without detergents. Biochem. Biophys. Res. Commun. **55:** 174–178.
18. WARREN, G. B., P. A. TOON, N. J. M. BIRDSALL, A. G. LEE & J. C. METCALF. 1974. Reconstitution of a calcium pump using defined membrane components. Proc. Nat. Acad. Sci. U.S.A. **71:** 622–626.
19. MARTONOSI, A., J. R. DONLEY, A. G. PUCELL & R. A. HALPIN. 1971. Sarcoplasmic reticulum. XI. The mode of involvement of phospholipids in the hydrolysis of ATP by sarcoplasmic reticulum membranes. Arch. Biochem. Biophys. **144:** 529–546.
20. RACKER, E. 1975. Carriers and channels in reconstitution systems. Ann. N.Y. Acad. Sci. This monograph.
21. SHAMOO, A. E. & D. H. MACLENNAN. 1974. A Ca^{++}-dependent and -selective ionophore as part of the $Ca^{++} + Mg^{++}$-dependent adenosinetriphosphatase of sarcoplasmic reticulum. Proc. Nat. Acad. Sci. U.S.A. **71:** 3522–3526.
22. THORLEY-LAWSON, D. A. & N. M. GREEN. 1973. Studies on the location and orientation of proteins in the sarcoplasmic reticulum. Eur. J. Biochem. **40:** 403–413.
23. MIGALA, A., B. AGOSTINI & W. HASSELBACH. 1973. Tryptic fragmentation of the calcium transport system in the sarcoplasmic reticulum. Z. Naturforsch. **28:** 178–182.
24. LOWEY, S., H. S. SLAYTER, A. G. WEEDS & H. BAKER. 1969. Substructure of the myosin molecule. I. Subfragments of myosin by enzymic degradation. J. Mol. Biol. **42:** 1–29.
25. LAEMMLI, U. K. 1970. Cleavage of structural proteins during the assembly of the head of bacteriophage T4. Nature **227:** 680–685.
26. SHAMOO, A. E. 1975. Isolation of ionophores from ion-transport systems. Ann. N.Y. Acad. Sci. This monograph.

27. WEBER, K. & M. OSBORN. 1969. The reliability of molecular weight determinations by dodecyl sulfate-polyacrylamide gel electrophoresis. J. Biol. Chem. **244:** 4406–4412.

Discussion

DR. SHAW: It seems to me that the particles you see inside are possibly proteolipids that have been pushed out by ATPase, and you're not seeing the ATPase as the globules inside.

DR. MACLENNAN: Why do you think that?

DR. SHAW: It's a possibility; the fact that you see that both fragments are at the surface may mean that it may be just lying at the surface and you're not seeing it inside.

DR. MACLENNAN: For one thing the proteolipid does not constitute enough protein to do what you're suggesting. A 40 by 60 angstrom particle would represent a lot of protein and would have a molecular weight of about 50,000. We don't know whether the proteolipid is stoichiometric in concentration with the ATPase, but it could be no more than stoichiometric. Its molecular weight is 10,000 or less. There simply isn't enough material in the membrane to account for the surface particles by the proteolipid. We'd rule that out without any problem.

DR. FINKELSTEIN: Is it known whether any other cations are transported against calcium or is an anion transported with calcium via the ATPase?

DR. MACLENNAN: Yes; there is kinetic data indicating that magnesium is transported as a counter ion, and that probably one magnesium and as many as two potassium ions could be countertransported for two calcium. . . .

DR. FINKELSTEIN: . . . . through the ATPase itself.

DR. MACLENNAN: I'm talking about kinetic data, not real measurement of magnesium moving out and calcium moving in; that is not known. In order to dephosphorylate the enzyme you require magnesium. Dr. Inesi has worked on this problem, and Drs. Yamamoto and Tonamura have shown that dephosphorylation is magnesium-dependent. Yesterday Dr. Racker mentioned the phosphorylation of the ATPase from inorganic phosphate, and that is a magnesium-dependent as is Pi-H_2O exchange. Also there is a change in affinity of a particular site for calcium and magnesium during transport.

DR. FINKELSTEIN: In these transport experiments in vesicles what is compensating operationally for the calcium movement? Is this just due to intrinsic leakage in the vesicles themselves?

DR. MACLENNAN: That's the weakest point of this countertransport argument. It's based on enzymatic data and we do not know how enough magnesium gets inside at all times to act in countertransport.

MECHANISM OF CALCIUM TRANSPORT IN SARCOPLASMIC RETICULUM*

W. Hasselbach, J. Suko,† M. H. Stromer,‡ and R. The
Max-Planck-Institut für Medizinische Forschung
Abteilung Physiologie
69 Heidelberg, Federal Republic of Germany

In striated muscle the sarcoplasmic membranes form a complicated network of tubules and cisternae around the myofibrils, covering a surface of approximately 3000–10,000 cm^2 per cm^3 of muscle.[1] Therefore the sarcoplasmic membranes can be separated from the contractile structures only after complete fragmentation. The small fragments heal over and perfectly sealed vesicles are formed whereby the sarcoplasmic surface remains outside.[2] It is the perfect tightness of the vesicles that makes it possible to analyze the translocation of calcium and the accompanying phosphate transfer reactions. The following functional characteristics of sarcoplasmic calcium transport have been established (see Reference 3):

1. Calcium-dependent phosphate liberation from various phosphate donors and calcium uptake are stoichiometrically related.
2. In the presence of membrane-permeable calcium precipitating anions, which keep the intravesicular calcium concentration low, calcium and the respective anions are taken up in stoichiometric amounts.
3. In the absence of calcium-precipitating anions, the internal calcium concentration rises steeply and produces transinhibition of calcium translocation and ATP splitting.
4. The maximal concentration ratio between the internal and the external calcium concentration reaches approximately 30,000 when phosphate is used as calcium-precipitating anion. Establishment of this high concentration ratio requires an energy expenditure of at least 6500 cal $\cdot$ mol^{-1}.
5. Transfer of the terminal phosphate of ATP to the membranal protein is catalyzed by calcium ions in the external medium and is independent of the intravesicular calcium concentration. It occurs with a comparatively high rate of 10 μmol $\cdot$ mg^{-1} $\cdot$ min^{-1} at 4°C.
6. External calcium catalyzes a fast phosphate exchange between ATP and ADP.
7. During the initial phase of calcium uptake, ADP induces a fast calcium-calcium exchange.
8. Calcium release is activated by the splitting products of ATP hydrolysis.
9. The ADP- and phosphate-dependent calcium release gives rise to the formation of stoichiometric amounts of ATP. The essential features of the calcium and phosphate transfer in the reaction sequence are illustrated in FIGURE 1.

It is the inherent property of sarcoplasmic membranes to form closed

* This work was supported in part by the Deutsche Forschungsgemeinschaft.

† Present address: Department of Pharmacology, University of Vienna, Vienna, Austria (Max-Planck-Stipendiat).

‡ Present address: Iowa State University, Ames, Iowa, USA (Humboldt-Stipendiat).

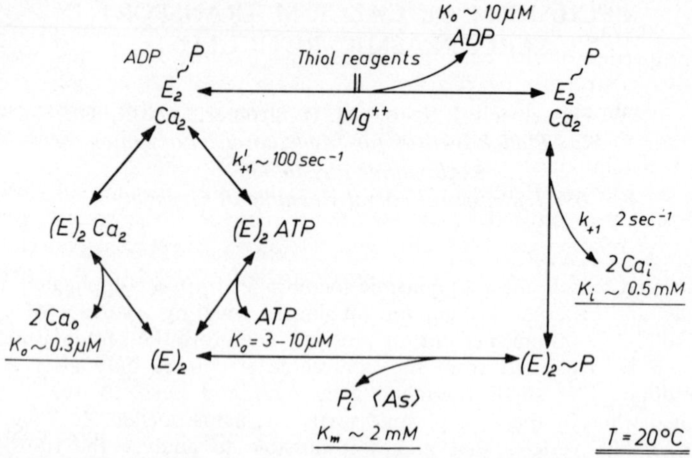

FIGURE 1. Reaction scheme of sarcoplasmic calcium transport. The figures in the scheme denote the approximate concentration of the respective reagents needed for half-saturation. k_{+1}' and k_{+1} are rate constants for phosphoprotein formation from the ATP protein complex and for the complete reaction cycle, respectively. The participation of magnesium ions is indicated only for the phosphorylation steps since its participation in dephosphorylation seems to be ambiguous. The liberation of calcium prior to that of phosphate is an essential feature of the scheme. The existence of an ADP-induced calcium-calcium exchange noted by Makinose[4] favors this order of the reaction sequence.

vesicles which provide the two-compartment system needed for the analysis of calcium translocation between the external solution and the vesicular space and the simultaneous phosphoryl transfer reaction. Although there is no compelling reason to assume that vesicle formation is directly related to the molecular events of calcium and phosphate transfer, the comparatively low calcium permeability of the sarcoplasmic membranes must be considered to be an inherent property of the calcium-translocating mechanism itself.

For instance, if the calcium ATPase molecule in the sarcoplasmic membrane would behave as a calcium carrier,[5] calcium stored inside the vesicles would rapidly escape even if movement of the carrier took place in a calcium-impermeable lipid layer. A free thermal motion of the calcium ATPase molecule in the sarcoplasmic membrane could accomplish a calcium efflux rate of 400 nmol $\cdot$ cm^{-2} $\cdot$ sec^{-1} at room temperature. At this temperature, however, the calcium efflux does not exceed 0.10 pmol $\cdot$ cm^{-2} $\cdot$ sec^{-1} (Reference 6) even if the internal calcium binding sites are completely saturated. One may object that phosphorylation of the transport protein, which occurs during calcium efflux, may prevent its motion in the lipid layer. However, even if phosphorylation of the transport protein is prevented by arsenate[7] or even if the phosphorylated protein is dephosphorylated as fast as possible by ADP,[8, 9] the release rate (3.0 pmol $\cdot$ cm^{-2} $\cdot$ sec^{-1}) never reaches the high value calculated for the freely moving calcium ATPase, nor does it reach the rate with which calcium is translocated from the external to the internal medium in the presence of ATP. Furthermore, it seems very unlikely that the calcium transporting molecule performs large amplitude movements. This notion is supported by the fact that neither the reaction of the ATPase with its antibody[10] nor selective cleavage by trypsin[11–14] interferes essentially with calcium transport. These properties of the calcium-translocating enzyme imply that very small changes in its structure can bring about calcium translocation across the sarcoplasmic membranes. The reassembly of the sarcoplasmic membranes from its partially dissociated components indicates that the ATPase molecule does not require additional components for calcium translocation.[15–18]

For the analysis of the activated calcium transport mechanism closed vesicles are mainly required if stoichiometric or energetic problems are considered, such as the relationship between calcium efflux, protein phosphorylation, or ATP formation. In contrast, the analysis of many aspects of the ATP-dependent forward reaction does not require the complete system, but rather requires decomposition of the system. In doing so, we have to presume that the phosphoryl transfer reactions are an inherent part of the calcium-translocating mechanism. Similar assumptions are tacit implications in transport chemistry.

It has long been known that activity of the calcium-dependent ATPase of the sarcoplasmic reticulum requires the presence of the lipid constituents of the membrane.[19, 20] Various procedures have tentatively been applied for the modification or the removal of these lipids with the aim of avoiding irreversible damage of the protein component.[21, 22] The lipid constituents are neither reduced nor selectively extracted by detergents like cholate or desoxycholate as long as they are applied in concentrations that do not produce irreversible inactivation.[23] Effective removal and modification of the lipids has been achieved by hydrolysis of the sarcoplasmic vesicles with phospholipase C or A in combination with removal of the degradation products.[22] The procedures yield highly calcium-permeable vesicular protein ghosts containing small amounts of the residual lipids like lysocephalin, lysolecithin, and monoglycerides. The diameter of these spherical ghosts is definitely smaller than that of the native vesicles and their size is relatively uniform[24] (FIGURE 2). In contrast to the case of native vesicles, the protein shell of the membrane ghosts has lost its ability to melt and to fuse, whereas native vesicles display fusing and melting of their membranes under a variety of conditions and form sheet-like streaks of lamellae (FIGURE 3a and b). These ordered areas may have given rise to the x-ray diffraction pattern observed by Dupont *et al.*[25]

The ghost protein of the vesicles is nearly completely deprived of all calcium-dependent phosphoryl transfer activities.[22, 26] However, it has retained its ability to bind calcium as well as ATP and ADP.[27, 28] Obviously the membrane lipids are neither part of nor do they directly influence the static structure of the calcium-translocating protein.

The lipid-deprived protein ghosts regain their ATP consuming power when they are supplemented by various lipid components. The most simple substituents are fatty acids provided that they fulfill the following requirements[22, 29]: (1) The fatty acids must be unsaturated. (2) The carbon chain of the fatty acids must at least be 14 carbon atoms long. (3) The double bonds should be in the middle of the chain. (4) The nonpolar end of the chain should not be longer than 8 carbon atoms (FIGURE 4).

According to this scheme, the most effective fatty acid is *cis*-vaccenate, while myristoleate produces only a barely detectable activation. *Cis* conformation of the fatty acid is not an essential requirement for activation; yet activation by the *trans* isomer of oleic acid—elaidinic acid—needs higher tempera-

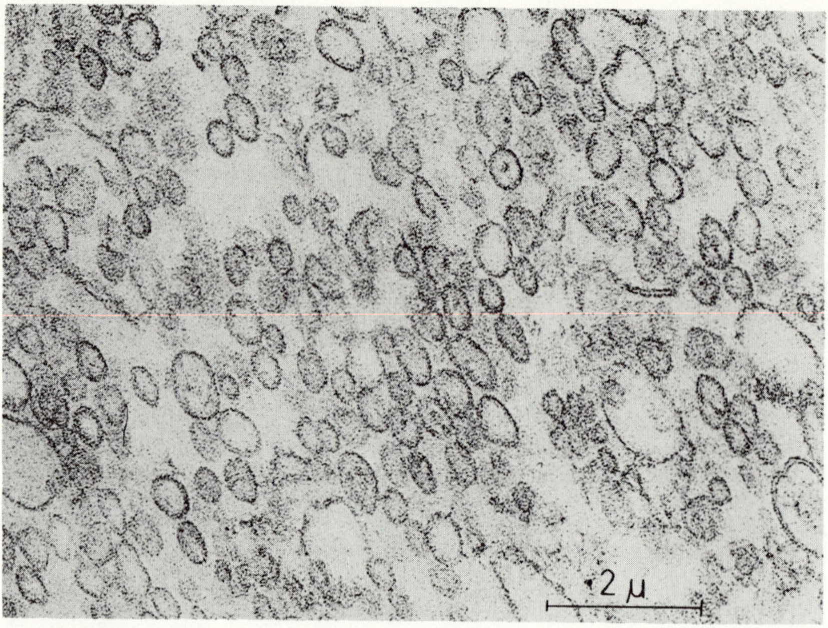

FIGURE 2. Electronmicroscopic picture of lipid-deprived vesicles. The vesicles were digested for one hour with 0.05 mg phospholipase A mg^{-1} protein, washing with serum albumin 50 $mg \cdot ml^{-1}$. Vestopal section from glutaraldehyde-fixed vesicles, lead citrate stain. Membranes of several vesicles are clearly asymmetric. Broken membrane fragments can be observed. The vesicles have become smaller and more uniform in diameter after treatment with phospholipase A and albumin. The residual lipids are mainly lysocephalin and, to a smaller extent lysolecithin. Reactivation of the calcium-dependent ATPase by oleate does not result in a displacement of the *lyso* compounds. (From Agostini and Hasselbach.[24] By permission of the publisher.)

tures, corresponding to an activation energy of approximately 48,000 $cal \cdot mol^{-1}$, which is twofold higher than that observed for preparations activated with oleate (FIGURE 5). Transition phenomena as observed with *cis*-vaccenate seem to be an exception. For none of the other acids tested have such breaks in the Arrhenius plot been observed. The calcium-dependent ATPase nearly regains the activity of the native preparations when supplemented with fatty acids and displays the same calcium dependence. However, the phosphoryl transfer reactions are not completely restored. While partial restoration of E-P formation can be achieved, the phosphoryl transfer from ATP to ADP cannot be restituted. Obviously, the intermediate states of the phosphoryl transfer reactions have become extremely labile so that they cannot be isolated nor can they give rise to the phosphate exchange reaction[30] (FIGURE 6).

Similarly incomplete is the restitution when delipidated preparations are reactivated by the addition of lysolecithin. Specific requirements which the lysolecithins have to meet are not known. Obviously, reconstitution of the calcium-dependent ATPase with unsaturated fatty acids or with lysolecithin

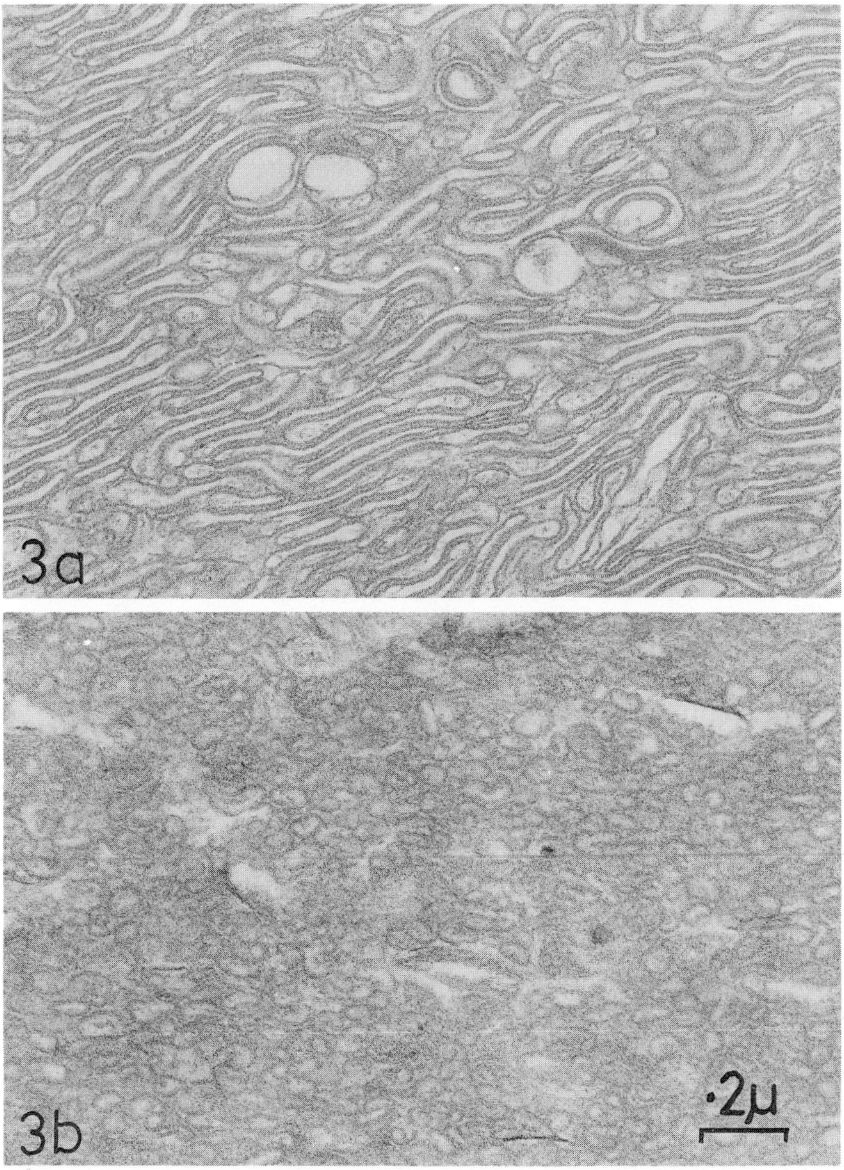

FIGURE 3. Electronmicroscopic picture of dehydrated native and lipid-deprived vesicles (a) Flattened sheet-like lamellae formed by exposing normal vesicles to high glycerol concentrations in a thin pellet. 60–70% gycerol, fixed in Karnovsky's fixative and embedded in epon-araldite. (b) Lipid-deprived vesicles treated under conditions identical to those in (a) do not melt or fuse.

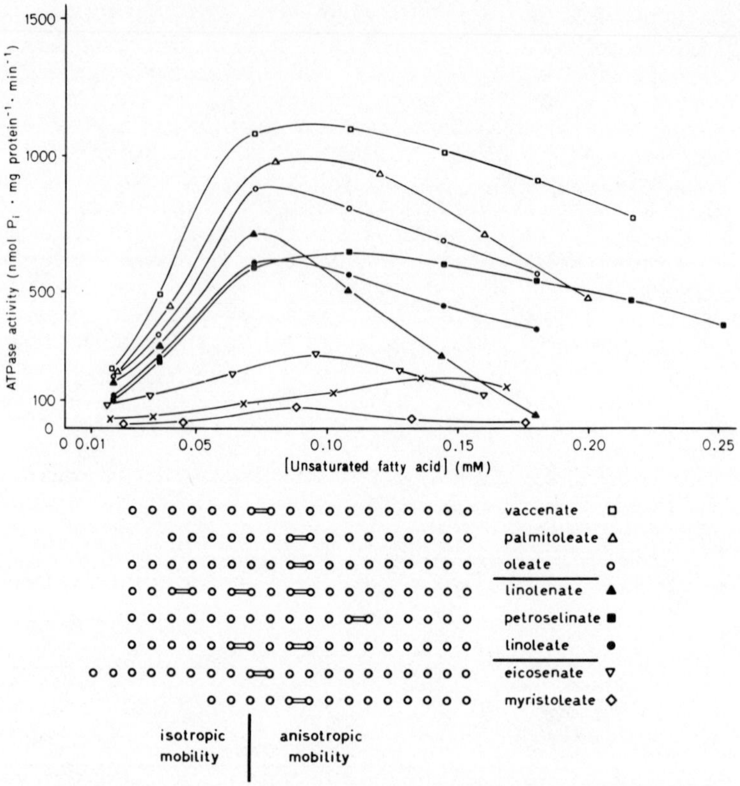

FIGURE 4a. Reactivation of lipid-deprived sarcoplasmic vesicles by various unsaturated fatty acids. Activity of the calcium-dependent ATPase has been measured at 20° C; pH 7.0; I = 0.1. Calcium 0.4 mM, 0.5 mM EGTA, 5 mg MgATP. (×) Ricinoleate was used as activating fatty acid; (○) the carbon chain proceeds from the left to the right. (From The and Hasselbach.[29] By permission of the *European Journal of Biochemistry*.)

results in degenerated enzyme species. To produce this kind of degeneration it is not necessary to remove the original lipid phase and to replace it by artificial lipids. Preparations whose phospholipids have been hydrolyzed with phospholipase A, but not removed from the protein matrix exhibit the same signs of degeneration. Even slighter alterations of the lipid-protein interaction in the membrane give rise to such changes of the activity pattern. For instance, treatment of the vesicles with low concentrations of halothane or ether, which do not remove essential lipids from the membranes, severely affects the ADP-ATP exchange reaction and simultaneously induces a distant rise in the rate of ATP hydrolysis[30, 31] (FIGURE 7). Similar changes occur after the vesicles have been treated with cholate or desoxycholate.

The most drastic changes of the energy transducing system, however, are produced by Triton X-100.[32, 33] Unlike treatment with cholate or desoxycholate, which does not remove the natural lipids, Triton treatment leads to a complete exchange of the natural lipids by Triton itself. Properties of the preparation depend on the conditions under which the lipid exchange takes place. Two different preparations can be obtained. If lipid replacement is performed in the presence of calcium, the preparation displays the calcium-dependent ATPase and, like the preparation mentioned above, it can neither be phosphorylated nor does it catalyze the phosphate exchange between ATP and ADP. In contrast, when the replacement is performed in the absence of calcium ions, completely inactive preparations are obtained. They can be recognized by their retarded elution pattern from tritonized DEAE columns, indicating a slightly changed protein structure produced by the exposure of a few acidic groups (FIGURE 8).

These preparations regain ATPase activity when they are supplemented with anionic lipids. In this case, even dodecylsulfate and phosphatidinic acid are effective activators. However, the resulting ATPase does not depend any longer on calcium, and affinity of the preparations for ATP and ADP is reduced. Like the other preparations, it is neither phosphorylated nor does it catalyze the phosphate exchange between ATP and ADP.

Obviously, the nature of the charge on the lipids is not essential for the described change in the activity pattern since it is produced not only by fatty acids, but also by lysolecithin and by Triton X-100 and to a smaller extent by oleylamine. These observations support the assumption that reactiva-

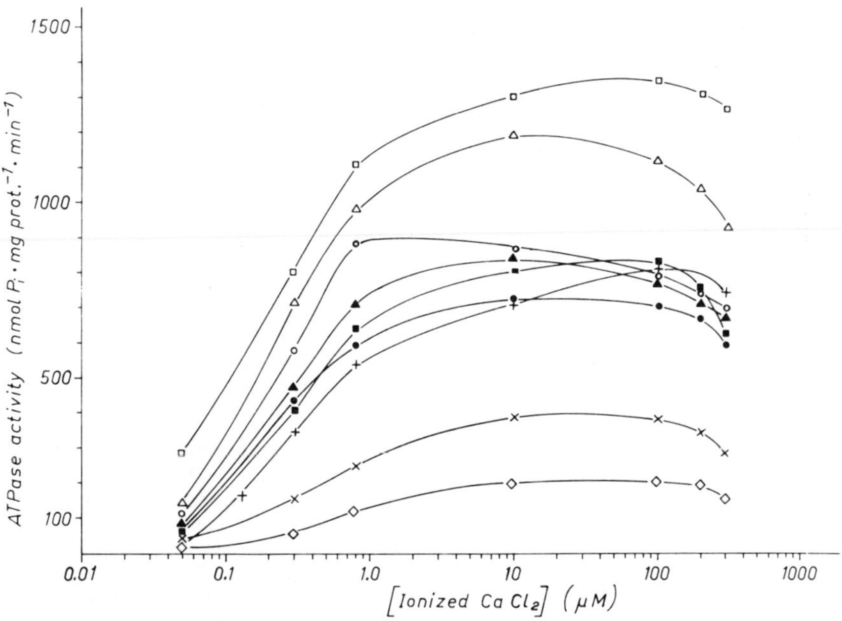

FIGURE 4b. Calcium requirement of the reactivated sarcoplasmic ATPase. At low concentrations of ionized calcium (<10 μM), calcium-EGTA buffers were used. Optimal concentrations of fatty acids were used. Other conditions as described in FIGURE 3a; symbols for the fatty acids as in FIGURE 4a. (From The and Hasselbach.[29] By permission of the *European Journal of Biochemistry*.)

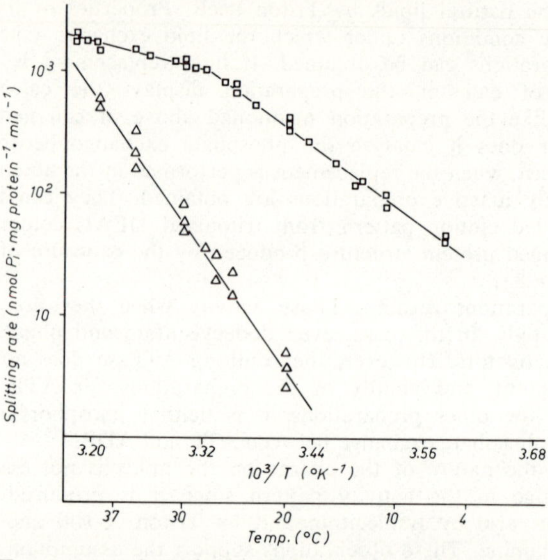

FIGURE 5. Temperature dependence of the reactivated sarcoplasmic ATPase. Activity was measured in the presence of 0.4 mM $CaCl_2$ and 0.5 mM EGTA; pH 7.0; I = 0.1; 5 mM MgATP. Reactivation with (□) cis-vaccenate 0.2 mg/mg protein and (△) elaidate 0.2 mg/mg protein.

tion is mainly based on hydrophobic interactions between superficial protein areas and the various lipid components. These interactions, however, must result in a complex allowing the lipids to move against each other. The importance of this flexibility of the complex is stressed by the finding that only those lipoprotein complexes in which spin-labeled fatty acids remain mobile or become mobile are enzymatically active.[34] Obviously, full and complete reactivation is based on a very delicate interaction between the hydrophobic protein and the hydrophobic region of the reactivating lipids. The charge on the hydrophilic residue of the agents contributes to this delicate balance by influencing micelle formation.[35]

As to the mechanism of this kind of activity modulation produced by lipid modifications, we are not able to attribute it to a distinct step in the

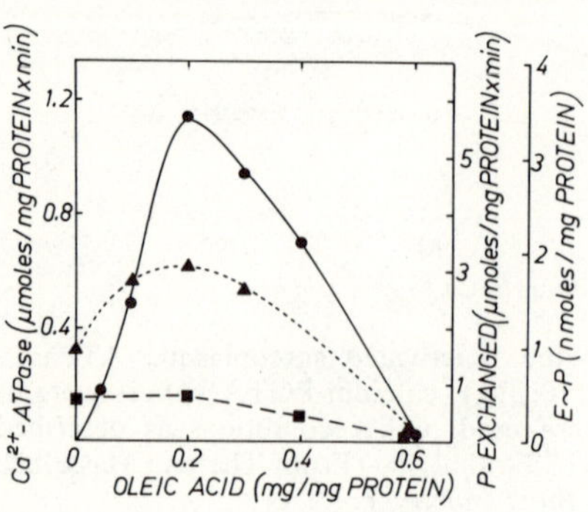

FIGURE 6. Reactivation of calcium-activated ATPase, phosphoprotein formation and ATP-ADP exchange by oleate. (●) ATPase activity; (▲) phosphoprotein formation; (■) ATP-ADP exchange. Note that the exchange reaction is not activated by oleate. pH 7.0; I = 0.1; 0.1 mM calcium; 7 mM Mg; 5 mM ATP. The exchange assay contained additionally 2 mM ^{32}P-ADP. The protein was phosphorylated with 5 mM ^{32}P-ATP.

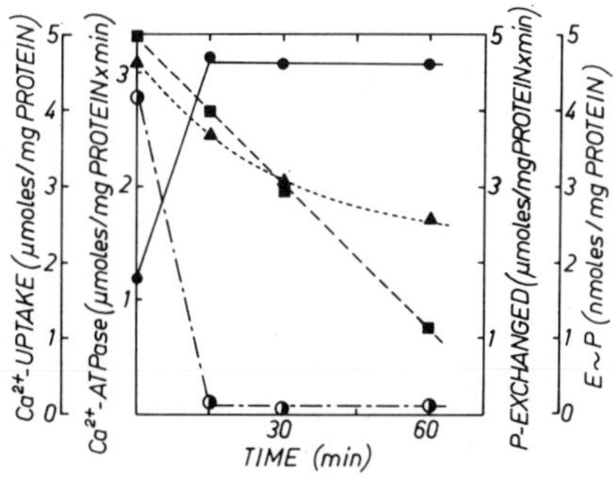

FIGURE 7. The effect of ether on calcium uptake, ATPase activity, phosphoprotein formation, and ATP-ADP exchange. T = 20° C; I = 0.1; pH 7.0; 5 mM ATP. The sarcoplasmic vesicles were treated with ether 7 v/v% for the time indicated at the abscissa. Protein concentration 1 mg·ml^{-1}. Subsequently the different activities were determined.[30] (●) ATPase activity; (▲) phosphoprotein formation; (■) ATP-ADP exchange; (◑) calcium uptake.

reaction chain. However, we can exclude a number of possibilities. For this purpose, properties of the various preparations are compared in TABLE 1. The results in the Table clearly demonstrate that functional degeneration occurs gradually and that the preparations can be separated into three groups.

Preparations whose lipid composition have been modified only mildly exhibit a high calcium-dependent ATPase activity at a significantly reduced exchange activity. Phosphoprotein formation is not or is only slightly affected. Activation of the calcium-dependent ATPase is especially pronounced for substrates like acetylphosphate.[36]

Protein ghosts whose lipid phase has been exchanged just reach or remain below the normal activity of the sarcoplasmic membranes at a considerably reduced exchange activity. The most highly degenerated species are preparations delipidated with Triton X-100 in the absence of calcium. No further consideration should be given to this enzyme species in its final state of degeneration

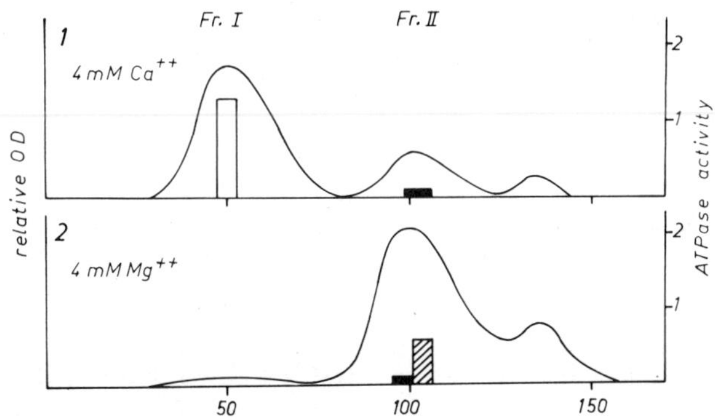

FIGURE 8. Elution pattern of Triton X-100-solubilized sarcoplasmic protein. The protein was solubilized by addition of a fivefold excess of Triton X-100. The DEAE column (2.5 × 50) was equilibrated with 0.2% Triton X-100; 5% glycerol; 50 mM Tris HCl; pH 8.1. Fraction I was eluted with the basic eluant. For the elution of fraction II, a linear NaCl gradient 0.0–0.3 M was superimposed. The proteins of fraction I and II move identically in SDS gel electrophoresis. (1) 4 mM $CaCl_2$ was present in all solutions. (2) 4 mM $MgCl_2$ was present in all solutions. (□) calcium-dependent ATPase activity; (▨) calcium-independent ATPase reactivated by oleic acid; (■) no ATPase activity.

TABLE 1

CHANGES OF CHARACTERISTIC PROPERTIES OF THE SARCOPLASMIC CALCIUM TRANSPORT ATPASE PRODUCED BY LIPID MODIFICATION

	Native Vesicles	Bile-Salt-Treated Vesicles	Ether-Treated Vesicles	Phospholipase-A-Treated Vesicles	Oleate-reactivated Vesicles	Triton (+Ca) Vesicles	Triton (–Ca) + Fatty Acids	Enzymatic Delipidated Vesicles
Ca ATPase (μmol · mg^{-1} · min^{-1})	0.7–1.0	2.0–3.0	2.0–3.0	2.0	0.9–1.2	0.8–3.0	0.8–1.5	0
Phosphoprotein (nmol · mg^{-1})	3–6	6	3	1.5–2.0	2.5	0	0	0.3–1.0
ATP–ADP exchange (μmol · mg^{-1} · min^{-1})	3.5	1.7	1.7	0.8	0.4	0	0	0.3
Ca affinity (μM)	0.3–0.5	0.3	0.3–0.5	0.3	0.3	0.15–0.30	0	0.3
ATP affinity (μM)	5–20	5	10–20	10	10	6	250	5
Authors	Hasselbach & Makinose[46,47] (1963, 1965); Makinose[48,49] (1966, 1969); Yamamoto *et al.*[50] (1967); Fiehn & Migala[27] (1970); Inesi *et al.*[31] (1967)	Meissner[44] (1973); Meissner & Fleischer[17] (1973); Hasselbach[41] (1975)	Inesi *et al.*[31] (1967); Suko *et al.*[32] (1975)	Fiehn & Hasselbach[23] (1970); Balzer *et al.*[45] (1968); The & Hasselbach[39] (1975)	The & Hasselbach[37] (1972, 1973)	Walter & Hasselbach[32] (1973); Hasselbach & Migala[33] (1972)	Walter & Hasselbach[32] (1973)	The & Hasselbach[37] (1972); Fiehn & Migala[27] (1970); The[38] (1975)

where only the hydrolytic mechanism of the phosphoryl transfer reaction has been retained.

If these initial states of transformation are considered, properties that reside in the protein moiety itself, such as ATP and calcium binding, are unaffected after reactivation. TABLE 1 shows that the calcium affinity found for native vesicles and lipid-deprived vesicles in binding studies coincides with the affinity found in splitting experiments with native and reactivated preparations.

Although the calcium affinity seems not to depend on the composition of the lipid phase, calcium-dependent properties of the enzyme emerge that are not present in the native form. In modified preparations, univalent cations can effectively compete directly or indirectly for calcium. Enzymatic activity is influenced in a complex manner by these ions.[37] For instance, potassium ions inhibit the ATPase at low calcium levels, where calcium ions produce activation; at high calcium levels, where the ATPase is inhibited by calcium, potassium ions produce activation (TABLE 2). The fact that the ion sensitivity is only slightly dependent on the charge of the reactivating lipids is an additional argument for the preponderance of lipid-protein interaction for these effects. One may assume that this kind of lipid modification makes anion binding sites accessible that are buried in the more rigid structure of the native preparation for direct or indirect competition with calcium ions.

Like the affinity of the lipoprotein system for calcium ions, ATP affinity is a property of the protein component alone. Binding studies with delipidated membranes revealed affinities similar or equal to those reported for native preparations.[27, 38] These affinities are in good agreement with those obtained in kinetic studies of intact and lipid-modified preparations. The insensitivity of ATP or ADP binding to lipid modification indicates that the nucleotide binding sites are presumably not in direct contact with the lipid phase of the membrane. This notion is supported by our recent finding concerning the severe but reversible inhibition of calcium transport and of calcium-dependent ATPase activity by low concentrations of chaotropic anions.[39] These anions do not measurably interfere with the cohesive forces between lipid and lipid or lipid and protein in the preparations. Furthermore, native and modified preparations are equally affected. The inhibition produced by chaotropic anions

TABLE 2

ACTIVATION AND INHIBITION OF THE CALCIUM-DEPENDENT ATPASE OF NATIVE AND MODIFIED SARCOPLASMIC VESICLES BY POTASSIUM CHLORIDE AT LOW AND HIGH CALCIUM CONCENTRATIONS*

Conditions		Native Vesicles (μmol · mg^{-1} · min^{-1})	Oleate-reactivated (μmol · mg^{-1} · min^{-1})	Triton-reactivated (μmol · mg^{-1} · min^{-1})	Reconstituted Vesicles (F) (μmol · mg^{-1} · min^{-1})
Ca^{++}	KCl				
0.2 mM + 0.4 mM EGTA	0.1 M	0.65	0.90	0.20	0.85
	0.5 M	0.50	0.50	0.05	0.30
0.2 mM	0.1 M	0.50	0.60	0.70	0.90
	0.5 M	0.35	1.00	0.80	0.85

* T = 20°C (1–7 min). Other conditions are as described in FIGURE 1.

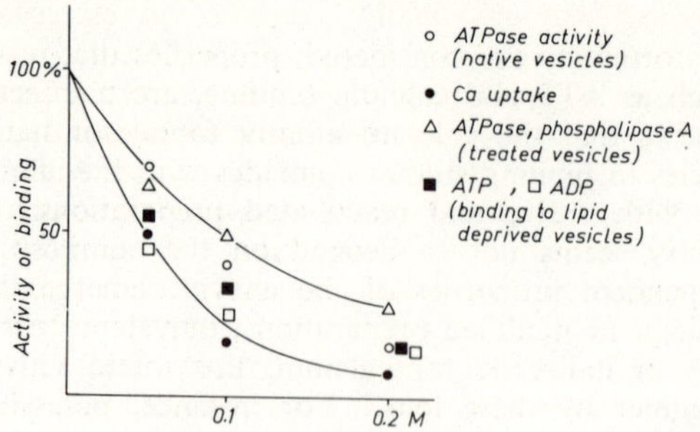

FIGURE 9. Inhibition of calcium uptake, calcium-dependent ATPase and of the binding of ATP and ADP to the sarcoplasmic membranes by chaotropic anions. Perchlorate and trichloracetate were used in these experiments (abscissa). ATP and ADP binding were measured with lipid-deprived preparations. Control binding values at concentrations of 10 μM of ATP or ADP; 1.75 nmol·mg^{-1} for ATP and 1.0 nmol·mg^{-1} for ADP.

can be explained by their competition for ATP and ADP in the protein matrix (FIGURE 9). The complicated activity profile of the calcium-activated ATPase, however, makes it difficult to quantify the apparent affinity reduction of ATP or ADP produced by the chaotropic agent and to distinguish between direct and indirect competition (FIGURE 10). One possibility is to assume that due to their low degree of solvation large anions penetrate into hydrophobic pockets of the molecule since such pockets have been identified as the most likely sites for nucleotide binding. This notion gets support from the fact that aldenylate kinase, like the transport ATPase, is severely inhibited by the chaotropic anions.[40]

Although the change of the affinity of the nucleotides for the protein seems not to be responsible for the degeneration of the transport ATPase, the considerable loss of activity in the Triton X-100 preparation indicates that ATP binding sites can be influenced by the lipid phase. That such an effect can be exerted by oleic acid must be deduced from the observation

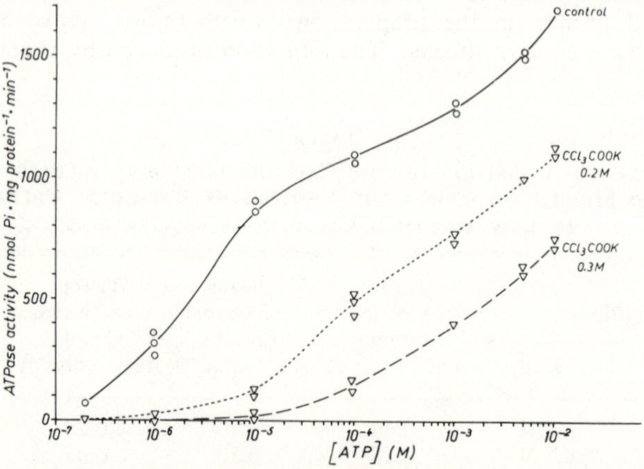

FIGURE 10. The calcium-dependent ATPase activity of phospholipase A digested sarcoplasmic vesicles at three trichloracetate concentrations. T = 25° C; pH 7.0; 5 mM Mg; 0.4 mM $CaCl_2$; 0.5 mM EGTA; 2.5 mM phosphoenolpyruvate and 50 μg·ml^{-1} pyruvate kinase.

that the protection of thiol groups of the transport protein by ATP in native and delipidated preparations is reduced and finally vanishes by increasing lipid replacement.[41]

Since neither nucleotide binding nor calcium binding are affected to such an extent as to explain the change in the activity pattern, the activation enthalpy of the different enzyme species should be considered as the last comparative criterion. The preparations represented by group one and two of our scheme are characterized by the same activation enthalpy as native preparations (TABLE 3). On the other hand, lipid exchange can produce changes in the activation enthalpy as demonstrated for elaidinic acid or for Triton X-100. The main characteristic of the modified preparations, especially that of the first group, is the considerable increase of the rate of hydrolysis connected with a decline of the exchange rate. In terms of the reaction rate theory, these changes imply that when membrane lipids have been modified, the formation of the reaction complex in the hydrolytic pathway is connected with a higher gain of entropy than the formation of the complex that leads to the transfer of phosphate to ADP. After affinity changes for the different substrates of the calcium transport system have been excluded as the cause of the changed enzyme activity pattern, our last statement concerning the change of activation entropy may give us a vague hint for further efforts to understand the interrelationship between phosphoryl transfer and calcium translocation.

In conclusion, the integrity of the membrane lipid phase of the sarcoplasmic vesicles is the essential prerequisite for the low calcium permeability and at the same time it is essential for the calcium-dependent phosphoryl transfer reaction. Even small alterations in the hydrophobic environment of the protein

TABLE 3

ACTIVATION ENTHALPY OF NATIVE AND MODIFIED SARCOPLASMIC PREPARATIONS FOR CALCIUM UPTAKE AND ATPASE ACTIVITY

Preparation	Activation Enthalpy (kcal/mol)	Authors
Native Vesicles		
Ca uptake	24–26	Inesi & Watanabe[51] (1967)
Extra ATPase	23–26	Hasselbach *et al.*[43] (1969)
Ether-treated		
Vesicles	23–28	Inesi & Watanabe[51] (1967)
ATPase		
Phospholipase-A-digested		
Ca-ATPase	30	The & Hasselbach[26] (1972)
Reactivated Ca-ATPase		
oleate	30	The & Hasselbach[26,29] (1972, 1973)
cis vaccenate	25 transition at 28°C to 8.5	
elaidate	48	
lysolecithine	29	

lead to a change of their reaction pattern with ATP and calcium. The exchange reaction proves to be the most sensitive step if it is inhibited at the expense of increased ATP hydrolysis. Nucletoide and calcium binding seem to be less vulnerable; they are affected only when, after severe structural alterations, the ATPase becomes calcium-insensitive.

References

1. PEACHEY, L. D. 1965. J. Cell. Biol. **25:** 209–231.
2. HASSELBACH, W. & L. G. ELFVIN. 1967. J. Ultrastruct. Res. **17:** 598–622.
3. HASSELBACH, W. 1974. Sarcoplasmic membrane ATPases. *In* The Enzymes, Vol. 10: 431–467. Academic Press. New York, N.Y.
4. MAKINOSE, M. 1973. FEBS Lett. **37:** 140–143.
5. SHAMOO, A. E. & D. H. MACLENNAN. 1974. Proc. Nat. Acad. Sci. U.S.A. **71:** 3522–3526.
6. HASSELBACH, W., W. FIEHN, M. MAKINOSE & A. J. MIGALA. 1969. Calcium fluxes across isolated sarcoplasmic membranes in the presence and absence of ATP. *In* The Molecular Basis of Membrane Function. Prentice Hall. Englewood Cliffs, N.J.
7. HASSELBACH, W., M. MAKINOSE & A. MIGALA. 1972. FEBS Lett. **20:** 311–1.
8. BARLOGIE, B., M. MAKINOSE & W. HASSELBACH. 1971. FEBS Lett. 267–268.
9. TEMPLE, D., W. HASSELBACH & M. MAKINOSE. 1974. Naunyn-Schmiedeberg's Arch. Pharmakol. **282:** 187–194.
10. MARTONOSI, A. & F. FORTIER. 1974. Biochem. Biophys. Res. Commun. **60:** 382–389.
11. MIGALA, A., B. AGOSTINI & W. HASSELBACH. 1973. Z. Naturforsch. **28:** 178–182.
12. THORLEY-LAWSON, D. A. & M. M. GREEN. 1973. Eur. J. Biochem. **40:** 403–413.
13. STEWART, P. S. & D. H. MACLENNAN. 1974. J. Biol. Chem. **249:** 985–993.
14. INESI, G. & D. SCALES. 1974. Biochemistry **13:** 3298–3306.
15. RACKER, E. 1972. J. Biol. Chem. **247:** 8198–8200.
16. RACKER, E. & E. EYTAN. 1975. Biochem. Biophys. Res. Commun. **55:** 174–178.
17. MEISSNER, G. & S. FLEISCHER. 1974. J. Biol. Chem. **249:** 302–309.
18. WARREN, G. B., P. A. TOON, N. J. BIRDSALL, A. E. LEE & J. C. METCALF. 1974. Proc. Nat. Acad. Sci. U.S.A. **71:** 622–626.
19. KIELLEY, W. W. & O. MEYERHOF. 1948. J. Biol. Chem. **176:** 591–601.
20. MARTONOSI, A. 1964. Fed. Proc. **23:** 913–921.
21. FLEISCHER, S. & B. FLEISCHER. 1967. Methods Enzymol. **10:** 406–433.
22. FIEHN, W. & W. HASSELBACH. 1970. Eur. J. Biochem. **13:** 510–518.
23. MACLENNAN, D. H., P. SEEMAN, G. H. ILES, C. C. YIP. 1971. J. Biochem. Chem. **246:** 2702–2710.
24. AGOSTINI, B. & W. HASSELBACH. 1971. Structural and cytochemical studies on fragmented sarcoplasmic reticulum following treatment with various lipolytic enzymes. *In* Basic Research in Myology: 226–234. Excerpta Medica. Amsterdam, The Netherlands.
25. DUPONT, Y., S. C. HARRISON & W. HASSELBACH. 1973. Nature **244:** 555–558.
26. THE, R. & W. HASSELBACH. 1972. Eur. J. Biochem. **28:** 357–363.
27. FIEHN, W. & A. MIGALA. 1971. Eur. J. Biochem. **20:** 245–248.
28. THE, R. 1975. Unpublished results.
29. THE, R. & W. HASSELBACH. 1973. Eur. J. Biochem. **39:** 63–68.
30. SUKO, J. & W. HASSELBACH. 1975. Unpublished results.
31. INESI, G., J. J. GOODMAN & S. WATANABE. 1967. J. Biol. Chem. **242:** 4637–4643.
32. WALTER, H. & W. HASSELBACH. 1973. Eur. J. Biochem. **36:** 110–119.
33. HASSELBACH, W. & A. MIGALA. 1972. FEBS Lett. **26:** 20–24.
34. SEELIG, J. & W. HASSELBACH. 1971. Eur. J. Biochem. **21:** 17–21.

35. Tanford, C. 1972. J. Mol. Biol. **67:** 59–74.
36. Friedman, Z. & M. Makinose. 1970. FEBS Lett. **11:** 69–72.
37. The, R. & W. Hasselbach. 1972. Eur. J. Biochem. **30:** 318–324.
38. The, R. 1975. Unpublished results.
39. The, R. & W. Hasselbach. 1975. In press.
40. Schulz, G. E., M. Elzinga, F. Marx & R. H. Schirmer. 1974. Nature **250:** 120–123.
41. Hasselbach, W. 1975. Unpublished results.
42. Inesi, G. & S. Watanabe. 1967. Arch. Biochem. Biophys. **121:** 665–671.
43. Hasselbach, W., M. Makinose, W. Fiehn. 1969. Activation and inhibition on calcium and cellular function. *In* Symposium on Calcium and Cellular Function: 75–84. Macmillan. London, England.
44. Meissner, G. 1973. Biochem. Biophys. Acta **298:** 906–926.
45. Balzer, H., M. Makinose, W. Fiehn & W. Hasselbach. 1968. Naunyn Schmiedeberg's Arch. Pharmak. Exp. Pathol. **260:** 456–473.
46. Hasselbach, W. & M. Makinose. 1963. Biochem. Z. **339:** 94–111.
47. Makinose, M. & W. Hasselbach. 1965. Biochem. Z. **343:** 360–382.
48. Makinose, M. 1966. Int. Congr. Biophys. (2nd): 276.
49. Makinose, M. 1969. Eur. J. Biochem. **10:** 74–82.
50. Yamamoto, T. & Y. Tonomura. 1967. J. Biochem. (Tokyo) **62:** 558–575.
51. Inesi, G. & S. Watanabe. 1967. Arch. Biochem. Biophys. **121:** 665–671.

Discussion

Dr. Sachs: If the calcium concentration gradient inside the vesicle is chemically or electrically coupled to ATP synthesis or to the movement of something like proton carried by oleic acid, then would you find some effect of protonophores like DNP, CCP, or S-13 on ATP synthesis?

Dr. Hasselbach: No; we did not find such an effect.

Dr. Sachs: The triton vesicle preparation which is active for calcium ATPase but makes no phosphoprotein: Would you comment on that as far as the phosphoprotein intermediate hypothesis is concerned.

Dr. Hasselbach: There might exist a phosphorylated intermediate. However, its steady-state concentration may be so low that it cannot be detected. That is one explanation.

ACTIVE TRANSPORT IN ISOLATED BACTERIAL MEMBRANE VESICLES: BINDING OF β-GALACTOSIDES TO THE *LAC* CARRIER PROTEIN

H. R. Kaback, G. Rudnick, S. Schuldiner, and S. A. Short

The Roche Institute of Molecular Biology
Nutley, New Jersey 07110

Bacterial cytoplasmic membrane vesicles provide a useful model system for the study of certain active transport systems.[1-4] These vesicles are devoid of the cytoplasmic constituents of the intact cell, and their metabolic activities are restricted to those provided by the enzymes of the membrane itself, constituting a considerable advantage over intact cells. Since transport by membrane vesicles per se is practically nil, the energy source for transport of a particular substrate can be determined by studying which compounds stimulate solute accumulation. In addition, metabolic conversion of the transport substrate and the energy source is minimal, allowing clear definition of the reactions involved.

Vesicles isolated from a variety of bacterial cells catalyze active transport of many different solutes by a respiration-dependent mechanism that does not involve the generation or utilization of ATP or other high-energy phosphate compounds.[1-7] In *Escherichia coli* and *Salmonella typhimurium* vesicles, most of these transport systems are coupled primarily to the oxidation of D-lactate or reduced phenazine methosulfate [PMS] (or pyocyanine) via a membrane-bound respiratory chain with oxygen or, under appropriate conditions,.[8-10] fumarate or nitrate as terminal electron-acceptors. On the other hand, in vesicles prepared from *Staphylococcus aureus,* active transport of amino acids is coupled exclusively to either L-lactate dehydrogenase or α-glycerol-P dehydrogenase depending upon the growth conditions of the parent cells.[11] In many cases the vesicles transport metabolites at rates that are comparable to those of the parent intact cells.[12, 13]

Energy-dependent Binding of β-Galactosides to the Lac *Carrier Protein*

Fluorescent β-galactoside [1′-(*N*-dansyl) amino-β-D-galactopyranoside (DG_0), 2′-(*N*-dansyl) aminoethyl-β-D-thiogalactopyranoside (DG_2), 2′-(*N*-dansyl) aminoethyl-β-D-galactopyranoside (oxy-DG_2), and 6′-(*N*-dansyl-amino-hexyl-β-D-thiogalactopyranoside (DG_6)] competitively inhibit lactose transport by membrane vesicles from *Escherichia coli* ML 308-225, but are not actively transported.[14-16] An increase in the fluorescence of these dansylgalactosides is observed upon addition of D-lactate, imposition of a membrane diffusion potential (exterior positive), or dilution-induced, carrier-mediated lactose efflux. The increase is not observed with 2′-(*N*-dansyl) aminoethyl-β-D-thio*gluco*pyranoside, nor with membrane vesicles lacking the β-galactoside transport system. Moreover, the D-lactate-induced fluorescence increase is blocked and/or rapidly reversed by addition of β-galactosides, sulfhydryl reagents, certain inhibitors of D-lactate oxidation, or uncoupling agents. The fluorescence increase exhibits an emission maximum at 500 nm and excitation maxima at 345 nm and at 292 nm. The latter excitation maximum is absent unless D-lactate is added, indicating that the bound dansylgalactoside molecules are excited by energy transfer from the membrane proteins. Titration of vesicles with dansylgalactosides in the presence of D-lactate demonstrates that the *lac* carrier protein constitutes 3 to 6% of the total membrane protein, and that the affinity of the carrier for substrate is directly related to the length of the alkyl chain between the galactosidic and the dansyl moieties of the dansylgalactosides. In addition, there is excellent agreement between the affinity constants of the various dansylgalactosides as determined by fluorimetric titration and their apparent K_i's for lactose transport (K_D's and/or apparent K_i's are approximately 550 μM, 30 μM, 40 μM, and 5 μM for DG_0, DG_2, oxy-DG_2, and DG_6, respectively).

Anisotropy of fluorescence measurements with 2′-(*N*-dansyl)aminoethyl-β-D-thiogalactopyranoside (DG_2) and 6′-(*N*-dansyl)aminohexyl-β-D-thiogalactopyranoside (DG_6) demonstrates a dramatic increase in polarization on addition of D-lactate, which is reversed by anoxia or addition of lactose.[15] These findings provide strong evidence for the contention that the fluorescence changes observed on "energization" of the membrane are due to binding of the dansylgalactosides per se, rather than binding followed by transfer into the hydrophobic interior of the membrane.

Fluorescence lifetime studies corroborate the results discussed above.[17] In the absence of D-lactate, DG_2 fluorescence exhibits a half-life of 3 nanoseconds. On addition of D-lactate to membrane vesicles containing the *lac* transport system, a percentage of the DG_2 molecule exhibits a half-life of 18 nanoseconds. Using the experimental values for anisotropy and excited-state lifetime derived from the above experiments, it can be calculated that the rotational relaxation time (ρ) of DG_2 is increased by approximately a factor of 200 when it is bound to the *lac* carrier protein.

Finally, it should be emphasized that Schuldiner *et al.*[18] have recently corroborated much of the fluorescence data both qualitatively and quantitatively by direct binding measurements using flow dialysis and high specific activity [^{3}H]DG_6.

The results are consistent with the suggestion that the *lac* carrier protein is inaccessible to the external medium unless energy is provided, and that energy is coupled to one of the initial steps in transport. It is proposed furthermore that at least one aspect of "energization" may be the generation of a membrane potential (exterior positive) resulting in increased accessibility of the *lac* carrier protein to the external solvent.

The strength of many of the conclusions discussed above rests heavily on observations that indicate that although the dansylgalactosides bind specifically to the *lac* carrier protein, they are not actively transported. Recent experimental evidence[16] provides definitive evidence that these compounds are not transported to any significant extent by either whole cells or isolated membrane vesicles: (i) Although DG_2 and oxy-DG_2 are competitive inhibitors of lactose transport in intact cells of *E. coli* and induce the synthesis of β-galactosidase *in vitro,* the compounds do not induce β-galactosidase *in vivo.* (ii) *p*-Chloromercuribenzenesulfonate (*p*-CMBS) does not cause efflux of lactose from the intravesicular pool, but causes rapid reversal of D-lactate-induced dansylgalactoside fluorescence. This difference between the effects of *p*-CMBS on β-galactoside transport and on the fluorescence behavior of the dansylgalactosides indicates that the fluorescence changes do not reflect events that occur subsequent to transport of the fluorescent galactosides into the vesicles. (iii) Dansylgalactosides inhibit dilution-induced, carrier-mediated efflux of lactose from the intravesicular pool. The results are consistent with the interpretation that the dansylgalactosides inhibit lactose efflux by preventing the unloaded carrier from returning to the inner surface of the membrane.

Studies with photoreactive β-galactosides, 2-nitro-4-azidophenyl-β-D-thiogalactopyranoside (APG_0)[19] and 2′-(2-nitro-4-azidophenyl)aminoethyl-β-D-thiogalactopyranoside (APG_2)[20] provide independent support for the conclusions derived from the dansylgalactoside experiments. The rationale behind the use of this class of compounds is that irradiation with visible light causes photolysis of the azido group to form a highly reactive nitrene which then reacts covalently with the macromolecule to which the azido-containing ligand is bound.

APG_0 is competitive inhibitor of lactose transport in membrane vesicles isolated from *E. coli* ML 308-225, exhibiting an apparent K_i of 75 μM. The initial rate and steady-state level of [^{3}H]APG_0 accumulation are markedly stimulated by the addition of D-lactate to vesicles containing the *lac* transport system, and kinetic studies reveal an apparent K_m of 75 μM. Membrane vesicles devoid of the *lac* transport system do not take up significant amounts of APG_0 in the presence or absence of D-lactate. When exposed to visible light in the presence of D-lactate, APG_0 irreversibly inactivates the *lac* transport system. Strikingly, photolytic inactivation is not observed in the absence of D-lactate. Kinetic studies of the inactivation process yield a K_D of 77 μM. Since lactose protects against photolytic inactivation and APG_0 does not inactivate amino acid transport, it is apparent that these effects are specific for the *lac* transport system. Analogous studies carried out with APG_2 demonstrate that this compound behaves similarly with respect to photoinactivation, but is not transported by the vesicles (like DG_2), and has a higher affinity for the *lac* carrier protein (i.e., the K_i for competitive inhibition of lactose transport and the K_D for photolytic inactivation in the presence of D-lactate are 35 μM). Moreover, it has been demonstrated that in addition to D-lactate oxidation, APG_2-dependent photolytic inactivation can also be induced by an artificially imposed membrane potential (exterior positive).

It seems clear from these studies with both dansyl- and azidophenylgalactosides that the *lac* carrier protein is not accessible from the outside surface of the vesicle membrane in the absence of energy-coupling. Moreover, all of the data are consistent with the notion that a membrane potential (exterior positive) causes the carrier to "move" to the outside of the membrane where it is able to bind ligand. Evidence in support of the hypothesis that D-lactate oxidation also generates such a membrane potential will be discussed in the *Mechanism of Active Transport* section. Clearly, in view of these findings, it is appropriate to postulate that the *lac* carrier protein itself, or a part of it, might carry a net negative charge.

Recently, it has been demonstrated[21, 22] that *p*-CMBS, a potent inhibitor of lactose transport which inhibits all carrier-mediated aspects of this transport system, does not block D-lactate-induced APG_2-dependent photoinactivation. Even more striking is the observation that *p*-CMBS induces APG_2 photoinactivation in the absence of D-lactate. The increase in susceptibility to APG_2 is reflected in the dissociation constant of APG_2 for *p*-CMBS- treated membranes, which is about 20 μM. This value is very similar to that determined for D-lactate-induced AGP_2-dependent photoinactivation. A reasonable interpretation of these results is that *p*-CMBS reacts with a sulfhydryl group which is exposed in the high-affinity form of the *lac* carrier, trapping the protein in that configuration and preventing substrate translocation. The covalent nature of this modification is demonstrated by the fact that the uncoupler carbonylcyanide *m*-chlorophenylhydrazone (CCCP) does not inhibit *p*-CMBS-induced APG_2 photoinactivation. Thus, a membrane potential is not required for the *p*-CMBS effect. It is also clear from the findings that *p*-CMBS does not block the binding site on the carrier since not only does APG_2 bind with high affinity, but also the *p*-CMBS-treated *lac* carrier protein is completely protected from APG_2 by lactose, thiodigalactoside, and mellibiose. Additional evidence that sulfhydryl groups are exposed upon conversion of the *lac* carrier protein to its high-affinity form is demonstrated by the observation that inactivation of lactose transport by *N*-ethylmaleimide is increased two- to fourfold in the presence of reduced phenazine methosulfate.[21, 22] The results as a whole indicate that in the presence of a membrane potential, exterior positive, a sulfhydryl group in the *lac* carrier protein that is not in the binding site becomes more accessible, and simultaneously, the carrier is converted into a form that has a high affinity for external substrate. Possibly, the sulfhydryl group in the *lac* carrier protein may exist in an ionized form in the hydrophobic milieu of the membrane and it may be this functional group in the protein that responds to the membrane potential.

Mechanism of Active Transport

There are certain fundamental observations that any proposed mechanism for energy-coupling in the vesicle system must take into account. (i) The apparent affinity of the carriers for substrate is much higher on the external surface of the membrane than on the internal surface, while the rate of "movement" from outside to inside and vice versa appears to be identical. (ii) There is no correlation between rates of oxidation of various electron donors and their relative effects on transport. Moreover, the cytochrome chain of *E. coli* vesicles is completely reduced by each substance that is oxidized by the vesicles, and essentially none of the vesicles in the preparations is inverted. (iii) Respiration-dependent transport is completely blocked by a variety of proton conductors, although these agents do not inhibit D-lactate oxidation. In a similar vein, mutants have been isolated that exhibit pleiotropic transport defects, and vesicles prepared from some of these mutants exhibit increased permeability to protons.[23, 24] (iv) Although all inhibitors of D-lactate oxidation block uptake, only those that act after the site of energy-coupling in the electron transfer chain cause efflux when added to preloaded vesicles. (v) Reduction of the respiratory chain from either side of the vesicle membrane is able to drive active transport. (vi) There is a large excess of carriers relative to D-LDH. (vii) Appropriate ion gradients result in binding of dansyl- and azidophenylgalactosides in the absence of D-lactate oxidation. Similarly, these gradients also result in uptake of lactose and amino acids.

An initial model proposed by Kaback and Barnes[25] depicts the carriers as electron-transfer intermediates in which a change from the oxidized to the reduced state results in translocation of the carrier-substrate complex to the inner surface of the membrane and a concomitant decrease in the affinity of the carrier for substrate. This model was intended merely to provide a working hypothesis that could account primarily for the observation that only certain electron-transfer inhibitors cause efflux of accumulated solutes. However, in view of a large number of observations that have appeared since the presentation of this model, it seems obvious that the model is insufficient to explain many of the other basic properties of the transport system.

A very different type of hypothesis, one that emphasizes the positioning of respiratory-chain components within the matrix of the membrane, was proposed by Mitchell.[26-30] As visualized by this chemiosmotic model, oxidation of electron donors is accompanied by expulsion of protons into the external medium, leading to a pH gradient and/or electrical potential across the membrane. This electrochemical gradient is postulated to be the driving force for inward movement of transport substrates by way of passive diffusion in the case of lipophilic cations such as dibenzyldimethylammonium ion,[31] via facilitated diffusion in the case of positvely charged substrates such as lysine or potassium ions or via coupled movement of protons with a neutral substrate such as lactose or proline (i.e., "symport"). In instances where sodium efflux is observed,[32] the chemiosmotic model invokes the concept of sodium-proton "antiport," which is postulated to catalyze electroneutral exchange of internal sodium with external protons and vice versa.[33] Moreover, the inhibitory effects of uncoupling agents on transport are attributed to the ability of these compounds to conduct protons across the membrane, thus short-circuiting the "proton-motive force" that drives transport.[34]

Recent evidence has provided direct support for the Mitchell hypothesis with regard to the vesicle system. During D-lactate or reduced PMS oxidation, lipophilic cations such as dimethyldibenzylammonium (in the presence of tetraphenylboron),[35-37] triphenylmethylphosphonium,[38] and safranine-0,[38] and rubidium (in the presence of valinomycin)[32] are accumulated. Moreover, there is a quantitative correlation between the steady-state level of accumulation of rubidium (in the presence of valinomycin), triphenylmethylphosphonium, and dibenzyldimethylammonium (in the presence of tetraphenylboron).[38] These observations are consistent with the interpretation of Harold and coworkers[35, 36] that D-lactate oxidation generates a membrane potential, interior negative.

Accumulation of these compounds is relatively specific for D-lactate or reduced phenazine methosulfate as electron donors. Since few, if any, of the vesicles in the preparations are inverted,[4] it is unlikely that the relative inability of other electron donors to drive uptake of these compounds can be attributed to an artifact of this nature. Thus, it appears likely that the machinery that is responsible for the generation of the membrane potential is located relatively specifically within the portion of the respiratory chain between D-LDH and cytochrome b_1.

Although all inhibitors of D-lactate oxidation block D-lactate-dependent solute accumulation, only those electron-transfer inhibitors that block the respiratory chain after the energy-coupling site cause efflux of accumulated solutes. These observations have been interpreted as being inconsistent with a chemiosmotic mechanism for active transport because it is not immediately obvious why only certain electron-transfer inhibitors should collapse the membrane potential and/or the pH gradient generated by D-lactate oxidation. Since precisely the same effects have been observed with a number of compounds that presumably equilibrate with the membrane potential and with D-lactate-induced proton gradients,[38] it seems reasonable to suggest that inhibition of electron flow in a manner that leads to reduction of the energy-coupling site results in dissipation of the membrane potential, while inhibition of electron flow in a manner that leads to oxidation of the energy-coupling site does not result in collapse of the potential. In other words, one explanation that would account for the data is that the membrane potential may be in equilibrium with the redox state of the respiratory chain at that site between D-LDH and cytochrome b_1 which generates the membrane potential. Assuming that the carriers are in equilibrium with a membrane potential, which in turn is in equilibrium with the redox potential of the energy-coupling site, a number of observations could be explained by a chemiosmotic mechanism.

Evidence is also available that provides a strong indication that the generation of a membrane potential, interior negative, is intimately related to the ability of the vesicles to catalyze active transport. As discussed above, binding of dansyl- and azidophenylgalactosides is induced by artificially imposed ion gradients that lead to a membrane potential, interior negative. Moreover, steady-state levels of lactose, proline, tyrosine, glutamic acid, and glycine accumulation in the presence of D-lactate and ascorbate-PMS are directly related to the magnitude of the membrane potential as determined by triphenylmethylphosphonium accumulation under identical conditions.[38] Finally, addition of lactose to membrane vesicles containing the β-galactoside transport system inhibits the uptake of both proline and triphenylmethylphosphonium.[38] A reasonable explanation for the last observation is that lactose transport partially dissipates the membrane potential in a manner similar to that described for glucose in *Neurospora crassa* by Slayman and Slayman,[39] who measured the potential directly. Despite the attractiveness of lactose/proton symport[40] as a mechanism for explaining these effects, however, all efforts to demonstrate this phenomenon directly in isolated membrane vesicles have been negative thus far.

Despite convincing evidence supporting a chemiosmotic mechanism for active transport in isolated membrane vesicles, a few inconsistencies remain that are not readily explained by the available data. According to the chemiosmotic theory, oxidation of any electron donor would presumably be associated with a unique value of the membrane potential, and this prediction is borne out by studies of triphenylmethylphosphonium and rubidium (in the presence of valinomycin) uptake.[32, 33] Accordingly, the order of effectiveness of the various electron donors in stimulating steady-state uptake would be expected to be the same for all respiration-linked transport systems, this order reflecting their relative effectiveness in generating a membrane potential. However, the order of effectiveness of various electron donors in stimulating uptake varies widely, depending on the transport system under study.[37] Moreover, in chemiosmotic coupling, the logarithm of the steady-state concentration ratio attained in the presence of a given electron donor relative to the logarithm of the ratio established in the presence of D-lactate would be expected to be constant for all the transport systems. These ratios, however, exhibit enormous variation among the different systems.

Membrane vesicles and ethylenediaminetetraacetic acid-treated cells catalyze lactose accumulation at least three to five times better than triphenylmethylphosphonium.[38] Based on a mechanism in which there is one positive charge (i.e., one proton) taken up per mole of lactose,[40] a potential of approximately —120 mV is required to achieve a lactose concentration gradient of 100. However, the vesicles generate a potential of only about −75 mV.[38] It seems apparent, therefore, that although the vesicles generate a membrane potential of the appropriate polarity, the magnitude of the potential is not sufficient to account for the phenomena observed. A similar situation has also been described in mitochondria[41] and in chloroplasts.[42, 43] In addition, triphenylmethylphosphonium accumulation in response to an artificially generated membrane potential approximates that observed during D-lactate or reduced PMS oxidation, while lactose accumulation in response to the artificially generated potential is considerably less than that observed with D-lactate or ascrobate-PMS.[38] These inconsistencies could be explained in a number of ways. First and most obvious, the accumulation of lipophilic cations or rubidium (in the presence of valinomycin) may not represent a truly quantitative estimate of the membrane potential. Second, there may be more than one positive charge (i.e., proton) taken up per mole of solute. Third, there may be another high-energy intermediate in addition to the membrane potential that is generated by D-lactate or reduced PMS oxidation, but not by an artificially imposed membrane potential. One such candidate would be the chemical gradient of protons; however, no significant pH gradient is observed when the vesicles are incubated in the presence of D-lactate or ascorbate-PMS and 5,5-dimethyloxazolidine-2,4-dione (DMO).[44] This observation is not surprising in view of the fact that the transport assays are carried out in highly buffered reaction mixtures. Hopefully, some of these problems will be resolved by more quantitative estimates of the membrane potential and by the development of techniques that allow an assessment of solute/proton symport in the vesicle system. In any case, it seems apparent that definitive answers regarding the intimate mechanism of active transport will have to await the isolation of functionally active carrier molecules.

References

1. KABACK, H. R. 1972. Biochim. Biophys. Acta **265:** 367–416.
2. KABACK, H. R. & J.-S. HONG. 1973. *In* CRC Critical Reviews in Microbiology. A. I. Laskin, Ed. Vol. 2:333–376. CRC Press. Ohio.
3. KABACK, H. R. 1973. *In* Bacterial Membranes and Walls. L. Leive, Ed. Vol. I.: 241 292, Marcel Dekker, Inc. New York, N.Y.
4. KABACK, H. R. 1974. Science **186:** 882–892.
5. PREZIOSO, G., J.-S. HONG, G. K. KABACK & H. R. KABACK. 1973. Arch. Biochem. Biophys. **154:** 575–582.
6. KONINGS, W. N. & FREESE. 1972. J. Biol. Chem. **247:** 2408–2418.
7. SHORT, S. A., D. C. WHITE & H. R. KABACK. 1972. J. Biol. Chem. **247:** 298–304.
8. KONINGS, W. N. & H. R. KABACK. 1973. Proc. Nat. Acad. Sci. U.S.A. **70:** 3376–3381.
9. BOONSTRA, J., M. T. HUTTUNEN, W. N. KONINGS & H. R. KABACK. 1975. J. Biol. Chem. In press.
10. KONINGS, W. N. & J. BOONSTRA. *In* Current Topics in Membranes and Transport. In press.
11. SHORT, S. A. & H. R. KABACK. 1974. J. Biol. Chem. **249:** 4275–4281.
12. LOMBARDI, F. J. & H. R. KABACK. 1972. J. Biol. Chem. **247:** 7844–7857.
13. SHORT, S. A., D. C. WHITE & H. R. KABACK. 1972. J. Biol. Chem. **247:** 7452–7458.
14. REEVES, J. P., E. SHECHTER, R. WEIL & H. R. KABACK. 1973. Proc. Nat. Acad. Sci. U.S.A. **70:** 2722–2726.
15. SCHULDINER, S., G. K. KERWAR, R. WEIL & H. R. KABACK. 1975. J. Biol. Chem. **250:** 1361–1370.

16. SCHULDINER, S., H. KUNG, R. WEIL & H. R. KABACK. 1975. J. Biol. Chem. **250:** 3679–3682.
17. SCHULDINER, S., R. D. SPENCER, G. WEBER & H. R. KABACK. J. Biol. Chem. In press.
18. SCHULDINER, S., R. WEIL & H. R. KABACK. Submitted for publication.
19. RUDNICK, G., R. WEIL & H. R. KABACK. 1975. J. Biol. Chem. **250:** 1371–1375.
20. RUDNICK, G., R. WEIL & H. R. KABACK. J. Biol. Chem. In press.
21. RUDNICK, G., R. WEIL & H. R. KABACK. 1975. Fed. Proc. **34:** 491.
22. RUDNICK, G., R. WEIL & H. R. KABACK. In preparation.
23. ROSEN, B. P. 1973. Biochem. Biophys. Res. Commun. **53:** 1289–1296.
24. ALTENDORF, K., F. M. HAROLD & R. D. SIMONI. 1974. J. Biol. Chem. **249:** 4587–4593.
25. KABACK, H. R. & E. M. BARNES, JR. 1971. J. Biol. Chem. **246:** 5523–5531.
26. MITCHELL, P. 1963. Biochem. Soc. Symp. **22:** 142.
27. MITCHELL, P. 1966. Biol. Rev. Cambridge Philos. Soc. **47:** 445–502.
28. MITCHELL, P. 1967. Fed. Proc. **26:** 1370.
29. MITCHELL, P. 1973. J. Bioenerg. **4:** 63–91.
30. HAROLD, F. M. 1972. Bacteriol. Rev. **36:** 172–230.
31. BAKEEVA, L. E., L. L. GRINIUS, A. A. JASAITIS, V. V. KULIENE, D. O. LEVITSKY, E. A. LIBERMAN, I. I. SEVERINA & V. P. SKULACHEV. 1970. Biochim. Biophys. Acta **216,** 13–21.
32. LOMBARDI, F. J., J. P. REEVES & H. R. KABACK, 1973. J. Biol. Chem. **248:** 3551–3565.
33. WEST, I. C. & P. MITCHELL. 1974. Biochem. J. **144:** 87–90.
34. MITCHELL, P. & J. MOYLE. 1967. Biochem. J. **104:** 588–600.
35. HIRATA, H., K. ALTENDORF, & F. M. HAROLD. 1973. Proc. Nat. Acad. Sci. U.S.A. **70:** 1804–1808.
36. ALTENDORF, K., H. HIRATA & F. M. HAROLD. 1975. J. Biol. Chem. **249:** 4587–4593.
37. LOMBARDI, F. J., J. P. REEVES, S. A. SHORT & H. R. KABACK. 1974. Ann. N.Y. Acad. Sci. **227:** 312–327.
38. SCHULDINER, S. & H. R. KABACK. 1975. Submitted for publication.
39. SLAYMAN, C. L. & C. W. SLAYMAN. 1974. Proc. Nat. Acad. Sci. U.S.A. **71:** 1935–1939.
40. WEST, I. C., & P. MITCHELL. 1973. Biochem. J. **132:** 587–592.
41. ROTTENBERG, H. 1970. Eur. J. Biochem. **15:** 22–28.
42. SCHULDINER, S., H. ROTTENBERG & M. AVRON. 1972. Eur. J. Biochem. **25:** 64–70.
43. ROTTENBERG, H., T. GRUNWALD & M. AVRON. 1972. Eur. J. Biochem. **25:** 54–63.
44. WADDEL, W. J. & T. C. BUTLER. 1959. J. Clin. Invest. **38:** 720–729.

BINDING PROTEINS AND MEMBRANE TRANSPORT*

Dale L. Oxender and Steven Quay

Department of Biological Chemistry
The University of Michigan
Ann Arbor, Michigan 48104

Bacterial cells have developed two basic types of active transport systems. One of these transport systems is a membrane-bound system and can be observed in membrane vesicle preparations prepared by the methods described in the previous presentation by Kaback.[1] Dr. Kaback described the transport properties of membrane vesicles from gram-negative bacteria that retain the membrane-bound transport system. He has shown that these membranes retain all components of the transport system except an energy source, which is necessary for the system to produce active transport. The membrane-bound systems derive their energy from an "energy-rich membrane state," which can be formed in a variety of ways such as respiration, ATP hydrolysis, or proton gradient-forming conditions.[3, 4]

A second type of active transport system that I would like to describe is a system that requires a shockable binding protein to produce active transport.[57–61] It appears that the binding protein transport systems obtain their cellular energy more directly from phosphate bound energy derived from glycolysis or oxidative phosphorylation.[2] These low molecular weight binding proteins can be removed from the bacterial cells by a cold osmotic shock treatment. In general osmotic shock treatment selectively destroys the type-two transport systems for which binding proteins can be isolated and has little effect on type-one transport systems in which the receptor site is tightly bound to the membrane.

The membrane of *E. coli* is a complicated triple-layer structure and the binding proteins appear to occupy the space between the outer and inner membranes of the cell envelope. The mild treatment of cold osmotic shock was worked out by Neu and Heppel.[5] The procedure is illustrated in FIGURE 1. In the first step cells are harvested in middle or late log phase and washed several times with Tris-HCl, pH 7. The washed cells are suspended in 40 to 80 volumes of room temperature 20% sucrose solution containing 0.1 mM EDTA. The cell suspension is stirred gently for 10 minutes and then centrifuged. The pellet is rapidly resuspended in 40 to 80 volumes of ice-cold distilled water containing 0.1 mM $MgCl_2$. The suspension is stirred for 10 minutes and centrifuged. The supernatant fluid is called the "shock fluid" and contains a number of hydrolytic enzymes as well as the binding proteins for various solutes. Usually about 4% of the total cellular protein is lost from the cells during this treatment. The procedure must be modified for certain strains that are more sensitive to osmotic shock treatment such as *Pseudomonas* species.

When we treated *E. coli* K12 by the osmotic shock treatment we found

* The work performed in our laboratory was supported in part by Public Health Service Grant GM11024 to D.L.O. from the National Institute of General Medical Sciences.

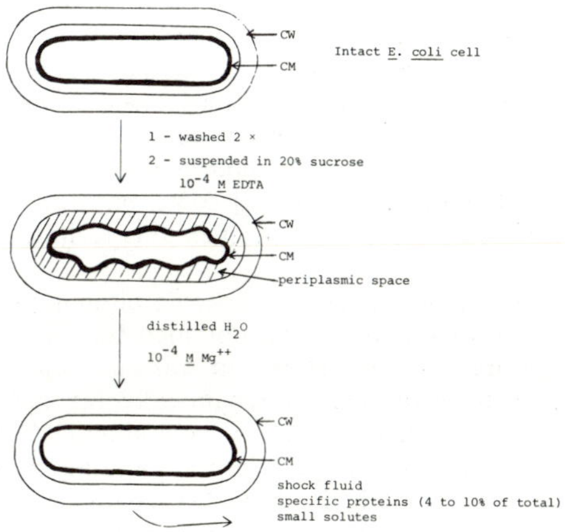

FIGURE 1. Schematic diagram of the process of osmotic shock and its effect on the bacterial cell.

that the transport activity for the branched-chain amino acids was greatly decreased. We examined the shock fluid and discovered leucine-binding activity by equilibrium dialysis.[6, 7]

Following concentration by ultrafiltration, the shock fluid was prepared for DEAE-cellulose by desalting on a Biogel-P10 column. The protein solution is applied to a DEAE-cellulose column equilibrated with 5 mM Tris-HCl, pH 7.6. The column was eluted with a linear gradient of sodium chloride. Three peaks with binding activity for branched-chain amino acids were found, corresponding to the LIV-binding protein, an isoleucine-preferring protein, and the leucine-specific protein.[8, 13] The elution pattern is illustrated in FIGURE 2. All three binding proteins crossreact with antibody prepared from the LIV-I binding protein, suggesting a considerable structural similarity. The bars in FIGURE 2 indicate crossreaction with anti-LIV-I antibody. The leucine-specific protein is purified to homogeneity by preparative polyacrylamide gel electrophoresis and crystallized from ammonium sulfate.[8] The isoleucine-preferring protein has not been extensively studied. The LIV-binding protein fraction from the DEAE-cellulose is purified with two steps, hydroxylapatite chromatography and isoelectric focusing. A final purification step can be achieved by

FIGURE 2. Elution profile of the DEAE-cellulose fractionation of the osmotic shock fluid. Crossreactive material is indicated by bars. (From Rahmanian *et al.*[13] By permission of the *Journal of Bacteriology*.)

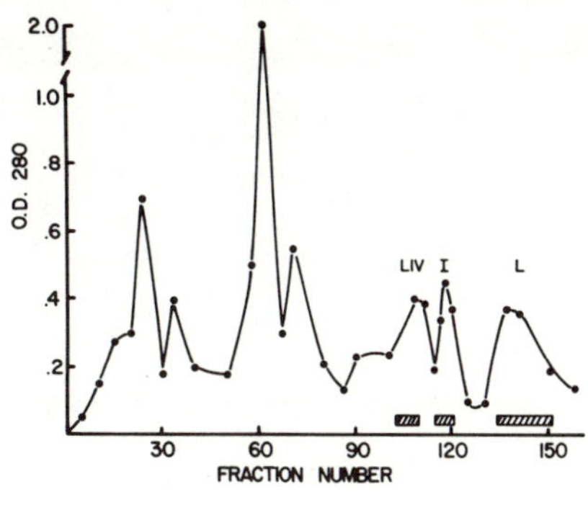

crystallization from 50–55% α-methyl-2,4-pentanediol.[7] Two crystal forms exist that depend on the pH at which crystallization takes place. At neutral pH values, needle-like crystals have been obtained,[7] while at pH 4 hexagonal columns occur (FIGURE 3). Initial x-ray diffraction studies have shown good resolution and suggest the possibility of doing structural studies by this technique. A large number of other binding proteins for amino acids, ions, sugars, and vitamins have been isolated by similar techniques from bacteria. TABLE 1 lists some of these binding proteins and some physical properties that have been measured. References for the purification technique have also been included.

As indicated in this Table, the leucine-specific binding protein has a molecular weight of 37,000,[8] very similar to that obtained for the LIV-binding protein.[8] The LIV-binding protein is very stable to various denaturation procedures. The active binding form is the thermodynamically favored conformation, and denaturation by heat treatment, urea, or guanidine-HCl is fully reversible. This is a general feature of the periplasmic binding proteins. The K_d for leucine is about 10^{-6} M for both the branched-chain amino acid binding proteins. The similarity of the two proteins is confirmed by the crossreactivity either protein exhibits when antibodies are prepared to leucine-specific or LIV-protein alone. Crossreactivity is also shown between the arabinose- and the galactose-binding proteins of *E. coli* B/r.[9, 10]

Specific antibodies have been used to localize the binding proteins to the cell envelope.[11, 12] In the study of Nakane *et al.*,[12] antibodies to the LIV-protein

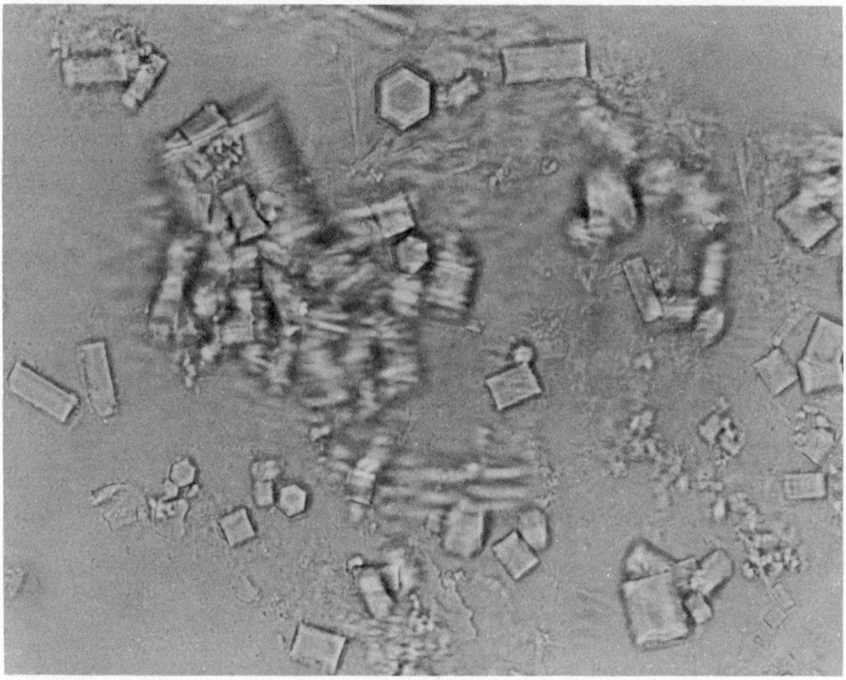

FIGURE 3. Crystals of the LIV-binding protein at pH 4.

TABLE 1
PHYSICAL PROPERTIES OF AMINO ACID-, SUGAR-, ION, AND VITAMIN-BINDING PROTEINS

Binding Protein	Organism	Molecular Weight	Cysteine Content	Dissociation Constant (10^{-6}M)	Specificity	References
Amino Acids						
Lysine, Arginine, Ornithine	*E. coli* K12	28,000	2	3, 1.5, 5	Specific	37
Arginine	*E. coli* K12	27,700	2	0.03	Specific	38
Histidine	*S. typhimurium* LT2	25,500	2	0.15	His > Arg > Lys > citrulline > azoserine	39
Cystine	*E. coli* W	27,000	3	0.01	Cystine > diaminopimelate	40
Glutamine	*E. coli* K12	27,000	0	0.3	Specific	41
Glutamate and Aspartate	*E. coli* K12	31,000	2	0.7, 1.2	Specific	42
Leucine, Isoleucine, Valine	*E. coli* K12	36,000	1	1	Leu = Ile > Val > Thr > Ala	7, 8
Leucine	*E. coli* 7	37,000	–	0.7	Leu > Trifluoroleucine	8, 13
Phenylalanine	*Comamonas*	–	–	0.2	–	43
Tryptophan	*Neurspora crassa*	–	–	80	Phe > Trp > Leu	44
Sugars						
Arabinose	*E. coli* B/r	39,000	2	0.3	Ara = Gal ≫ Glu	9
Galactose	*E. coli* K12	36,000	0	0.5 and 10	Gal = Glu > Ara	32, 45, 46
Ribose	*S. typhimurium* LT2	30,400	0	0.33	Specific	47
Maltose	*E. coli* K12	40,500	–	1.5 and 10	Maltose > several maltodextrins	48
Ions						
Calcium	Chick, intestinal mucosa	24,700	3	26	Calcium > strontium > barium	49, 50
Phosphate	*E. coli* K10	42,000	–	0.8	–	51
Sulfate	*S. typhimurium* LT2	32,000	0	20	Sulfate > chromate	52
Vitamins						
Thiamine	*E. coli* W	–	–	0.03	Thiamine > thiamine monophosphate > thiamine diphosphate	53, 54
Cyanocobalamine	*E. coli* B	22,000	–	0.006	–	55

were used in an enzyme-labeled technique. Late log phase cells were fixed in acetone, washed with phosphate buffer, and treated with rabbit antibody to the binding protein. The cells were then washed and allowed to react with sheep anti-rabbit globulin that had been conjugated to horseradish peroxidase. The fixed cells were stained cytochemically for peroxidase. The location of the insoluble product, which reacts with osmium tetraoxide, could be observed in the electron microscope. Using unimmunized rabbit serum as control, it was possible to localize the reaction to the cell wall or cell membrane. Using the fluorescent reagent, diazo-7-amino-1,3-naphthalene disulfonate, Pardee and Watanabe[11] obtained evidence that the sulfate-binding protein was located internal to the cell wall, but external to the plasma membrane. This reagent penetrates the cell wall but not the cellular membrane.

Multiplicity of Branched-Chain Amino Acid Systems

When the kinetics of leucine transport were examined over a wide concentration range the reciprocal plots of the initial rates of uptake were distinctly biphasic, indicating heterogeneity in leucine transport.[13] The multiplicity of branched-chain amino acid systems has been confirmed in other laboratories.[8, 13–15, 56] We have used a combination of kinetic, genetic, and biochemical studies to describe the number of transport systems that serve for leucine, isoleucine, and valine.

High-Affinity Uptake Systems

The high-affinity component of the kinetic plot is composed of at least three transport systems. The major high-affinity system is called the LIV-I system. This system has a broad specificity since it serves for leucine, isoleucine, valine, threonine, alanine, homoserine,[16] cysteine, serine, methionine, and a variety of analogs.[15, 17] Additional minor high-affinity systems are represented by a leucine-specific system (L-S system)[8, 13] and an isoleucine-preferring system (I-system).[18] These three systems each have approximately the same Km value for uptake of their preferred substrate (0.2 μM). Thus, they appear collectively as one component in the kinetic plots of uptake of the respective substrate. All three of these systems are repressible by leucine in the growth media and are sensitive to osmotic shock treatment. The three proteins (isolated as described and shown in FIGURE 2) have specificities that correspond to the three high-affinity transport systems described above. In wild type *E. coli* the LIV-I system constitutes about 75 to 85% of the high-affinity uptake of leucine and isoleucine, while the specific uptake systems constitute the balance of the high-affinity transport of these amino acids.

Low-Affinity Transport of Leucine, Isoleucine, and Valine

The low-affinity component of the kinetic plot of leucine uptake is referred to as the LIV-II transport system. This component is not sensitive to osmotic shock treatment and can be observed in shocked cells. It is not subject to repression by growth of the cells on leucine. Kinetic plots of leucine uptake in shocked cells or cells grown in the presence of leucine are essentially linear and provide a good measure of the LIV-II transport component. The Km values for leucine and isoleucine are from 2 to 4 micromolar.[13] The Km value for valine is around 10 to 20 micromolar.

In contrast to the LIV-I system, the LIV-II transport system is much more specific for leucine, isoleucine, and valine. Some analogs of the branched-chain amino acids such as norleucine[56] have low affinity for the LIV-II system. Since the LIV-II system is not sensitive to osmotic shock treatment, it can also be observed in membrane vesicle preparations of bacteria as described in the previous presentation by Kaback. In wild-type *E. coli* K12 the LIV-II system ordinarily constitutes about 10 to 15% of the uptake of 1×10^{-5} M leucine. At this concentration of leucine both high- and low-affinity transport systems would be saturated.

Regulation of Leucine Transport

We have previously shown that when cells are grown in nutrient medium or in media containing leucine the transport activity and binding activity is repressed.[7] The represson by leucine of both transport and shockable-binding activity is coordinate, supporting the view that the binding protein is involved in the rate-limiting step in the high-affinity leucine transport system (LIV-I).

Since the biosynthesis of leucine is also repressed by external leucine, we examined the possibility that biosynthesis and transport of leucine shared common regulatory components. We examined the transport activity in mutant strains with derepressed leucine biosynthetic enzymes obtained from Dr. Umbarger. We found that the transport was normal and still regulated by leucine in the growth media. In addition a mutant with derepressed leucine transport activity, strain E0312, showed normal regulation of the biosynthetic enzymes for leucine.[19] These data are summarized in TABLE 2.

TABLE 2

THE SEPARATION OF BRANCHED-CHAIN AMINO ACID TRANSPORT AND BIOSYNTHESIS

		Percent of Parent	
Strain	Phenotype	Biosynthetic Enzymes	LIV Transport System
CU5	wildtype	100	100
CU5001	derepressed *leuABCD, ilvB*	700*†	100†
CU5002	derepressed *leuABCD, ilvB, ilvADE*	439*†	119†
EO303	wildtype	100	100
EO312	derepressed branched-chain amino acid transport and binding proteins	79†	200‡

* Data are for the *ilvB* gene product, acetohydroxy acid synthetase. Similar results were obtained for the *leuB* gene product, isopropyl malate dehydrogenase.

† Data from Reference 19.

‡ Data from Reference 13.

Transport Mutants

Many of the mutant selection procedures are designed to use amino acid auxotrophs. The high-affinity transport systems that require binding proteins are greatly reduced in leucine auxotrophs since growth on leucine leads to their repression.

In order to obtain mutants that were no longer repressed by the addition of leucine, we plated an auxotroph on D-leucine. D-Leucine is not transported efficiently enough to provide for a source of L-leucine via racemization in an auxotroph unless a derepression in transport occurs. When a wild-type *E. coli* K12 strain that has a lesion in leucine biosynthesis is plated on 200 mg/L D-leucine, D-leucine-utilizing mutants (DLU) were obtained. Most of these mutants had normal transport; however, about 3% of these mutants had greatly increased D- and L-leucine and isoleucine transport activity.[20, 21] These mutants showed corresponding increases in the levels of the various binding proteins for leucine, isoleucine, and valine.

One class of these mutants, represented by strain E0312,[13] shows derepressed levels of the high-affinity transport systems, and the addition of leucine to the growth media no longer represses the level of the high-affinity transport and the level of the three binding proteins. The kinetics of the high-affinity transport of leucine shows a two- to threefold increase in maximal capacity without significant changes in the Km value. This mutation has been partially mapped by genetic techniques and appears to be about 2% linked to the *pyrD* locus at minute 21 of the *E. coli* chromosome.[22]

A second class of D-leucine utilizing mutants is represented by strain E0318. This mutant, like the previous mutant, also has a two- to threefold increase in leucine transport activity; however, the major increase in leucine transport results from an increased level of the leucine-specific system. The leucine-specific system makes up about 50% of the leucine uptake in this strain. The leucine-specific binding protein is also increased severalfold.[13] The LIV-I transport system appears unchanged in the E0318 strain and is still repressed by leucine. This regulatory mutation has been mapped and is 60 to 70% linked to the *aroA* locus at minute 20 on the *E. coli* chromosome.[22] These regulatory mutants show increased transport activity as a result of the increased binding proteins, again indicating that the binding proteins are involved in the rate-limiting step in high-affinity branched-chain amino acid transport systems.

Defective Transport Mutants

The D-leucine-utilizing strains E0311 and E0318 are able to utilize D-leucine because of increased transport activity for D-leucine. Using these strains, we applied a penicillin selection in the presence of D-leucine to isolate transport mutants among the survivors. Several mutants were isolated that showed greatly decreased D- and L-leucine transport activity.[13] The transport defect in these mutants was shown to be in the high-affinity transport systems and the LIV-II low-affinity membrane-bound system was relatively unaffected. We observed three types of transport mutants in these studies. One type proved to be defective in a large number of transport systems and was therefore considered to have a general membrane defect. The parental DLU strain E0311, which had the high-affinity systems derepressed, was used to obtain a second type of mutant. This mutant, strain E0323, showed a loss in all the high-affinity systems, but appeared to have normal binding protein levels. Strain E0323 apparently has a defect in a common component for the binding protein transport systems. A third type of transport mutant was derived from the parental DLU strain E0318, which has high leucine-specific activity. This mutant strain, E0321, has lost only the specific leucine transport system. When we examined the shock fluid for leucine-specific binding activity, none was found. A protein fraction corresponding to the leucine-specific protein was eluted from DEAE columns at a slightly slower rate than authentic leucine-specific binding protein. This protein fraction could be detected in the eluate by its crossreactivity with antibody to the LIV-I binding protein. This strain, containing a mutation in the structural gene of the leucine-specific binding protein, is being mapped to establish the position of the structural gene on the *E. coli* chromosome.

Analog-Resistant Mutants

The use of analogs of leucine such as azaleucine and trifluoroleucine have not been effective in obtaining mutants with defective binding proteins. The inability to obtain such mutants may have several causes. First, these analogs are not very toxic to wild-type strains. Second, they are ineffective for selection methods using leucine auxotrophs because the requirement for exogenous leucine in these strains prevents uptake of the analogs. In addition many strains are able to develop resistance to moderate levels of the analog by derepressing the leucine biosynthetic pathway. The DLU mutant strains with increased leucine uptake described above are much more sensitive to low levels of azaleucine or trifluoroleucine.

An examination of azaleucine transport activity in these strains shows that it can enter the cell by the LIV-I system and by the general aromatic amino acid transport system called *aroP*.[17] Although the derepressed transport system can increase the sensitivity of a cell to azaleucine, extensive studies have indicated that a variety of cellular processes can alter the azaleucine toxicity.[17]

A large number of azaleucine-resistant mutants have been examined for leucine transport activity. Approximately 50% of the mutants have defective leucine transport. Some of these mutants have membrane defects and are thus pleiotropic transport mutants. The rest of the transport mutants have a specific defect in the high-affinity leucine transport system. An examination of several hundred has failed to yield a binding-protein negative mutant; therefore, we conclude that the predominant transport mutant that derives from the analog resistance selection is defective in a common component of the high affinity systems other than the binding protein component. These mutants will be useful for characterizing the nonbinding protein components of the high-affinity leucine transport system.

Valine-Resistant Mutants

E. coli K12 is extremely sensitive to growth inhibition by low levels of L-valine since valine causes feedback inhibition of acetohydroxy acid synthetase, a common enzyme in valine and isoleucine biosynthesis. Valine-induced growth inhibition is relieved by the addition of isoleucine to the medium. The most common class of valine-resistant mutants obtained is the regulatory mutant in the biosynthetic pathway for isoleucine and valine. Various mutations leading to the valine-resistant phenotype have been mapped at six different chromosomal loci in *E. coli.*[23] Guardiola and Iaccarino have used valine resistance in the presence of leucine and methionine to obtain several types of mutants.[15, 24] In addition to finding regulatory mutants for the *ilv* pathway they identified transport mutants. Some of these mutants have been genetically mapped, but the biochemical characterization is not sufficient at the present time to assign the defect to a specific branched-chain amino acid binding protein.

Requirement of Binding Proteins in Chemotaxis

Chemotaxis by *E. coli* has been the subject to extensive studies[25-29] and will not be covered now except to indicate the relationship of certain binding proteins to chemotaxis. Although not all chemo receptors are shockable binding proteins, a number of the sugar-binding proteins appear to be absolutely required for chemotaxis toward the respective sugar. Chemotaxis of *E. coli* toward galactose,[26, 27] ribose,[28] and maltose[29] requires the presence of the specific sugar-binding protein. There are additional components required for both transport and chemotaxis even though the binding protein serves as a common element. The relationship of the binding proteins to transport and chemotaxis can be diagrammed as follows:

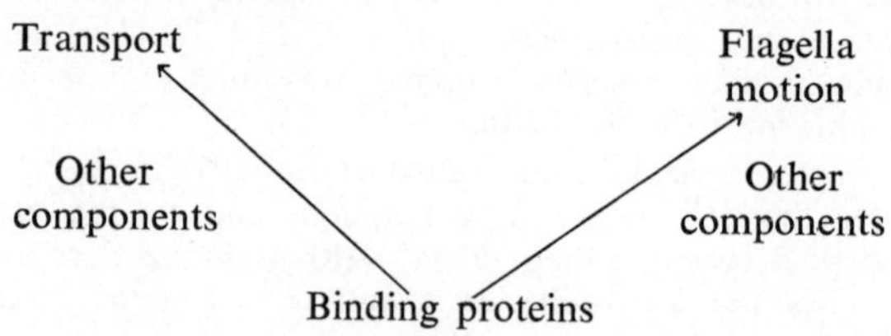

There are transport-specific components not required for chemotaxis and chemotaxis-specific components not required for transport as illustrated in this diagram. These additional components have been identified by genetic techniques.[25] Galactose transport mutants have been obtained that show normal chemotaxis for galactose and certain chemotactic mutants have normal transport. These latter types of mutants both have normal levels of galactose-binding protein.

Role of Binding Proteins in Transport

A variety of indirect evidence has accumulated to support a direct role of the binding protein in solute transport by the high-affinity shock-sensitive transport systems. These lines of evidence include: (a) osmotic shock treatment causing a parallel loss in transport activity and binding protein; (b) similarity of the kinetic constants; (c) parallel regulation of transport and binding protein activities; (d) localization of binding proteins to the periplasmic space; and (e) reversion of transport negative mutants that do not contain functional binding proteins, which always results in a recovery of both transport and binding protein activities.

Reconstitution studies to find direct evidence for a role of binding proteins have been generally discouraging, although genetic studies have provided direct evidence for a role of certain binding proteins.

Mutants with Altered Binding Proteins

Starting with strains that can utilize D-histidine, Ames and Lever were able to isolate mutants with the histidine-binding protein (J) missing.[30, 31] Revertants of these histidine transport mutants (induced by the frameshift mutagen ICR191 in a *his*J mutant that was itself induced with ICR191) were obtained that contained *his*J proteins with altered chemical and physical properties. The revertant strain was more temperature-sensitive for both growth and histidine transport activity than was the parental strain. The histidine-binding protein isolated from the temperature-sensitive revertant was also more sensitive to heat treatment than normal histidine-binding protein. These data provide direct evidence for a role of the histidine-binding protein in the high-affinity histidine transport system.

Direct evidence that the galactose-binding protein plays a role in β-methyl-D-galactoside transport was provided by obtaining mutants of *E. coli* that contained altered galactose-binding proteins that resulted in an altered chemotaxis[26] and an altered transport of galactose.[32] The amino acid structure of the galactose-binding protein was found to be altered leading to a lower affinity for galactose which was reflected simultaneously in a higher Km value of transport and a higher threshold for the detection of positive chemotaxis specifically for galactose.[33]

As indicated in an earlier section we have also isolated a mutant that appears to be in the structural gene of the leucine-specific binding protein. This mutation produces a CRM protein that results in the complete loss of leucine-specific transport activity.

Recently Kustu and Ames[34] obtained a histidine transport mutant from *Salmonella typhimurium* that had lost the high-affinity histidine transport, although it retained normal levels of the histidine-binding protein. This mutation proved to be in the structural gene of the histidine-binding protein, suggesting that a site on the binding protein other than the recognition site may serve for binding additional membrane components.

Extensive genetic studies[35, 36] on galactose transport have shown that mutants of the β-methyl-D-galactoside transport system can all be placed in one of three classes using genetic complementation analysis.[35] The three genes have been called *mglA, mglB,* and *mglC* and are genetically linked to the *his* loci of the *E. coli* chromosome. The *mglB* was determined to be the structural gene for the galactose-binding protein, and the products corresponding to *mglA* and *mglC* appear to be additional membrane-bound components of the galactose transport system. Robbins and Rotman[36] found that the galactose-binding protein was required for active transport of galactose, although *mglB* mutants could grow on high levels of galactose. They concluded that the binding protein was not involved in the translocation step in transport.

FIGURE 4 presents a diagram of the possible role of the binding protein in galactose transport. Components A and C are believed to be membrane

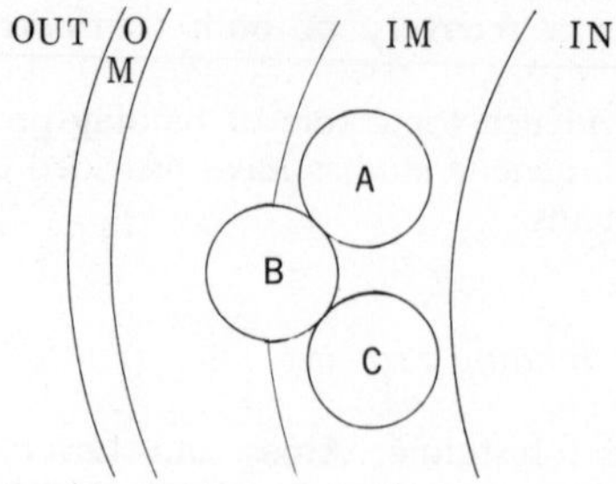

FIGURE 4. The role of the galactose-binding protein in galactose transport. OUT, outside; OM, outer membrane; IM, inner membrane; IN, inside of cell.

components of the translocation process and the binding protein (B) creates an active transport system by increasing the affinity of the membrane-bound systems for galactose. Robbins and Rotman[36] suggest that the affinity of the system may be as much as 1000 times greater in the presence of the binding protein, although the capacity of the system is not altered by the presence of the binding protein.

Silhavy *et al.*[32] have prepared an extensive review of the evidence for the role of the galactose-binding protein. In agreement with Robbins and Rotman,[36] they do not believe that the binding protein for galactose is involved in the translocation step. They review the evidence for two sugar binding sites on the galactose-binding protein. These authors suggest that one site is the recognition site for galactose and that the other is a glycoprotein recognition site. Their model predicts that either or both of components A and C in the membrane may be a glycoprotein, and they have experiments under way to test this possibility.

The results from our laboratory on the role of the leucine-binding protein in transport are not in disagreement with those found for galactose. At the present time, most of our data suggest that LIV-I and LIV-II are parallel systems and not connected by a series of steps, although some mutants appear to be defective in both LIV-I and LIV-II transport systems, which suggests they may share a common component.

Conclusions

The recent studies have clearly established two types of active transport systems. One type is membrane-bound and can be observed in membrane vesicles and the other type is osmotic-shock-sensitive and requires binding proteins to produce active transport. It appears that the membrane-bound systems derive cellular energy from an energy-rich membrane state which can be formed from respiration or ATP-hydrolysis, while the binding protein systems are more directly coupled to phosphate bond energy derived from glycolysis or oxidative phosphorylation.

The following conclusions concerning the role of the binding proteins are offered:

1. The binding proteins are present in relatively large amounts ($\sim 10^{-6}$ or 10^{-5} M) and appear to reside in the periplasmic space.

2. They do not appear to be involved in solute translocation steps, although they contain a second binding site that could interact with membrane components.

3. The binding proteins appear to increase the affinity of the transport system for the solute by interacting with a membrane component. This may result from the possibility that the solute-binding protein complex is the actual substrate for the membrane transport system.

References

1. KABACK, H. R. 1971. Bacterial membranes. Methods Enzymol. **22:** 99.
2. BERGER, E. A. & L. A. HEPPEL. 1974. Different mechanisms of energy coupling for the shock-sensitive and shock-resistant amino acid permeases of *Escherichia coli.* J. Biol. Chem. **249:** 7747.
3. HAROLD, F. M. 1972. Conservation and transformation of energy by bacterial membranes. Bacteriol. Rev. **36:** 172.
4. KABACK, H. R. 1974. Transport studies in bacterial membrane vesicles. Science **186:** 882.
5. NEU, H. C. & L. A. HEPPEL. 1965. The release of enzymes from *Escherichia coli* by osmotic shock and during the formation of spheroplasts. J. Biol. Chem. **240:** 3685.
6. PIPERNO, J. R. & D. L. OXENDER. 1966. Amino acid-binding protein released from *Escherichia coli* by osmotic shock. J. Biol. Chem. **241:** 5732.
7. PENROSE, W. R., G. E. NICHOALDS, J. R. PIPERNO & D. L. OXENDER. 1968. Purification and properties of a leucine-binding protein from *Escherichia coli.* J. Biol. Chem. **243:** 5921.
8. FURLONG, C. E. & J. H. WEINER. 1970. Purification of a leucine-specific binding protein from *Escherichia coli.* Biochem. Biophys. Res. Commun. **38:** 1076.
9. PARSONS, R. G. & HOGG, R. W. 1974. Crystallization and characterization of the L-arabinose-binding protein of *Escherichia coli* B/r. J. Biol. Chem. **249:** 3602.
10. PARSONS, R. G. & R. W. HOGG. 1974. A comparison of the L-arabinose- and D-galactose-binding proteins of *Escherichia coli* B/r. J. Biol. Chem. **249:** 3608.
11. PARDEE, A. B. & K. WATANABE. 1968. Location of sulfate-binding protein in *Salmonella typhimurium.* J. Bacteriol **96:** 1049.
12. NAKANE, P. K., G. E. NICHOALDS & D. L. OXENDER. 1968. Cellular localization of leucine-binding protein from *Escherichia coli.* Science **161:** 182.
13. RAHMANIAN, M., D. R. CLAUS & D. L. OXENDER. 1973. Multiplicity of leucine transport systems in *Escherichia coli* K12. J. Bacteriol. **116:** 1258.
14. GUARDIOLA, J., M. DE FELICE, T. KLOPOTOWSKI & M. IACCARINO. 1974. Multiplicity of isoleucine, leucine, and valine transport systems in *Escherichia coli* K12. J. Bacteriol. **117:** 382.
15. GUARDIOLA, J., M. DE FELICE, T. KLOPOTOWSKI & M. IACCARINO. 1974. Mutations affecting the different transport systems for isoleucine, leucine, and valine in *Escherichia coli* K12. J. Bacteriol. **117:** 393.
16. TEMPLETON, B. A. & M. A. SAVAGEAU. 1974. Transport of biosynthetic intermediates: Regulation of homoserine and threonine uptake in *Escherichia coli.* J. Bacteriol. **120:** 114.
17. HARRISON, L. I., H. N. CHRISTENSEN, M. E. HANDLOGTEN, D. L. OXENDER & S. C. QUAY. 1975. The transport of L-4-azaleucine in *Escherichia coli* K12. J. Bacteriol. **122:** 957.
18. FURLONG, C. E., C. CIRAKOGLU, R. C. WILLIS & P. A. SANTY. 1973. A simple preparative polyacrylamide disc gel electrophoresis apparatus: Purification of three branched-chain amino acid binding proteins from *Escherichia coli.* Anal. Biochem. **5:** 297.
19. QUAY, S. C., D. L. OXENDER, S. TSUYUMU & H. E. UMBARGER. 1975. Separate regulation of transport and biosynthesis of leucine, isoleucine, and valine in bacteria. J. Bacteriol. **122:** 994.

20. Rahmanian, M. & D. L. Oxender. 1972. Derepressed leucine transport activity in *Escherichia coli.* J. Supramol. Struc. **1:** 55.
21. Oxender, D. L. & M. Rahmanian. 1972. Leucine transport in *Escherichia coli. In* The Molecular Basis of Biological Transport. J. F. Woessner, Jr. and F. Huijing, Eds. : 271–289. Academic Press. New York, N.Y.
22. Anderson, J. J., S. C. Quay & D. L. Oxender. 1975. Mapping of two D-leucine utilization mutations affecting the regulation of branched-chain amino acid transport systems in *Escherichia coli.* Abst. Ann. Meet. Am. Soc. Microbiol. **:** 169.
23. Glover, S. W. 1962. Valine-resistant mutants of *Escherichia coli* K12. Genet. Res. Camb. **3:** 448.
24. Guardiola, J. & M. Iaccarino. 1971. *Escherichia coli* K12 mutants altered in the transport of branched-chain amino acids. J. Bacteriol. **108:** 1034.
25. Adler, J. 1975. Chemotaxis in bacteria. Ann. Rev. Biochem. **44:** 341.
26. Hazelbauer, G. L. & J. Adler. 1971. Role of the galactose binding protein in chemotaxis of *Escherichia coli* toward galactose. Nature New Biol. **230:** 101.
27. Ordal, G. & J. Adler. 1973. Isolation and complementation of mutants in galactose taxis and transport. J. Bacteriol. **117:** 509.
28. Aksamit, R. R. & D. E. Koshland, Jr. 1974. Identification of the ribose binding protein as the receptor for ribose chemotaxis in *Salmonella typhimurium.* Biochemistry **13:** 4473.
29. Adler, J., G. L. Hazelbauer & M. M. Dahl. 1973. Chemotaxis toward sugars in *Escherichia coli.* J. Bacteriol. **115:** 824.
30. Ames, G. F. & J. Lever. 1970. Components of histidine transport: Histidine-binding proteins and *hisP* protein. Proc. Nat. Acad. Sci. U.S.A. **66C:** 1096.
31. Ames, G. F. & J. E. Lever. 1972. The histidine-binding protein J is a component of histidine transport. Identification of its structural gene, *hisJ.* J. Biol. Chem. **247:** 4309.
32. Silhavy, T. J., W. Boos & H. M. Kalckar. 1974. The role of the galactose-binding protein in galactose transport and chemotaxis. Moosback Colloquium. Moosback, West Germany. In press.
33. Boos, W. & M. O. Sarvas. 1970. Close linkage between a galactose binding protein and the β-methyl galactoside permease in *Escherichia coli.* Eur. J. Biochem. **13:** 526.
34. Kustu, S. G. & G. F-L. Ames. 1974. The histidine-binding protein J, a histidine transport component, has two different functional sites. J. Biol. Chem. **249:** 6976.
35. Ordal, G. W. & J. Adler. 1974. Properties of mutants in galactose taxis and transport. J. Bacteriol. **117:** 517.
36. Robbins, A. R. & B. Rotman. 1975. Evidence for binding protein-independent substrate translocation by the methylgalactoside transport system of *Escherichia coli* K12. Proc. Nat. Acad. Sci. U.S.A. **72:** 423.
37. Rosen, B. 1971. Basic amino acid transport in *Escherichia coli.* J. Biol. Chem. **246:** 3653.
38. Rosen, B. P. 1973. Basic amino acid transport in *Escherichia coli.* II. Purification and properties of an arginine-specific binding protein. J. Biol. Chem. **248:** 1211.
39. Lever, J. E. 1972. Purification and properties of a component of histidine transport in *Salmonella typhimurium.* The histidine-binding protein J. J. Biol. Chem. **247:** 4317.
40. Berger, E. A. & L. A. Heppel. 1972. A binding protein involved in the transport of cystine and diaminopimelic acid in *Escherichia coli.* J. Biol. Chem. **247:** 7684.
41. Weiner, J. H. & L. A. Heppel. 1971. A binding protein for glutamine and its relation to active transport in *Escherichia coli.* J. Biol. Chem. **246:** 6933.

42. WILLIS, R. C. & C. E. FURLONG. 1975. Purification and properties of a periplasmic glutamate-aspartate binding protein from *E. coli* K12 strain W3092. J. Biol. Chem. **250:** 2574.
43. GUROFF, G. & K. E. BROMWELL. 1971. Phenylalanine uptake and phenylalanine binding material in *Comamonas* sp. Arch. Biochem. Biophys. **137:** 379.
44. WILEY, W. R. 1970. Tryptophan transport in *Neurospora crassa:* A tryptophan-binding protein released by osmotic shock. J. Bacteriol. **103:** 656.
45. ANRAKU, Y. 1968. Transport of sugars and amino acids in Bacteria. II. Properties of galactose- and leucine-binding proteins. J. Biol. Chem. **243:** 3123.
46. ANRAKU, Y. 1968. Transport of sugars and amino acids in bacteria I. Purification and specificity of the galactose- and leucine-binding proteins. J. Biol. Chem. **243:** 3116.
47. AKSAMIT, R. & D. E. KOSHLAND, JR. 1972. A ribose binding protein of *Salmonella typhimurium.* Biochem. Biophys. Res. Commun. **48:** 1348.
48. KELLERMAN, O. & S. SZMELCMAN. 1974. Active transport of maltose in *Escherichia coli* K12. Involvement of a "periplasmic" maltose binding protein. Eur. J. Biochem. **47:** 139.
49. WASSERMAN, R. H., R. A. CORRADINO & A. N. TAYLOR. 1968. Vitamin D-dependent calcium-binding protein: Purification and some properties. J. Biol. Chem. **243:** 3978.
50. WASSERMAN, R. H. & A. N. TAYLOR. 1968. Vitamin D-dependent calcium-binding protein: Response to some physiological and nutritional variables. J. Biol. Chem. **243:** 3987.
51. MEDVECZKY, N. & H. ROSENBERG. 1969. The binding and release of phosphate by a protein isolated from *Escherichia coli.* Biochim. Biophys. Acta **192:** 369.
52. PARDEE, A. B. 1966. Purification and properties of a sulfate-binding protein from *Salmonella typhimurium.* J. Biol. Chem. 241: 5886.
53. IWASHIMA, A., A. MATSUURA & Y. NOSE. 1971. Thiamine-binding protein of *Escherichia coli.* J. Bacteriol. **108:** 1419.
54. NISHIMUNE, T. & R. HAYASHI. 1971. Thiamine-binding protein and thiamine uptake by *Escherichia coli.* Biochim. Biophys. Acta **244:** 573.
55. TAYLOR, A. L., S. A. NORREL & M. L. HANNA. 1972. Uptake of cyanocobalamine by *Escherichia coli* B: Some characteristics and evidence for a binding protein. Arch. Biochem. Biophys. **148:** 366.
56. WOOD, J. M. 1974. A comparison of L-leucine, L-isoleucine, and L-valine transport in *Escherichia coli* strains 7 (K12) and ML308-255. Fed. Proc. **33:** 1394.
57. OXENDER, D. L. 1972. Membrane transport. Ann. Rev. Biochem. **41:** 777.
58. OXENDER, D. L. 1974. Membrane transport proteins. *In* Biomembranes. L. A. Manson, Ed. Vol. 5, Chapt 2: 25–79. Plenum Press. New York, N.Y.
59. OXENDER, D. L. & S. C. QUAY. 1976. Isolation and characterization of membrane binding proteins. *In* Methods of Membrane Biology. E. D. Korn, Ed. Vol. 6. Plenum Press. New York, N.Y. In press.
60. BOOS, W. 1974. Bacterial transport. Ann. Rev. Biochem. **43:** 25–79.
61. ROSEN, B. & L. A. HEPPEL. 1973. Present status of the binding proteins that are released from gram-negative bacteria by osmotic shock. *In* Bacterial Membranes and Walls. L. Leive, Ed.: 209–240. Marcel Dekker, Inc. New York, N.Y.

Discussion

DR. TOSTESON: I wonder whether binding proteins function so poorly as transporters because their affinities for the transported substances are too high and the rate coefficient for release is too low? To that end it would be interesting to know whether there is any evidence either in bacterial systems or in vesicles that binding proteins can actually go all the way through the bilayer.

DR. OXENDER: The binding proteins don't appear to have measurable affinities; also we've sent protein to Thompson at Virginia, and he has attempted to see whether they would have any affinity for lipid bilayers, and his experimental results were negative. We've also taken liposomes and loaded them with radioactive leucine and then added the binding protein to see if binding proteins will stimulate the efflux of leucine; again the results were negative.

Part VIII. Permeation Mechanism across Biological Membranes: II

BIOLOGICAL APPLICATIONS OF IONOPHORES: THEORY AND PRACTICE*

Berton C. Pressman and Norberto T. deGuzman

Department of Pharmacology
University of Miami School of Medicine
Miami, Florida 33152

The discovery of the ionophores[1-3] has provided the means for studying a wide range of ion-selective processes. These compounds of moderate molecular weight (ca. 500–1500) form lipid-soluble complexes which provide a vehicle for a wide variety of cations to transverse lipid barriers, hence the term *ionophore* signifying *ion-bearer*.[3] A series of representative ionophores is shown in Figure 1. Ionophore-mediated transport has not only been employed to perturb biological systems for study, but has also provided clues to the molecular mechanisms by which the biological membranes themselves transport and

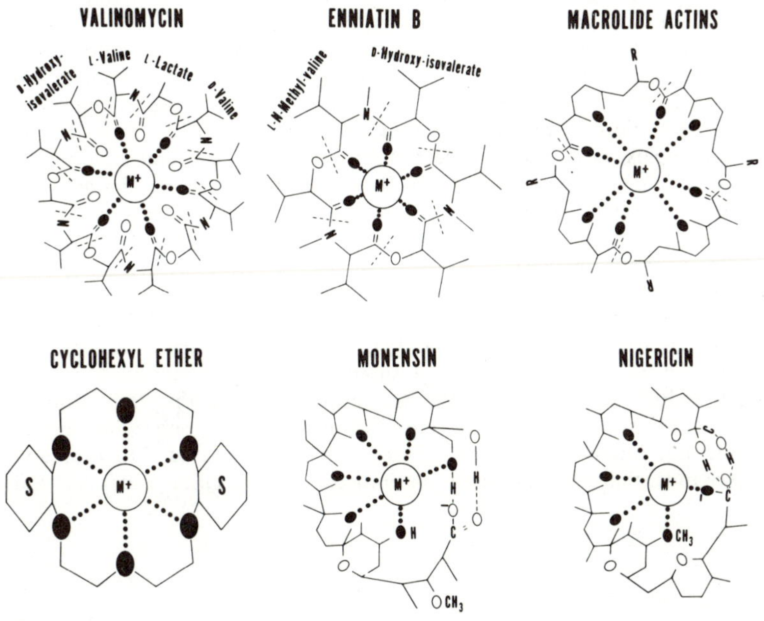

Figure 1. Structures of representative neutral (valinomycin, enniatin B, macrolide actins and dicyclohexyl-18-crown-6) and carboxylic (monensin, nigericin) ionophores.

* This work was supported by gifts from Hoffman La-Roche and Eli Lilly and by Grant No. M73A64 from the Heart Association of Greater Miami and Grants No. HL-14434 and HL-16,117 from the National Institutes of Health.

discriminate between ions. Accordingly, ionophore transport and complexation have been intensively studied in a multitude of bulk phase and lipid bilayers systems. A practical offshoot of this work has been the development of ionophore-based liquid-exchange ion-selective electrodes, which, in their standard and micro variations, have further extended the impact of ionophores in biology. General reviews of this subject are contained in References 4 and 5.

The first clue to the unique properties of ionophores was the observation that valinomycin induces the energy-dependent uptake of K^+, but not Na^+, by mitochondria.[1] Those ionophores whose existence was earlier revealed by the mitochondrial test system interact primarily with alkali ions,[3, 6] although they exhibit a diversity of cationic discrimination patterns.[6–8] More recently it was recognized that the ionophore A-23187 is highly selective for divalent cations[9] and that X-537A and several of its derivatives are broad-spectrum ionophores, capable of complexing and transporting not only alkali ions but polyvalent inorganic ions and complicated organic amines as well.[10–12] Its ability to transport such key biological control agents as Ca^{++} and the catecholamines led us to explore the effects of X-537A on the isolated heart and ultimately on the intact organism. Initial results were encouraging and indicated that ionophores have unprecedented physiological effects with pharmacological implications which, we are encouraged to believe, will develop into yet another major area of ionophore-related studies. This paper will focus on the pharmacological properties of X-537A and related ionophores and examine how these correlate with their properties in other systems.

X-537A (FIGURE 2) belongs to the class of *carboxylic* ionophores that contain a single carboxyl group which is usually deprotonated in its cationic complexes. In most cases the resulting negative charge of the ionophore offsets that of the cation complexed so that for 1:1 complexes with monovalent ions the complex as a whole is electrically neutral. In parallel fashion X-537A forms a 2:1 neutral complex with Ba^{++}, which has been characterized by x-ray crystallography. The cation is sandwiched unequally between two deprotonated ionophore molecules termed *"unprimed"* (six ligands; see FIGURE 2) and *primed* (three ligands).[13]

X-537A contains an aromatic group that provides several interesting molecular properties. It gives rise to three absorption bands at 315 nm (aromatic hydroxyl), 245 nm (aromatic carboxyl), and 210 nm (aromatic ring proper). Although the 315-nm band shows a small shift on protonation-deprotonation, no useful shift occurs during the transition between the free anionic and complexed forms. The 315-nm and 245-nm bands, along with the weakly absorbing but highly asymmetric ketonic carbonyl peak at 290 nm, give rise to CD bands that can be used as indices of molecular conformation and

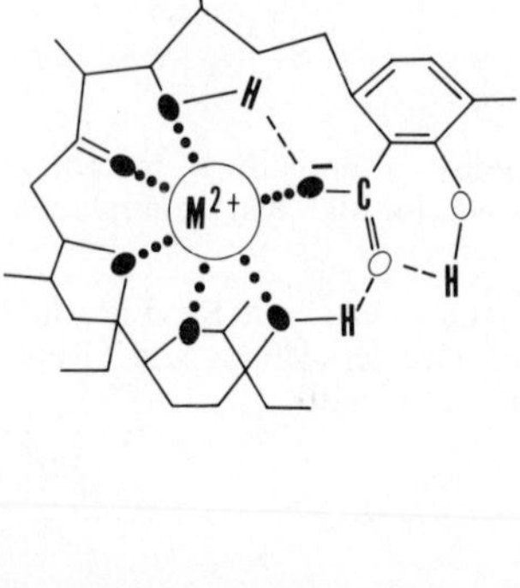

FIGURE 2. Structure of the "unprimed" moiety of X-537A as it occurs in the barium salt. It is assumed that the liganding pattern for monovalent cations is similar.

TABLE 1

RELATIVE AFFINITIES OF IONOPHORES FOR DIVALENT CATIONS OBTAINED FROM TWO-PHASE DISTRIBUTION STUDIES*

Ionophores Cation	Acetyl-X-537A	X-537A	Br-X537A	Dianemycin
Mg^{++}	.26	.38	.54	.01
Ca^{++}	.29	1.0	2.8	.08
Sr^{++}	1.8	8.5	18	.20
Ba^{++}	72	2600	5600	

* Complex formation was determined from the migration of the test cation from aqueous Tricine buffer (pH 9.0) into 70% toluene: 30% *n*-butanol. The [ionophore] was 0.5 mM and the two phase K_A calculated from the following expression which assumes 2:1 complex formation: $K_A = [M^{++}\ \text{Ionophore}^{-}{}_{2\ org}]/[M_{H_2O}] \cdot [\text{Ionophore}_{org}]^2$. All values were normalized by dividing by the two phase K_A for the Ca^{++}-X-537A complex.

that are unique for each complexation species.[12, 14] The aromatic hydroxyl and carboxyl groups collectively constitute a fluorophore, exciting at 313 nm and emitting at 420 nm.[15] Each complexation species has its own characteristic fluorescent yield that is sufficiently different from that of the uncomplexed anion species to serve as a basis for affinity titrations.

X-537A is readily susceptible to chemical modification. The aromatic ring may be substituted *para* to the phenolic hydroxyl (Br, Cl, I, NO_2), the phenolic hydroxyl may be acetylated or the ketonic carbonyl reduced,[16] all of which produce functional ionophores with altered cationic selectivities.

In order to ascertain whether the ability of X-537A to complex Ba^{++} extends to the biological trigger ion Ca^{++}, complexation affinities were initially measured by two-phase distribution to establish ionophore affinities.[6] The principle involved is the redistribution of the inherently hydrophilic test cation into a standard lipid phase upon the formation of a lipid-soluble complex with the test ionophore.[6, 7] TABLE 1 shows that although complexation affinity falls off with decreasing ionic radius of the cation complexed, X-537A does form readily detectable complexes with Ca^{++}.

For the sake of comparison, the divalent cation affinities of other carboxylic ionophores were measured. The next most effective one is dianemycin, which is lower in affinity for divalent ions by an order of magnitude; monensin and nigericin show still lower affinities for Ca^{++}.

In the course of studying the effects of X-537A on biological systems, we observed that this ionophore forms a complex with the buffer Tris, which is a primary amine. In order to assay the ability of X-537A to complex amines in general by the two-phase partition procedure, ethanolamine was chosen as a model amine since it is highly hydrophilic, remains in the aqueous phase unless complexed, and is available isotopically labeled. As seen in TABLE 2, it readily forms lipid-soluble complexes with X-537A and to a considerably lesser extent with other carboxylic ionophores.

The structural analogy between ethanolamine (H—CHOH—CH_2NH_2) and the catecholamines such as norepinephrine [$(HO)_2C_6H_3$—CHOH—CH_2NH_2] is readily apparent and directed us to discover the important ability of X-537A to complex catecholamines. The strong preference observed of X-537A for

TABLE 2

K_A OF ORGANIC AMINE-IONOPHORE COMPLEXES*

Ionophore	Ethanolamine	Norepinephrine	Epinephrine
X-537A	415	163	9.8
Dianemycin	65	31	4.2
Monensin	3	9	0.9
Nigericin	1.2	5	1.8

* The same system described in TABLE 1 was used. K_A was calculated on the assumption of 1:1 complexes with the monovalent cations.

norepinephrine over epinephrine, presumably due to steric hindrance for complexation by the *N*-methyl group of the latter, parallels the selectivity of the biological α-adrenergic receptor for which it might be considered a model.

The fluorescence properties of X-537A and its derivatives ultimately provided a nonisotopic method applicable for determining complexation affinities in a single phase. The effects of low concentrations of the test ions on the change in ionophore fluorescence were plotted as double reciprocal Langmuir saturation isotherms; the slopes of the resulting plots yielded the complexation affinities, while extrapolation of the linear isotherms to infinite cation concentration indicated the true fluorescent yield of the ionophore complexation species under saturation conditions. Often, for example, for Li^+, Mg^{++}, Ca^{++}, and Sr^{++}, these isotherms displayed abrupt changes in slope, indicative of the formation of different species of complexes, an interpretation also supported by CD spectra.[12] It should be noted that the two-phase distribution procedure for measuring complex formation selects for electrically neutral species; any tendency to form electrically charged complexes would set up interphasic potentials which would oppose further migration of cation to the nonpolar phase. The formation of electrically charged cation-ionophore complexes, the existence of which is implied by the ability of X-537A to carry current through a lipid bilayer,[10] is under no such restriction in a homogeneous bulk phase. Preliminary titrations indicate that at low cation concentrations the first complex detectable in ethanol is 1:1, ionophore:ion, for ions of *all valences.* It should be noted that the 1:1 complexes of all polyvalent cations are charged and therefore could carry current. Thus, the complexes formed during single-phase fluorescence titrations may be different species from those governing cation migration in the two-phase test system described above, and those responsible for carrying current through the lipid bilayer.

Haynes and Pressman have already reported that the fluorescence quantum efficiency of anionic uncomplexed X-357A in ethanol is about 0.75.[11] Ketodihydro-X-537A is about equally fluorescent while the decrease in fluorescence quantum efficiency of Br-X-537A by a factor of ten does not introduce undue difficulties into the titration procedure. From TABLE 3 we see that the alkali cation complexes of all three ionophores show qualitatively similar preferences for ions. Although by the partition method the most favored ion is Cs^+ as opposed to Rb^+,[6] this could reflect differences in cation desolvation energies between water and ethanol. The dihydro derivative shows considerably less affinity for alkali ions than the other two ionophores, a condition most marked for the nonspherical "pseudoalkali" ion Ag^+.

In the alkaline earth series increasing ionic radius from Ca^{++} to Ba^{++} greatly increases cation affinity for X-537A. It is noteworthy that the affinity of dihydro-X-537A for Ca^{++} is 66% of that of the parent compound; for K^+ it is 15%, while for ethanolamine it is less than 3%. The differential effects of modifications of a parent ionophore on its ability to translocate K^+, Ca^{++}, and catecholamines could be optimized for a given pharmacological application by a systematic and comprehensive survey of derivatives.

The wide variety of amines that complex X-537A provides a means for obtaining information about the ionophore binding site. The progressive gradual loss of affinity from methylamine to octylamine could be ascribed to increasing molecular "jiggling" of the hydrocarbon chain from thermal agitation. Introducing a second *N*-methyl group into either methylamine or ethanolamine is equivalently unfavorable for complexation, presumably due to steric hindrance (cf. norepinephrine vs. epinephrine, TABLE 2).

For the sake of comparison K_A values for some higher valent complexes of X-537A are also given. All inorganic cations tested gave definite complexes with X-537A including *all* available lanthanides, Cu^{++}, Zn^{++}, Fe^{++}, etc. Net transport velocities across membranes depend on a more complex reaction sequence than is involved in equilibrium affinity measurements, hence the latter predict only roughly what potential biological effects a given ionophore is capable of. An ion that forms a poor complex with an ionophore is not likely to

TABLE 3

K_A OF IONOPHORE COMPLEXES IN ABSOLUTE ETHANOL*

	X-537A	Br-X537A	H_2-537A
Li^+	.1	.4	.2
Na^+	6.7	17	.5
K^+	67	40	10
Rb^+	77	53	11
Cs^+	59	38	10
NH_4^+	5.3	10	.4
Tl^+	50	25	.7
Ag^+	11	24	.1
$HO-CH_2-CH_2-NH_3^+$	11	8	.3
Mg^{2+}	7.4	45	4
Ca^{2+}	7.6	20	5
Sr^{2+}	25	83	5.5
Ba^{2+}	1000	67	6.5
Mn^{2+}	83	33	22
La^{3+}	170		
Th^{4+}	67		
$CH_3NH_3^+$	8.0		
$C_2H_5NH_3^+$	6.9		
$C_3H_7NH_3^+$	6.5		
$C_4H_9NH_3^+$	6.1		
$C_8H_{17}NH_3^+$	5.6		
$(CH_3)_2{=}NH_2^+$	1.3		
$(CH_3)_3{\equiv}N^+$	.3		

* 10 mM triethanolamine partially neutralized with 5 mM $HClO_4$ present as buffer. All K_A expressed as mM^{-1}.

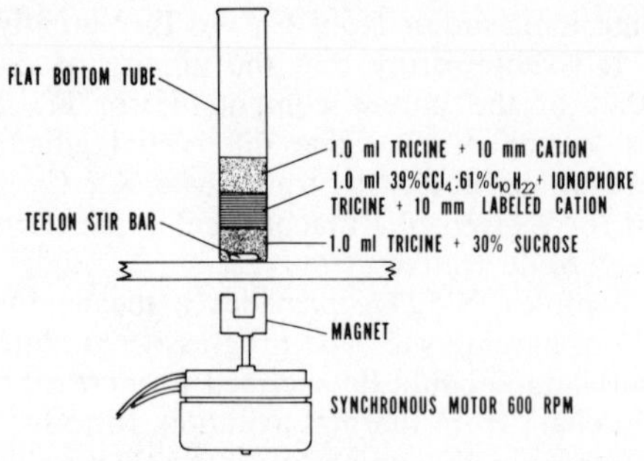

FIGURE 3. Vertically stacked bulk phase transport system. Transport is followed by measuring the migration of isotope from the lower to the upper phase. Further details about this system are given in Reference 12.

be transported effectively by it; one that forms so tight a complex that it cannot be readily dissociated is an equally poor candidate for transport.

Accordingly we devised the vertically stacked bulk transport test system shown in FIGURE 3 which stimulates more closely the dynamic conditions that occur during ionophore-mediated membrane transport. The system has been described previously[12] and is dependably reproducible when replicate determinations are carried out, either simultaneously or in sequential runs.

TABLE 4 reveals that the alkali ion-transport rates follow the lyotropic series $Cs^+ > Rb^+ > Na^+$ (for convenience these experiments are restricted to the readily available isotopically labeled cations). The poorer affinity of dihydro-X-537A shown in monovalent cation affinities in TABLE 3 is more apparent under the dynamic conditions of transport in TABLE 4. On the other hand all three ionophores are about equally effective in transporting divalent ions, despite the extraordinarily high equilibrium affinity of X-537A for Ba^{++}. This supports the earlier-stated inference that the complexation affinities seen in TABLE 3 may not be those of the same complex species which most easily cross lipid barriers. Conversely Ca^{++} appears to be transported unusually efficiently by comparison with its K_A values.

TABLE 4

BULK PHASE TRANSPORT RATES FOR CATION COMPLEXES OF IONOPHORES*

	X-537A	Br-X537A	Dihydro-X-537A
Na^+	39		1.4
Rb^+	45	46	2.1
Cs^+	118	132	3.1
Ca^{2+}	119	128	104
Sr^{2+}	54	57	52
Ba^{2+}	89	88	65
Tm^{3+}	17		
Ethanolamine	55	47	5.6
Norepinephrine	29		1.2
Epinephrine	9		0.5
Isopropylnorepinephrine	6		

* Solvent system employed is shown in FIGURE 3. Rates are expressed as nmol/hr cation appearing in the upper phase. Turnover numbers may be calculated from the information that the 1-ml organic barrier layer (which was approximately 1 cm in height) contained 100 nmol ionophore.

Transport in the amine series parallels the affinity constants seen in TABLE 2, except that discrimination between norepinephrine and epinephrine is less sharp. Determinations with other biogenic amines establish the rank order of transport capability of X-537A: dopamine > serotonin > ethanolamine norepinephrine > epinephrine > isopropylnorepinephrine.

The collective *in vitro* studies of X-537A and its derivatives described indicate a diversity of complexation species and an amenability to examination by a wide variety of techniques. It seemed to us remarkable if these ionophores could not produce profound and interesting biological perturbations, particularly in view of the potentiality of X-537A for transporting the key biological control cations, Ca^{++} and the biogenic amines, and for shifting the resting potentials of excitable membranes by altering alkali ion gradients. Because of the well-established effects of Ca^{++} and catecholamines on the cardiovascular system we chose to test X-537A with a standard heart preparation.

When the ionophore is perfused through spontaneously beating rabbit or guinea pig hearts (modified Langendorff preparation), increases in force of contraction with some increase in rate are observed in the dose range 10^{-6}M — 10^{-5}M.[10-12] These results were confirmed in the rat heart by the Williamson group.[18] Preparations from animals "reserpinized" to deplete catecholamines showed markedly diminished contractility responses to X-537A.[12] At a constant perfusion pressure the perfusion rate was increased by the ionophore. Since in this preparation the perfusion fluid normally flows through the coronary arteries, which regulate flow by their hydraulic resistance, this indicated that the ionophore induces a dilation of the coronary vasculature that could even be observed in animals that had previously been heavily reserpinized.

These preliminary observations encouraged us to extend our studies to the standard subject for pharmacological evaluation of cardiovascular drugs, the anesthetized dog.[19-21] The time course of the response of a dog to X-537A is shown in FIGURE 4. The effect of X-537A upon the maximal rate of pressure

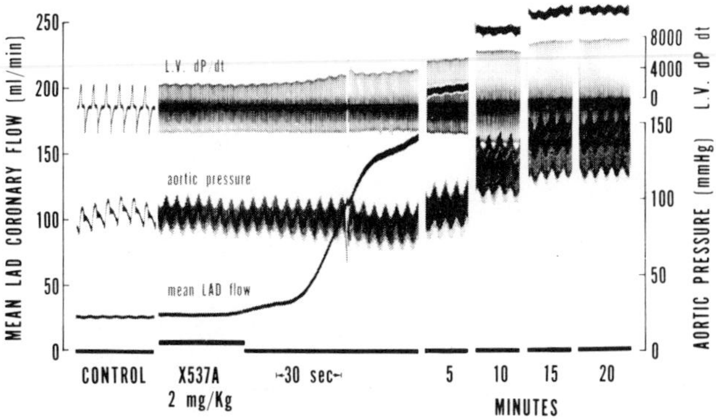

FIGURE 4. Continuous monitoring of the response of an anesthetized dog to X-537A (2 mg/kg). (*Upper trace*) left ventricular dP/dt is the rate of rise of pressure of the left ventricle as obtained from a pressure transducer tipped catheter. (*Middle trace*) aortic pressure. This tracing merges with the left ventricular dP/dt trace after 15 minutes. (*Lower trace*) mean blood flow in the left anterior descending coronary artery obtained from an electromagnetic flow probe. This trace becomes the uppermost one 10 minutes after the administration of X-537A.

development (dP/dt) in the left ventricle, which is conventionally regarded as an index of the contractile state of the heart, is shown in the upper trace of FIGURE 4. This function begins to rise almost immediately upon administration of a large dose of X-537A (2 mg/kg), attaining a value of almost threefold the control value within twenty minutes of the drug injection. The middle trace, aortic pressure, eventually undergoes a marked elevation after an initial transient decrease. This and other evidence (data from cardiac output) indicate that there is an initial drop in systemic vascular resistance paralleling the previously cited drop in coronary resistance that was noted in the isolated heart. As contractility rises, accompanied by an increased cardiac output, the drop in aortic pressure is gradually replaced by an increase. The lower tracing shows the blood flow through a major coronary artery (left anterior descending) as measured continuously by means of an electromagnetic flow probe. Coronary flow rises much more sharply and earlier than contractility or aortic pressure and in this experiment, after twenty minutes, was fully ten times higher than the control value, indicative of a striking drop in coronary vascular resistance.

The hemodynamic responses to X-537A over an extended period of time in a series of six dogs are presented in FIGURE 5. The upper trace shows a definite but minimal increase in heart rate. The rise in contractility, left ventricular dP/dt, is only slightly less than that shown for the representative dog in FIGURE 4 and attenuates slowly and gradually during the four-hour observation period. The initial rise in aortic (arterial) pressure is less sustained, returning to the control value after two hours. Cardiac output, a complicated function of contractility, systemic resistance and other factors that govern the venous return to the heart, shows an initial doubling which attenuates during the first hour and then persists at a mildly elevated level for four hours. Total peripheral resistance, calculated by dividing aortic pressure by flow, i.e., cardiac output, shows a decided initial drop during the first half hour, followed by a minimal decrease during the remainder of the observation period. These results collectively indicate a stimulation of the pumping characteristics of the heart and a decrease in vascular resistance which might be

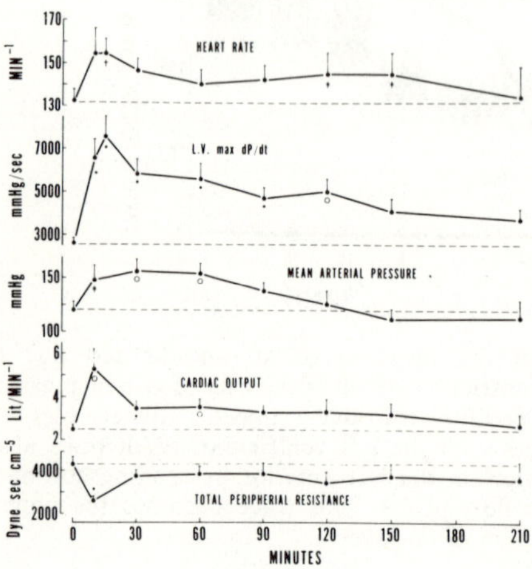

FIGURE 5. Hemodynamic effects of X-537A (2 mg/kg) on a series of six dogs. Measurements were made on records such as represented in FIGURE 4. Values are the series means with the standard error indicated.

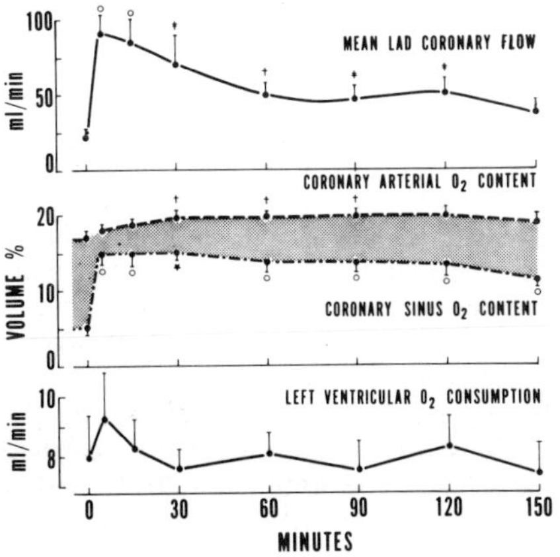

FIGURE 6. Effects of X-537A (2 mg/kg) on the consumption of O_2 by the left ventricle. pO_2 measurements were made on blood entering (aorta) and leaving (coronary sinus) the left ventricle. These were converted to actual O_2 content of the basis of the hemoglobin content and pH of the blood. Left ventricular O_2 consumption was obtained as the product of the atrioventricular $\triangle O_2$ across the ventricle and the flow of blood through the coronary artery, and extrapolated to the whole ventricle from the measured ratio of myocardium supplied by this vessel to the total mass of the left ventricle.

useful during pathological cardiovascular conditions such as those that occur in acute pump failure or shock.

FIGURE 5 depicts some of the effects of X-537A on the circulatory parameters of the heart itself. The upper trace shows the series average of the increase in coronary blood flow, which is somewhat more modest than the example chosen in FIGURE 4. The O_2 content of the coronary sinus blood draining the left ventricle rises markedly as a consequence of the increased flow rate through the myocardium (FIGURE 6). The stippled area represents the extraction of O_2, i.e., the atrioventricular difference, from which the left ventricular O_2 consumption is calculated by multiplying it by the coronary blood flow through the ventricular muscle. The O_2 consumption shows a slight, but significant rise, which returns to the control values within thirty minutes.

The implications of these data on cardiac work efficiency are shown in FIGURE 7. The small increases in ventricular O_2 consumption have been contrasted with the marked increase in the hydraulic work calculated as the product of cardiac output and effective pressure (mean aortic pressure). This

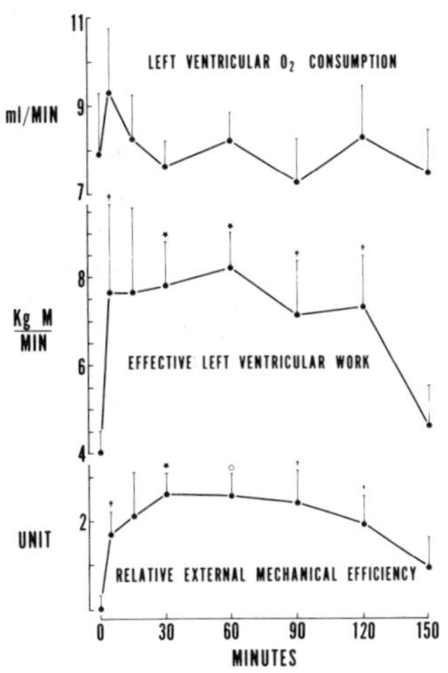

FIGURE 7. Effects of X-537A on cardiac work efficiency.

derived index of external cardiac work shows an approximately 100% increase, which was sustained over a period of two hours. If we divide cardiac work by O_2 consumption we see that X-537A induces a considerable increase in cardiac work output per unit O_2 consumed, i.e., a marked improvement in cardiac efficiency. This indicates that it is possible to improve cardiac performance under conditions of limited O_2 supply, such as occur in coronary vessel disease (often accompanied by angina), or in selected regions of the heart following an infarction. This is in sharp contrast to the effect of catecholamines, the administration of which increases cardiac work, but only at the expense of a considerable augmentation of tissue O_2 consumption, which would be contraindicated under conditions with accompanying ischemia.

The doubling of the work efficiency of the heart by ionophores might appear implausible. However, reasonable values for the thermodynamic cardiac efficiency of a barbiturate-anesthetized dog would be about 15–20%, and a doubling of this value would not call into question any thermodynamic principles. The heart is known to increase its O_2 consumption in response to increased heart rate and pressure; since the major component of the ionphore-induced work increase is *flow* (a consequence of decreased peripheral resistance) rather than *pressure,* the minimal increase in myocardial respiration is understandable.

Our original rationalization for administration of X-537A to dogs was the hope that the mobilization of calcium and catecholamines would prove advantageous for cardiac performance.[10–12] We have recently surveyed a series of other carboxylic ionophores and were startled to observe that they duplicated most of the cardiovascular properties of X-537A; even those ionophores are virtually incapable of transporting calcium and catecholamines *in vitro* (especially true of monensin and X-206).[21] In FIGURE 8 five of these have been compared with X-537A in terms of the dose necessary to produce a doubling of left ventricular dP/dt. They are all more potent than X-537A; e.g., only one seventh the dose of A-204, is required to produce the same effect as 1 mg/kg X-537A. Their potency to increase cardiac output roughly parallels their effects on contractility. The carboxylic ionophores show qualitative differences in response patterns as well. Thus, the dose of monensin that doubles left ventricular dP/dt appears to be more effective than an equivalent inotropic dose of X-537A in increasing coronary blood flow. The maximal responses

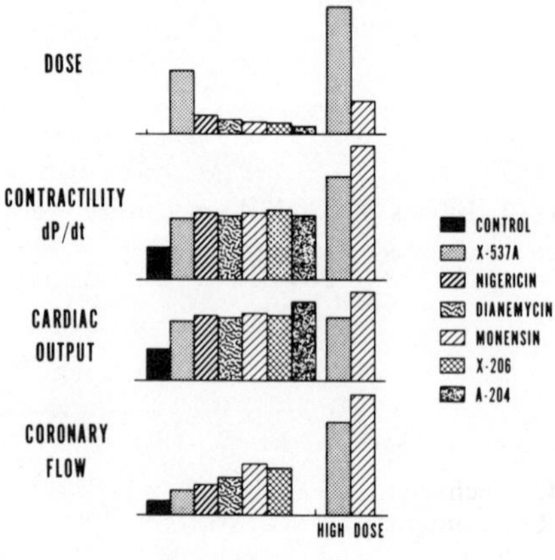

FIGURE 8. Comparison of the cardiovascular effects of various carboxylic ionophores. Dosages were adjusted to produce comparable effects on cardiac contractility, left ventricular dP/dt for X-537A this represented 1 mg/kg. Total cardiac output (less the coronary flow) was measured by the thermal dilution technique. Coronary flow was monitored by means of an electromagnetic probe placed on the left anterior descending coronary artery.

in all three parameters induced by subtoxic doses of monensin seem to be somewhat greater than those produced by equivalent doses of X-537A.

It thus emerges that despite our original rationalization, the ability to transport calcium and catecholamines well is not essential for a carboxylic ionophore to exert pharmacologically useful cardiovascular effects. A multitude of physiological effects have been ascribed to the calcium-mobilizing effects of X-537A (and A23187), e.g., coagulation of platelets,[22–24] release of histamine from mast cells,[25, 26] fertilization of eggs,[27, 28] and a variety of calcium-mediated exocytotic processes.[29] The direct mobilization of catecholamines has been found to release catecholamines from chromaffin granules[30] and peripheral adrenergic neurons.[31] Accordingly, there might be a distinct advantage in avoiding unwanted side effects by selecting the ionophore with the narrowest cation-selectivity spectrum compatible with cardiovascular activity for further therapeutic development.

Earlier reports of at least partial inhibition of the cardiovascular effects of ionophores by catecholamine-depleting or β-blocking agents implicate catecholamines in their action.[19, 32] The discovery that carboxylic ionophores virtually devoid of *in vitro* catecholamine transport capacity can have potent cardiovascular activity suggests that whatever the involvement of catecholamines is, it is indirect. These effects are not shared by all types of ionophores since the neutral subclass, e.g., valinomycin and the macrotetrolide actins, are cardiovascular depressants and exceedingly toxic. At this juncture we surmise that the essential property of an ionophore required for useful cardiovascular applications is the ability to induce some reequilibration of alkali ions, i.e., Na^+ for K^+ exchange, across certain membranes in critical control centers without causing undue changes in membrane potentials (as would the neutral ionophores). Their mechanism of action now seems much more complicated than was previously assumed.

We thus see how the concept of ionophores, conceived in the mitochondrion and matured in the test tube, has reemerged in a new biological context, that of providing a means of controlling biological processes for therapeutic purposes. The stimulation of cardiovascular function by carboxylic ionophores might be expected to find applications in chronic heart failure, reversal of shock of various etiologies, limiting the damage from myocardial ischemia and coronary insufficiency (encompassing angina). Finally it is hoped that these studies will serve as the prototype for a chemotherapeutic tactic of wide applicability: the alteration of membrane permeability by appropriately designed ionophores.

Acknowledgments

The technical assistance of Frank Lattanzio, Rick Barrios, and Valeria Gomez is gratefully acknowledged. We are also indebted to Dr. Peter Somani for many stimulating and helpful discussions.

References

1. Moore, C. & B. C. Pressman. 1964. Mechanism of action of valinomycin on mitochondria. Biochem. Biophys. Res. Commun. **15:** 562–567.

2. PRESSMAN, B. C. 1965. Induced transport of ions in mitochondria. Proc. Nat. Acad. Sci. U.S.A. **53:** 1076–1083.
3. PRESSMAN, B. C., E. J. HARRIS, W. S. JAGGER & J. H. JOHNSON. 1967. Antibiotic-mediated transport of alkali ions across lipid barriers. Proc. Nat. Acad. Sci. U.S.A. **58:** 1949–1956.
4. PRESSMAN, B. C. 1973. Alkali metal chelators—the ionophores. *In* Inorganic Biochemistry. G. L. Eichhorn, Ed.: 203–226. Elsevier. New York, N.Y.
5. OVCHINNIKOV, YU. A., V. T. IVANOV, & A. M. SHKROB. 1974. Membrane-active complexones. B.B.A. Library, Vol. 12. Elsevier. New York, N.Y.
6. PRESSMAN, B. C. 1968. Ionorphorous antibiotics as models for biological transport. Fed. Proc. **27:** 1283–1288.
7. PRESSMAN, B. C. & D. H. HAYNES. 1969. Ionophorous agents as mobile ion carriers. *In* the Molecular Basis of Membrane Function. D. C. Tosteson, Ed.: 221–246. Prentic Hall, Inc. Englewood Cliffs, N.J.
8. PRESSMAN, B. C. & M. J. HEEB. 1972. Permeability studies on erythrocyte ghosts with ionphorous antibiotics. *In* Symposium on Molecular Mechanisms of Action of Antibiotics which Affect Membrane Permeability. E. Munoz, F. Garcia-Ferrandiz and D. Vazquez, Eds.: 603–614. Elsevier. New York, N.Y.
9. REED, P. W. & H. A. LARDY. 1972. A23187: A divalent cation ionphore. J. Biol. Chem. **247:** 6970–6977.
10. PRESSMAN, B. C. 1973. Properties of ionphores with broad range cation selectivity. Fed. Proc. **32:** 1698–1703.
11. PRESSMAN, B. C. 1972. Carboxylic ionphores as mobile carriers for divalent ions. *In* The Role of Membranes in Metabolic Regulation. M. A. Mehlman and R. W. Hanson, Eds.: 149–164. Academic Press. New York, N.Y.
12. PRESSMAN, B. C. & N. T. DE GUZMAN. 1974. New ionophores for old organelles. Ann. N.Y. Acad. Sci. **227:** 380–391.
13. JOHNSON, S. M., J. HERRIN, S. J. LIU & I. C. PAUL. 1970. The crystal and molecular structure of the barium salt of an antibiotic containing a high proportion of oxygen. J. Am. Chem. Soc. **92:** 4428–4430.
14. DEGANI, H., H. L. FRIEDMAN, G. NAVON & E. M. KOSSOWER. 1973. Fluorometric complexing constants and circular dichroism measurements for antibiotic X-537A with univalent and bivalent cations. Chem. Commun.: 431–432.
15. CASWELL, A. H. & B. C. PRESSMAN. 1972. Kinetics of transport of divalent cations across sarcoplasmic reticulum vesicles induced by ionophores. Biochem. Biphys. Res. Commun. **49:** 292–298.
16. WESTLEY, J. W., E. P. OLIVETO, J. BERGER, R. H. EVANS, JR., R. GLASS, A. STEMPEL, V. TOOME & T. WILLIAMS. 1973. Chemical transformations of antibiotic X-537A. J. Med. Chem. **16:** 397–403.
17. HAYNES, D. H. & B. C. PRESSMAN. 1974. X-537A: A Ca^{2+} ionophore with a polarity-dependent and complexation-dependent fluorescent signal. J. Membr. Biol. **16:** 195–205.
18. SCHAFFER, S. W., B. SAFER, A. SCARPA & J. R. WILLIAMSON. 1974. Mode of action of the calcium ionophores X-537A and A23187 on cardiac contractility. Biochem. Pharmacol. **23:** 1609–1617.
19. DEGUZMAN, N. T. & B. C. PRESSMAN. 1974. The inotropic effects of the calcium ionophore X-537A in the anesthetized dog. Circulation **49:** 1972–1977.
20. DEGUZMAN, N. T. & B. C. PRESSMAN. 1974. Cardiovascular effects of the inotropic ionophore X-537A. Circulation **50** Suppl. **III:** 36.
21. DEGUZMAN, N. T. & B. C. PRESSMAN. 1975. The carboxylic ionophores—a new group of inotropic agents. Am. J. Cardiol. **35:** 131.
22. MASSINI, P. & E. F. LUSCHER. 1974. Some effects of ionophores for divalent cations on blood platelets: Comparison with the effects of thrombin. Biochem. Biophys. Acta **372:** 109–117.

23. White, J. G., H. R. Rao Gundu & J. N. Gerrard. 1974. Effects of the inonophore A23187 on blood platelets. Am. J. Pathol. **77:** 135–144.
24. Feinstein, M. B. & C. Fraser. 1975. Secretion and aggregation of human platelets induced by calcium ionophores. Inhibition by PGE, and dibutyryl cyclic AMP. J. Gen. Physiol. In press.
25. Forman, J. C., J. L. Mongar & B. D. Gomperts. 1973. Calcium ionophores and movement of calcium ions following the physiological stimulus to the secretory process. Nature (London) **245:** 249–251.
26. Cochrane, D. E. & W. W. Douglas. 1974. Calcium-induced extrusion of secretory granules (exocytosis) in mast cells exposed to 48/80 or the ionophores A23187 and X-537A. Proc. Nat. Acad. Sci. U.S.A. **71:** 408–412.
27. Steinhardt, R. A., D. Epel, E. J. Carroll, Jr. & R. Yanagimachi. 1974. Is calcium ionophore a universal activator for unfertilized eggs? Nature **252:** 41–43.
28. Chambers, E. L., B. C. Pressman, & Birgit Rose. 1974. The activation of sea urchine eggs by the divalent ionophores A23187 and X-537A. Biochem. Biophys. Res. Commun. **60:** 126–132.
29. Douglas, W. W. Exocytosis and the exocytosis-vesiculation sequence: with special reference to neurohypophysis, chromaffin and mast cells, calcium and calcium ionophores. *In* Secretory Mechanisms of Exocrine Glands. N. A. Thorn and O. H. Petersen, Ed.: 116–136. Munksgaard. Copenhagen, Denmark.
30. Johnson, R. G. & A. Scarpa. 1974. Catecholamine equilibration gradients of isolated chromaffin vesicles induced by the ionophore X-537A. FEBS Lett. **47:** 117–121.
31. Thoa, N. B., J. L. Costa, J. Moss & I. J. Kopin. 1974. Mechanism of release of norepinephrine from peripheral adrenergic neurones by the calcium ionophores X-537A and A23187. Life Sci. **14:** 1705–1719.
32. Schwartz, A., R. M. Lewis, H. G. Hanley, R. G. Munson, F. D. Dial & M. V. Ray. 1974. Hemodynamic and biochemical effects of a new positive inotropic agent antibiotic R02-2985. Circulation Res. **34:** 102–111.

Discussion

Dr. Green: A very intriguing question is posed by these remarkable studies, and it has to do with the mechanism of entry of calcium into the cardiac cell. Presumably there are active processes physiologically involved in the movements of calcium; that is, calcium just doesn't flow passively across the cell membrane. And yet, these remarkable effects are presumably achieved by a passive mechanism. It is hard to believe that these ionophores are acting in any way other than in a passive capacity. How can you combine both active and passive processes in the regulation of cardiac function?

Dr. Pressman: Cardiology is a rather arcane field, although not quite as much as cardiologists would have us believe. There is a well-known property of cardiac cells: They seem to have inherently a calcium for sodium exchange system. So if we bring sodium into a cell, the increased sodium is now capable of driving an uptake of calcium. Now, since the contractility of the heart is improving, ultimately somewhere at the bottom there has got to be an increase in the amount of calcium that can be delivered to the myofibril during contraction. But it now appears not to be getting there directly, as we originally thought, by the ionophore carrying the calcium in. However, it may turn out that the movement of calcium is secondary to the movement of sodium. The curious thing is that we got the right results, but for the wrong reason: that is, the calcium is getting in *indirectly* and not *directly*.

All of these ionophores move sodium. There is a gradient of sodium across the cell that is greater on the outside than on the inside. So if we induce a passive leak of sodium into the cell, we will raise the intracellular sodium level, which can then cause an increase passively in the amount of calcium within the cell.

DR. YUMAN: It is interesting that X-537A, either in your studies or in Schwartz's, seems to have rather prolonged effects. If X-537A were continually releasing catecholamines, the effects certainly wouldn't be that long-lasting.

It is intriguing that X-537A causes some effects that are similar to those of dopamine; at least these effects are seen on the renal blood flow and the cardiovascular system, aren't they?

DR. PRESSMAN: Yes, but dopamine would give rise to a response like that of the other catecholamine-like substances. It would stimulate the myocardial oxygen consumption. Ionophores have some effects that are similar to those of dopamine, and resemble dopamine more than they do the other catecholamines. Some differences still remain, however.

RECONSTITUTION OF ACTIVE TRANSPORT BY KIDNEY AND BRAIN $(Na^+ + K^+)$-ATPase*

Stanley M. Goldin† and Kathleen J. Sweadner‡

The Biological Laboratories
Harvard University
Cambridge, Massachusetts 02138

INTRODUCTION

Reconstitution of membrane proteins is a technique that potentially can provide information about membrane function. The first convincing demonstration that a purified membrane protein could be inserted into a model membrane system was Racker's reconstitution of ATP-dependent Ca^{++} uptake by the purified Ca^{++}-ATPase from sarcoplasmic reticulum[1]; the results obtained in that study could be consistent with either ATP-dependent ionophoresis or active uptake of calcium.

Employing a similar technique, we have reconstituted ouabain-inhibitable, ATP-dependent Na^+ transport by the purified $(Na^+ + K^+)$-ATPase (sodium and potassium ion-stimulated adenosine triphosphatase) from canine renal red medulla[2]: Hilden *et al.* reproduced these findings with purified $(Na^+ + K^+)$-ATPase from *Squalus acanthias* rectal gland.[3] It has been demonstrated that the Na^+ transport observed was active, and that the kidney enzyme actively transports Cl^- in a 1:1 ratio along with Na^+. This unexpected result could be due to the purification or reconstitution techniques or to a tissue-specific difference between the function of $(Na^+ + K^+)$-ATPase in kidney versus its function in nerves[4] and red cells.[5]

We summarize here recent findings, to be published elsewhere in detail,[6] to the effect that in this reconstitution system the brain $(Na^+ + K^+)$-ATPase actively transports $K^+(Rb^+)$ counter to Na^+ in a 1.8:3 ratio. This is close to the 2:3 ratio observed in red cells and nerves. This paper further deals with the assumptions involved in interpreting the transport measurements made with this system and their mathematical consequences.

RECONSTITUTION OF THE $(Na^+ + K^+)$-ATPASE AND OBSERVATION OF Na^+, Cl^-, AND K^+ TRANSPORT

The methodology summarized in this section has been more thoroughly described elsewhere.[2, 6]

Microsomal $(Na^+ + K^+)$-ATPase is cosolubilized with phosphatidyl choline

* This work was supported by Grant No. HL 08893 from the National Institute of Health and Grant No. GB 36827 from the National Science Foundation (to Guido Guidotti).

† Supported by United States Public Health Service Training Grant No. GM 138.
Correspondence should be addressed to Mr. Goldin.
‡ Supported by United States Public Health Service Training Grant No. GM 36.

using 2% sodium cholate. After a high-speed centrifugation that pellets protein resistant to detergent solubilization, vesicles are formed when the resulting supernatant is dialyzed against histidine or imidazole buffer containing 30 mM NaCl, 20 mM KCl, and 0.25 M sucrose. The high osmolarity of sucrose in the buffer ensures that quantification of intravesicular concentration changes due to active transport are not complicated by a significant change in vesicle size because of osmotically driven swelling or shrinking. Vesicle formation is established by the ability of the preparation to trap radioactive solutes (e.g., [^{3}H]-glucose, $^{22}Na^+$, $^{36}Cl^-$) after dialysis, which removes 99% of the detergent. Both phospholipid and trapped [^{3}H]-glucose elute in the void volume of a Sepharose 4B column, indicating that these vesicles are substantially larger than the 250-Å diameter phosphatidyl choline "Huangosomes"[7] formed by sonication.

Vesicles have been made from the Kyte microsomal renal ($Na^+ + K^+$)-ATPase by this technique. They contain only the two polypeptides of the renal red medullary ($Na^+ + K^+$)-ATPase of the purest preparations of this enzyme obtained so far.[8, 9] Microsomal ($Na^+ + K^+$)-ATPase was purified from brain gray matter by the first two steps of the procedure of Nakao *et al.*[10] and reconstituted into vesicles by this technique as well.

The vesicles incorporating either brain or kidney ($Na^+ + K^+$)-ATPase have similar monovalent ion permeability characteristics, and are similar to vesicles formed by the same technique but without added enzyme. $^{22}Na^+$, $^{86}Rb^+$, or $^{42}K^+$ added to the outside of the vesicles equilibrates with the interior with a $t_{1/2}$ of approximately 5 minutes at 37° C, as shown by rapidly chilling small aliquots of these vesicles after incubation, and separating free from trapped isotope on small Sephadex columns at 2° C. $^{36}Cl^-$ equilibrates roughly twice as fast, to the same equilibrium level as the cations (FIGURE 1a and b).

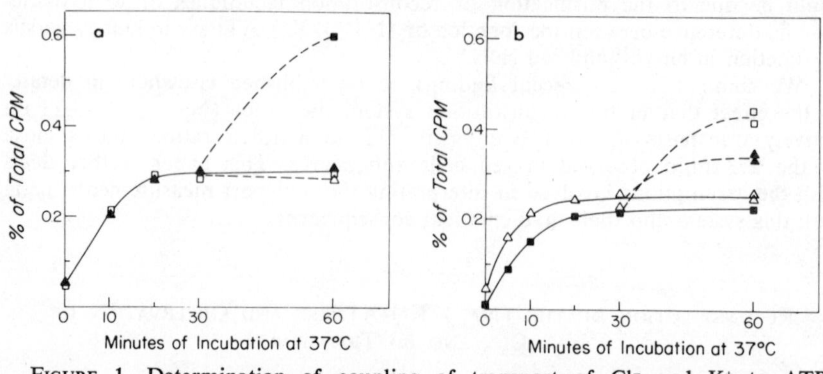

FIGURE 1. Determination of coupling of transport of Cl^- and K^+ to ATP-dependent active uptake of Na^+, using "low-K^+ vesicles" formed as described in text. (a) Double-label determination of $^{42}K^+$ and $^{22}Na^+$ transport equilibrium of $^{42}K^+$ (□) and of $^{22}Na^+$ (▲) with vesicles in absence of ATP. ATP-dependent $^{22}Na^+$ active uptake (■) and ATP-dependent $^{42}K^+$ transport (△), both after 30-minute pre-equilibration without ATP. See text for details. (b) Parallel determination of $^{36}Cl^-$ and $^{42}K^+$ transport. Equilibration of $^{36}Cl^-$ (△) and of $^{22}Na^+$ (■) with vesicles in absence of ATP. ATP-dependent active uptake of $^{22}Na^+$ (□/□) and of $^{36}Cl^-$ (▲/▲), both after 30-minute preequilibration without ATP.

Trapping these isotopes in the vesicles during their formation by dialysis gives the same trapping capacity as when the vesicles are passively equilibrated with externally added isotope, showing that there are no aqueous compartments that are not accessible to externally added isotope. Addition of ATP to the outside of the vesicles after thirty minutes of passive preequilibration with radioactive isotope results in a twofold increase in trapped $^{22}Na^+$, almost no change in trapped $^{42}K^+$, and a 1.4-fold increase in trapped $^{36}Cl^-$; taking into account the relative concentrations of Na^+ and Cl^-, the ratio of Na^+ to Cl^- ion movement is seen to be 1:0.92. This ATP-dependent uptake is resistant to 0.3 mM externally added ouabain, an inhibitor of the $(Na^+ + K^+)$-ATPase, but is completely inhibited by 0.3 mM ouabain trapped inside the vesicles during their formation. ATP-dependent uptake does not occur in vesicles formed without enzyme.

Reconstitution of the brain enzyme by this method shows a similar pattern of passive and ouabain-inhibitable, ATP-dependent $^{22}Na^+$ transport, but also shows a marked (greater than twofold) decrease in internal $K^+(^{86}Rb^+)$ (FIGURE 2A). Another feature of this system, true for both kidney and brain enzyme reconstitutions, is demonstrated in FIGURE 2: when 50 mM fructose is present in the reconstitution buffer, rapid depletion of the energy source (ATP) can be achieved by using it to phosphorylate fructose when hexokinase is added to the system (FIGURE 2B). This results in eventual reequilibration of all actively transported ions back to the levels observed before ATP is added (FIGURE 2A). By simultaneously measuring ATP hydrolysis together with ATP-dependent $^{22}Na^+$ uptake, 3.1 Na^+ ions are transported per ATP hydrolyzed in our most active preparations during their first minute in the presence of ATP. Corrections were made for the Mg^{++}-ATPase activity present in the brain preparation.

Qualitative Interpretation of These Results

The foregoing results can be qualitatively interpreted in only one clear and simple way. In reconstituted preparations of both brain and kidney, a portion of the enzyme is oriented in vesicle membranes in such a way as to catalyze active uptake of $^{22}Na^+$ coupled to ATP hydrolysis on the vesicles' outside surface, and is inhibitable by internally trapped ouabain. In the case of the brain enzyme, active efflux of $K^+(Rb^+)$ coupled to this ATP hydrolysis is observed, but not in the case of the kidney enzyme. That this ATP-dependent transport is active and not passive is evident because the ATP-dependent measurements are taken *after* the vesicles are preequilibrated with isotope; under these circumstances any passive change in the permeability of the vesicles to these ions would not be observed. Added confirmation of this is that this uptake is reversible—ATP depletion allows diffusion of these ions back to their passive equilibrium levels.

FIGURE 3 illustrates the behavior of each enzyme in the reconstitution system.

Quantitative Analysis of the Ion Movements

To obtain more quantitative insight into the mechanism of action of these reconstituted enzyme preparations, we must decide what assumptions about

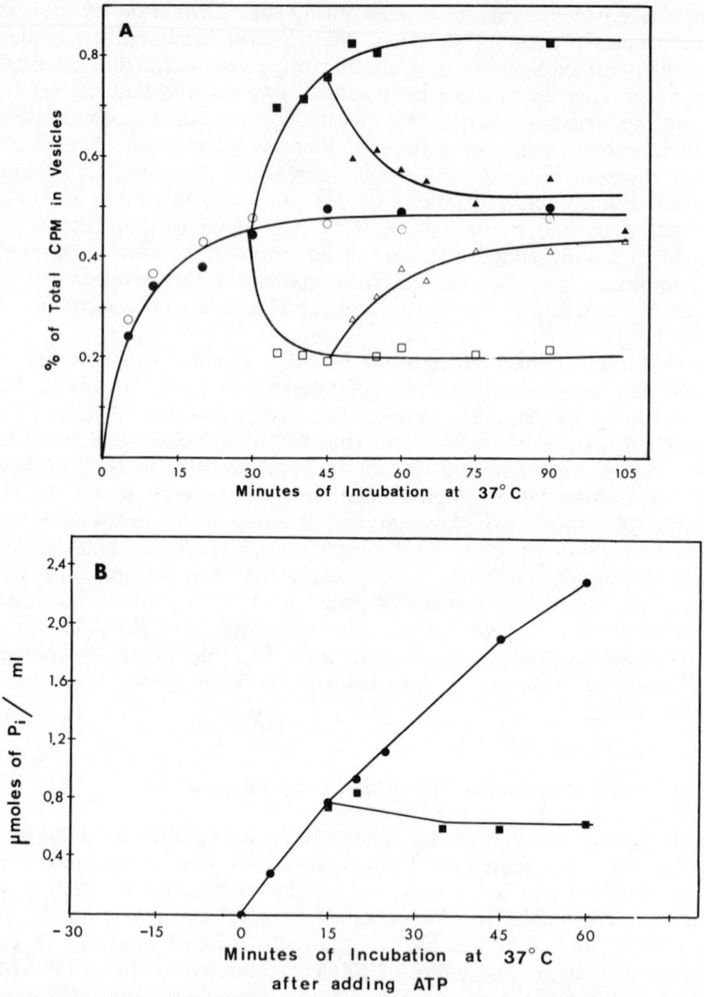

FIGURE 2. (A) Active transport of Na^+ and Rb^+. $^{22}Na^+$ or $^{86}Rb^+$ was allowed to passively equilibrate with the vesicles from outside (*circles*). At 30 minutes, Tris ATP was added, to 4 mM (*square s*). At 45 minutes, hexokinase was added (*triangles*). Closed symbols, $^{22}Na^+$; open symbols, $^{86}Rb^+$. The reconstitution buffer contained 30mM NaCl and 20mM (KCl + RbCl). (B) Pi produced from the hydrolysis of ATP under the conditions of the experiment in FIGURE 1a. Tris ATP was added at zero minutes (*circles*), and hexokinase was added at 15 minutes (*squares*).

the reconstitution system are reasonable to make. Ideally, we would like to not have to assume that these vesicles are a completely monodisperse, homogeneous preparation with a precisely defined radius. All of the mathematical manipulations shown below make no such assumption.

What we wish to see is whether the ATP-dependent movements of K^+ and Cl^- in both the kidney and the brain ($Na^+ + K^+$)-ATPase reconstitution systems can be explained by a merely passive response to active transport of Na^+. Active transport of Na^+ would create a membrane potential; the magnitude of this potential, and the relative permeabilities of K^+ and Cl^- would determine their movements in response to the membrane potential.

Derivation of Relative Permeabilities of K^+ and Cl^-

Let us first obtain an expression that enables us to calculate P_{K^+}/P_{Cl^-}, the ratio of the K^+ and Cl^- permeabilities. Consider a vesicle of radius r, permeable to both K^+ and Cl^-. If radioactive K^+ is added to the outside of this vesicle, it will enter the vesicle according to the following expression:

$$\frac{dN_{K^+}}{dt} = (P_{K^+})(4\pi r^2)(C_{K_o} - C_{K_i})$$

where dN_{K^+}/dt is the flux, in moles per second, of labeled K^+ into the vesicle, P_{K^+} is the membrane permeability coefficient for K^+, and $C_{K_o} - C_{K_i}$ are the outside and inside K^+ isotope concentrations, respectively.

Dividing both sides of the equation by $4/3\pi r^3$, the volume of the vesicles, we get

$$\frac{dC_{K_i}}{dt} = P_{K^+}(3/r)(C_{K_o} - C_{K_i})$$

If $C_{K_i} = 0$ initially, and C_{K_o} is a constant, we get, upon integration:

$$\ln \frac{C_{K_o}}{C_{K_o} - C_{K_i}} = P_{K^+}(3/r)t$$

When $C_{K_i} = \frac{1}{2}C_{K_o}$, then $0.693 = P_{K^+}(3/r)t_{\frac{1}{2}K^+}$, where $t_{\frac{1}{2}K^+}$ is the time needed for half of the isotope to equlibrate with the vesicles.

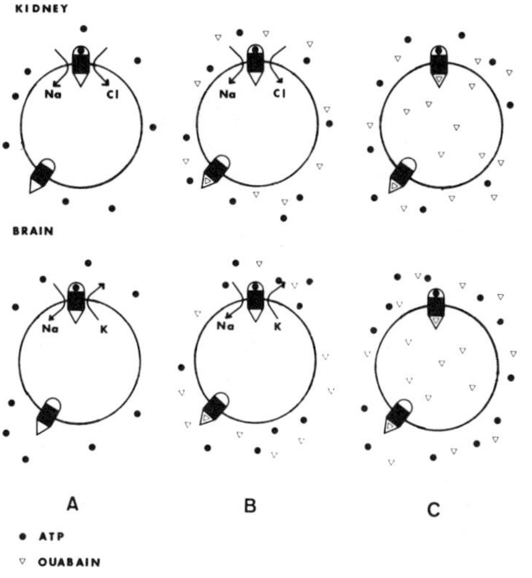

FIGURE 3. Diagrammatic representation of that fraction of $(Na^+ + K^+)$-ATPase that is located in sealed vesicles at the end of the reconstitution. Some enzyme will be oriented with its ATP-hydrolyzing site facing out, some facing in. In (A) only that fraction with the site facing out is activated. The kidney enzyme transports Na^+ and Cl^- into the vesicles. The brain enzyme exchanges K^+ for Na^+ in a 2:3 ratio. In (B) ouabain is added from the outside of the vesicles, and it has no effect on the fraction of the enzyme that is activated by external ATP. In (C) ouabain is trapped inside during dialysis, and inhibits all active transport.

By the same treatment we analogously obtain

$$0.693 = P_{Cl^-}(3/r)t_{½Cl^-}$$

So we see that the ratio of the relative isotope permeabilities,

$$\frac{P_{K^+}}{P_{Cl^-}} = \frac{t_{½Cl^-}}{t_{½K^+}}$$

Thus the relative Cl^- and K^+ permeabilities can be obtained without knowledge of the vesicle radius by the measurement of the time course of equilibration of $^{42}K^+$ and $^{36}Cl^-$ with the vesicle preparation. For the kidney enzyme preparation, $P_{K^+}/P_{Cl^-} = 0.56$.

Prediction of Movements of K^+ and Cl^- in Response to Active Pumping of Na^+

Hodgkin and Katz[11] present an equation for the movement of K^+ ion across a membrane under the influence of diffusion and an electric field which involves now generally accepted assumptions about the behavior of the cell membrane:

$$-I_K = RT\mu_K \frac{dC_K}{dx} - C_K\mu_K F \frac{V}{a} \qquad (1)$$

where the above quantities are defined as usual. §

Now, for a vesicle of radius r,

$$\left(\frac{dN_K}{dt}\right)\left(\frac{1}{4\pi r^2}\right) = -I_K$$

This converts a current measurement to a flux measurement, and

$$\frac{d[K^+]}{dt}\frac{(4/3 \cdot \pi r^3)}{4\pi r^2} = -I_K = \frac{r}{3}\frac{d[K^+]}{dt}.$$

P_K is the K^+ permeability coefficient defined as

$$P_K = \frac{\mu_K \beta_K RT}{aF}$$

Inserting the above into Equation 1 and rearranging, we get

$$\frac{d[K^+]_i}{dt} = \frac{3}{r} FP_K \left[[K]_o \left(1 + \frac{FV}{2RT}\right) - [K]_i \left(1 - \frac{FV}{2RT}\right)\right] \qquad (2)$$

Rearranging for integration:

$$\int_{[K_o]}^{[K_i]} \frac{d[K_i]}{[K_o]\left(1 + \frac{FV}{2RT}\right) - [K_i]\left(1 + \frac{FV}{2RT}\right)} = \frac{3}{r} FP_K \int_0^t dt \qquad (3)$$

Before we activate the enzyme by adding ATP, the vesicles have been preequilibrated with isotope so

$$[K_i] = [K]_o \text{ at } t = 0$$

§ Definitions of terms used in the Hodgkin-Katz equation are as follows: I_K is the K^+ current density through the membrane; C_K is the K^+ concentration in the membane; μ_K is the K^+ mobility; a is membrane thickness; V is the membrane potential; dC_K/dx is the K^+ concentration gradient across the membrane; and F, R, and T have their usual meanings

Evaluating this integral, we have:

$$\frac{-1}{\left(1-\frac{FV}{2RT}\right)}\left(\ln\left[\frac{[K]_o\left(1+\frac{FV}{2RT}\right)-[K]_i\left(1-\frac{FV}{2RT}\right)}{[K]_o\left(\frac{FV}{RT}\right)}\right]\right)=\frac{3}{r}FP_Kt \quad (4)$$

Employing the same procedure for the movement of Cl^-, we get:

$$\frac{-1}{\left(1+\frac{FV}{2RT}\right)}\left(\ln\left[\frac{[Cl]_i\left(1+\frac{FV}{2RT}\right)-[Cl]_o\left(1-\frac{FV}{2RT}\right)}{[Cl]_o\left(\frac{FV}{RT}\right)}\right]\right)=\frac{3}{r}FP_Kt \quad (5)$$

To eliminate vesicle radius, as we promised we would, simply divide Equation 5 by Equation 4:

$$\frac{\left(1-\frac{FV}{2RT}\right)}{\left(1+\frac{FV}{2RT}\right)}\left(\frac{\ln\left[\frac{[Cl]_i\left(1+\frac{FV}{2RT}\right)-[Cl]_o\left(1-\frac{FV}{2RT}\right)}{[Cl]_o\left(\frac{FV}{RT}\right)}\right]}{\ln\left[\frac{[K]_o\left(1+\frac{FV}{2RT}\right)-[K]_i\left(1-\frac{FV}{2RT}\right)}{[K_o]\frac{FV}{RT}}\right]}\right)=\frac{P_{Cl}}{P_K} \quad (6)$$

Now in our system, if Na^+, K^+ and Cl^- are the only ions that move, conservation of charge dictates that at any given time:

$$[Na]_i - [Na]_o = [Cl]_i - [Cl]_o + [K]_o - [K_i] \quad (7)$$

Using this to replace $[Cl]_i$ in Equation 6, we get:

$$\frac{\ln\left[\frac{([Na]_i-[Na]_o+[Cl]_o+[K]_i-[K]_o)\left(1+\frac{FV}{2RT}\right)-[Cl]_o\left(1-\frac{FV}{2RT}\right)}{[Cl]_o\frac{FV}{RT}}\right]}{\ln\left[\frac{[K]_o\left(1+\frac{FV}{2RT}\right)-[K_i]\left(1-\frac{FV}{2RT}\right)}{[K_o]\frac{FV}{RT}}\right]}=\frac{P_{Cl}}{P_K}\frac{\left(1+\frac{FV}{2RT}\right)}{\left(1-\frac{FV}{2RT}\right)} \quad (8)$$

$[Cl]_o$, $[K]_o$, and $[Na]_o$ as well as P_{Cl}/P_K are all known; $[Na]_i$ is experimentally determined. The membrane potential, V, must be less than the membrane breakdown voltage but greater than the minimal voltage necessary to cause $[Na]_i$ to increase by the observed value (we will show how to determine this V_{min} later.) Equations 7 and 8 put rigid constraints on the change in $[Cl]_i$ and $[K]_i$ that must occur if these movements are passive responses. Solving this set of equations using computerized numerical methods, we see that with the twofold increase in $[Na]_i$ exhibited in FIGURE 4 an 18–25% decrease in $[K]_i$ and a 34–36% increase in $[Cl]_i$ would be expected. In actuality, what is observed is a 40% increase in $[Cl]_i$ and only a 6.5% decrease in $[K]_i$, less that one-third of the predicted value. This result suggests that the enzyme selectively, actively translocates Cl^- along with Na^+ rather than letting K^+ and Cl^- passively distribute themselves according to their permeabilities in the vesicle membrane.

Demonstration of Active K^+ Transport in the Brain

Now, in the case of the brain enzyme reconstitution (FIGURE 2) the time course of ATP-dependent Na^+ uptake is shown to reach a steady state, where the rate of active uptake of $^{22}Na^+$ equals its rate of back-leakage. In this situation, if we make the assumption that the K^+ movement is a passive response to the active Na^+ transport, the left half of Equation 2 equals zero, and we can then say that

$$0 = [K]_o\left(1 + \frac{FV}{2RT}\right) - [K]_i\left(1 - \frac{FV}{2RT}\right),$$

so

$$\frac{[K]_i}{[K]_o} = \frac{\left(1 + \frac{FV}{2RT}\right)}{\left(1 - \frac{FV}{2RT}\right)}$$

It is this value of FV/2RT that is the minimum necessary to produce a given passive movement of an ion. A value larger than this would dictate that the ion movement has not yet reached steady state. Analogously, it can be shown that:

$$\frac{[Cl]_i}{[Cl]_o} = \frac{\left(1 - \frac{FV}{2RT}\right)}{\left(1 + \frac{FV}{2RT}\right)},$$

and so:

$$\frac{[K]_i}{[K]_o} = \frac{[Cl]_o}{[Cl]_i}$$

This relationship, together with the requirement of conservation of bulk electrical neutrality, allows us to solve for predicted $[K]_i$ and $[Cl]_i$ for a given observed increase in $[Na]_i$. This enables us to compare the predicted enzyme-catalyzed movements with those actually observed.

FIGURE 4 shows the results of ATP-dependent K^+ and Na^+ measurement for a series of experiments identical in protocol to that of FIGURE 2A. Three different starting Na^+ and K^+ concentrations were employed—30:20, 20:30,

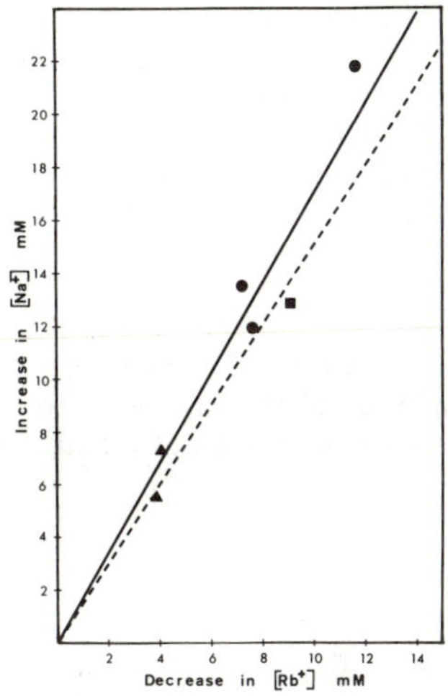

FIGURE 4. The increase in [Na^+] and decrease in [Rb^+] inside the vesicles was calculated from the levels attained at steady state in the presence of ATP. The solid line is a least-squares fit; the dotted line is a ratio of 3:2 of Na^+ to Rb^+ transported. *Circles:* 30 mM Na^+, 20 mM K. *Square:* 20 mM Na^+, 30 mM K^+. *Triangles:* 40 mM Na^+, 10 mM K^+. The point at 11.6 mM K^+ is from vesicles made by the four-day procedure.[2] All others were from vesicles made by the 16-hour procedure.[6]

and 40:10 (mM). The ratio of Na^+ to Rb^+ transported at steady state, taken from the slope of the line fitted by the method of least squares, is 3:1.8. All three starting Na^+:K^+ concentration ratios gave essentially the same result. The small deviation from the expected ratio of 3:2 may have been due to the substitution of $^{86}Rb^+$ for K^+.

These experimentally observed changes in [K^+] are three- to sixfold higher than the predicted passive response, indicating that most of the observed outward K^+ transport in the reconstituted brain ($Na^+ + K^+$)-ATPase is active and not merely a passive response to an electrogenic sodium pump.

CONCLUDING REMARKS

The studies described here were performed primarily to achieve two goals. The first was to demonstrate what was suspected for almost two decades[12]: that a purified ($Na^+ + K^+$)-ATPase is indeed capable of catalyzing ATP-dependent, ouabain-inhibitable active cation transport. The reconstitution of the purified ($Na^+ + K^+$)-ATPase from renal red medulla has confirmed this.

The human brain is a complex, highly organized entity containing many cell structures that have no known analog elsewhere. The composition and behavior of the cell membranes of these structures may give us important clues about the functions of these structures and, indeed, of the brain as a whole. A computer, for example, may appear complex, but is really made up of a finite number of simple switching and storage elements; a person unfamiliar with how the computer is put together might do well to first understand the function of the individual elements it consists of.

With such thoughts in mind, our second goal has been to begin to develop a methodology to obtain, by reconstitution, detailed information about the mechanism of active transport proteins. The unexpected result that the reconstituted kidney ($Na^+ + K^+$)-ATPase actively pumps both Na^+ and Cl^- has been rendered more credible by the fact that the reconstituted brain ($Na^+ + K^+$)-

ATPase behaves precisely as physiological studies on nerves would lead one to expect. Further confirmation of the active Cl^- transport finding comes from recent electrophysiological studies on isolated perfused thick ascending limbs of Henle's loop. It is this part of the kidney that is richest in $(Na^+ + K^+)$-ATPase[13] and is the source of the purified $(Na^+ + K^+)$-ATPase used for this reconstitution work. These studies[14, 15] demonstrate that the thick ascending limb is capable of ouabain-inhibitable active transport of Cl^- as well as of Na^+.

We hope to see, in the near future, further development and use of membrane reconstitution methods: not just to confirm what we already know from physiological studies, but also to obtain useful and reliable information about structures that are not readily accessible to such physiological techniques.

Acknowledgment

This work was performed in the laboratory of Professor Guido Guidotti. We wish to thank him for his generous support, encouragment, and advice.

References

1. Racker, E. 1972. J. Biol. Chem. **247:** 8198–8200.
2. Goldin, S. M. & S. W. Tong. 1974. J. Biol. Chem. **249:** 5907–5915.
3. Hilden, S., H. M. Rhee & L. E. Hokin. 1974. J. Biol. Chem. **249:** 7432–7440.
4. Thomas, R. C. 1969. J. Physiol. (London) **201:** 495–514.
5. Post, R. L. & P. C. Jolly. 1957. Biochim. Biophys. Acta **25:** 118–128.
6. Sweadner, K. J. & S. M. Goldin. 1975. J. Biol. Chem. **250:** 4022–4025.
7. Huang, C. 1969. Biochemistry 8: 344–352.
8. Kyte, J. 1971. J. Biol. Chem. **246:** 4157–4165.
9. Jorgensen, P. L. 1974. Biochim. Biophys. Acta **356:** 36–52.
10. Nakao, T., M. Nakao, N. Mizuno, Y. Komatsu & M. Fujita. 1972. J. Biochem. (Tokyo) **73:** 609–619.
11. Hodgkin, A. L. & B. Katz, 1949. J. Physiol. (London) **108:** 37–77.
12. Skou, J. C. 1965. Physiol. Rev. **45:** 596.
13. Schmill, V. & V. C. Dubeck. (1971). Eur. J. Clin. Invest. **1:** 307–312.
14. Rocha, A. S. & J. P. Kokko. 1973. J. Clin. Invest. **52:** 612–619.
15. Burg, M. B. & N. Green. 1973. Am. J. Physiol. **224:** 659–668.

Discussion

Dr. Finkelstein: Couldn't chloride transport with sodium in the kidney preparation be demonstrated less ambiguously by treating the vesicles with valinomycin so that the passive permeability to potassium was made much larger than that of chloride or sodium? So then, without any algebra or comparing 18% effects with 3% effects, you would have a clear-cut demonstration of whether or not that chloride was obligatorily transported along with the sodium.

MR. GOLDIN: I see your point; we intend to perform that experiment.

DR. SACHS: If you substitute with a nonpermeant anion, do you affect the sodium uptake?

MR. GOLDIN: This is something that had been done by Michael Lincoln, an undergraduate working under my supervision. If you replace chloride in the kidney reconstitution system by an impermeant anion, such as sulfanilic acid, you do see the potassium efflux. But the work is very preliminary.

DR. GREEN: It seems that the studies you carried out in kidney and brain ATPase provide a completely unexpected new insight into the mechanism. From your data it appears that you have separated the two parts of the process. One part has to be an active transport of sodium; that's true in both the kidney and in the brain preparation. But this can be coupled either to a symport movement of chloride in kidney or to an antiport passive movement of potassium. By passive movement, I mean a specialized system where the active transport of sodium is coupled to a moving back of potassium. It's not a random process; it requires a special system. And in the case of the kidney that system is missing and you have only a corresponding movement of chloride. So it would suggest that the mechanism of sodium potassium ATPase must involve a primary active transport of sodium, coupled either to a symport movement of chloride or an antiport movement of potassium. The point is: if you don't move the potassium, you're going to run into problems if you can't move the chloride.

MR. GOLDIN: The ions always have the option of moving passively.

A COMPARISON OF RECENT SUGGESTIONS FOR THE FUNCTIONAL ORGANIZATION OF RED-CELL SUGAR-TRANSPORT SITES BASED ON KINETIC OBSERVATIONS

Paul G. LeFevre

Department of Physiology and Biophysics
Health Sciences Center
State University of New York at Stony Brook
Stony Brook, New York 11794

The above title, submitted to our Conference Chairman some months ago, does not place quite the proper emphasis on what I now wish to define as my intention in this presentation. In the last few years a number of diverse propositions have been offered as more adequate alternatives to the classical symmetrical mobile-carrier picture of the mediated permeation of human erythrocyte membranes by glucose and other simple sugars. Originally I had thought to catalog here the essential postulates of these several newer formulations, taking note of the kinetic basis for proposing them, and briefly tabulating what seem to be their comparative successes and shortcomings. However, I am aware of the generally low tolerance toward this brand of model-building among those who have not themselves indulged, and I suspect that many may view the inclusion of this type of review in our program as something of a paradigmatic lapse; I would not want to aggravate this misunderstanding by parading the details of a series of formalized daydreams, especially when I would have to conclude that no one of these seems remarkably more satisfactory than any of the others.

Accordingly, I will not attempt here to detail the kinetic analyses of the respective models, and will in fact identify only briefly the nature of the crucial kinetic observations that they have been designed to encompass. Rather, my prime emphasis will be on the relatively small number of *conceptual distinctions* regarding the sequential steps in the "carrier" operation that crop up among the new proposals, and that bear on the problems of matching the kinetic peculiarities of the biological system. It is rather critical for our communication that the reader take note of some diagrammatic shorthand that I propose to use to facilitate these comparisons; I hope to explain this symbolism as we proceed through the figures. I must confess that in some instances I have taken the liberty of modifying significantly the diagrammatic presentations of the original authors in order to emphasize the comparative basic concepts rather than the specifics of the various formulations; I apologize in advance if this should inadvertently result in any distortion of the original propositions.

This system, which facilitates the movement of monosaccharides through red-cell membranes, is almost universally chosen for the textbook illustration of "facilitated diffusion," presumably the simplest type of mediated transport. The picture generally presented is that of Widdas,[1] as illustrated in FIGURE 1. Widdas originally developed this schema to interpret his observations on placental sugar transfer,[2] but its main application has been to the erythrocyte

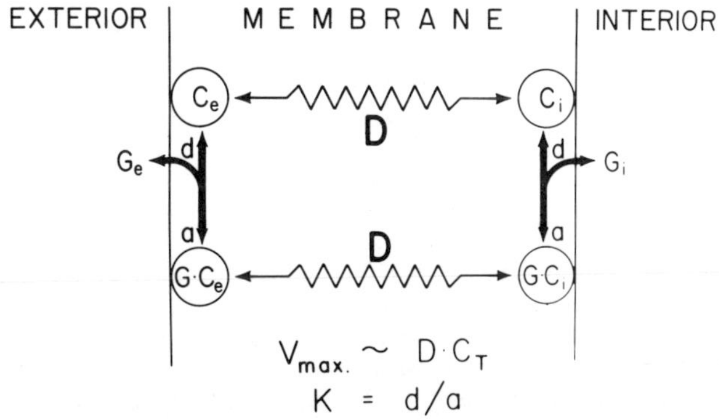

FIGURE 1. The classical symmetrical mobile-carrier model of Widdas[1] for facilitated diffusion. Carrier C at either membrane interface reacts reversibly with substrate G in adjacent compartment to produce complex G·C; heavy arrows indicate relative rapidity of this association-dissociation (a, d) equilibrium. Both free carrier and complex diffuse (process D) within membrane between exterior and interior interfaces (denoted by subscripts (e, i). Relative slowness of transmembranal diffusion is indicated by *resistance* symbol in arrows denoting D process and by size of D symbol. Free mobility of carrier and complex across membrane is indicated by circular enclosures. These conventions apply also to other figures. C_T expresses total carrier content of membrane. K is equilibrium (dissociation) constant for process a, d.

system. For many years it provided a remarkably satisfactory, quantitatively consistent framework for systematizing a wide variety of kinetic findings, for both net transfers and unidirectional tracer fluxes, in either direction through human red-cell membranes. Although the symbolism in this diagram is rather obvious, I want to call attention to certain conventions here that will persist through all the other figures. The critical membrane sites are designated by the capital letter C (for carrier); these can combine reversibly at either membrane interface with the sugar, usually designated as G (for glucose) in the adjacent aqueous compartment, thus forming G · C, a "complex" or an "occupied site." This occurs by some sort of fairly specific adsorption-desorption or association-dissociation (designated by lowercase a and d). The capital K designates the equilibrum constant for this process, expressed as a dissociation constant (hence d/a). On any of these symbols, the subscripts e and i refer respectively to the *e*xterior and *i*nterior compartments or interfaces. In this "classical" schema, the carrier is *truly* a carrier, acting as a simple ferryboat for the substrate G in its traversal of the barrier. Here and elsewhere, this type of function will be indicated by diagramming the sites as *circles,* and by use of the capital letter D (for diffusion) to label the *transmembranal* event (or its rate constant), whereby the sites are interchanged within the membrane between the e and i interfaces. It is essential to Widdas's kinetic treatment that this transmembranal diffusion be the bottleneck step in the reaction sequence, setting the overall pace for the sugar transfer. This is indicated in the figure by the large, thick lettering for this step "D," and also by the jagged path of the corresponding arrows, to suggest considerable *resistance* in this step. Furthermore, this assumption in the classical model means that the interfacial reactions are viewed as *always virtually at equilibrium;* this is symbolized by the thick, heavy arrows designating the a, d processes.

It is noteworthy that for a transfer in one direction, this set of assumptions parallels the postulates of classical enzymology, so that the *unidirectional fluxes* follow Michaelis-Menten kinetics. Thus, the translocation behavior of a particular sugar at a given temperature in this system is completely specifiable by statement of only two constants, as indicated at the foot of FIGURE 1: a transmembranal *rate* parameter, or maximal flux, V_{max} (set by the mobility constant, D, multiplied by some expression of the carrier's total population density, C_T, in the membrane), and the *affinity* constant, K (i.e., d/a).

The difficulties with this classical model became gradually more apparent about ten years ago, and were particularly well crystallized experimentally in a systematic series of studies in the late 1960s by D. M. Miller[3-6] at the Research Institute of the Canada Department of Agriculture in London, Ontario. Additional related problems[7-13] have since been documented from tests of critical predictions from some of the newer models we will discuss. Time limitations preclude the reiteration here of the specifics of these failings in the classical model; and in any case, a number of excellent presentations of this are already available.[6, 14, 15] Suffice it to say that the acute problems arise primarily when quantitative data from diverse kinetic procedures are considered together. In particular, the apparent magnitude of the nominal constants varies systematically according to which of several equally justifiable procedures is involved. The affinity factor K behaves especially badly, varying for glucose at 20° C by more than 20-fold[4]; at lower temperatures the problem is even worse by reason of opposed temperature-dependencies in the respective apparent values.[16, 17] This vastly surpasses the range of possible technical uncertainty, or the calculable systematic divergences that are sometimes inherent in the methods of data analysis. Yet the elegantly simple, symmetrical, two-factor Widdas model generally proves admirably consistent for any *homogeneous* series of tests involving a particular class of kinetic measurement. In fact, in one such investigation comparing a series of different sugars, all the V_{max}'s appropriately appeared roughly equivalent, while the affinity constants differed by 400-fold.[18]

The attempt in the last few years to retain these consistencies while adjusting to the discordant findings has brought about a rather quiet revolution, still not clearly resolved, in our ideas regarding how the system might basically be organized. In seeking to survey the proposals that have been made, one could readily be dismayed both by the widely varying specifics and by the frequently elusive vagueness in the expression of the critical presumptions. However, it appears to me that *three primary conceptual revisions* keep cropping up, singly or in combinations:

(a) replacement of the fully mobile carriers with "fixed sites" or "hemiports," which *exchange* the adsorbed substrate but *do not accompany* it in its move across the membrane;

(b) insertion of an appreciable *resistance in series* with the carrier system proper, either as an "unstirred layer" adjacent to the membrane or within the membrane's superficial structures, through which the sugar must diffuse to approach or depart from the reactive sites; and

(c) abandonment of Widdas's simplifying restrictions regarding the relative speed of the several steps, thus allowing (1) various loci of the rate limitation, (2) quantitative alteration of carrier behavior when the complex with substrate is formed, and particularly (3) *asymmetry* in the interior and exterior affinities and in the movement through the membrane in the two directions.

It is not really feasible to segregate completely these three issues in our disussion here, as they recur in an overlapping way in the several models that we should look at. However, let us first consider the conceptual significance of a proposal of type (a). This harks back to the possibility raised about 20 years ago by Stein and Danielli[19] that the selective adsorption sites might be aligned along the walls of narrow polar channels through the otherwise continuous lipid membrane, as in FIGURE 2. These channels might be actual aqueous pores, but are now more likely to be identified with membrane-spanning protein units. For contrast with the circular mobile carriers, here and in all later figures, sites of such restricted mobility are diagrammed as *squares* (or rectangles). Following an initial adsorption of the sugar or other substrate at one of the interfacial positions (as with the formation of $G \cdot C_e$ at the left of the diagram), transfer proceeds by a *basically random jumping* between adjacent sites—a process that has been referred to as "polar creep" or "one-dimensional diffusion."[20, 21] (It makes very little difference in the kinetics how many sites are presumed to make up the linear chain; the use of five in FIGURE 2 is simply in deference to the alphabetical subscripts, e ↔ i.) For all fixed-site models here, we will designate the creep between adjacent sites in either direction as process "x," which switches the occupancy state of the pair of sites between upper and lower lines in a diagram such as that of FIGURE 2. In analogy with the diffusion step in the classical mobile-carrier model, this *x*-process is here made the rate-limiting (high-resistance) step; thus it is easy to appreciate that for many situations the predicted kinetics will parallel closely those of the Widdas model, and may be characterized by analogous constants, as indicated at the foot of the diagram.

However, the "single-file" characteristic[20, 21] makes this model as it stands incompatible with the observed phenomenon[22] of *counter-transport,* predicted from the Widdas model: under appropriate conditions, the downhill flow of

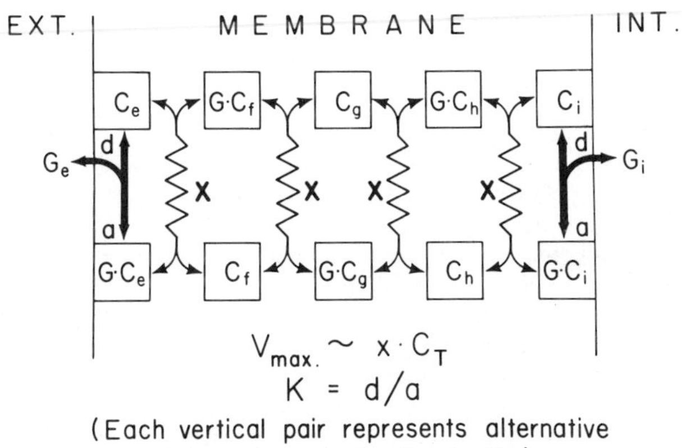

FIGURE 2. Single-file fixed-site "polar-creep" model of Stein and Danielli.[19] Interfacial sites C_e and C_i react with G as in FIGURE 1, but remain in essentially fixed position (fixity indicated by square enclosures). Substrate is passed by process x between adjacent fixed sites in chain spanning membrane (subscripts e, f, g, h, i).

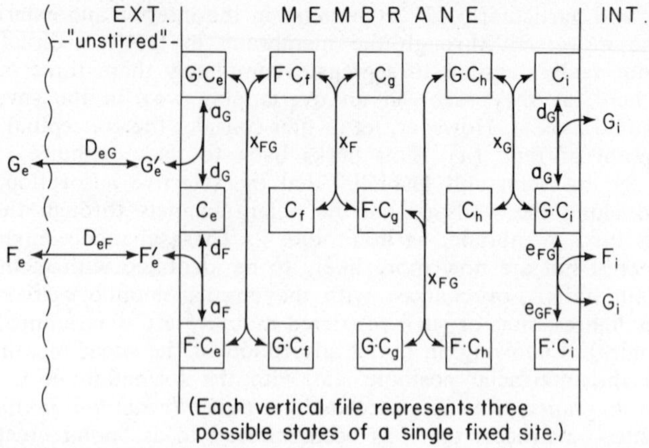

FIGURE 3. Two-way-traffic fixed-site model of Naftalin.[23] Two substrates F and G interact with C as in FIGURE 2, but can also *exchange* position, both in interfacial reactions (process e) and in transmembranal migration (process x), as indicated by respective subscripts. External to cells is "unstirred layer" imposing resistance to substrate's diffusion (D_e), so that concentrations of (F_e') and (G_e') directly in contact with exterior interface may differ from bulk-phase levels of F_e and G_e.

a sugar in one direction through the membrane can produce a transient net flow uphill of a tracer component, or of a second substrate, in the opposite direction. This finding clearly suggests that an unevenly distributed substrate can compete for the carrier transfer in one direction, while not similarly interfering with the movement of the second sugar in the other direction. A fixed-site "bucket-brigade" mechanism seems to pose a mutual interference between fluxes in the two directions; consequently, for many years little serious consideration was given to such hypotheses.

The issue was reopened in 1970 by Naftalin's effort[23] to provide a basis for the kinetic anomalies to which Miller's work had called special attention. The critical innovation in Naftalin's proposal, illustrated in FIGURE 3, was the allowance of *exchange of two occupants* between adjacent sites, so that the movement of sugars through the site-lattice need no longer be strictly "single-file." Thus the creep of one substrate may be coupled with the counter-creep of another (or of an isotopic tracer component of the same species), shown in two places in FIGURE 3 as the x_{FG} exchange process. In its general form, Naftalin's schema does not stipulate which step in the process be rate-limiting; but he stresses that the likelihood of an x_{FG} event may well exceed that of an x_F or x_G migration into a vacant site. Similarly, he tends to assign a greater probability to the surface exchange process "e" indicated at the lower right in FIGURE 3, than to the independent a, d events involving a single substrate. It is also emphasized that e_{FG} need not equal e_{GF}.

It is not surprising that the postulating of such exchanges (both e and x) allowed the reconciliation of the polar-creep concept with the occurrence of counter-transport. However, there is another type of coupled-flow behavior observed experimentally which made it necessary for Naftalin to postulate also the significant external unstirred-layer diffusion barrier[23, 24] indicated in FIGURE 3. The phenomenon in question, referred to as *trans*-stimulation or "accelerated exchange," involves the enhancement of unidirectional efflux when a substrate is added to a previously sugar-free external medium (a point we will return to later on). Interposing this diffusion barrier obviously shifts the local concentrations of substrates immediately in contact with the cells, F_e' and G_e', somewhat away from the levels in the bulk medium, in accordance with the prevailing flows into or out of the cells. Given this complication, Naftalin's computer simulation behaved like the actual transport system in many significant respects. In particular, it provided for the large differences in the operationally defined K's obtained with the several recognized procedures; and qualitatively, it permitted the curious pattern seen in the comparison of *trans*-stimulation interactions between different pairs of sugars. However, both Miller[25] and Lieb and Stein[26] have, on independent grounds, challenged the realism of assigning significance to unstirred layers in this system, calculating that the proposition leads to quantitatively absurd or self-contradictory conclusions.

As mentioned earlier, the kinetics predicted for such fixed-site models as Naftalin's do not depend significantly on the number of intra-membrane stations lined up in the bucket brigades between the interfacial layers. The crucial distinction between our "circles" and "squares" is simply that the squares *at the opposite interfaces* remain distinctly segregated from each other; each acts, as implied by Stein's term *hemiport,*[27] to mediate the substrate transfer *not all the way* to the opposite side, but only between one of the aqueous phases and *the interior of the membrane.* Thus a simple *bilayer* of juxtaposed hemiports, within which substrate may migrate or exchange from one monolayer to the other, will behave essentially like Naftalin's more extended matrix.

It was in the pursuit of a simplified computer simulation of this condensed version of Naftalin's schema a few years ago that I stumbled upon a possible new emphasis in our conception of the critical steps in a hemiport operation[28]; this is diagrammed in FIGURE 4. Naftalin had utilized in his computer-model program a random variable to signal either "go-right" or "go-left" instructions

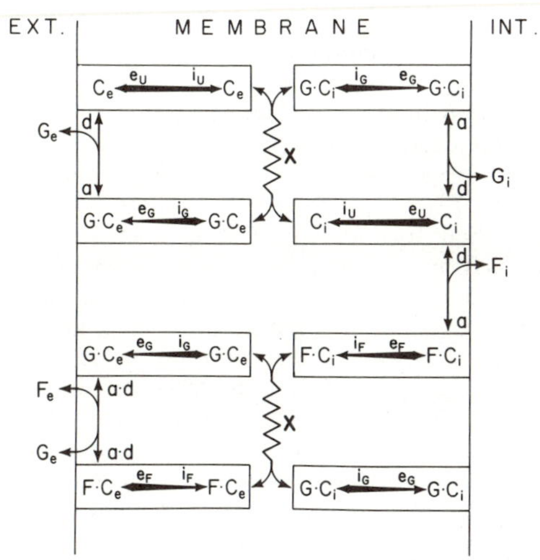

FIGURE 4. Bilayer "introversion" model of LeFevre.[28] Rectangular C_e and C_i units operate like squares of FIGURES 2 and 3, but oscillate in rapid conformational equilibrium between introverted orientation (allowing event x) and extroverted orientation (allowing a, d events). Varied tapering of arrows denoting the introversion-extroversion process (e, i) denotes shifting of its equilibrium according to the state of occupancy of site: subscript U denotes unoccupied state; F and G denote occupancy by corresponding substrate. At lower left, simultaneous events a, d are equivalent to exchange process e of FIGURE 3, but do not entail its independent rate constant.

for each site in the lattice at each iteration of the program. In application to a simple bilayer, this reduces to a sort of orientative or conformational choice at each site, disposing on the one hand toward interaction with the adjacent aqueous compartment, and on the other hand toward interaction with any apposed hemiport. I termed this shift an *introversion* or *extroversion*. The reason for placing any emphasis on this seemingly rather inconsequential step in the transfer process was the unexpected finding that if the left-right decision were *not* random, the overall kinetics were dramatically altered. In fact, if just this introversion-extroversion equilibrium were presumed to be governed by the identity of the substrate occupying the site, no other distinctive parameter need be postulated to account for essentially all the kinetic observations and for the large differences among the several substrates (classically taken to indicate widely differing "affinities"). Thus, as FIGURE 4 indicates, the interfacial events (a and d) and the hemiport transfer (x) may be assigned fixed values independent of which substrate or combination of substrates is involved. The single parameter conditioned by substrate occupancy is the statistical distribution of the hemiport's time between the introverted and extroverted states, in a rapid equilibrium; variation in this parameter is represented by the varying taper of the arrows in the balance of the i, e processes symbolizing intro- and extroversion in FIGURE 4. For example, the observed kinetics suggest that an unoccupied site spends somewhat more time in the extroverted than in the introverted state (e_U appreciably exceeds i_U); occupation by glucose very heavily favors the introverted state ($e_G \ll i_G$), while fructose occupation has the opposite effect ($e_F \gg i_F$). Again, as with the classical model, the transmembranal step, x, is taken as rate-limiting. Full specification of the transport operation thus requires assigning the three fixed constants (a, d, and x) plus an i-e equilibrium constant for the unoccupied site, and one for each type of occupancy (substrate) involved.

This schema, which comes fairly close to kinetic accord with all the facts available, is clearly reasonably consonant with many aspects of current thought regarding cell membrane structure. Let me note again that the principal conceptual distinction from other models is the status indicated by the *inner portions of the hemiport rectangles* in FIGURE 4: in this introverted state, the adsorbed sugars are neither necessarily committed to the transmembranal passage, nor yet available for discharge or exchange at the adjacent interface. It is the substrate-conditioned variation in the prominence of this "limbo"-like state that especially characterizes this schema.

An altogether different sort of organization of introverting hemiports is found in a series of reports from Lieb and Stein.[14, 29, 30] There are several points that specifically distinguish their formulation. Taking note of the wide variation in apparent K's derived from different procedures, Lieb and Stein looked into the possibility that the sugar-adsorbing sites actually are heterogeneous in this respect. They thus arrived at the model illustrated in FIGURE 5; here, sites of *two affinity classes* are assembled into *tetrameric groups*. (To those of you familiar with the original Liebs-Stein diagrams of this model, the present format must appear a bit strange; but I believe this view will help to bring out the fundamental analogies to other models, while underscoring the unique elements.) The tetramer is represented in the diagram by any horizontal row spanning the membrane, and consists of two high-affinity sites and two low-affinity sites, one of each type associated in hemiport-fashion with each side of the membrane. The hemiports in this case, however, do

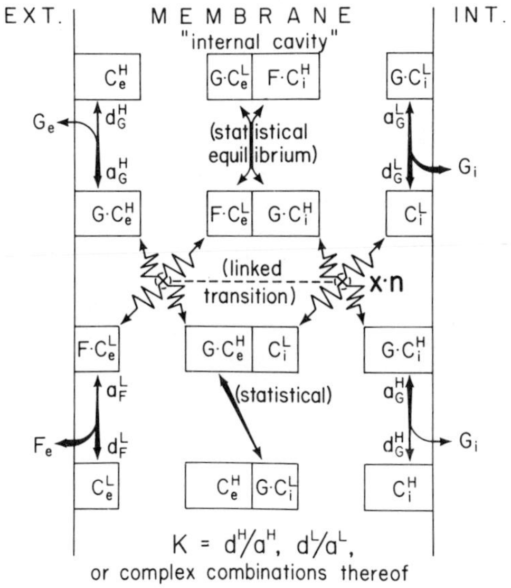

FIGURE 5. Tetrameric "internal-transfer" model of Lieb and Stein.[14, 29, 30] Tetramer unit consists of two high-affinity sites (C^H) and two low-affinity sites (C^L) assembled with one of each type at opposite interfaces, where a, d events occur as in previous figures. Other pair of sites is apposed at "internal cavity," within which any occupant sugars (F, G) readily migrate or exchange between the sites, according to relative affinities. Transition process x rearranges entire tetramer so that within each half of membrane, the introverted and extroverted sites are switched in position. Frequency of transition event is proportional to n, the number of sites (0, 1, 2, 3, 4) currently occupied.

not operate independently; they are arranged in such a way that at any instant the high-affinity site on one side and the low-affinity site on the other side are presented at the membrane interfaces, while the other member of each pair is introverted so that they are mutually apposed in an "internal cavity" within the barrier. So long as none of the sites in the tetramer is occupied, this conformation remains static. But any occupancy by substrate induces a tendency for the tetramer to undergo a "transition" whereby the monomeric units on each side are reversed in conformation. This generates an exactly equivalent tetrameric arrangement, but with each hemiport having undergone an intro- or extroversion from its previous position, so that the high- and low-affinity orientation is reversed. Lieb and Stein find it a workable postulate that the frequency of this transition is simply proportional to the number of sites occupied (the factor *n* in the diagram).

Within the internal cavity, it is presumed that any temporarily captured sugars are free to migrate between the two introverted hemiports, and will become distributed statistically according to the relative affinities at the two loci. Let us very briefly consider the operational results of such an arrangement. With a single trapped substrate (as at the bottom of FIGURE 5), the equilibrium will favor occupation of the high-affinity site, so that at the instant of transition, the occupant will almost certainly leave in that direction (in the instance diagrammed, toward the exterior). In contrast, when two occupants are present (upper part of FIGURE 5), the distribution and the route of departure may be about 50:50. Accordingly, when a one-way transfer from a finite sugar level into an empty compartment is examined, it will be the low-affinity site that will define the apparent K; but the high-affinity site will predominate when the saturation of the back-flux is studied, and various combinations will be seen under other circumstances—in keeping with the observed "anomalies."

Let us now turn to the matter of relaxation of the Widdas model's con-

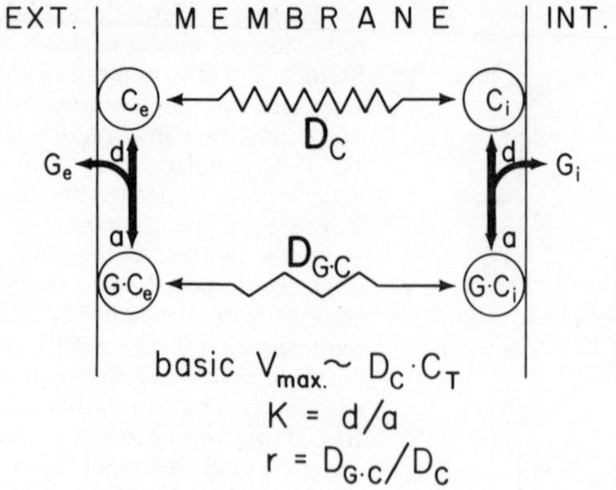

FIGURE 6. Modification (by Mawe and Hempling[31] and Levine *et al.*[32]) of classical model by substrate conditioning of carrier mobility. Symbols as in FIGURE 1, except that C and G·C here have different transmembranal mobilities: process D_C is slower than process $D_{G\cdot C}$; r is ratio of the two rate constants, expressing the activation of mobility by substrate occupancy.

straints regarding the relative rates of the several steps in the transfer process. One aspect of this is that the migration of the carrier across the membrane may be conditioned by its *occupancy* by the substrate, and perhaps by the identity of the occupant sugar, just as we have seen to be the useful case with the introversion of a hemiport. The first definitive evidence of this type of complication of the classical model was brought out independently in two laboratories[31, 32] in 1965. As given diagrammatically in FIGURE 6, the system behaved as if it were more difficult to move the naked carrier than to move the glucose-carrier complex. This conclusion resulted from the finding that the unidirectional efflux of tracer glucose was decidedly enhanced by adding ordinary glucose to the outside medium—i.e., on the side *toward* which the flux was directed. This so-called *trans*-stimulation effect seemed to establish that the return of carrier by process $D_{G\cdot C}$ is intrinsically faster than by process D_C, as indicated symbolically in FIGURE 6. This has the formal effect of adding a third constant to the classical two; it amounts to a second V_{max}, but is generally expressed in the form of a ratio of the two (r, at the foot of the figure). Perhaps the most meaningful consequence of this complication is that now the carrier does not maintain a homogeneous distribution through the membrane, but may pile up on one side when the G_e and G_i concentrations differ sufficiently to produce significant differences in the ratio, C/G · C, at the two interfaces. The interdependence of influx and efflux may thus give rise to either *trans*-stimulation or *trans*-inhibition (phenomena not encompassed by the classical schema).

It may also be that the transmembranal step is not always rate-limiting, and thus the interfacial adsorption-desorption process may be quite far from equilibrium whenever a transfer of sugar is in progress. Such a proposition is at the heart of a model developed in 1972 by Wilbrandt,[33] which accommodates a number of the established anomalies. It was also implicit in the earliest model designed specifically to fit the glucose transfer kinetics in these cells, which was quickly displaced by the Widdas model. I refer to the schema which my wife and I[34] proposed in 1952, which also presumed an *asymmetry in reaction rates* at the two sides (but retained symmetry in the *equilibria*). This primitive model is diagrammed in FIGURE 7. We postulated that the adsorption-desorption at the cell surface is *catalyzed,* presumably enzymatically (by a factor, E, in the diagram); and that the transmembranal step constituted an insignificant resistance in comparison to the uncatalyzed events at the interior interface. Even this simple schema gave remarkable quantitative accord with a variety of measurements then available on net sugar transfer in both directions. The most compelling argument for its early demise in favor of the classical model was the previously mentioned demonstration[22] of the reality of countertransport, whereby (as predicted by Widdas) the flow of one substrate may drive an opposing flow of another, even transiently uphill; in effect, each substrate serves to accompany the carrier in the return lap of what might have been the other substrate's round-trip. The occurrence of this behavior appeared to establish that the carrier sites on the two sides of the membrane form a single pool in that they constantly undergo mutual exchange, but two distinct pools in that at any instant they may differ significantly in their state of occupancy by the substrates. Hence, Widda's assignment of the primary transport resistance to step "D" appeared confirmed.

The other deviant aspect of the LeFevres' 1952 schema, the asymmetry consideration, has in the last few years been resurrected independently by a host of investigators (though strongly challenged by others[35]); but the newer asymmetry is of a more radical sort, dealing *not merely with rates but with equilibria.* For a picture of this complete removal of all the relative-rate restrictions adopted by Widdas, we may consider the completely generalized mobile-carrier presentations given some ten years ago by Regen and Morgan[36] and by Britton[37], as shown in FIGURE 8. With this wide-open view of the matter, we have eight potentially independent constants to define the transport behavior of a single substrate—the a and d on each side, and the transmembranal D in each direction for empty and for loaded carrier. This may be quickly reduced to seven constants for the particular system we are discussing, because of its rather well established inability to "pump" the sugars uphill. The absence of an energizing link to a metabolic process imposes the restriction noted at the bottom of the figure: when G_e and G_i are at the same concentration, the net-inflow cycle has to be as likely as the net-outflow cycle, so that the product

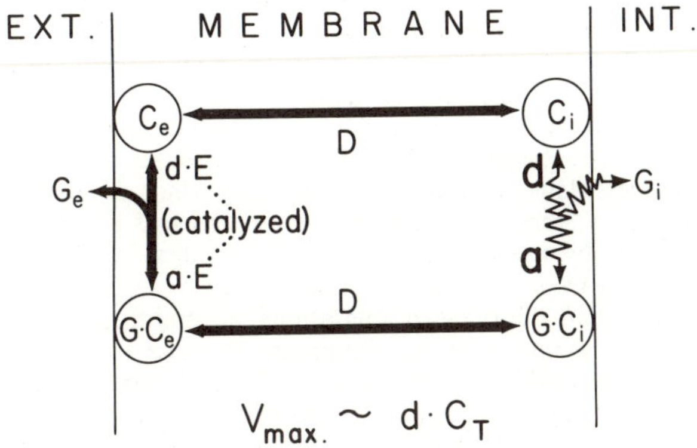

$$V_{max.} \sim d \cdot C_T$$
$$K = d/a$$

FIGURE 7. Early model by LeFevre and LeFevre[34] with rate limitation by asymmetric interfacial process. Symbols as in FIGURE 1, but with catalysis of a, d equilibrium by factor E at exterior interface only. In absence of catalysis at interior surface, a, d process on this side becomes rate-limiting for overall transfer, and process D is essentially at equilibrium.

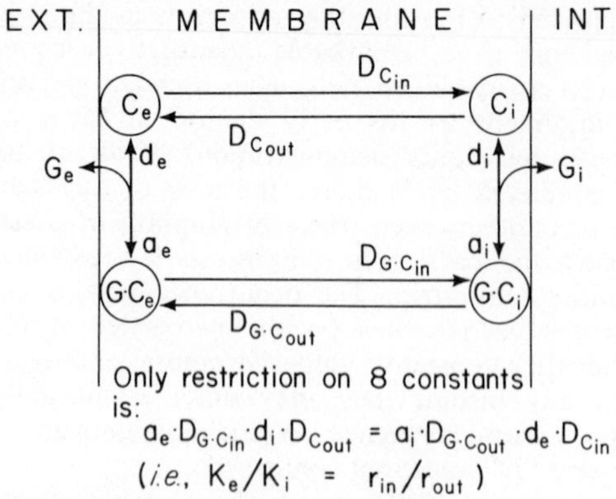

FIGURE 8. Generalized mobile-carrier model (of Regen and Morgan[36] and Britton[37]) without classical restrictions on relative rates of steps. Symbols as in previous mobile-site figures, but with independence of rate constants at the two interfaces and in the oppositely directed transmembranal movements, as identified by subscripts. The single restriction stated reflects absence of link to a power source: no work can be performed.

of four of the constants equals the product of the other four. Note that this does not mean that the saturation characteristics are equivalent at the two interfaces; the a's and d's may relate in any way, so that K_e and K_i may differ. The restriction is that this difference must be balanced by a proportional difference in the factor that we have previously called r (the ratio of the *D* for loaded carrier to that for empty carrier) in the two directions. Thus, if $K_e \neq K_i$, there must be a *rectification* of some kind in the D-process: either $D_{C_{in}} \neq D_{C_{out}}$, or $D_{G \cdot C_{in}} \neq D_{G \cdot C_{out}}$, or (more probably) both.

Such asymmetry in the transmembranal rate constants *does not suggest a diffusion* process, and thus perhaps again argues for the greater realism in picturing the carrier sites as hemiports (squares) rather than truly mobile elements (circles). It was in terms of the squares that Baker and Widdas[38] couched their proposal two years ago for an asymmetric model as given in FIGURE 9. As diagrammed in the tapering of the arrows for the a, d equilibria at the two interfaces, Baker and Widdas suggest substantially *tighter association at the exterior* surface ($K_i > K_e$). With no work performance, it is then a thermodynamic necessity (as indicated in the final line at the bottom of the figure) that the rate constants for outward and inward activation differ by the same factor; thus, with the higher affinity on the exterior side, the x_{in} step poses the greater resistance.* In order to accommodate the observations relating to greater rapidity in exchange than in net transfers, Baker and Widdas added a third sort of x-process (lower portion of FIGURE 9) by which occupants are swapped between apposed sites; the illustration here is in terms of a glucose:glucose exchange, recognizable by use of isotopically labeled sugar (G*). In analogy with taking $D_{G \cdot C}$ to exceed D_C in the earlier instance (i.e., $r > 1$), this $x_{exch,}$ process must occur more readily than the one-way migrations, x_{iu} and x_{out}. Baker and Widdas retain the classical assignment of rate limitation to these transmembranal steps. Altogether, their model adds two parameters to the classical two: one expresses the in:out asymmetry, and the other the independent rate of the exchange process. This model proves very satisfactory for matching quantitatively the peculiarities of the erythrocyte system, especially as regards the time-course of transient uphill counterflow of labeled glucose—a matter that has seriously strained other models.

* Geck[7] and Wilbrandt[33] had arrived at essentially identical interpretations of the situation in their slightly different asymmetric models; however, Batt and Schachter[13] find it necessary to suggest that the fundamental constants are altered by the presence of intracellular sugar.

However, it appears that a *combination* of the revisions we have discussed may have to be considered—that, in itself, the setting aside of the classical restrictions on the relative magnitude of the several rate constants is not quite adequate to allow fitting of all the facts. In an especially comprehensive and trenchant analysis of the applicability of the completely generalized mobile-carrier model to this biological system, Regen and Tarpley[15] have recently concluded that something analogous to the "unstirred layer" must also be incorporated into the picture. In FIGURE 10 I have taken some minor liberties with Regen and Tarpley's schematic presentation in order to focus attention on the fundamental relation of this proposal to the others we have considered. The relative slowness of glucose entry, in comparison with its exit or its isotopic exchange, is identified with the assigning of an especially high resistance to *the return of the empty carrier to the outside* (the $D_{C_{out}}$). A similar asymmetry applies to the movement of the carrier-substrate complex, G · C, but as we have previously noted this complex moves much more readily than the empty carrier. Another major resistance is a diffusion barrier lying at, or just within, the membrane interfaces, particularly the inner one, which must be navigated by the substrate in order to reach the carrier mechanism or to escape from it to one of the pools. Consequently, the concentrations of substrate encountered by the carriers will be intermediate between the concentrations in the two pools, G_e and G_i. Regen and Tarpley calculate the most appropriate distribution of this extraneous resistance to be on the order of 15% as D_e, 85% as D_i.

Regen and Tarpley's analysis[15] seems to be the most detailed now available, and to accommodate virtually all the data. As we have just seen, it necessitates assumption of transmembranal asymmetry in both the affinity and the mobility† of the carrier, substrate-conditioning of this mobility, and also a significant

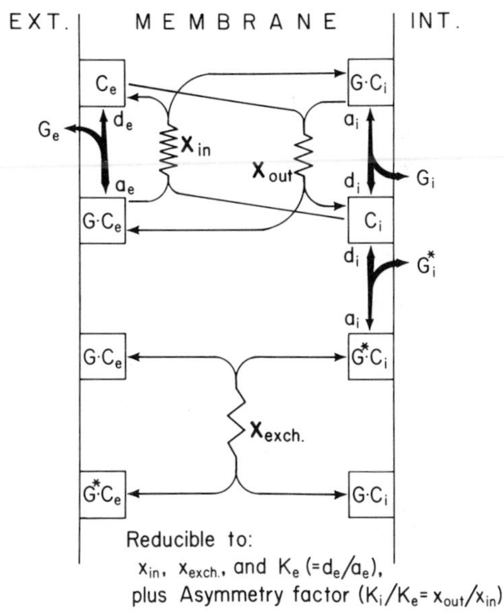

FIGURE 9. Asymmetric hemiport exchange model of Baker and Widdas.[38] Symbols as in previous fixed-site figures, but with independence of rate con- in the transmembranal (x) pro- cesses of exchange and of stants at the two interfaces and migration in the two directions. G* represents isotopically labeled G.

† However, Regen and Tarpley tend to favor the suggestion that this in:out asymmetry is referable to the presence of a competitive inhibitor restricted to the interior side, rather than to an *inherent* asymmetry in the carrier mechanism itself.

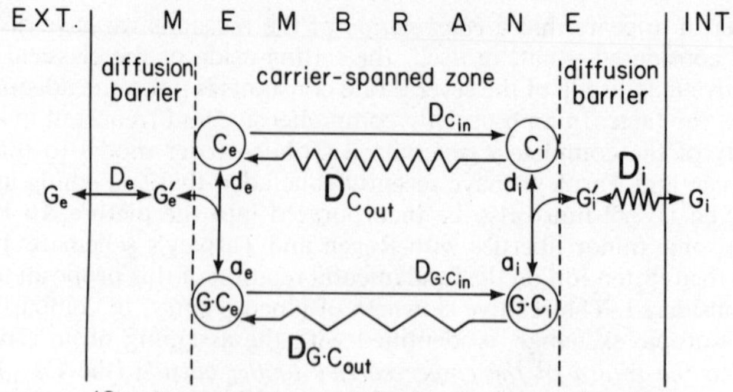

FIGURE 10. Asymmetric mobile-carrier model slightly modified from Regen and Tarpley.[15] Symbols as in previous figures, except that carrier proper lies within extraneous diffusion barriers constituting part of superficial membrane structure. Primary overall resistances appear in processes D_i and $D_{C_{out}}$.

series resistance, again asymmetric, hindering the interfacial access to the carrier. Except for the latter point, the analysis by Baker and Widdas[38] in terms of a hemiport system leads to closely parallel conclusions, giving essentially identical asymmetry factors and rate ratios in comparing exchange to net transfer. This underscores the generality that an imaginary hemiport mechanism can readily (and reasonably) be endowed with properties allowing it to emulate in detail any overt behavior that may be foreseen for the equally imaginary mobile-carrier apparatus. Apparently, kinetic study will not allow a decision between the circles and the squares or rectangles of our diagrams here. However, it does allow us, at this stage, to circumscribe in some detail the manner in which the steps in the facilitated-diffusion process in this particular biological system may be interrelated; and it is reassuring to note the marked similarities between the pictures we thus come up with for the cell membrane assembly and those presented by our colleagues here on totally different grounds.

References

1. WIDDAS, W. F. 1954. Facilitated transfer of hexoses across the human erythrocyte membrane. J. Physiol. (London) **125:** 163–180.
2. WIDDAS, W. F. 1952. Inability of diffusion to account for placental glucose transfer in the sheep and consideration of the kinetics of a possible carrier transfer. J. Physiol. (London) **118:** 23–39.
3. MILLER, D. M. 1965. The kinetics of selective biological transport. II. Equations for induced uphill transport of sugars in human erythrocytes. Biophys. J. **5:** 417–423.
4. MILLER, D. M. 1968. The kinetics of selective biological transport. III. Erythrocyte-monosaccharide transport data. Biophys. J. **8:** 1329–1338.
5. MILLER, D. M. 1968. The kinetics of selective biological transport. IV. Assess-

ment of three carrier systems using the erythrocyte-monosaccharide transport data. Biophys. J. **8:** 1339–1352.

6. MILLER, D. M. 1969. Monosaccharide transport in human erythrocytes. *In* Red Cell Membrane Structure & Function. G. A. Jamieson and T. J. Greenwalt, Eds. : 240–290. J. B. Lippincott Co. Philadelphia, Pa.
7. GECK, P. 1971. Eigenschaften eines asymmetrischen Carrier-Modells für den Zuckertransport am menschlichen Erythrozyten. Biochim. Biophys. Acta **241:** 462–472.
8. MILLER, D. M. 1971. The kinetics of selective biological transport. V. Further data on the erythrocyte-monosaccharide transport system. Biophys. J. **11:** 915–923.
9. LACKO, L., B. WITTKE & H. KROMPHARDT. 1972. Zur Kinetik der Glucose-Aufnahme in Erythrocyten. Effekt der Trans-Konzentration. Eur. J. Biochem. **25:** 447–454.
10. KARLISH, S. J. D., W. R. LIEB, D. RAM & W. D. STEIN. 1972. Kinetic parameters of glucose efflux from human red blood cells under zero-*trans* conditions. Biochim. Biophys. Acta **255:** 126–132.
11. EILAM, Y. & W. D. STEIN. 1972. A simple resolution of the kinetic anomaly in the exchange of different sugars across the membrane of the human red blood cell. Biochim. Biophys. Acta **266:** 161–173.
12. HANKIN, B. L. & W. D. STEIN. 1972. On the temperature dependence of initial velocities of glucose transport in the human red blood cell. Biochim. Biophys. Acta **288:** 127–136.
13. BATT, E. R. & D. SCHACHTER. 1973. Transport of monosaccharides. I. Asymmetry in the human erythrocyte mechanism. J. Clin. Invest. **52:** 1686–1697.
14. LIEB, W. R. & W. D. STEIN. 1972. Carrier and non-carrier models for sugar transport in the human red blood cell. Biochim. Biophys. Acta **265:** 187–207.
15. REGEN, D. M. & H. L. TARPLEY. 1974. Anomalous transport kinetics and the glucose carrier hypothesis. Biochim. Biophys. Acta **339:** 218–233.
16. LEVINE, M. & W. D. STEIN. 1966. The kinetic parameters of the monosaccharide transfer system of the human erythrocyte. Biochim. Biophys. Acta **127:** 179–193.
17. LEVINE, M., S. LEVINE & M. N. JONES. 1971. The effect of temperature on the competitive inhibition of sorbose transfer in human erythrocytes by glucose. Biochim. Biophys. Acta **225:** 291–300.
18. LEFEVRE, P. G. 1962. Rate and affinity in human red blood cell sugar transport. Am. J. Physiol. **203:** 286–290.
19. STEIN, W. D. & J. F. DANIELLI. 1956. Structure and function in red cell permeability. Disc. Faraday Soc. **21:** 238–251.
20. HECKMANN, K. 1965. Zur Theorie der "Single File"-Diffusion I. Z. Physik. Chem. Neue Folge **44:** 184–203.
21. HECKMANN, K. 1965. Zur Theorie der "Single File"-Diffusion, II. Z. Physik. Chem. Neue Folge **46:** 1–25.
22. ROSENBERG, T. & W. WILBRANDT. 1957. Uphill transport induced by counterflow. J. Gen. Physiol. **41:** 289–296.
23. NAFTALIN, R. J. 1970. A model for sugar transport across red cell membranes without carriers. Biochim. Biophys. Acta **211:** 65–78.
24. NAFTALIN, R. J. 1971. The role of unstirred layers in control of sugar movements across red cell membranes. Biochim. Biophys. Acta **233:** 635–643.
25. MILLER, D. M. 1972. The effect of unstirred layers on the measurement of transport rates in individual cells. Biochim. Biophys. Acta **266:** 85–90.
26. LIEB, W. R. & W. D. STEIN. 1972. The influence of unstirred layers on the kinetics of carrier-mediated transport. J. Theoret. Biol. **36:** 641–645.
27. STEIN, W. D. 1969. Intra-protein interactions across a fluid membrane as a model for biological transport. *In* Membrane Proteins. Proceedings of a Symposium Sponsored by the New York Heart Association. : 81–90. Little, Brown. Boston, Mass.

28. LeFevre, P. G. 1973. A model for erythrocyte sugar transport based on substrate-conditioned "introversion" of binding sites. J. Membr. Biol. **11:** 1–19.
29. Lieb, W. R. & W. D. Stein. 1970. Quantitative predictions of a noncarrier model for glucose transport across the human red cell membrane. Biophys. J. **10:** 585–609.
30. Lieb, W. R. & W. D. Stein. 1971. New theory for glucose transport across membranes. Nature New Biol. **230:** 108–109.
31. Mawe, R. C. & H. G. Hempling. 1965. The exchange of C^{14} glucose across the membrane of the human erythrocyte. J. Cell. Comp. Physiol. **66:** 95–104.
32. Levine, M., D. L. Oxender & W. D. Stein. 1965. The substrate-facilitated transport of the glucose carrier across the human erythrocyte membrane. Biochim. Biophys. Acta **109:** 151–163.
33. Wilbrandt, W. 1972. Carrier diffusion. *In* Biomembranes, Vol. 3. Passive Permeability of Cell Membranes. F. Kreuzer and J. F. G. Slegers, Eds. : 79–99. Plenum Press, New York & London.
34. LeFevre, P. G. & M. E. LeFevre. 1952. The mechanism of glucose transfer into and out of the human red cell. J. Gen. Physiol. **35:** 891–906.
35. Hankin, B. L., W. R. Lieb & W. D. Stein. 1972. Rejection criteria for the asymmetric carrier and their application to glucose transport in the human red blood cell. Biochim. Biophys. Acta **288:** 114–126.
36. Regen, D. M. & H. E. Morgan. 1964. Studies of the glucose-transport system in the rabbit erythrocyte. Biochim. Biophys. Acta **79:** 151–166.
37. Britton, H. G. 1964. Permeability of the human red cell to labelled glucose. J. Physiol. (London) **170:** 1–20.
38. Baker, G. F. & W. F. Widdas. 1973. The asymmetry of the facilitated transfer system for hexoses in human red cells and the simple kinetics of a two component model. J. Physiol. (London) **231:** 143–165.

Discussion

Dr. Bender: Is there any kind of chemical-physical description pertaining to sites or hemiports?

Dr. LeFevre: No; we really don't know what they are. We are currently attempting to get them labelled by various combinations of irreversible inhibitors. Those results are not promising at the present time. There has been extended effort with various proteinaceous extracts to indicate binding, but these efforts have all been totally fruitless over a period of ten years.

Dr. Bender: Are you still looking at this as a hydrophobic or hydrocarbon sea in which you have these ports?

Dr. LeFevre: I think so.

Dr. Holtz: What are the major objections that have come up to a simple carrier model or to a carrier that makes you believe you are dealing mainly with a channel rather than a carrier type of mechanism?

Dr. LeFevre: As emphasized in my report, I don't think that we can necessarily distinguish the two on kinetic grounds; we can readily fit the kinetics to either case.

Dr. Chase: Dr. Keston found an enzyme commonly called mutarotase. Recently this enzyme was found also in the erythrocyte membrane. Do you have any comments?

DR. LEFEVRE: Only that anyone must acknowledge the extreme parallelism in much of the behavior between that site and the sugar-carrier site. There certainly is a lot of analogy in the spectrum of substrate specificity, for example, and in its sensitivity to various inhibitors. There may very well be some connection.

DR. HEINZ: Some time ago a paper appeared by Edwards, who also came to the conclusion that inside there is a quasi-unstirred layer, which of course need not be a real unstirred layer but which acts as an unstirred layer, and maybe it is due to an activation energy for getting on the carrier. Is this model in any way related to that?

DR. LEFEVRE: Yes. That factor is certainly indicated by Regen and Tarpley's analysis as necessary to impose as an added complication on any of the symmetrical models that I described. But this is a very recent development.

INTESTINAL SUGAR TRANSPORT: STUDIES WITH ISOLATED PLASMA MEMBRANES*

U. Hopfer

Department of Anatomy
School of Medicine, and
Developmental Biology Center
Case Western Reserve University
Cleveland, Ohio 44106

K. Sigrist-Nelson

Laboratory for Biochemistry
Federal Institute of Technology
Zurich, Switzerland

H. Murer

Max Planck Institute for Biophysics
Frankfurt, Federal Republic of Germany

INTRODUCTION

The absorptive epithelium of the small intestine is specialized in the transfer of nutrients from the lumen to the blood and lymph vessels. Traditionally, transport studies have been carried out in intestinal preparations with intact epithelial cells and precautions were taken to keep cells viable during the time of measurements. The information obtained in this way led to concepts about the nature of transport processes, specificity of transport systems, input of energy, etc., and there are several recent review articles on these matters.[1-3] Unfortunately, because of the complexity of eukaryotic cells, ambiguities often exist in the interpretation of transport data, and this has resulted in considerable controversies, especially with respect to the nature of the energy supply for uphill D-glucose transport and the role of Na^+ in this process.[4-6] Is chemical energy, in the form of ATP for example, directly involved in D-glucose uptake by epithelial cells or is the energy to drive sugar transport provided indirectly via cation gradients? Can useful energy be gained for the concentrative uptake of nutrients from ion gradients?

In order to define the mechanisms in sugar transport more clearly we have gone to a subcellular system of isolated plasma membranes. Deliberate efforts were made to break up the intestinal epithelial cell and purify separately its two types, i.e., the luminal brush border (microvillous) and the contraluminal lateral-basal plasma membrane. This approach, by necessity, involves disruption of the surface membrane and thereby destruction of its barrier property. Since any assay for transport depends on the latter property of membranes, it became necessary to find conditions under which the membrane fragments

* This research was supported by Grants No. 3.511.71 and 3.0090.73 from the Swiss Nationalfonds.

would form vesicles and reseal after they had been isolated from the cell. This paper will discuss some of the considerations that enabled us to carry out the studies on intestinal sugar transport in isolated plasma membranes as well as the conclusions derived from them. Many of the experimental details and results have been published.[7–11]

Isolation of Plasma Membranes

In intestinal epithelial cells the luminal and contraluminal region of the surface membrane have distinctly different chemical and enzymatic composition as well as appearance in the electron microscope.[12–15] Therefore, is was desirable not only to separate plasma membrane from intracellular membranes and organelles, but also to brush borders from lateral-basal membrane. Several methods are available for the isolation of intestinal plasma membranes.[7, 9, 16–22] Figure 1 illustrates the methods that we found convenient and effective in yielding highly purified plasma membrane fractions from rat small intestine. Briefly, for isolation of microvillous membranes intact brush borders are sheared off epithelial cells and isolated by low-speed centrifugation. Subsequently, most of the microfilaments are removed whereby smaller membrane fragments are formed. These fragments are collected by centrifugation at $7.2 \times 10^5 \times$ g min.

The purification of plasma membrane enzymes is given in Table 1. The specific activities of typical brush border enzymes like sucrase (E.C. 3.2.1.48), leucine amino peptidase (E.C. 3.4.11.1), and alkaline phosphate (E.C. 3.1.3.1) are 36–41 times higher in the final membrane as compared with the starting homogenate of intestinal scrapings. If isolated cells are taken as basis for comparison the purification of brush border enzymes is about 20-fold. This difference is caused by the presence of other cell types besides enterocytes in the scrapings. Typical enzymes of the lateral-basal plasma mem-

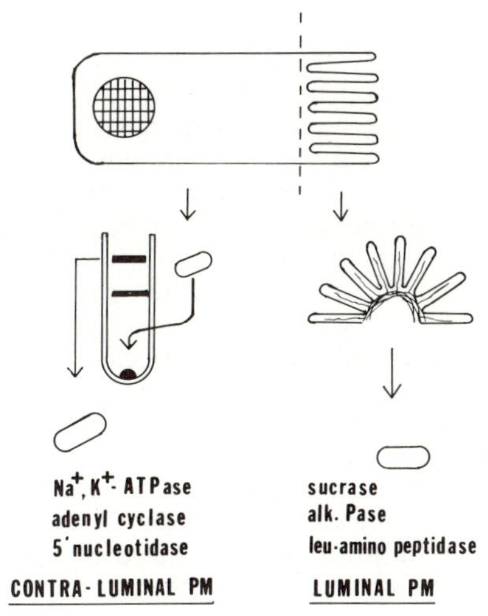

Figure 1. Preparation of luminal and contraluminal plasma membrane.

TABLE 1
ENZYMATIC COMPOSITION OF BRUSH BORDER MEMBRANE

	Intestinal Scrapings	Brush Border Membrane	Purification
Sucrase*	0.09	3.33	37
Leucine amino peptidase*	0.16	5.70	36
Alkaline phosphatase*	0.08	3.18	41
K^+-stimulated phosphatase*	0.054	0.030	0.6
Adenyl cyclase†	36	32	0.9

NOTE: All enzyme activities were measured at 37°C. Sucrase, leucine-aminopeptidase, and Na^+, K^+-ATPase were determined according to References 7, 58, and 17, respectively. K^+-stimulated phosphatase was assayed in the presence of 10 mM $MgCl_2$, 5 mM EDTA, and 6 mM *p*-nitrophenyl phosphate (di-Tris salt) at pH 7.8; the K^+-stimulation was calculated from the difference in phosphate release with either 90 mM KCl or 90 mM NaCl. 5′-nucleotidase was measured in a medium containing 10 mM $MgCl_2$, and either 5 mM 5′-AMP or 5 mM 3′-AMP at pH 7.6; the difference in phosphate release from the two nucleotides was contributed to specific 5′-nucleotidase activity. Alkaline phosphatase was assayed in the presence of 2.5 mM $MgCl_2$, 2.5 mM $ZnSO_4$ and 5.5 mM *p*-nitrophenyl phosphate at pH 10.5. Adenyl cyclase was determined in the presence of 4 mM $MgCl_2$, 0.1 mM EDTA, 3 mM ATP, 10 mM theophylline, 10 mM creatine phosphate, 0.1 mg/ml creatine kinase, and 10 mM NaF at pH 7.4.

* Activity in μmoles substrate hydrolyzed per min per mg protein.

† Activity in pmoles product synthesized per min per mg protein.

brane, e.g., K^+-stimulated phosphatase (E.C. 3.1.3.) and F^--activated adenyl cyclase (E.C. 4.6.1.1.), are not enriched in the purified brush border membrane (TABLE 1). Their low levels indicate little crosscontamination from the lateral-basal plasma membrane.

The physical properties of membrane fragments from the contraluminal region resemble those of smooth intracellular membranes and may be similar to the plasma membranes of other cell types. Therefore, preparation of this membrane starts out with collection of intestinal epithelial cells. The isolated cells are then homogenized and a "crude" plasma membrane fraction is prepared by differential centrifugation. Most of the plasma membranes settle slightly slower than the mitochondria, as described previously.[9] Further purification of the surface membranes is achieved by flotation of the membranes through a discontinuous gradient of 20–50% sucrose at $18 \times 10^6 \times g$ min. Purified lateral-basal plasma membranes are collected at the 20/30% interface and brush border membrane at the 40/50% interface. TABLE 2 shows enzyme activities of the lighter membrane faction (20/30% sucrose). Na^+,K^+-ATPase (E.C. 3.6.1.3), K^+-stimulated phosphatase, 5′nucleotidase (E.C. 3.1.3.5.), and F^--activated adenyl cyclase are enriched 10–14 times over the cell homogenate. All these enzymes are considered to be plasma membrane markers.[23] Very little activity of brush border enzymes is associated with this fraction, as indicated by sucrase or alkaline phosphatase in TABLE 2. Hence most of the membrane has to be derived from the lateral-basal region. Stirling[24] actually showed by autoradiography that ouabain, a specific inhibitor of Na^+, K^+-ATPase, binds specifically to this part of the epithelial cell. A membrane fraction enriched in brush border enzymes can be collected at the 40/50% interface from the same gradient (data not shown).

TABLE 2

ENZYMATIC COMPOSITION OF LATERAL-BASAL PLASMA MEMBRANE

	Isolated Cells	Lateral-Basal Plasma Membrane	Purification
Na^+, K^+-ATPase*	0.067	0.69	10
K^+-stimulated phosphatase*	0.017	0.23	14
5′-nucleotidase*	0.080	0.93	12
Adenyl cyclase†	41	454	11
Sucrase*	0.16	0.18	1.1
Alkaline phosphatase*	0.23	0.12	0.5

NOTE: For determination of enzyme activities see TABLE 1.

* Activity in μmoles substrate hydrolyzed per min per mg protein.

† Activity in pmoles product synthesized per min per mg protein.

TRANSPORT STUDIES IN ISOLATED MEMBRANES

Any assay for transport of solutes is contingent upon membranes enclosing a topographical closed space and effectively separating two aqueous compartments. Fortunately, many plasma membranes form vesicles during homogenization. Depending on experimental conditions, the isolated intestinal membranes contain a space from about 0.5 to 10 μl per mg membrane protein. The magnitude of the space has been estimated from the uptake of permeant solutes equilibrating between the suspension medium and the intravesicular space. Size of membrane fragments appears to be one of the factors determining the space. It has been noted that lateral-basal membranes form larger vesicles than the brush border membranes as judged by electronmicroscopy of negatively stained preparations, and this correlates with a greater equilibrium uptake of permeant solutes in lateral-basal membranes (unpublished observation).

Osmotic forces also influence the size of the intravesicular compartment. To prevent collapse of this space during an experiment, it is necessary to trap an impermeant solute during formation of the membrane vesicles. We have employed extensively either metabolically inert polyols, like mannitol and sorbitol, or carbohydrates like cellobiose. The osmotic behavior of membrane vesicles can be followed by measuring the equilibrium uptake of permeant solutes such as D-glucose. At equilibrium the intravesicular concentration equals the medium concentration. Hence, the amount of permeant solute per unit weight of the membrane is proportional to the volume when only this parameter is varied. FIGURE 2 shows the equilibrium uptake of labeled D-glucose by different aliquots of the same membrane preparation suspended in media with different osmolarity. The change in the medium osmolarity is achieved by increasing concentrations of the impermeant solute cellobiose. FIGURE 2 demonstrates that D-glucose uptake is indirectly proportional to the medium osmolarity with no uptake on extrapolation to infinity (zero on the inverse osmolarity scale). Thus, the membrane vesicles behave as predicted from osmotic considerations. The equilibrium uptake of other permeant sugars (D-fructose, L-glucose) or neutral amino acids (D- and L-alanine, L-valine) shows identical behavior. However, we have observed that under certain experimental conditions binding of labeled solutes to the membranes takes place in addition to transport into the membrane space.[11]

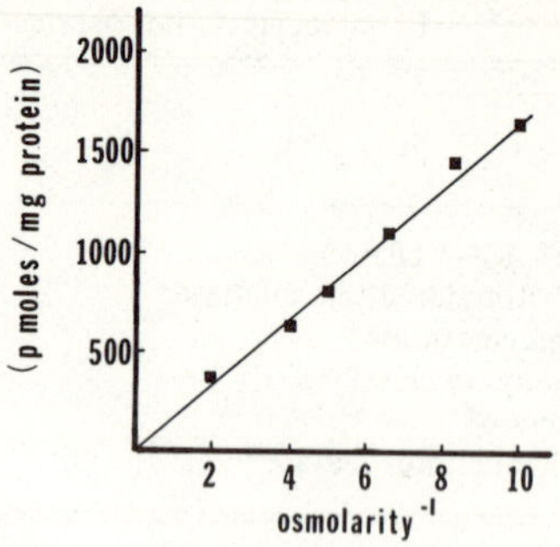

FIGURE 2. Effect of medium osmolarity on D-glucose uptake by brush border membranes. Preparation of membranes (in 100 mM cellobiose, 1 mM HEPES adjusted with Tris to pH 7.5, and 0.1 mM $MgSO_4$) has been described.[7] Uptake of D-glucose was measured in aliquots of the same membrane preparation from media with the following final composition: NaSCN (25 mM), Tris-HEPES (1 mM), $MgSO_4$ (0.1 mM), D-(1-^{3}H)glucose (1 mM) and enough cellobiose to give the indicated osmolarity (shown as inverse osmolarity). Length of incubation: 10 minutes; temperature: 25°C. The line was calculated by regression analysis (correlation coefficient 0.99).

SUGAR TRANSPORT IN BRUSH BORDER MEMBRANE

The existence of transport systems or carriers in biological membranes is usually inferred from such observations as differential rates of transport for structurally similar compounds, saturation of transport rates with increasing substrate concentration, uptake of solutes against a concentration gradient, and enhancement of solute flow across the membrane by the flow of other solutes in the same direction (cotransport) or related compounds in the other direction (countertransport). All of these phenomena have been shown to occur with monosaccharides in the intestinal membrane vesicles, testifying to the existence of intact sugar carriers after membrane isolation.[7, 10]

FIGURE 3 shows the uptake of D-glucose, D-fructose, and L-glucose by brush border membranes. The sugar transport is followed using labeled monosaccharides and quenching transport with ice-cold buffer (for details see reference 10). Aliquots of membranes, containing the transported sugars, are collected on membrane filters and the retained sugars are determined with a scintillation counter. Other radioactive compounds can be added to the quench solution to get an estimate of the contamination from the incubation medium. Retention of medium sugars is usually negligible. One of the advantages of working with isolated membranes, namely, absence of sugar metabolism, becomes evident at this point. Under our conditions we have never seen any evidence for conversion of the employed labeled sugars to other compounds, and thus radioactivity on the filter is proportional to the amount of the compound originally containing the label.

It is obvious from FIGURE 3 that all three monosaccharides are taken up to the same extent on prolonged incubation. This finding is expected for solutes equilibrating into the same membrane space and in the absence of metabolic energy. More interestingly, the transport rates differ considerably for the various monosaccharides and have the following order: D-glucose > D-fructose > L-glucose. It is difficult to draw any conclusions about possible factors determining L-glucose permeability of the membrane since it exhibits the slowest rate. However, specific agencies (carriers) must exist in the membrane that catalyze the higher rates for other sugars in comparison to L-glucose. Inhibition and countertransport experiments with brush border membranes have established that separate carriers exist for D-fructose and for D-glucose

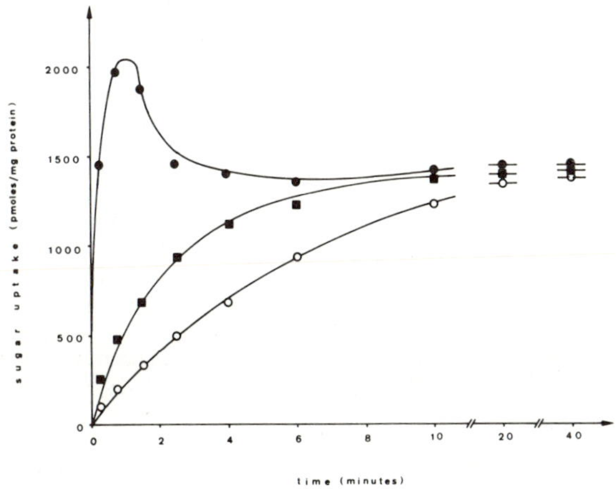

FIGURE 3. Time course of D-glucose (●), D-fructose (■), and L-glucose (○) uptake by brush border membranes. Membranes were prepared in 100 mM D-mannitol, 1 mM HEPES adjusted with Tris to pH 7.5, and 0.1 mM $MgSO_4$. Aliquots of the membrane were incubated in media containing (final concentration): 100 mM D-mannitol, 1 mM Tris-HEPES, pH 7.5, 0.1 mM $MgSO_4$, 100 mM NaSCN and 1 mM of either D-(1-^{3}H)glucose, D-(1-^{3}H)fructose, or L-(1-^{14}C)glucose. Temperature: 25°C. For further details see Reference 10. (From Sigrist-Nelson and Hopfer.[10] By permission of *Biochimica et Biophysica Acta.*)

and glucalogs including D-galactose.[7, 10] The properties of the two carriers are discussed separately.

D-*Glucose Transport*

Intestinal glucose transport differs from that of most other tissues in that it is dependent on the presence of Na^+,[25–27] is active in the sense of absorption against a concentration gradient,[28–30] and is quite specifically inhibited by the glucoside phlorizin.[31, 32] Interestingly, all these phenomena have been demonstrated in the isolated brush border membrane,[7, 8] and these similarities have helped to ascertain that the glucose carrier is retained intact through the preparative procedure.

Conditions under which D-glucose is transported against a concentration gradient have contributed considerably to our understanding of the mechanism of active nutrient transport in animal cells. Examples of such experiments are given in the upper curves of FIGURES 4 and 5 where the overshooting uptake of D-glucose upon addition of NaSCN represents transient concentrative uptake. In order to understand the forces responsible for the D-glucose movement into the vesicles it is necessary to describe the experiments in some detail: The brush border membranes are preincubated with labeled D-glucose (1 mM) in the absence of Na^+. The sugar equilibrates, although slower than in the presence of Na^+, between the medium and the intravesicular space as evidenced by constant D-glucose levels after about 10 minutes of the preincubation period. On addition of NaSCN (final concentration 0.1 M) further uptake of D-glucose takes place before a new, lower equilibrium value is

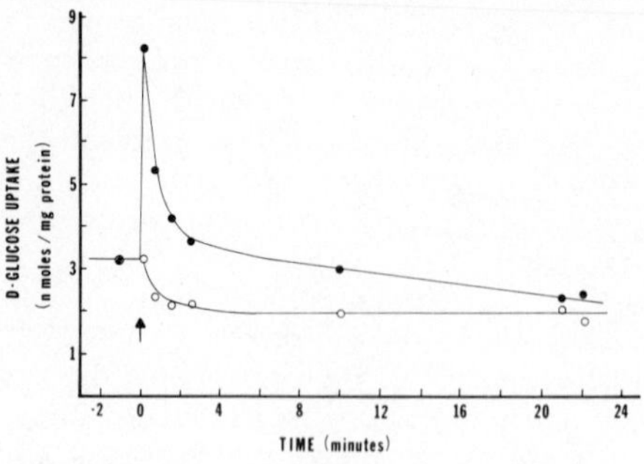

FIGURE 4. Effect of monactin on active D-glucose uptake by brush border membrane. D-Glucose uptake was measured in the absence (●) or presence of 12.5 μg/ml monactin (○). Membranes were preincubated at 25°C for 12 minutes in the following medium: 100 mM D-mannitol, 0.1 mM $MgSO_4$, 1 mM Tris-HEPES, pH 7.5, 1% ethanol, and 1 mM D-(1-^{3}H)glucose. The arrow indicates the addition of a 1-M solution of NaSCN (final concentration: 0.1 M). Uptake measurements were made as usual.[8]

reached. The overshooting part is dependent on the establishment of a NaSCN gradient.[8]

The interpretation of the uptake data is facilitated by the absence of D-glucose metabolism and osmotic behavior of membrane vesicles. In the experiments of FIGURE 4 and 5 the amount of D-glucose in the membranes increased over the equilibrium level after addition of a hypertonic 1 M NaSCN solution, i.e., in the absence of an increase in vesicle volume. The overshoot must therefore represent a higher intravesicular D-glucose concentration as compared to equilibrium, and therewith to medium, since the two are identical for permeant nonelectrolytes. An estimate of the D-glucose gradient (intravesicular/medium concentration) can be obtained from the ratio of the overshooting to the equilibrium uptake. Under optimal conditions, in the presence of a proton gradient instead of a NaSCN gradient, transient accumulation of D-glucose up to ten times the medium concentration has been observed.[8] Although the concentration gradients are small in comparison to those established by bacterial membranes they are in line with usual findings for intact animal tissue.[31]

Energy for this active transport must have been derived from the salt gradient present at time zero of the incubation period. It is a limited amount and dissipated within about 2 minutes as measured by the equilibration of labeled NaSCN between medium and membranes.[33] The time curve of active D-glucose transport is consistent with the limited supply of energy.

The relevant findings with respect to the transport mechanism of D-glucose can be summarized as follows[7, 8]: (i) Active D-glucose transport is observed

FIGURE 5. Effect of gramicidin D on active D-glucose uptake by brush border membrane. D-Glucose uptake was measured in the absence (●) or presence of 18 μg/ml gramicidin D (○). Membranes were treated as explained in the legend to FIGURE 4. The arrow indicates the addition of a 1-M solution of NaSCN (final concentration: 0.1 M).

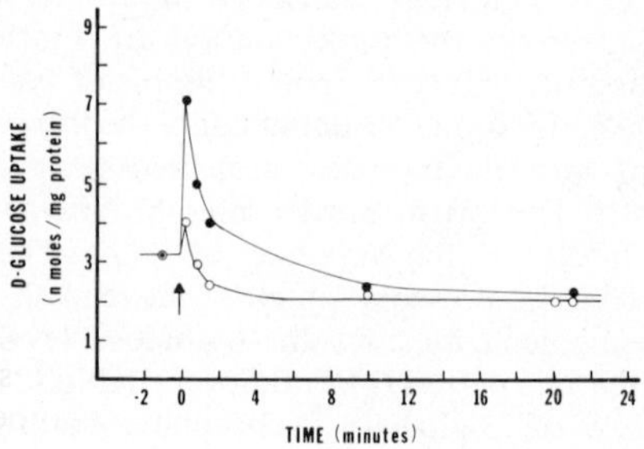

in the absence of ATP and in membranes virtually free of Na^+, K^+-ATPase. (ii) The presence of Na^+, but not of K^+ is required for active transport. (iii) An electrochemical Na^+ gradient across the brush border membrane produces concentrative D-glucose uptake if the gradient is oriented such that the inside has the lower Na^+ potential. The electrochemical Na^+ gradient can consist of either a Na^+ concentration gradient, provided the membrane potential is maintained close to zero by the presence of permeant ions other than Na^+, or an electrical potential, or a combination of the two.

Membrane potentials have been produced experimentally by fluxes of ions other than Na^+ down their concentration gradients in the presence of specific ionophores (K^+ in the presence of valinomycin; H^+ in the presence of dinitrophenol or derivatives of carbonyl cyanide phenylhydrazone and salicyl anilide).[8] The efficacy of SCN^- in producing active D-glucose uptake in FIGURES 4 and 5 can be explained on the same basis, namely a contribution of the SCN^- flow to the electrical component of the Na^+-gradient. SCN^- permeates biological membranes predominantly in the ionic form at the pH of the experiments (pH 7.5).[34] Thus, a SCN^- concentration gradient (medium > intravesicular space) would produce a diffusion potential negative on the inside of the vesicles.

The uphill movement of D-glucose in the presence of Na^+ and a membrane potential indicates the existence of Na^+-sugar cotransport through the membrane, which may be formulated as follows:

$$Na^+_{lu} + \text{D-glc}_{lu} \rightleftharpoons Na^+_{in} + \text{D-glc}_{in}$$

where lu = incubation medium and in = intravesicular in the case of the isolated membranes. The 1:1 stoichiometry of Na^+ and D-glucose is suggested by studies with intact rabbit mucosal sheets.[35]

The equation illustrates that the direction of net nonelectrolyte flow depends not only on the D-glucose concentration gradient, but also on the electrochemical Na^+ gradient. The latter amounts to a driving force for D-glucose transport. The close association of D-glucose flow with an electrical current has important consequences since all electrical currents across the same membrane are coupled. Since macroscopic electroneutrality is required for fluxes into the vesicles, the Na^+-D-glucose cotransport is dependent on charge compensation through sites other than the glucose carrier.

In order to observe active sugar transport an electrochemical Na^+ gradient has to be set up either artificially through ion gradients, as in the case of isolated membrane vesicles, or through ion "pumps" as in the intact cell. However, the magnitude of the Na^+ gradient is only partially determined by the salt gradients or the "pumps." The other important factor is dissipative Na^+ flow, i.e., Na^+ translocation through the membrane down its gradient without coupling to other solute flow. Agents that increase the Na^+ conductance of membranes allow rapid equilibration of Na^+ and would therefore be expected to act as uncouplers in Na^+-dependent D-glucose transport. Several antibiotics are known to increase the Na^+ conductance of biological and lipid bilayer membranes,[36–40] and two of them, monactin and gramicidin D, have been tested in the isolated vesicle system. The results are shown in the lower curves of FIGURES 4 and 5 for monactin and gramicidin D, respectively. Both experiments demonstrate that the ionophores lower the active D-glucose transport produced by a NaSCN gradient. It appears to be a specific effect since neither of the antibiotics affects the ability of the membrane to discriminate between D- and L-glucose uptake in the absence of a driving force (unpublished observation).

D-*Fructose Transport*

In contrast to D-glucose, the rate of D-fructose transport is not influenced by the presence or absence of Na^+, K^+, SCN^-, Cl^-, or phlorizin. Also Na^+ gradients have no effect. The net flow of D-fructose appears to always follow its concentration gradient, thus equilibrating this monosaccharide between the medium and the intravesicular compartment. The existence of a fructose-carrier can be inferred from higher transport rates for D-fructose as compared to L-glucose and D-mannitol, saturation of the transport rates above 0.2 M D-fructose, and countertransport of labeled and unlabeled forms of this sugar. In summary, the available evidence from experiments with isolated brush border membranes suggests that D-fructose is transported by a mechanism classically referred to as "facilitated diffusion."[10]

D-Glucose Transport in Lateral-Basal Plasma Membranes

Previously, D-glucose transport had been investigated in vesicles of lateral-basal plasma membrane prepared by an electrophoretic method.[9] More extensive studies have been carried out recently with the lateral-basal plasma membrane fraction isolated by the sucrose gradient technique mentioned above. Results of both studies agree very well and the existence of a glucose transport system can be demonstrated through such phenomena as stereospecificity, saturability, and countertransport of labeled and unlabeled forms of D-glucose. The uptake rate appears to saturate around 50 mM D-glucose and preloading of the isolated membranes with this sugar enhances the subsequent uptake of labeled D-glucose from the medium. In contrast to the glucose transport system of the brush border, Na^+, K^+, SCN^-, Cl^- or their gradients do not exert a noticeable effect on D-glucose transport in the lateral-basal plasma membrane. Furthermore, net flow of D-glucose follows the concentration gradient as evidenced by equilibrating in- or efflux. Thus, D-glucose transport across the lateral-basal plasma membrane resembles the "facilitated diffusion" in erythrocytes.

The view that D-glucose transport is similar for erythrocytes and lateral-basal plasma membrane, is strengthened by a comparison of inhibitors of D-glucose transport in brush border and contraluminal plasma membrane. Inhibitors effective in erythrocytes also inhibit D-glucose translocation across the lateral-basal plasma membrane. Table 3 shows the effects of phlorizin, phloretin, and cytochalasin B. A comparison of the 15-sec uptake of D-glucose in Table 3(a) demonstrates that phlorizin is more effective than phloretin in inhibiting D-glucose transport in brush border membranes, while cytochalasin B does not inhibit (at least in the presence of a driving force). Phlorizin and phloretin only affect the rate of transport and not the final uptake as evidenced by similar uptake values after 16 minutes of incubation when equilibrium has been reached. In the lateral-basal membrane (Table 3(b)), however,

TABLE 3

INHIBITION OF D-GLUCOSE TRANSPORT BY PHLORIZIN, PHLORETIN, AND CYTOCHALASIN B

Length of Incubation	Control	Phlorizin (0.5 mM)	Phloretin (0.1 mM)	Cytochalasin B (0.1 mM)
(a) D-*Glucose Uptake by Brush Border Membranes**				
15 sec	740	340 (−55)	620 (−17)	790 (+ 6)
16 min	1575	1760 (+11)	1670 (+ 6)	1690 (+ 7)
(b) D-*Glucose Uptake by Lateral-Basal Plasma Membranes**				
15 sec	3704	2610 (−30)	1607 (−57)	1740 (−54)
16 min	10380	10235 (− 2)	10340 (− 1)	10380 (0)

NOTE: Brush border and lateral-basal plasma membranes were isolated from the same sucrose gradient and homogenized in 100 mM D-mannitol, 0.1 mM $MgSO_4$, 1 mM Tris-HEPES buffer, pH 7.5. Inhibitors (in 50% ethanolic solution) were added to different aliquots of the membranes and the uptake reaction started by the addition of D-glucose and NaSCN. The control received the same amount of ethanol, which did not affect D-glucose uptake. Final composition of the incubation medium: 100 mM D-mannitol, 0.1 mM $MgSO_4$, 1 mM Tris-HEPES, pH 7.5, 100 mM NaSCN and 1 mM D-(1-^{14}C)glucose. Uptake was measured as described.[9] Temperature: 25°C.

* In pmol/mg membrane protein; percentage of stimulation = (+); percentage of inhibition = (−).

phloretin is more potent than phlorizin and cytochalasin B inhibits strongly. The relative inhibitory power of phloretin and phlorizin and the inhibition by cytochalasin B agree well with studies in erythrocytes and plasma membranes in other tissues.[32, 41–43] Together with the enzyme data these findings suggest that the plasma membrane of the lateral-basal region is similar to those of other cell types containing Na^+, K^+-ATPase, adenyl cyclase, and a Na^+-independent glucose transport system. In contrast, the brush border membrane represents an unusual, specialized surface membrane with a Na^+-dependent glucose transport system, disaccharidases and dipeptidases.

D-GLUCOSE TRANSPORT ACROSS THE CELL

In the intact cell D-glucose exhibits a vectorial flow from lumen to blood. However, transport experiments with isolated plasma membranes have provided no indications of unidirectional transport. On the contrary, the Na^+-independent as well as the Na^+-dependent glucose transport system appear to enhance the equilibration of their substrates equally well in either direction. In the case of the lateral-basal plasma membrane, the direction of net D-glucose flow is dependent solely on its concentration gradient. In the brush border membrane D-glucose flow is a function not only of the D-glucose gradient, but also of the Na^+ concentration gradient and the membrane potential. Thus, to a first approximation, asymmetry properties are not found in the individual glucose transport systems; rather asymmetry lies in the distribution pattern of two different glucose transport systems within the plasma membrane: a Na^+-dependent system operating in the brush border membrane and a Na^+-

independent system in the lateral-basal plasma membrane. In addition, the electrochemical potential of Na^+ appears to be lower inside the cell as compared to the lumen under physiological conditions, and this ensures a virtually unidirectional inflow of D-glucose across the brush border membrane. Measurements of Na^+ concentration gradient and electrical potential difference with microelectrodes indicate that under physiological conditions the Na^+ gradient is approximately 60–70 mV (inside negative). About half of this force is due to an electrical potential difference and the other half is caused by the Na^+ concentration gradient.[44–47] The Na^+, K^+-ATPase in the lateral-basal plasma membrane probably constitutes the biochemical counterpart to the ion "pump" that is responsible for establishing and maintaining the Na^+ gradient.

With some simplifying assumptions transepithelial D-glucose transport can be reconstructed from the transport events across the isolated plasma membranes. Overall transport can be viewed as the sum of the following four transport reactions:

	Reaction	*Location*	*Inhibitor*
(1)	$ATP_{in} + 3Na^+_{in} \rightleftharpoons ADP_{in} + P_{in} + 3Na^+_{se}$	lateral-basal PM	ouabain
(2)	$3Na^+_{se} \rightleftharpoons 3Na^+_{lu}$	tight junction	
(3)	$3D\text{-}glc_{lu} + 3Na^+_{lu} \rightleftharpoons 3D\text{-}glc_{in} + 3Na^+_{in}$	brush border PM	phlorizin
(4)	$3D\text{-}glc_{in} \rightleftharpoons 3D\text{-}glc_{se}$	lateral-basal PM	cytochalasin B
sum:	$ATP_{in} + 3D\text{-}glc_{lu} \rightleftharpoons ADP_{in} + P_{in} + 3D\text{-}glc_{se}$	cell	all of the above

where in = intracellular, se = serosal (extracellular, but not luminal), lu = luminal, and PM = plasma membrane.

The assumptions concern the stoichiometry of the reactions and the omission of K^+. They can be justified from transport data with intact intestinal tissue,[35] suggesting the 1:1 stoichiometry of Na^+-D-glucose influx, and from investigations of erythrocyte Na^+, K^+-ATPase where evidence for a 3:1 stoichiometry between Na^+ and ATP was obtained.[48, 49] Although K^+ can be assumed to be involved in the ATP-dependent step, it does not appear to play a role in sugar transport. K^+ may recycle between cell and extracellular space in the steady state, and thus would not enter in the equations of D-glucose transport. Very good evidence for a high Na^+ conductance of the intercellular tight junctions in the small intestine has come forth within the last years.[3, 50, 51]

In conclusion, the studies with the isolated intestinal plasma membranes confirm the general concepts that had been suggested for active nonelectrolyte transport in animal cells by Christensen and his colleagues,[52] Crane,[4] and Schultz and Curran.[1] More importantly, they have shown association of electrical currents with nonelectrolyte transport. Consequences of an electrogenic translocation mechanism are coupling of all electrical fluxes across the same membrane and the possibility of circular Na^+ fluxes from cell → intercellular space → lumen → cell in intact epithelium. Oversight of such possible consequences may form the basis for controversies concerning the question of a common sugar-amino acid carrier in the brush border membrane[53–55] or of the Na^+-dependence of D-glucose transport *in vivo*.[4–6, 56, 57]

Summary

The luminal and the contraluminal plasma membranes of small intestinal epithelial cells were isolated separately and in the form of membrane vesicles. In the luminal membrane distinct transport systems exist for D-fructose and for D-glucose and glucalogs as evidenced by stereospecificity for the substrate, counter- and cotransport phenomena, and inhibitor specificity. D-Glucose is taken up against a concentration gradient when energy is supplied in the form of an electrochemical Na^+ gradient between medium and intravesicular membrane space. The contraluminal membrane also possesses a glucose transport system. However, it is sensitive to different inhibitors than the glucose transport in the luminal membrane. D-Glucose uptake by the contraluminal membrane is not enhanced by Na^+ gradients and can be characterized as "facilitated diffusion." The studies with the isolated plasma membranes allow a description of transcellular D-glucose transport in terms of the translocation events across single membranes.

Acknowledgments

We would like to thank Professor G. Semenza for his interest in this work and Dr. R. Przybylski for reading the manuscript. We are greatly indebted to Mrs. E. Ammann and Mr. J. Biber who helped carry out part of the investigations during their undergraduate training.

References

1. Schultz, S. G. & P. F. Curran. 1970. Physiol. Rev. **80:** 637–718.
2. Kimmich, G. A. 1973. Biochim. Biophys. Acta **300:** 31–78.
3. Schultz, S. G., R. A. Frizzell & N. H. Nellans. 1974. Ann. Rev. Physiol. **36:** 51–91.
4. Crane, R. K. 1965. Fed. Proc. **24:** 1000–1006.
5. Frodtran, J. S., F. C. Recktor Jr. & N. W. Carter. 1968. J. Clin. Invest. **47:** 884–900.
6. Kimmich, G. A. 1970. Biochemistry **9:** 3669–3677.
7. Hopfer, U., K. Nelson, J. Perrotto & K. J. Isselbacher. 1973. **248:** 25–32.
8. Murer, H., U. Hopfer. 1974. Proc. Nat. Acad. Sci. U.S.A. **71:** 484–488.
9. Murer, H., U. Hopfer, E. Kinne-Saffran & R. Kinne. 1974. Biochim. Biophys. Acta **345:** 170–179.
10. Sigrist-Nelson, K. & U. Hopfer. 1974. Biochim. Biophys. Acta **367:** 247–254.
11. Hopfer, U. & K. Sigrist-Nelson. 1974. Nature **252:** 422.
12. Trier, J. 1968. Handbook of Physiology. Section 6. Vol. 3: 1125–1174. American Physiological Society. Washington, D.C.
13. Johnson, C. F. 1967. Science **155:** 1670–1677.
14. Benson, R. L., B. Sacktor & J. W. Greenawalt. 1971. J. Cell. Biol. **31:** 186–193.
15. Fujita, M., K. Kawai, S. Asano & M. Nakao. 1973. Biochim. Biophys. Acta **307:** 141–151.
16. Eichholz, A. & R. K. Crane. 1974. Methods in Enzymology. S. Fleischer & L. Packer, Eds. Vol. **31A:** 123–133. Academic Press, New York, N.Y.
17. Quigley, J. P. & S. G. Gotterer. 1969. Biochim. Biophys. Acta **173:** 456–468.
18. Fujita, M., H. Matsui, K. Natano & M. Nakao. 1971. Biochim. Biophys. Acta **233:** 404–408.

19. FUJITA, M., H. MATSUI, K. NATANO & M. NAKAO. 1972. Biochim. Biophys. Acta **274:** 336–347.
20. DOUGLAS, A., R. KERLEY & K. J. ISSELBACHER. 1972. Biochem. J. **128:** 1329–1338.
21. SCHMITZ, J. H. PREISER, D. MAESTRACCI, B. GHOSH, J. CERDA & R. K. CRANE. 1973. Biochim. Biophys. Acta **323:** 98–112.
22. LOUVARD, D., S. MAROUX, J. BERRATTI, P. DESNUELLE & S. MUTAFSCHIEV. 1973. Biochim. Biophys. Acta **291:** 747–763.
23. WALLACH, D. F. H. 1972. The Plasma Membrane: Dynamic Perspectives, Genetics and Pathology. Springer-Verlag. New York, N.Y.
24. STIRLING, C. E. 1972. J. Cell. Biol. **53:** 704–714.
25. RIKLIS, E. & J. H. QUASTEL. 1958. Can. J. Biochem. Physiol. **36:** 347–362.
26. CSAKY, T. Z., & M. THALE. 1960. J. Physiol. **151:** 59–65.
27. SCHULTZ, S. G. & R. ZALUSKY. 1964. J. Gen. Physiol. **47:** 1043–1059.
28. MCDOUGAL, D. B., K. D. LITTLE & R. K. CRANE. 1960. Biochim. Biophys. Acta **45:** 483–489.
29. KINTER, W. G. & T. H. WILSON. 1965. J. Cell Biol. **25:** 19–39.
30. LEESE, H. 1974. Nature **251:** 512.
31. CRANE, R. K. 1968. Handbook of Physiology Section 6. Vol. 3: 1323–1352. American Physiological Society. Washington, D.C.
32. STEIN, W. D. 1967. The Movement of Molecules Across Cell Membranes. : 283. Academic Press. New York, N.Y.
33. SIGRIST-NELSON, K., H. MURER & U. HOPFER. 1975. J. Biol. Chem. **250:** 5674–5680.
34. MICHELL, P. & J. MOYLE. 1969. Eur. J. Biochem. **9:** 149–155.
35. GOLDNER, A. M., S. G. SCHULTZ & P. F. CURRAN, 1969. J. Gen. Physiol. **53:** 362–383.
36. SZABO, G., G. EISENMAN & S. CIANI. 1969. J. Membr. Biol. **3:** 346–354.
37. GRAVEN, S. N., H. A. LARDY & S. ESTRADA-O. 1967. Biochemistry **6:** 365–371.
38. HENDERSON, P. J. F., J. D. MCGIVAN & J. B. CHAPPELL. 1969. Biochem. J. **111:** 521–535.
39. GOODALL, M. C. 1970. Biochim. Biophys. Acta **219:** 471–478.
40. MYERS, V. B. & D. A. HAYDON. 1972. Biochim. Biophys. Acta **274:** 313–322.
41. ESTENSEN, R. D. & P. G. W. PLAGEMANN. 1972. Proc. Nat. Acad. Sci. U.S.A. **69:** 1430–1432.
42. ZIGMOND, S. H. & J. G. HIRSCH. 1972. Science **176:** 1432–1434.
43. TAVERNA, R. D., & R. G. LANGDON. 1973. Biochim. Biophys. Acta **323:** 207–219.
44. ROSE, R. C. & S. G. SCHULTZ. 1971. J. Gen. Physiol. **57:** 639–663.
45. WHITE, J. F. & W. MCD. ARMSTRONG. 1971. Am. J. Physiol. **221:** 194–201.
46. CURRAN, P. F., J. J. HAJJAR & J. M. GLYNN. 1970. J. Gen. Physiol. **55:** 297–308.
47. SCHULTZ, S. G., R. E. FUISZ & P. F. CURRAN. 1966. J. Gen. Physiol. **49:** 849–866.
48. SEN, A. K. & R. L. POST. 1964. J. Biol. Chem. **239:** 345–352.
49. WHITTAM, R. & M. E. AGER. 1965. Biochem. J. **97:** 214–227.
50. FROEMTER, E. & J. M. DIAMOND. 1972. Nature New Biol. **35:** 9–13.
51. MORENO, J. H. & J. M. DIAMOND. 1975. Membranes. G. Eisemann, Ed. Vol. 3. Marcel Dekker, Inc. New York, N.Y.
52. RIGGS, T. R., L. M. WALKER & H. N. CHRISTENSEN. 1958. J. Biol. Chem. **233:** 1479–1484.
53. ALVARADO, F. 1966. Science **151:** 1010–1013.
54. ROBINSON, J. W. L. & F. ALVARADO. 1971. Pfluegers Arch. **326:** 48–75.
55. CHEZ, R. A., S. G. SCHULTZ & P. F. CURRAN. 1966. Science **153:** 1012–1013.
56. SALTZMAN, D. A., F. C. RECTOR & J. S. FORDTRAN. 1972. J. Clin. Invest. **51:** 876–885.
57. FOERSTER, H. & J. HOOS. 1972. Z. Physiol. Chemie. **353:** 88–94.
58. WACKER, H., P. LEHKY, E. FISCHER & E. STEIN. 1971. Helv. Chim. Acta. **54:** 473–485.

Discussion

DR. SACHS: You said that the system was symmetric with respect to each membrane. Have you checked the electrical symmetry of the carrier system by invoking the potential and showing a stimulated efflux?

DR. HOPFER: We have looked at the efflux, although not extensively. Future investigations may turn up some asymmetry features. But I think the statement holds to a first approximation.

Part IX. Permeation Mechanism across Biological Membranes: III

DRIVING FORCES OF AMINO ACID TRANSPORT IN ANIMAL CELLS*

E. Heinz, P. Geck, and C. Pietrzyk

Abteilung für Physikalische Biochemie
Gustav-Embden-Zentrum der Biologischen Chemie
Johann Wolfgang Goethe Universität
Frankfurt am Main, Federal Republic of Germany

Amino acids, neutral ones as well as charged ones, are accumulated in most animal cells, especially in tumor cells. There is now little doubt that this accumulation is by active transport by the usual definition of this term. There is, on the other hand, still much controversy as to the immediate driving force for this transport: Is it primary active, i.e., immediately coupled to a metabolic chemical reaction such as ATP hydrolysis; or is it secondary active, i.e., coupled immediately to the thermodynamically downhill movement of electrolyte ions, by co- or countertransport? In the first case the immediate driving force would derive directly from the affinity of the chemical reaction, and in the second case, from the electrochemical potential gradient of one or more electrolyte ions. Since the driving ion gradient is continuously restored at the expense of metabolic energy, also secondary amino acid transport is ultimately but indirectly coupled to a chemical reaction of metabolism. In the last years, considerable evidence has been accumulated showing that the transport of amino acid is indeed coupled to the inward movement of Na^+ ions and that energy of the Na^+ gradient does drive amino acid uphill. Also, the (inverse) gradient of potassium ions appears to contribute some energy for this transport, but since under certain conditions valinomycin stimulates this effect, the potassium efflux is probably not directly linked to the amino acid influx via countertransport, so that the energy from the K^+ gradient is contributed *via* the electrical potential difference (PD) across the cell membrane, thereby enhancing the electrochemical potential gradient of Na^+ ions.[1] In view of all evidence, the question can no longer be whether the ion gradient is a driving force for amino acid transport or not. Rather, the question presently whether the ion gradient is the only driving force for amino acid transport, or whether in addition a major part of this force stems directly from a metabolic reaction. There are significant observations that indeed appear to support the latter alternative and that have been interpreted to indicate that part of the amino acid transport is primary, i.e., directly linked to metabolism.

Even though the ion gradient may be effective in the absence of metabolic activity, the converse has not unequivocally been shown. Hence the two hypothetical pathways of energy transfer could not likely be separate and parallel, but would have to be geared to each other in a more or less intricate way.

The major arguments that have been raised against the gradient hypothesis are the following:

* This work has been supported by Grant No. He 102/12 from the Deutsche Forschungsgemeinschaft.

1. The total force available from the electrochemical potential gradient (in the following simply called "gradient") of Na^+, the stoichiometric coupling ratio taken to be unity, appears to be grossly inadequate to account for the amino acid accumulation observed.[2] Even the sum of both the Na^+ and the (inverted) K^+ gradients would barely be fully adequate. As mentioned before, however, the K^+ gradient is probably not fully available as the driving force for amino acid transport, neither via direct linkage nor indirectly via the electrical PD, so that the true driving force seems nontheless far from adequate.

2. Respiring cells accumulate glycine about three times more effectively than do metabolically inhibited cells at the same ion distribution.[3] So it seems that the ion gradient contributes only a minor, perhaps merely accidental portion of the total transport energy.

3. Active uptake of amino acids, although at a reduced rate, may still take place in respiring cells, if the Na^+ and K^+ gradients are abolished or even inverted.[4] This transport appears to be against an opposing driving force, at least as long as the latter does not exceed a limiting value of 4000 to 5000 J/mol.

We shall now scrutinize these arguments and see that each of them is based on certain assumptions which can be refuted or at least weakened.

1. As to the energetic inadequacy of the ion gradients, previous calculations were based on two assumptions: (1) that the activity coefficient of Na^+ and K^+ is about the same inside and outside the cell; and (2) that the ions are evenly distributed over the whole intracellular space. Whereas assumption (1) may not be too far off, assumption (2) is wrong. By fractionating the cells, we found that under normal conditions Na^+ and Cl^- ions are strongly sequestered in the cell nucleus, which covers about one-third of the cellular volume, whereas amino acid and K ions appear to be evenly distributed inside the cell.[5]

It turned out that the effective electrochemical potential gradients of Na^+, i.e., after correction for nuclear sequestration and in consideration of the fact that the electrical component, as determined by various methods, is at least 25 mV (Reference 6), should under these conditions account energetically for the highest accumulation ratios of amino acid.[5, 7] From this fact it does of course not necessarily follow that this driving force is actually used in full by the transport mechanism. We could meanwhile show, however, applying the principles of irreversible thermodynamics, that the direct coupling between amino acid transport and Na^+ influx is tight enough to warrant about adequate energy transfer from the ion gradient to amino acid transport.[8] Analogous investigations on the direct coupling between amino acid transport and ATP hydrolysis did not demonstrate any coupling at all, so that on these grounds ATP is very unlikely to contribute directly to the transport energy[9] (TABLE 1).

The apparent stimulation of active transport by cellular metabolism can at least in part also be explained by the nuclear sequestration of both Na^+ and Cl^-. Though no direct experimental evidence is available, we may safely presume that during metabolic inhibition, owing to a paralyzed Na^+ pump, the cytoplasmic Na^+ and Cl^- will rise without much change of the nuclear portion of these ions. It can easily be shown that under these circumstances the overall, i.e., uncorrected, distribution of Na^+ and Cl^- between cell and medium will be affected by metabolic inhibition much less than the effective distribution of these ions between cytoplasm and medium. This difference may not account for the full discrepancy between metabolizing and nonmetabo-

TABLE 1

COUPLING OF AIB TRANSPORT TO NA INFLUX AND ATP HYDROLYSIS

		J_i	
		$i = Na^+$ (Na influx)	$i = ATP$ (ATP hydrolysis)
Degree of coupling	q	0.5 – 0.6	Not detectable
Efficacy of accumulation	$-\frac{X_A}{X_i}$	≈0.6	0
Efficiency of energy transfer	η	$\approx \frac{\nu A}{\nu Na} 0.6$	0

lizing cells. There is likely to be an additional effect, however, which may concern the electrical membrane PD, as will be discussed later.

3. The last of the above arguments rests on cases of apparent incongruity in direction between amino acid transport and ion gradients; that is, active amino acid transport and the driving gradients appear to oppose each other. This is the hardest argument to disprove. Nuclear sequestration of Na^+ was found to vanish if the extracellular Na^+ was reduced to the low value required to invert the gradient.[5] As far as the concentration gradients of Na^+ and K^+ are concerned, their inversion thus appears to be real under these circumstances, and certainly cannot be explained away by nuclear sequestration. With inverted gradients active transport of aminoisobutyrate (AIB) is still possible, even after correction for nuclear sequestration, provided that the opposing driving force does not exceed about 5000 J/mol (FIGURE 1). The gradient hypothesis would in the present case still hold if the electrical potential across the cellular membrane were high enough to outweigh the opposed chemical potential gradients, so that the total driving force would still be in the direction congruent to the amino acid transport. Since active transport is observed as long as the opposing driving force is less than 4000 to 5000 J/mol, the electrical potential difference, in order to overcome this force, has to be higher than 40 to 50 mV, inside negative. In other words, with inverted gradients this PD should be substantially higher than that with normal gradients. Before looking into the experimental evidence on this point, let us first consider its theoretical basis and ask whether, and under what conditions, we could reasonably expect a PD of the postulated magnitude here from what we know about the origin of the PD in these cells. Clearly any reasonable prediction as to this p.d. depends on the alleged mechanism of electrical PD generation. There is little known about this mechanism, but there are several hypothetical possibilities, which we shall briefly discuss:

Most arguments on the electrical PD in these cells appear to have been more or less explicitly based on the assumption that the electrical PD in these cells represents a membrane-diffusion potential resulting from differential passive permeability through the cell membrane of Na^+ and K^+ ions, which are maintained at a disequilibrium distribution by an electroneutral exchange pump (FIGURE 2). The Cl^- ions, which are not subject to any pumping, should instead attain equilibrium distribution, from which the electrical PD can be derived directly.[7, 10] In view of these assumptions the electrical PD

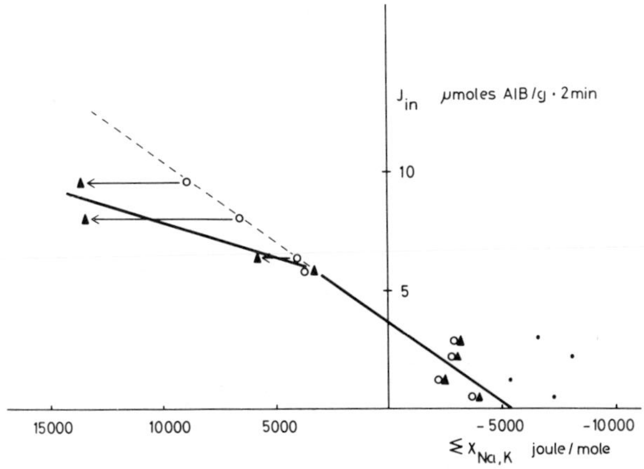

FIGURE 1. The net influx of 2-aminoisobutyrate as a function of the ionic driving forces. The final results of eight typical experiments with varying Na^+ and K^+ gradients are summarized. *Abscissa:* $X_{Na^+} - X_{K^+}$, the sum of the ionic driving forces of Na^+ and K^+, in J/mol. *Ordinate:* Influx of 2-aminoisobutyrate (J_{AIB}) in μmol/g per 2 min. (○) uncorrected driving forces; (●) corresponding driving forces corrected for nuclear sequestration of the alkali ion. The corresponding values are connected by a horizontal arrow. Since in the experiments with inverted gradients (right side of the ordinate) the intracellular concentrations of Na^+ and K^+ changed considerably towards normalization of the gradients during the 2-minute incubation, the initial values are also given (●). Hence the intermediate values are somewhere in between, and are in any case more towards the right from the final (corrected) values. Each pair of values represents a whole experiment. (From Pietrzyk and Heinz.[5] By permission of *Biochimica et Biophysica Acta.*)

and its dependence on ion concentrations and mobilities should be satisfactorily represented by the Goldman-Hodgkin equation.

From this equation it is clear that with normal Na^+ and K^+ distribution an electrical PD has to be expected if P_K is greater than P_{Na}. The distribution of Cl^-, after it has reached equilibrium, does not enter this equation. According to this equation, what will happen if the gradients of Na^+ and K^+ are inverted?

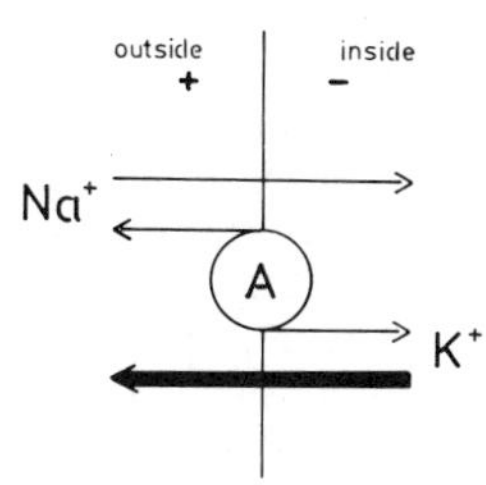

$$\Delta\Psi = -\frac{RT}{zF} \ln \frac{P_K[K^+]_i + P_{Na}[Na^+]_i}{P_K[K^+]_o + P_{Na}[Na^+]_o}$$

FIGURE 2. Kinetic model of an electroneutral Na^+/K^+ pump. The electrical potential is due to differential permeability coefficients of Na^+ and K^+ as under a simplifying assumption derived from the Goldman-Hodgkin equation. $\Delta\Psi$ is the electrical potential difference between the inside and outside of the cell, in mV. R and T are the gas constant and the absolute temperature, respectively. z is the electric valency, F the Faraday constant and ln the natural logarithm. P_K and P_{Na} are the permeability coefficients of K^+ and Na^+, respectively, through the cellular membrane, here in ml per gram (dry weight) per minute. The subscripts i and o refer to inside and outside the cell, respectively.

TABLE 2
LACK OF CORRELATION BETWEEN POTENTIAL DIFFERENCE AND K DISTRIBUTION DURING PUMP ACTIVATION

K_e^+ (mM)	Distribution Ratio Cell/Medium	
	K^+	R_{TPP^+}*
0.16	63.1	78
0.90	31.3	90
2.29	29.6	159
8.70	10.8	169

* R_{TPP^+} = tetraphenylphosphonium distribution ratio in percentage of distribution ratio in normal cells.

In one typical experiment taken from the paper of Schafer and Heinz,[4] active transport of AIB was clearly observed, while Na^+ and K^+ had the following distribution: Inside the cell Na^+ was 96 mM, and K^+ was 134 mM; whereas in the medium Na^+ was 40 mM, and K^+ 149 mM. At this low extracellular Na^+ no appreciable sequestration of this ion in the nucleus takes place, so that the intracellular Na^+ concentration should be approximately equal to the cytoplasmic Na^+ concentration. We see that both ion gradients are moderately inverted, that of Na^+ ions more strongly than that of K^+ ions. According to the above equation the electrical PD should be strongly reduced or even reversed, being positive inside the cell and negative outside. Measurements of this potential with microelectrodes have not been carried out under these conditions. From the Cl^- distribution, however, the PD was calculated to be strongly reduced, to about —4 mV, though not reversed.[4] In short, both the expected and the measured PD are obviously far too low to outweigh the inverted chemical potential gradients of Na^+ and K^+, unless both prediction and measurement are wrong. There is indeed evidence that they are, because the underlying assumptions are probably not correct.

The abolishment or even inversion of the electrical potential, as predicted above from the Goldman-Hodgkin equation, rests on the assumption that the Na^+-K^+ pump is electroneutral. The membrane potential would behave differently if this pump were instead electrogenic (FIGURE 3). There is indeed evidence that it is, in line with many other Na^+-K^+ pumps in animal cells. At the present juncture it is not important to know which ion is electrogenically transported. It could be that, as in other tissues, the ratio of Na^+ pumped outwards over the K^+ ions pumped inward is higher than unity. This would mean that the Na^+-K^+ pump is partly electrogenic. There is also evidence that H^+ ions are transported actively out of these cells and also this process could be electrogenic. Any electrogenic pump will make a contribution to the membrane potential of its own. The magnitude of this potential depends in a somewhat complicated way on the concentrations and permeabilities of the other permeant ions. The effect of the electrogenic Na^+ pump on the PD can best be illustrated by a modification of the Goldman-Hodgkin equation, introducing another term for the active net transport of Na^+ ions, A_{Na} (FIGURE 3). (P_{Cl} is so small that Cl^- can be considered to be virtually nonpenetrating in this context, as shown below.) A_{Na} is not a constant. Like the pump activity it can be expected to in-

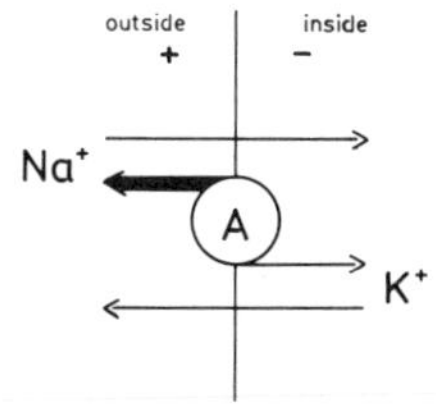

$$\Delta\Psi = -\frac{RT}{zF}\ln\frac{P_K[K^+]_i + P_{Na}[Na^+]_i + A_{Na}}{P_K[K^+]_o + P_{Na}[Na^+]_o}$$

FIGURE 3. Model of an electrogenic pump, based on a stoichiometry greater than one between Na^+ extrusion and K^+ uptake. The resulting electrical membrane potential is in part directly due to the pump, depending on the concentrations and permeability coefficients of other permeant ions present. Under simplifying assumptions it can be predicted by an expanded Goldman-Hodgkin equation, as given below. A_{Na} is to denote the electrogenic net transport of Na^+. Its magnitude must not be considered constant since it may vary with the degree of activation of the pump. The other symbols are the same as in FIGURE 2. Further explanations are given in the text.

crease accordingly as the distribution of Na and K is removed from the physiological value. It can easily be seen from this equation that if A_{Na} is sufficiently greater than $P_{Na}[Na^+]_o$, then the inverted distribution of Na^+ and $K,^+$ as exemplified above, should give rise to an electrical potential. So, in the presence of an electrogenic Na pump, the electrical PD need not fall; it should even rise considerably if the Na^+ and K^+ distributions are inverted. It could also be predicted that upon approaching normal steady-state distribution of Na^+ and K^+, the electrical potential should decrease since A_{Na} decreases as the $Na,^+$ K^+-ATPase is likely to become less activated if the cellular K^+ concentration rises.[13]

Why does the Cl^- distribution not show the predicted increase? Calculation of the PD from this distribution rests on the assumption of thermodynamic equilibrium distribution, and that is evenly distributed all over the cell. The first assumption seemed to be justified in view of the rapid equilibration of labeled Cl^- between the inside and outside of the cell, having a half-time of about two minutes.[7] These data, however, do not distinguish between exchange diffusion and net movement. The importance of such a distinction is most dramatically illustrated with human red cells in which the net movement of Cl^- is about 10^4 times slower than the Cl^- exchange.[12] We carried out similar tests for the Cl^- penetration in Ehrlich cells with qualitatively similar results. FIGURE 4 shows a typical experiment carried out according to Hunter.[12] The good fit between experimental points and the predicted curve appears to indicate that the permeability coefficient of Cl^- net transport is correctly chosen. It is also seen that the presence of Na^+ does not appreciably alter the results. The omission of Na^+ from the calculation appears therefore to be justified. Similar values for P_{Cl} were obtained in different experiments, in which the exit of labeled Cl^- into a Cl^--free medium was measured. This P_{Cl} should refer to the true net penetration of Cl^-. It is about 20 times smaller than that derived previously from isotopic Cl^- exchange. The half-time of Cl^- equilibration, previously thought to be two minutes, is in reality about one hour. It is clear that the Cl^- ions could not even approximately reach equilibrium distribution in the previous experiments, which lasted only five minutes.[4] On this basis the too-low distribution, ratios can plausibly be explained as follows:

During the extended pretreatment in the cold, Cl^- may have entered the cell, reducing the distribution ratio. If later the cells were suddenly exposed to a medium low in Na^+ and high in K^+, and if under these circumstances

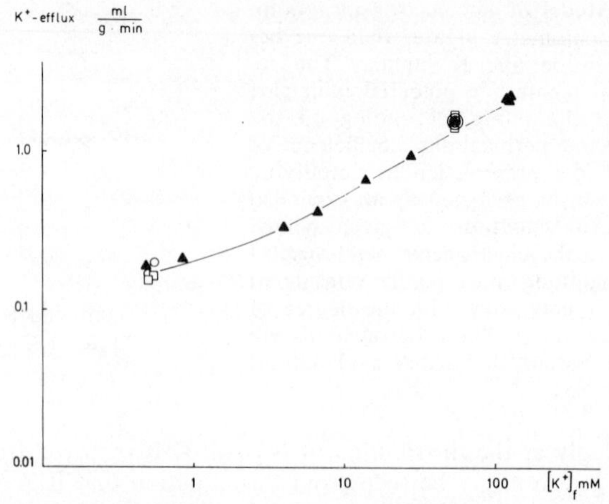

FIGURE 4. Determination of permeability coefficient of Cl^- (net movement) according to Hunter. Cells preloaded with radioactive potassium were suspended in media with varying K^+ concentration. In most experiments all extracellular Na^+ was replaced by tetraethylammonium (TEA) (▲). In some experiments also Na^+ was present in small concentrations: (□) Na^+ = 13 mM; (○) Na^+ = 23 mM. The points give experimental values of K^{42} efflux and the solid line gives the expected values calculated under the assumption of a constant electrical field using the equation of Hunter (see below). The best fit is obtained if P_K = 2,4 ml/gram/min and P_{Cl} = 0,057 ml/gram/min.

$$K^+\text{-efflux} = P_K\,[K^+]_i\,\frac{\ln B}{B-1};\quad B = \frac{P_{Cl}\,[Cl^-]_o + P_K\,[K^+]_i}{P_{Cl}\,[Cl^-]_i + P_K\,[K^+]_o}$$

The symbols are the same as in FIGURE 2.

the electrical PD were to rise again, the subsequent incubation should have been far too short for the Cl^- distribution to reach a new equilibrium.

Also the second assumption is wrong in view of the demonstrated sequestration of Cl^- between nucleus and cytoplasm. We have shown that with normal concentrations of K^+ and Na^+ inside and outside the cell, the nuclear sequestration of Cl^- in the nucleus is such that the true cytoplasmic Cl^- concentration is about half the overall cellular Cl^- concentration. This means that previous p.d. determination on the basis of Cl^- distribution, even if there were equilibrium, must have given values that are by 15 to 20 mV too low. In short, the Cl^- distribution can be dismissed as a reliable indicator of the true membrane potential.

In order to test the behavior of the electrical PD under various conditions, a more reliable method than that based on the Cl^- distribution has to be applied. For this purpose we developed a method based on the distribution of a lipid-soluble cation. The widely used dibenzyldimethylammonium (DDA^+) enters the cells too slowly to reach equilibrium distribution in a reasonable time. Among various such cations tested, tetraphenylphosphonium (TPP^+) was found to reach equilibrium distribution within about five minutes (FIGURE 5). This cation is toxic to cell metabolism in higher concentrations, but rather innocuous if applied in radioactively labeled tracer concentrations. The distribution ratio between cell and medium reached by this compound is considerably higher than expected, so that without appropriate standardization the absolute PD values derived from this distribution may be treated with caution. On the other hand, the distribution ratio of this compound responds linearly and in the predicted direction upon alterations of the electrical cell PD by ionophores (FIGURE 6). The compound therefore appears to be useful to monitor relative PD changes, which in the context are sufficiently informative. Using this compound, we now tried to test the behavior of the PD under conditions in which Na^+ and K^+ concentrations inside the cell were inverted. For this purpose cells were depleted of K^+, which was replaced by Na^+. In the absence of extracellular K^+ the PD, as monitored by the TPP^+ distribution, was about the same as or slightly lower than that of normal cells. If, however, the Na^+-K^+ pump was activated by the addition of K^+ to the medium, the TPP^+ distribution strongly increased (FIGURE 7). This should be expected if the activated pump is, at least in part, electrogenic, as has been shown for other cells and tissues.[14] With 10 mM K^+ the TPP^+ distribution reached a value that was about three times higher than was found with cells of normal ion content. It can also be seen on FIGURE 7 that the distribution of TPP^+ decreases with time, as should be expected as the ion pump slows down if the ion distribution changes towards the normal steady-state distribution.

These results can thus be interpreted as indicating an electrogenic pump that upon activation by external K^+ is capable of raising the PD to such an extent that active amino acid transport could be energized against opposing ion gradients. This interpretation, however, is weakened to the extent that the described observations can also be explained in terms of an electroneutral

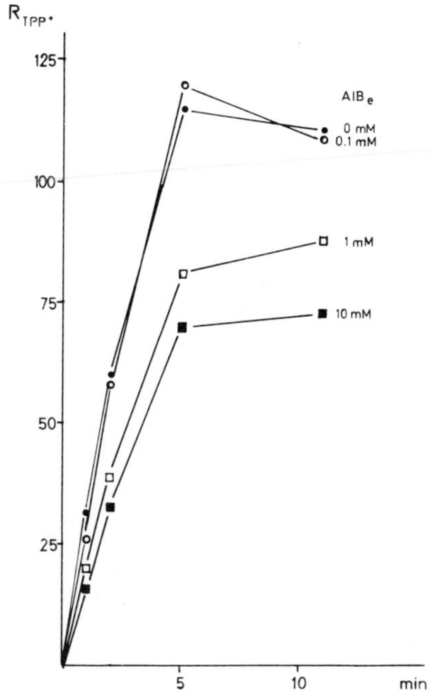

FIGURE 5. Distribution of ^{3}H-tetraphenylphosphonium ions between normal cell and medium. Ehrlich cells without special pretreatment were suspended in Ringer's solution containing 10 μM ^{3}H-labeled TPP^+ at 37° C. *Abscissa:* time after TPP addition. *Ordinate:* distribution ratio of TPP^+ (R_{TPP}). In some experiments the nonmetabolizable amino acid aminoisobutyrate (AIB) was present at the concentrations indicated.

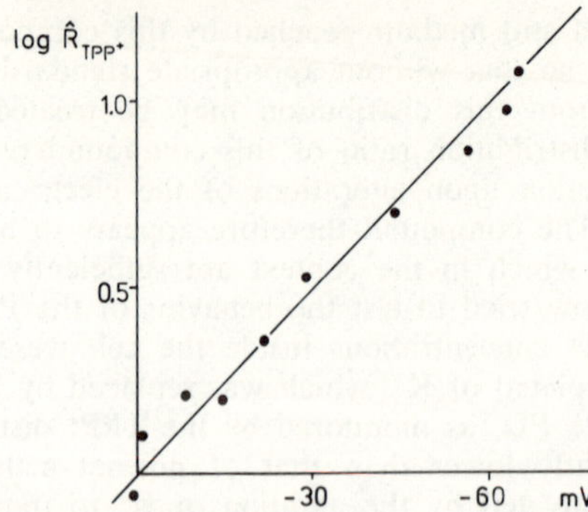

FIGURE 6. Distribution of TPP^+ as a function of membrane potential. Normal Ehrlich cells were incubated in cholin-K-Ringer-phosphate buffer at 37° C in the presence of 25 μg/ml valinomycin and 10 μM TPP^+. The extracellular $[K^+]$ was varied with complementary changes of $[choline^+]$. The electrical potential was estimated using the simplified Goldman-Hodgkin equation, inserting P_{Cl} and P_K as determined in the experiments described above. The TPP^+ distribution ratio was corrected for adsorbed TPP^+ by subtracting the extra uptake of TPP^+ at zero PD. *Ordinate:* corrected distribution ratio of TPP^+. *Abscissa:* estimated PD in mV.

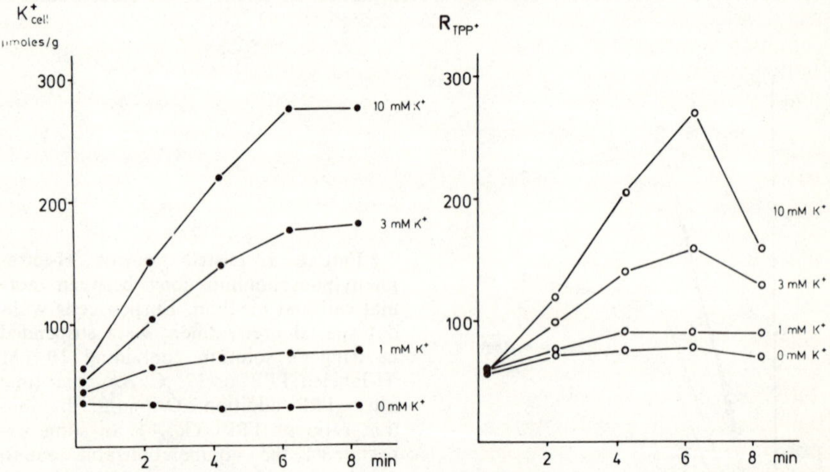

FIGURE 7. Uptake of K^+ and accumulation of TPP^+ in K^+-deficient Ehrlich cells as a function of extracellular K^+. Cells were depleted of K^+ by preincubation in K^+-free Ringer's solution at 1° C. The cellular K^+ drops to about 15 to 20 mM by this procedure whereas the Na^+ rises accordingly. In the main incubation, the uptake of K^+ (left side) and the distribution ratio of TPP^+ were measured in the presence of extracellular K^+ at the concentrations indicated. *Ordinate* (left): cellular content of K^+ in μM/g (dry weight). *Ordinate* (right): distribution ratio (R_{TPP}). *Abscissa* (both sides): time in minutes. The extracellular $[Na^+]$ was adjusted to make total extracellular $[Na^+] + [K^+] = 150$ mM.

pump, in line with the alternative hypothesis discussed before. In that case the electrical PD would entirely be due to the distribution and differential permeabilities of K^+ and Na^+, presumably mostly to the K^+ distribution, according to the unmodified Goldman-Hodgkin equation. So the question arises whether, for instance, in FIGURE 7 the K^+-dependent rise of the PD could also be attributed to the rising K^+ distribution rather than to an electrogenic pump. Since, as seen from FIGURE 7, the distributions of both K^+ and TPP^+ rise with time for each given K^+ concentration, a direct linkage between the two seems to suggest itself. There are, however, three significant points in which the response of the TPP^+ distribution contradicts such as interpretation.

1. As shown in TABLE 2, which is based on the data of FIGURE 7, the distribution ratio of K^+ strongly decreases with increasing extracellular K^+ concentration, as should be expected from a saturable ion pump. Consequently, the electrical PD due to this K^+ distribution should also decrease with increasing extracellular K^+. By contrast, all PD values, estimated from the TPP^+ distribution, strongly increase with increasing extracellular K^+. This increase, while it contradicts the hypothesis of an electroneutral pump, is very compatible with an electrogenic pump which can be expected to be increasingly activated by increasing extracellular K^+.

2. If the Na^+-K^+ pump were neutral, the PD should predominantly be determined by the K^+ distribution, as has been stated before. In the present experiments with extracellular $K^+ = 10$ mM the maximal distribution ratio for K^+ is only 10. This is considerably lower than the static head distribution ratio with normal cells. One would expect that the electrical PD should also be lower in the present experiments than under normal conditions. The contrary, however, was found: the peak distribution of TPP^+ under the present condition is about three times higher for K^+-depleted cells than for normal cells. Since the response of the TPP^+ distribution, owing to the limited penetration rate of this compound, appears to lag behind the activation PD change by about five to six minutes, one may presume that the real PD maximum, which should appear right after the activation, may exceed the static head PD by still more than three times. The relative magnitude of the PD estimated from these experiments thus appears to be at variance with that predicted for an electroneutral pump. It is, on the other hand, well consistent with the operation of an electrogenic pump, which, as has been discussed before, is likely to be more powerful in K^+-depleted than in normal cells.

3. Since the K^+ distribution after six minutes incubation does not decrease (indeed it rather slightly increases), an electrical potential if it were due to this K^+ distribution should not decrease either. In reality, however, the TPP^+ distribution markedly dropped after six minutes. This drop, while hardly compatible with a K^+-diffusion PD is very compatible with an electrogenic pump, which is likely to become less activated as the cellular K^+ rises.

Under the condition that the changes in TPP^+ distribution, as discussed above, reliably monitor direction and magnitude of the underlying changes in PD, all three of the above points argue strongly against an electroneutral pump and in favor of an electrogenic pump which is activated by extracellular K^+ and becomes less active as the intracellular K^+ level rises. In the following a few more experimental results are reported which are consistent with this view, but which would have argued against this pump, if they had come out differently. FIGURES 8 and 9 show that ouabain added in increasing concentration reduces the uptake of K^+ and AIB and the distribution ratio of TPP^+

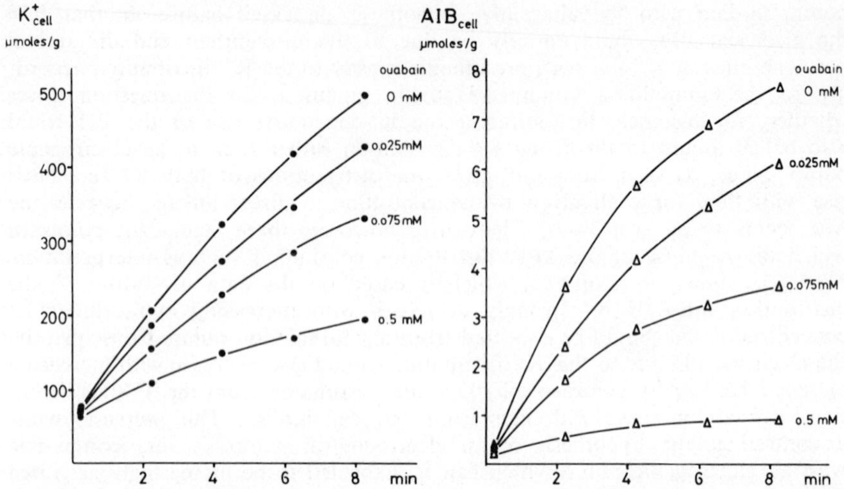

FIGURE 8. Inhibition of the uptake of AIB and of K^+ in K^+-depleted Ehrlich cells by ouabain. The cells were depleted of K^+ as described in FIGURE 7. The main incubation was in modified Ringer's phosphate buffer, containing 18 mM K^+ and 132 mM Na^+ at 37° C. Initial AIB concentration was 0.1 mM. *Ordinate* (left): cellular content of K^+ in $\mu M/g$ (dry weight); (right): cellular content of AIB in $\mu M/g$ (dry weight). *Abscissa:* time in minutes.

in K^+-depleted cells. To the extent that the K^+ uptake indicates the activity of the electrogenic pump, it can be inferred that this pump is directly inhibited by this drug, as seems to be supported by the apparent inhibition of PD generation, as monitored by the TPP^+ distribution. It is true that also in the absence of an electrogenic pump an inhibition of K^+ uptake could be expected, since ouabain may also inhibit the electroneutral pump, and the PD would in that case depend on the K^+ distribution (K^+ battery). Without an electrogenic pump, however, the electrochemical potential gradient of Na^+ is under the present conditions not high enough to account for the AIB transport observed. In the beginning the Na^+ distribution is near unity, and the PD of the K^+ battery is too low. Only with time should the driving Na^+ gradient slowly grow as the ions redistribute towards the normal. The uninhibited uptake of AIB, however, appears to be equally powerful from the beginning, as if an additional driving force were supplied directly from metabolism. The strong inhibition of the AIB transport by ouabain as seen in the Figure argues against direct coupling to metabolism, unless the glycoside attacks the amino acid transport directly, but favors the assumption of an electrogenic pump. If AIB is cotransported with Na^+, as postulated by the gradient hypothesis, and if this cotransport under the present condition is energized by the electrical PD, the addition of amino acid to the system should shunt this PD. FIGURES 5 and 10 show that this is actually the case: the TPP^+ distribution markedly dropped with increasing AIB addition. Again, if under the present condition amino acid transport were predominantly driven directly by metabolic energy, this shunting effect would not necessarily occur. Although a negative result of this test would have strongly argued against the gradient hypothesis, the present result is highly consistant with it. It may be also inferred that with normal distribution of Na^+ and K^+ the electrogenic pump accounts at least in part for the electrical PD, even though this contribution, for reasons discussed above, may be smaller than that in the state of inverted Na^+ and K^+ distribution. As a consequence, metabolic inhibition should abolish this part, so that even with the same distribution ratios for Na^+ and K^+ the electrical PD, and hence the electrochemical potential difference of Na^+ ions, should be smaller by this amount. This would give a further explanation for the observation that during metabolic inhibition the active transport is considerably smaller than it is during respiration, even if the Na^+ distribution ratios are the same.

It is admitted that the evidence so far presented of the existence of an electrogenic pump and its decisive function in amino acid transport in cells with inverted alkali ion gradients is not yet conclusive. To make it so, more experiments are required. Especially the relation between TPP^+ distribution and electrical potential has to be tested more rigorously, preferably in conjunction with direct PD measurements by micropuncture. This could not be done yet, among other things because none of the PD measurements in these cells so far applied can be considered reliable beyond doubt. It should also be shown that the stoichiometry between Na^+ extrusion and K^+ uptake in K^+-depleted cells is greater than one under the critical condition. The direct dependence of the electrical PD on the rate of excess Na^+ extrusion should be demonstrated. Moreover it has to be made sure that the permeability coefficients for alkali ions and Cl^- do not drastically change under any of the critical experimental conditions applied. Nevertheless, the evidence available at this juncture appears to be strongly suggestive of an electrogenic pump,

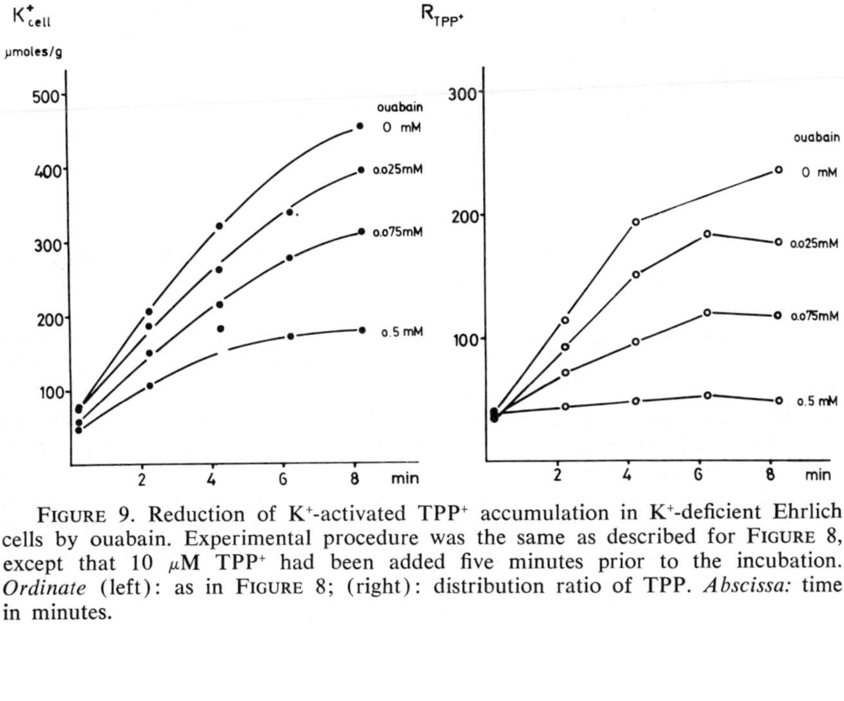

FIGURE 9. Reduction of K^+-activated TPP^+ accumulation in K^+-deficient Ehrlich cells by ouabain. Experimental procedure was the same as described for FIGURE 8, except that 10 μM TPP^+ had been added five minutes prior to the incubation. *Ordinate* (left): as in FIGURE 8; (right): distribution ratio of TPP. *Abscissa:* time in minutes.

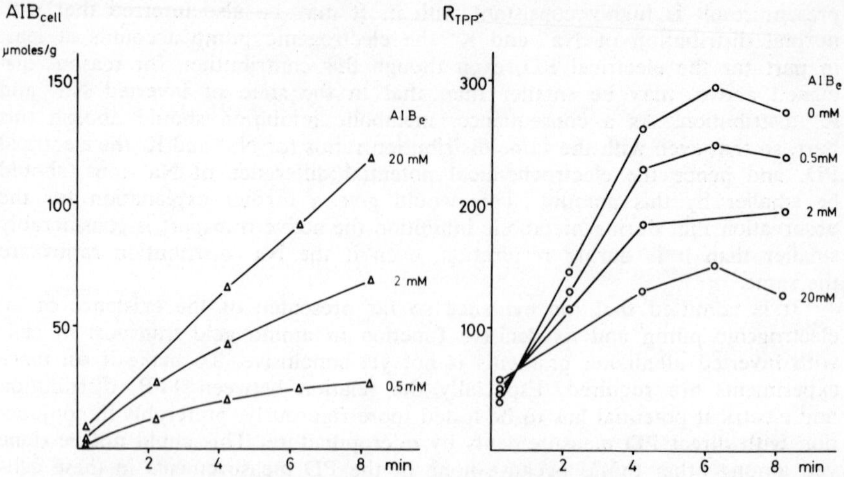

FIGURE 10. Reduction of K^+-activated TPP^+ accumulation in K^+-deficient Ehrlich cells by simultaneous AIB uptake. Cells were depleted of K^+ as described for FIGURE 7, and subsequently incubated in modified Ringer's phosphate buffer containing 18 mM K^+ and 132 mM Na^+ at 37° C, with the initial concentrations of AIB as indicated. 10 μM TPP^+ had been added five minutes prior to incubation. *Ordinates* and *abscissa* as in FIGURE 7.

also in analogy with electrogenic Na^+ pumps shown in other cells and tissues.[14] In short, the assumption of a direct energetic coupling between metabolism and amino acid transport in these cells, in the sense of primary active transport, is no longer mandatory and in view of the bulk of evidence is very unlikely. The problem appears no longer to be intractable and a complete solution of the problem of energy supply for amino acid transport in these cells in terms of the gradient hypothesis seems to be in reach.

Conclusions

Our observations on the response of TPP^+ uptake into K^+-depleted cells on the extracellular K^+ concentration and on the addition of ouabain and amino acid appear to provide suggestive evidence of, or are at least consistent with, the operation of a powerful, K^+-activated electrogenic pump. They are also consistent with the assumption that in these K^+-depleted cells the active uptake of amino acid is energized by the electric PD presumably generated by this electrogenic pump. It follows that an energy source other than the electrochemical potential gradient of Na^+ ions need not be invoked to explain active amino acid transport with inverted Na^+ and K^+ distribution. In the presence of an electrogenic cation pump one should expect that owing to the activity of this pump the electrical PD is raised under these conditions to a value outweighing the opposed chemical potential gradient of Na^+. The experimental evidence so far available is clearly consistent with such an assumption.

Acknowledgments

The competent technical assistance of Miss B. Pfeiffer and Miss G. Werner is gratefully acknowledged. We also thank Miss B. Pfeiffer and Mrs. E. Kemsley for preparing the manuscript.

References

1. GIBB, L. E. & A. A. EDDY. 1972. Biochem. J. **129:** 979–981.
2. JACQUEZ, J. A. & J. A. SCHAFER. 1969. Biochim. Biophys. Acta **193:** 368–383.
3. EDDY, A. A. 1968. Biochem. J. **108:** 489–498.
4. SCHAFER, J. A. & E. HEINZ. 1971. Biochim. Biophys. Acta **249:** 15–33.
5. PIETRZYK, C. & E. HEINZ. 1974. Biochim. Biophys. Acta **352:** 397–411.
6. LASSEN, U. V., A.-M. T. NIELSEN, L. PAPE & L. O. SIMONSEN. 1971. J. Membr. Biol. **6:** 269–288.
7. HEINZ, E. 1974. *In* Current Topics in Membranes and Transport. F. Bronner and A. Kleinzeller, Eds. Vol. 5: 137. Academic Press. New York, N.Y.
8. HEINZ, E. & P. GECK. 1974. Biochim. Biophys. Acta **339:** 426–431.
9. GECK, P., E. HEINZ & B. PFEIFFER. 1974. Biochim. Biophys. Acta **339:** 419–425.
10. GROBECKER, H., H. KROMPHARDT, H. MARIANI & E. HEINZ. 1963. Biochem. Z. **337:** 462.
11. HODGKIN, A. L. 1958. Proc. Roy. Soc., Ser. B **148:** 1.
12. HUNTER, M. J. 1971. J. Physiol. **218:** 49 p.
13. CHIPPERFIELD, A. R. & R. WHITTAM. 1974. **187:** 269–280.
14. RITCHIE, J. M. 1971. *In* Current Topics in Bioenergetics. D. R. Sanadi Ed. Vol. 4: 327. Academic Press. New York, N.Y.

Discussion

DR. HOPFER: In the intestine where you also have active amino acid transport, we can show for the neutral amino acid at least that you do have a mechanism for concentrated uptake supported by electrochemical gradient of sodium. I could have told this whole story, which I told for sugar, for glucose and also for neutral amino acids, which supports your conclusion.

DR. HEINZ: Thank you.

IMPLICATIONS OF THE NEUTRAL CARRIER Cl^--HCO_3^- EXCHANGE MECHANISM IN GASTRIC MUCOSA*

Warren S. Rehm and Sue S. Sanders

Department of Physiology and Biophysics
University of Alabama in Birmingham
The Medical Center
Birmingham, Alabama 35294

In keeping with the goals of this conference we will focus our attention in this paper on the mechanisms of ion transport across the limiting membranes of the gastric cells. As a result of the analysis of an aspect of gastric HCl secretion we recently have had an interesting spin-off. We found that the established criteria for passive ion transport are not sufficient and we will give a brief account of this later.

FIGURE 1 shows the overall problem of ion transport in the stomach. In this figure SEC refers to the secretory or lumen membrane and NUTR refers to the nutrient membrane (the membrane facing the muscularis mucosa or blood side). We will assume initially that there is only one cell type and hence only a single nutrient membrane and a single secretory membrane. Later we will present a more realistic histologic picture. In addition to H^+ and Cl^- secretion small amounts of Na^+, K^+ and trace amounts of other ions are transported across the gastric mucosa and secreted into the lumen. For simplicity we will in general ignore the problem of Na^+ and K^+ transport across the whole cell layer and confine our attention to H^+, Cl^- and HCO_3^- transport. It has been amply demonstrated that Cl^- is actively transported and that a metabolic machine (or machines) is required for the secretion of H^+ and Cl^-.[1-3]

Hydrogen ion and Cl^- are secreted into the lumen and Cl^- enters the cells via the nutrient membrane and HCO_3^- appears in the nutrient bathing fluid. The equality of the rate of H^+ secretion and HCO_3^- appearance in the nutrient bathing solution has been experimentally demonstrated numerous

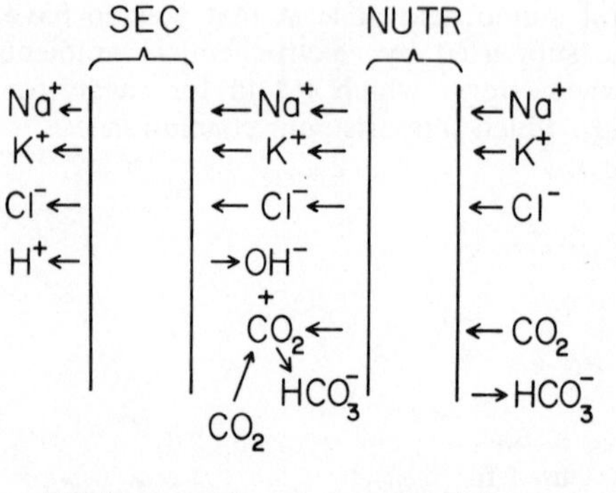

FIGURE 1. Summary of the transport of ions across the cell layer of the gastric mucosa. SEC indicates the secretory or lumen-facing membrane and NUTR the nutrient membrane, which faces the blood side. CO_2 is needed to maintain the acid-base balance of the cells and more H^+ is secreted than CO_2 produced so exogenous CO_2 is needed. (See text for details.)

* This research was supported by grants from the National Science Foundation (GBMS-72-02508) and National Institutes of Health (5 RO1 EY00456).

times.[4-7] This equality also follows on the basis of simple theoretical considerations.[8]

Our main theme is concerned with whether the active transport of H^+ and Cl^- is by electrogenic mechanisms or by a unitary or neutral mechanism. In our analysis we will use well-established findings and conclusions. These findings have been presented in detail previously and an in-depth presentation is not appropriate here. Consequently, we will base our analysis on findings and conclusions that we believe have been previously established. First I will briefly describe our experimental technique.

Most of the experimental work forming the basis for our present picture was performed on either the *in vitro* amphibian stomach or on a flap of the dog stomach with an intact blood supply. FIGURE 2 shows a method developed in the early 1940s in which a flap of dog stomach was mounted between two lucite chambers in such a way as to maintain an intact blood supply.[9, 10] Two electrodes are used for potential difference (PD) measurements and two for current-sending. The resistance is determined as the ΔPD per unit current. The ΔPD is taken after charging of the dielectric capacitors of the membranes and before there is a change in the PD due to changes in ionic gradients between the cytoplasm and external fluids.[11, 12] The secretory rates are easily measured; in the case of the dog stomach by changing the secretory solution every five or ten minutes or in the case of the *in vitro* stomach by the pH stat method.[13] With the frog stomach there is no blood flow; it is a truly *in vitro* method.

FIGURE 3A illustrates by means of an equivalent circuit a theory of HCl production formulated in the 1940s, the gist of which is that there are separate electrogenic mechanisms for Cl^- and H^+ in the secretory membrane.[2] Obviously in the absence of an external current the mechanisms must be electrically coupled. We will refer to this theory as the separate-site or electrogenic theory of HCl formation. FIGURE 3B represents the alternative to the electrogenic theory, i.e., a neutral or unitary theory.

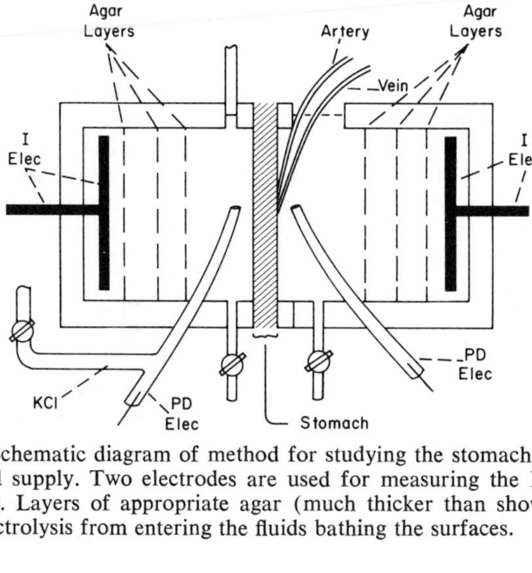

FIGURE 2. Schematic diagram of method for studying the stomach of the dog with an intact blood supply. Two electrodes are used for measuring the PD and two for current-sending. Layers of appropriate agar (much thicker than shown) prevent the products of electrolysis from entering the fluids bathing the surfaces.

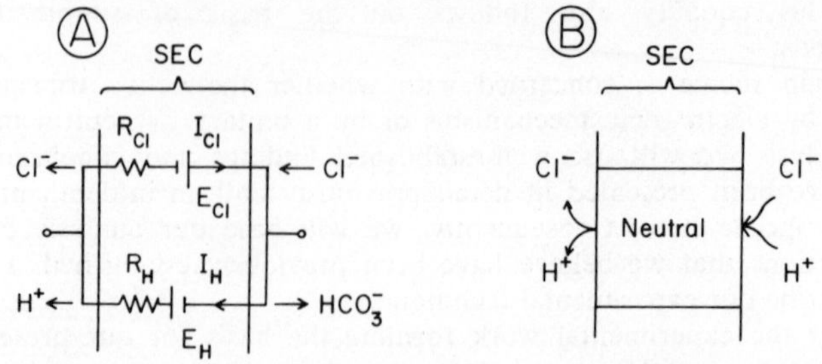

FIGURE 3. (A) The H^+-Cl^- electrogenic theory presented by an equivalent circuit. (B) The alternative to the electrogenic theory, i.e., a neutral HCl mechanism. SEC refers to secretory membrane.

We use the term "electrogenic" to mean that in the active transport mechanism itself there is a net transport of charge across the membrane. Some workers use the term for schemes in which the active transport mechanism is neutral, but results in ion gradients across the limiting membranes which in turn give rise to PD's.

In 1945 we showed that the H^+ rate could be changed by the application of electric current from an external source. Current sent from the nutrient or blood side to the secretory side should increase the current in the H^+ limb and hence the H^+ secretory rate. Current sent in the opposite direction should decrease the H^+ rate. This prediction was borne out experimentally in our studies on the flap of dog stomach[9] and Davies and his colleagues confirmed this in studies on the *in vitro* frog stomach.[14] This finding was one of the major reasons why we proposed the separate-site or electrogenic theory.

In the late 1940s, following the discovery that the H^+ rate could be increased or decreased by the application of electric current, a number of workers proposed a theory of HCl secretion that involves an oriented oxidation in the secretory membrane of a hydrogen carrier with the expulsion of H^+ into the lumen and of electrons to an electron-acceptor in the cytoplasm. Within a year five different laboratories presented models that were basically the same from the point of view of the production of H^+ ions.[15-19] Conway coined the phrase "redox pump" for this type of mechanism. All of the models except for ours ignored the fact that the potential difference is opposite to that predicted on the basis of an electrogenic H^+ pump. With an electrogenic H^+ pump and with the Cl^- being a passive ion the secretory side would be positive to the nutrient side. Experimentally it has been found that the orientation is exactly the opposite. The nutrient or blood side is positive to the secretory side and in the dog stomach is about 60-mv positive.[1, 10] Since Cl^- is moved up both a concentration gradient and an electrical gradient, it is clear that Cl^- is actively transported. I believe we have now convinced everyone in the field that Cl^- is actively transported. The electrogenic theory is illustrated in FIGURE 4, in which a redox scheme is presented for the electrogenic H^+ pump. The upper left hand circuit represents an equivalent circuit for the H^+ pump. In parallel with the H^+ pumps are Cl^- pumps, which are represented by a resistance, and a Cl^- emf as shown in the upper right hand side. The Cl^- emf would tend to make the nutrient or blood side positive, while the

H^+ emf would tend to make it negative. Since the nutrient side is positive, then within the framework of the theory the parameters of the two pumps must be such as to result in this orientation. With this scheme the emf of the H^+ mechanism is a function of the mass action constants and the concentrations of the reactants and the products.[20] The equation for the H^+ emf E_H is shown in FIGURE 4 and for purposes of illustration is divided into two terms. The first term, $E_H{}^*$, is a function of the mass action constants and the concentrations of the reactants and products; the second term shows that the emf is a logarithmic function of the H^+ concentration in the lumen. With everything held constant except the H^+ concentration of the secretory fluid, the H^+ emf will decrease in magnitude with increasing H^+ concentration. Although we are not presenting a detailed scheme for the active transport of Cl^-, such schemes show that the emf is a function of an $E_{Cl}{}^*$ and the ratio of the Cl^- concentrations between the lumen and the cytoplasm.[20] It can be shown that any formulation of an electrogenic pump for Cl^- can be resolved into two terms, one of which, shown on the right, is the Nernst equation. Actually these equations are formulated for equilibrium conditions, i.e., for electrogenic mechanisms in the absence of a completed circuit. During current flow they are only approximations of the magnitudes of the emfs. Again the equivalent circuit for the electrogenic scheme is shown in the lower left hand corner.

All that the electrogenic theory postulates in its general form is that there are separate electrogenic mechanisms for the H^+ and Cl^-. An electrogenic H^+ mechanism could also be driven by ATP (adenosine triphosphate), in which the net transport of charge was due to the movement of ions and not electrons. This possibility is illustrated in FIGURE 5 and the ATPase is essentially similar to what Mitchell[21] calls ATPase I. With the switches, shown in the upper part of this figure to the Cl^- and H^+ electrogenic mechanisms open and with no H^+ gradient across the membrane this ATPase model would

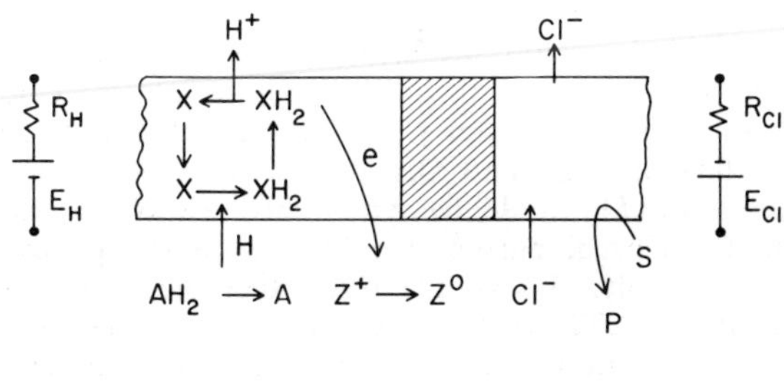

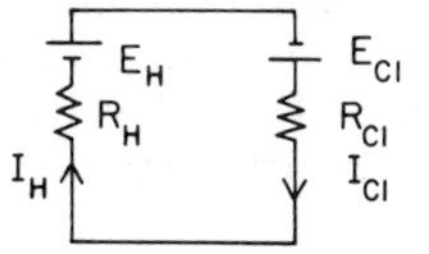

$$E_H = E_H^* - \frac{RT}{F} \log [H^+] \text{Lumen}$$

$$E_{Cl} = E_{Cl}^* - \frac{RT}{F} \log \frac{[Cl^-] \text{ Lumen}}{[Cl^-] \text{Cell}}$$

FIGURE 4. Upper portion illustrates a redox scheme for electrogenic H^+ pump. Hydrogen atoms are donated to the carrier X in the membrane and H^+ secreted into lumen and electrons (e) donated to an electron acceptor in cytoplasm. Active Cl^- electrogenic pump is in parallel with H^+ pump. The upper left and right circuits are equivalent ones for the two pumps; the lower circuit shows the electrical coupling. (See text and Rehm.[8])

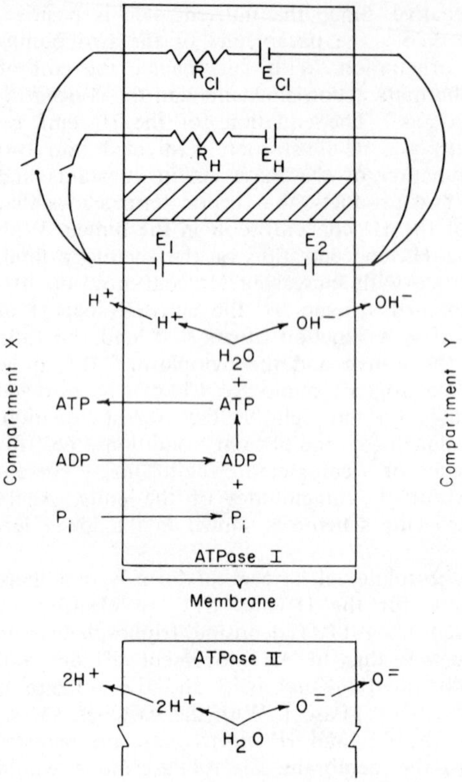

FIGURE 5. An ATPase scheme for the production of H^+ ions. ATPase I, in the absence of completed circuit (switches on left hand side are open) and with no H^+ gradient across the membrane, would yield a PD of over 400 mv. With the E_H switch closed and with E_H greater than the emf of the ATPase I system current flows so as to produce movement of ions in the direction indicated by the arrows, with the consequence that ATP is synthesized. With the E_H switch open and the E_{Cl} closed, the ions would move in directions opposite to those indicated by the arrows, ATP would be hydrolyzed, and OH ion would move from the lumen (right side, side Y) into the cell; this is the same end result as would occur if H^+ were secreted into the lumen. ATPase II in place of ATPase I could only decrease H^+ concentration to pH of about 3.5. (From Rehm.[8] By permission of the publisher.)

result in a PD of over 400 mv with lumen side (side Y, the right hand side) being positive.[8] Now if the switch to the electrogenic H^+ mechanism were closed and with E_H greater than the emf of the ATPase system, then current would flow and H^+ and OH^- would move in the direction indicated by the arrows and ATP would be synthesized. However, with the E_H switch open and the E_{Cl} closed, current would flow in the opposite direction and ATP would be hydrolyzed with the result that the lumen (right hand side) would be acidified.[8] ATPase I could produce H^+ ions at a sufficient high concentration to account for the H^+ concentration of gastric juice. Mitchell's ATPase II[22], the ATPase that is probably present in subcellular microsomes, would only reduce the pH to about 3.5 and by itself is not a viable mechanism for H^+ secretion. The magnitude of the emf with ATPase I would be a function of E_H* and a Nernst term for the ratio of the H^+ concentration in the lumen to that of the cell.

My colleague, Dr. George Sachs, originally planned to deal with (1) the problem of the relationship between the state of the electron transport system in the gastric cells and the secretory rates[23-25] and (2) the gastric ATPases and the evidence for and against ATP being the source of energy for the H^+ and Cl^- mechanisms.[26-29] However, he will deal primarily with the interesting recent work on the effect of ATP on the gastric microsomes.[30] I believe that, apart from the recent work on the gastric microsomes, the other evidence does not warrant the conclusion that ATP is the source of energy for the transport mechanisms.

Now we return to the main theme: the electrogenic theory vs. the neutral theory. Since the evidence for the electrogenic or separate-site theory has been presented previously[31] I will not go into detail here.

The most unequivocal evidence for an electrogenic H^+ pump in the gastric mucosa is found in the studies on the frog gastric mucosa with bathing solutions that contain no Cl^-. Heinz and Durbin found that H^+ secretion occurs in the absence of Cl^- in the bathing solution.[32] The PD in the absence of Cl^- in the bathing solution is inverted; normally the nutrient side is positive, but in chloride-free media it is negative. This is the orientation expected from an electrogenic H^+ pump. However, this evidence by itself does not warrant the conclusion that the H^+ pump is electrogenic. You recall that in the original scheme proposed by the Ussing school for active Na^+ transport in the frog skin the PD is oriented in the direction for an electrogenic Na^+ pump, but they presented a scheme (consistent with the data available at the time) in which the active process was neutral. The scheme had a neutral forced exchange mechanism for Na^+ and K^+.[33] Marion LeFevre and I extended the work of Heinz and Durbin and found, as shown in FIGURE 6, that in Cl^--free media there was a precise linear relationship between the PD and the H^+ rate.[34] In this representative experiment the magnitude of the PD before the addition of an inhibitor was about 30 mv and the H^+ rate (given in μa cm^{-3}) was about 25. Following the addition of small amounts of dinitrophenol the magnitude of the PD and the H^+ rate decreased. The rather startling finding is the precise linear relationship between the PD and H^+ rate. In my opinion, there is no way to explain this linear relationship other than on the basis that the H^+ mechanism (under these conditions) is electrogenic. The analysis of this finding has been presented in detail at a previous conference.[35] The linear relationship found with dinitrophenol was confirmed with other inhibitors by Sachs and his colleagues[36] and by Schwartz and MacKrell.[37, 38] In other words, the experimental basis for the conclusion that the H^+ mechanism is electrogenic under Cl^--free conditions has been confirmed many times with a variety of inhibitors. I should point out parenthetically that in the intact gastric mucosa dinitrophenol produces a marked increase in electrical resistance. I mention this because dinitrophenol produces a decrease in resistance

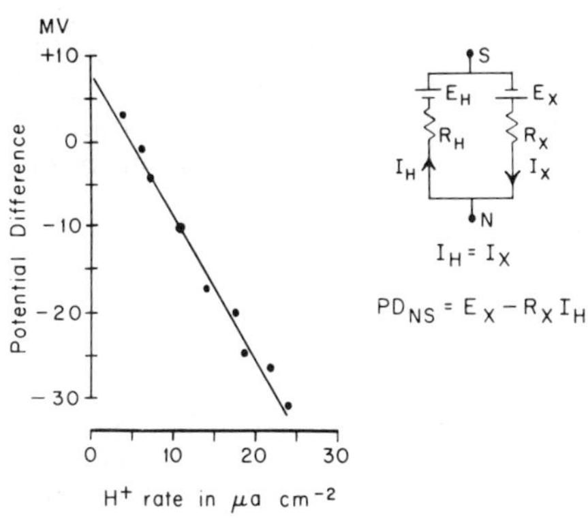

FIGURE 6. PD vs the H^+ secretory rate in Cl^--free media; -PD means nutrient side negative. Farthest point to right gives PD and H^+ rate before addition of dinitrophenol to nutrient fluid; with addition of aliquots of dinitrophenol the PD decreases linearly with the decrease in H^+ rate. Equivalent circuit on the right used for analysis of linear relationship.[34]

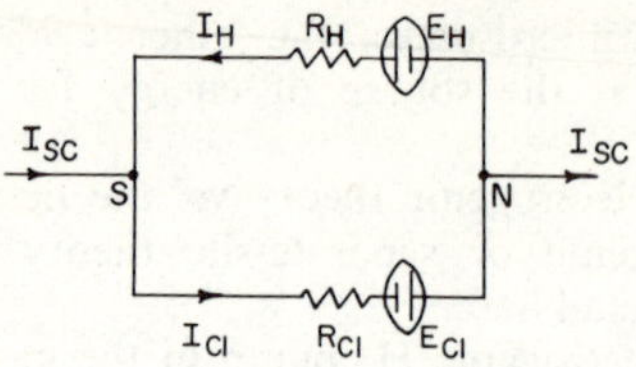

FIGURE 7. Single membrane, two-limb circuit for analysis of relationship between short circuit current (I_{SC}) and H^+ rate (I_H). R and E refer to resistances and emfs, and the meaning of the subscripts is self-evident.

$$I_{SC} = I_{Cl} - I_H$$

$$\Delta I_{SC} = -\Delta I_H$$

in subcellular and bilayer membranes,[39, 40] and one might infer that the primary action of dinitrophenol in decreasing H^+ secretion was the result of a high conductance H^+ leak; but such is not the case.

Since the H^+ mechanism is clearly electrogenic in the absence of Cl^- it seems reasonable to assume that it is also electrogenic in the presence of Cl^- bathing media and since there must be a completed circuit, the active Cl^- transport mechanism must also be electrogenic. On the other hand, in view of other findings one is forced to consider the possibility that with Cl^- media the H^+ mechanism is no longer electrogenic and that the HCl mechanism is a neutral one.

It is a truism that theories are never proved—theories are for disproving and not for proving. So I will not marshal the evidence in favor of the electrogenic theory when the bathing media contains Cl^-. Instead I will concentrate on the evidence that seemingly disproves the electrogenic theory when Cl^- is present in the bathing media. The really challenging evidence against the electrogenic theory was presented by Heinz and Durbin.[41, 42] Their analysis has seemingly convinced most of my colleagues in the field of gastric secretion as well as workers outside the field who write reviews on the subject.[43] However, only recently we realized there is an implicit assumption in the Heinz-Durbin analysis that turned out to be unjustified on the basis of relatively recent findings.[44] This assumption is that the ion pathways in the nutrient membrane which are involved in HCl secretion are conductive pathways. (A similar implicit assumption is present in the criteria for passive ion transport that had been developed by the Ussing school,[45] and we will touch on this briefly later.)

FIGURE 7 illustrates the Heinz-Durbin argument. They used a two-limb circuit to represent the whole mucosa; starting with the electrogenic assumption, they attempt to show that it is incompatible with certain experimental findings and conclude that the HCl mechanism is therefore a neutral one. Their argument is as follows: Under short circuit conditions and in the absence of H^+ secretion $I_{SC} = I_{Cl}$ and during H^+ secretion the I_{SC} is equal to $I_{Cl} — I_H$. Heinz and Durbin assumed that gastric stimulants turn on the H^+ mechanism without changing the values of the resistance and emf of the electrogenic Cl^- mechanism. With the voltage-clamp at zero the chloride current in the chloride limb would be invariant, and hence with the onset of H^+ secretion the short circuit current must decline by an amount equal to the increase in H^+ rate, i.e., $\Delta I_{SC} = —\Delta_H$. They found for a large number of mucosae that a plot of the short-circuit current vs. the H^+ ion rate did not reveal the predicted relationship, hence the electrogenic theory was not tenable and therefore the HCl mechanism must be neutral. One might argue that the two-limb circuit was too simplistic and that with a double-membrane model the relationship would be different. However, with a two-membrane model, as shown in FIGURE 8, and with all ion pathways conductive, it is easily shown that on the basis of the electrogenic theory the short-circuit current should also decrease by an amount equal to the increase in H^+ rate, i.e., $\Delta I_{SC} = -I_H$. The Heinz-Durbin analysis was based on the assumption that the parameters of the chloride limb or limbs were not changed by the gastric stimulant. Later we showed that with the onset of H^+ secretion, the short-circuit current transiently increases rather than decreasing. The simplest interpretation is that the gastric stimulant, histamine, not only turns on the H^+ mechanism, but increases the potency of the Cl^- mechanism. As a result of this finding of a transient increase in short-circuit current, it was possible to show that the relationship between the short-circuit current and the H^+ rate can be explained on the basis of the electrogenic theory. However, after the H^+ mechanism was partially turned on with small concentrations of histamine, increasing the concentration of histamine often increased the H^+ rate and produced only small changes in the short-circuit current.[13] So the basic contention of Heinz and Durbin was still applicable. We argued, but not too convincingly, that the electrogenic theory was still viable, but that there was a type of biochemical coupling between the pumps that resulted in equal changes in the H^+ and Cl^- rate.[13] It was only after we found that a neutral mechanism was present in the nutrient membrane for the entrance of Cl^- into the cells and for the appearance of HCO_3^- in the nutrient fluid that we realized that in the Heinz-Durbin analysis the implicit assumption was made that all ion pathways are conducive. The present picture of the nutrient membrane is shown in FIGURE 9. The evidence for a neutral carrier mechanism for Cl^- and HCO_3^- in the nutrient membrane has been presented in detail previously,[31] but I will briefly review some of the evidence. The first inkling of a neutral Cl^--HCO_3 mechanism in the nutrient membrane came from our Ba^{++} studies. As shown in FIGURE 10 addition of Ba^{++} to the nutrient fluid markedly increases the electrical resistance without producing much change in the H^+ rate. The H^+ rate continues at a high level for many hours. We showed that the site of the Ba^{++} action was on the nutrient membrane.[47] Simple calculations reveal that the resistance of the nutrient membrane was much too high to account for the rate of Cl^- and HCO_3^- movement across this membrane via passive conductive pathways.[31] Other evidence also led to the same conclusion.

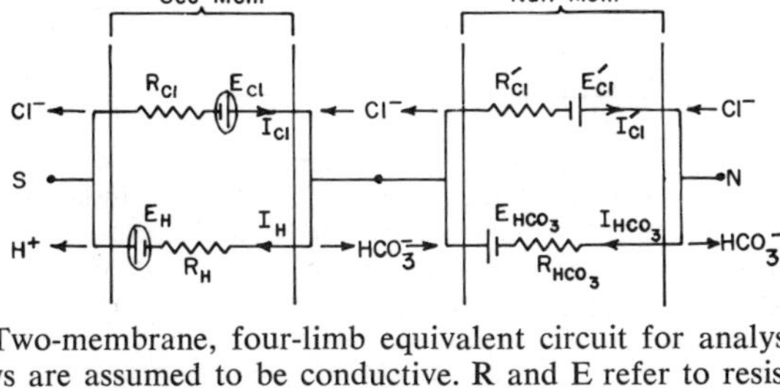

FIGURE 8. Two-membrane, four-limb equivalent circuit for analysis of I_{SC} vs. I_H; all ion pathways are assumed to be conductive. R and E refer to resistances and emfs and the meaning of the subscripts is self-evident.

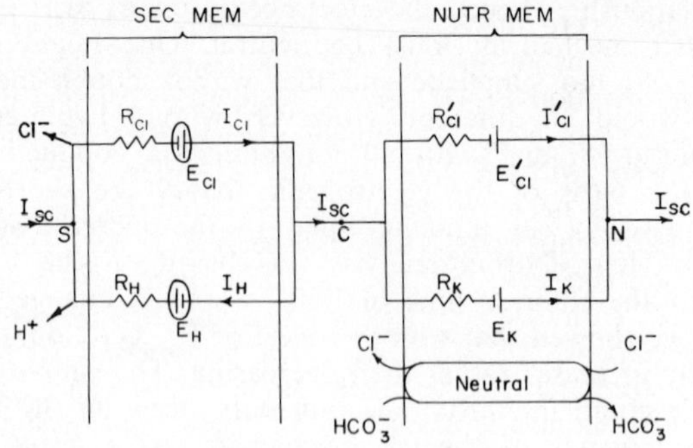

$$\Delta I_{SC} = -\frac{R_{Cl}}{R_{Cl} + R'_{Cl}} \Delta I_H$$

FIGURE 9. Secretory membrane has electrogenic Cl^- and H^+ mechanisms. Nutrient membrane has a neutral Cl^--HCO_3^- mechanism and the total conductance of the nutrient membrane equals the sum of the K^+ and Cl^- conductances. With the voltage across the system clamped at zero, the relationship between the change in I_{SC} and I_H is given by the equation.

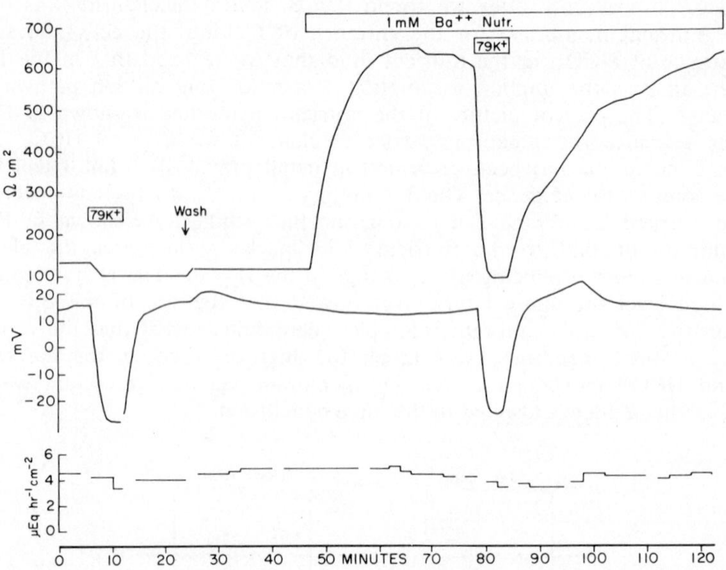

FIGURE 10. Effect of increasing K^+ in nutrient fluid from 4 to 79 mM in presence and absence of Ba^{++}. The bar at top, 1 mM Ba^{++} NUTR, indicates period in which Ba^{++} was in nutrient fluid. The bars labeled 79 K^+ indicate periods in which nutrient K^+ was 79 mM (K^+ replacing Na^+). During period indicated by second 79 K^+ bar, Ba^{++} was 1 mM in this solution. The resistance, PD, and H^+ plotted vs. time. The arrow labeled wash indicates washing of nutrient side with regular nutrient fluid to insure that the K^+ concentration was 4 mM. (From Pacifico *et al.*[47] By permission of the *American Journal of Physiology*.)

Systematic studies were made of the effects on the PD of changing the concentrations of the ions in the nutrient fluid, and it was found that only changes in the concentrations of Cl^- and K^+ resulted in changes in the potential difference.[48–50] Changes in Na^+, bicarbonate, H^+ and OH^- concentrations of the nutrient fluid were without effect on the PD. The evidence shows that the conductance of the nutrient membrane is equal to the sum of the K^+ and Cl^- conductances. Thus, from this approach it appears that bicarbonate cannot move out of the cell via a conductive pathway, and even though there is a conductive pathway for Cl^-, Cl^- must enter via a neutral mechanism or else electroneutrality would be violated. The entrance of H^+ or exit of OH^- via the nutrient membrane would result in the appearance of bicarbonate in the nutrient fluid, but again movement of these ions must be via neutral mechanisms.[51]

As illustrated in FIGURE 9, it is easily shown that instead of the short-circuit current decreasing by an amount equal to the increase in H^+ current, the decrease in short-circuit current is equal to $R_{Cl}/(R_{Cl} + R_{Cl}')$ times the increase in H^+ current.[42] So if the chloride resistance of the nutrient membrane (R_{Cl}') is very much greater than the chloride resistance of the secretory membrane (R_{Cl}), then within the framework of the electrogenic theory there would be essentially no change in the short-circuit current with changes in the hydrogen ion current. *Therefore the Heinz-Durbin argument does not disprove the electrogenic theory.* However, if the chloride resistance of the nutrient membrane was of the same magnitude as that of the secretory membrane, we might be able to disprove the theory that the HCl mechanism in the presence of Cl^- is electrogenic. It is therefore obvious that it is of critical importance to determine the Cl^- resistances in the two opposing membranes.[52] A substantial amount of our recent efforts has been devoted to this problem. We find it a rather formidable one. FIGURE 11 shows in a diagram-

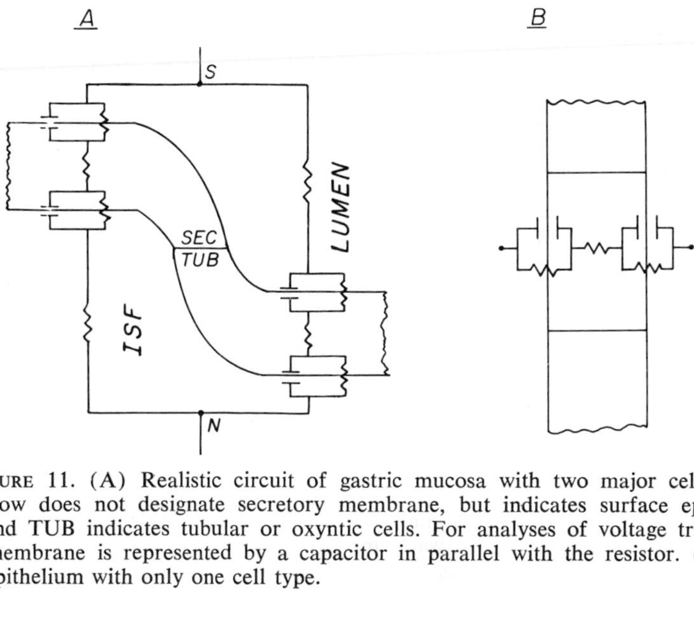

FIGURE 11. (A) Realistic circuit of gastric mucosa with two major cell types. SEC now does not designate secretory membrane, but indicates surface epithelial cells and TUB indicates tubular or oxyntic cells. For analyses of voltage transients each membrane is represented by a capacitor in parallel with the resistor. (B) An ideal epithelium with only one cell type.

matic way a more realistic circuit of the frog gastric mucosa.[11] There are two major cell types, the surface cells, which line the surface and the pits, and the tubular or oxyntic cells, which line the tubules. So we have four membranes, and a first goal is the determination of the magnitude of the conductances of each membrane. The next goal is to determine the conductive pathways in each membrane in terms of the specific ion conductances. We have made some progress by the further use of Ba^{++}, microelectrodes,[53] analysis of transient voltage responses to step currents, and very hypotonic bathing media.[54] But a presentation of the current status of our attempt to determine the Cl^- resistances of the two opposing membranes of the oxyntic cells is beyond the scope of this paper. Before considering FIGURE 12 I would like to summarize the foregoing:

The evidence for a H^+ electrogenic mechanism in the frog gastric mucosa with Cl^--free media is quite conclusive. The prevailing main argument against the separate-site or electrogenic theory has been the relationship in the presence of Cl^- between changes in short-circuit current and the H^+ rate. In the original Heinz-Durbin analysis it was implicitly assumed that Cl^- and HCO_3^- transport across the nutrient membrane is via conductive pathways. Since we showed that there is a neutral Cl^--HCO_3^- mechanism in the nutrient membrane, the Heinz-Durbin argument within the framework of our present knowledge does not disprove the electrogenic theory. It should be apparent that a knowledge of the ratio of the chloride resistance of the two opposing membranes is of crucial importance in further testing the theory.

FIGURE 12 illustrates a spin-off of the foregoing analysis. After we became aware of the implicit assumption in the Heinz-Durbin analysis we realized that the establishment of the criteria for passive ion transport by the Ussing school also made use of the same implicit assumption; that is, all ion pathways were considered conductive or could be represented by equivalent circuit conductive pathways. The model shown in FIGURE 12 was not designed as a viable model for any particular tissue, but primarily to illustrate the conclusion that the accepted criteria for passive ion transport are not sufficient to justify the conclusion that a given ion is passively transported. In this model there

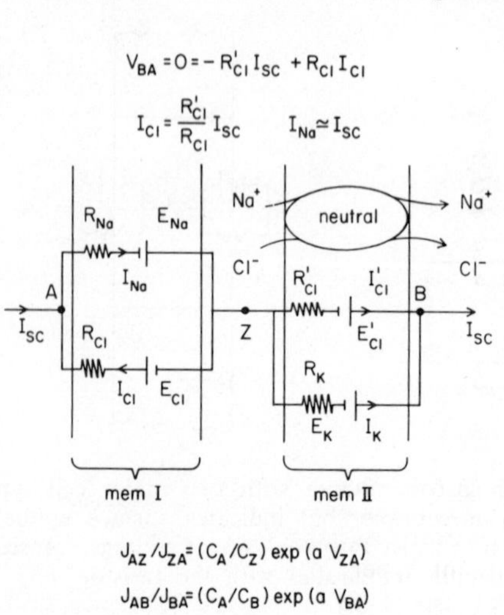

FIGURE 12. Conceptual model with two membranes to illustrate that the classical criteria for the establishment of passive ion transport are not sufficient. (See text for details.)

are two membranes and the membrane labeled membrane I contains passive Na^+ and Cl^- conductive pathways and is impermeable to all other ions. Membrane II has passive K^+ and Cl^- conductive pathways and in addition contains a neutral active pump that pumps NaCl from the cytoplasm to the bathing fluid on the right, and no other ions can enter or exit via membrane II.

Let us take a look at the situation in the model under open-circuit conditions. The neutral NaCl pump pumps equal amounts of Na^+ and Cl^- from the cytoplasm to the right-hand bathing media and maintains a low concentration of Na^+ and Cl^- in the cytoplasm. Na^+ and Cl^- move from the left-hand bathing solution via the passive conductive pathways in membrane I into the cytoplasm. The potassium current in membrane II under steady-state open-circuit conditions must be zero since potassium cannot move across membrane I. With the potassium current zero, it then follows on the basis of the electroneutrality principle that the chloride current in membrane II must also be zero. Therefore under steady-state open-circuit conditions the rate of Na^+ and Cl^- entry into the cytoplasm via membrane I equals the rate of pumping of Na^+ and Cl^- via the neutral NaCl pump.

We now examine the model under steady-state Ussing-Zerahn short-circuit conditions. The current flow in the K^+ limb in membrane II must obviously be zero. With the K^+ current zero all of the short-circuit current via membrane II is via the Cl^- conductive pathway. The magnitude of the chloride current in membrane I must equal the net transport of chloride across the whole tissue. The bathing solutions are the same on the two sides of the model so that the two chloride ion diffusion potentials E_{Cl} and E_{Cl}' are equal in magnitude and opposite in orientation. Therefore, with the voltage between B and A clamped at zero, it follows that the magnitude of the RI drops in the two Cl^- limbs must be equal. Therefore, the chloride current in membrane I equals the ratio of chloride resistances times the short circuit current. I hope that it is clear that the chloride current in membrane I must equal the total net transport of chloride across the system. Therefore, if R_{Cl}' is very much less than R_{Cl} the net transport of chloride becomes vanishingly small and the short-circuit current is essentially equal to the net transport of sodium, that is, equal to the sodium current in membrane I. It is obvious that if experimentally there is no net transport of a given ion across a tissue under the Ussing-Zerahn conditions, one is not justified in concluding that the ion is a passive ion—other information is needed, such as the information found in Cl^--free solution.

One might argue that this model would not be applicable to tissues like the frog skin since in the absence of Cl^- the short-circuit current equals the net transport of Na^+. If one argues like this our point is missed. However, if one wishes, one can modify the model to take account of certain other experimental observations. If one postulates that in the presence of Na^+ and Cl^- the pump is a neutral pump, but in the absence of one of the ions it can function as an electrogenic pump for the other ion, then it is more difficult to eliminate this model as a viable model for certain epithelial tissues.

It can also be shown on the basis of this model that with the appropriate assumptions concerning the chloride resistances of the two opposing membranes and the characteristics of the neutral pump, the flux ratio equation for a passive ion would be satisfied.[52] However, a detailed consideration of this aspect lies outside the scope of the present paper. This extension of our analysis to epithelial tissues in general with respect to the problem of the criteria for passive ion transport is of recent development. However, we have presented it on several occasions in the last few months and we usually get rather violent reactions. However, the reactions do not result from the analysis of short-circuit current, which is the crucial part of the argument, but instead come from the flux ratio analysis, which I have not presented here. So it is possible that the elements of our argument with respect to the flux ratio have been pointed out by other investigators. However, when we read the recommended papers, we have not yet found any clear statements to the effect that the absence of net transport under the Ussing-Zerahn short-circuit conditions is not sufficient to conclude that the ion is passive.

References

1. Rehm, W. S. 1944. Am. J. Physiol. **141:** 537–548.
2. Rehm, W. S. 1950. Gastroenterology **14:** 401–417.
3. Hogben, C. A. M. 1955. *In* Electrolytes in Biological Systems. A. M. Shanes, Ed.: 176–204. Amer. Physiol. Soc. Washington, D.C.
4. Hanke, M. E., R. E. Johannesen & M. M. Hanke. 1931. Proc. Soc. Exp. Biol. Med. **28:** 698–700.
5. Davies, R. E. 1951. Biol. Rev. Cambridge Phil. Soc. **26:** 87–120.
6. Teorell, T. 1951. J. Physiol. (London) **114:** 267–276.
7. Sanders, S. S., V. B. Hayne, Jr. & W. S. Rehm. 1973. Am. J. Physiol. **225:** 1311–1321.
8. Rehm, W. S. 1972. *In* Metabolic Pathways, 3rd edit. L. E. Hokin, Ed. Vol. **6:** 187–241. Academic Press. New York, N.Y.
9. Rehm, W. S. 1945. Am. J. Physiol. **144:** 115–125.
10. Rehm, W. S. 1953. Am. J. Physiol. **172:** 689–699.
11. Noyes, D. H. & W. S. Rehm. 1970. Am. J. Physiol. **219:** 184–192.
12. Kidder, G. W., iii & W. S. Rehm. 1970. Biophys. J. **10:** 215–236.
13. Rehm, W. S. 1962. Am. J. Physiol. **203:** 63–72.
14. Crane, E. E., R. E. Davies & N. M. Longmuir. 1948. Biochem. J. 366–342.
15. Conway, E. J. 1951. Science **113:** 270–273.
16. Patterson, W. B. & D. Stetten. 1949. Science **109:** 256–258.
17. Robertson, R. N. 1960. Biol. Rev. **35:** 231–264.
18. Davies, R. E. & A. G. Ogston. 1950. Biochem. J. **46:** 324–333.
19. Rehm, W. S. 1949. Am. J. Physiol. **159:** 586.
20. Rehm, W. S. 1965. Fed. Proc. **24:** 1387–1395.
21. Mitchell, P. 1961. Nature (London) **191:** 144–148.
22. Mitchell, P. 1967. Fed. Proc. **26:** 1370–1379.
23. Kidder, G. W., iii, P. F. Curran & W. S. Rehm. 1966. Am. J. Physiol. **211** 513–519.
24. Hersey, S. J. & F. F. Jobsis. 1969. Biochem. Biophys. Res. Commun. **36:** 243–250.
25. Hersey, S. J. 1974. Biochim. Biophys. Acta **344:** 157–203.
26. Forte, J. G., P. H. Adams & R. E. Davies. 1963. Nature **197:** 874–876.
27. Durbin, R. P. 1968. J. Gen. Physiol. 51(Suppl): 233–239.
28. Kasbekar, D. K. & R. P. Durbin. 1965. Biochim. Biophys. Acta **105:** 472–482.
29. Forte, J. G. 1970. *In* Membranes and Ion Transport. E. E. Bittar, Ed. Vol. 3: 111–165. Wiley. New York, N.Y.
30. Lee, J., G. Simpson & P. Scholes. 1974. Biochem. Biophys. Res. Commun. **60:** 825–832.
31. Rehm, W. S. 1967. Fed. Proc. **26:** 1303–1313.
32. Heinz, E. & R. P. Durbin. 1959. Biochim. Biophys. Acta **31:** 246–247.

33. USSING, H. H. 1965. Harvey Lectures **59:** 1–30.
34. REHM, W. S. & M. E. LEFEVRE. 1965. Am. J. Physiol. **208:** 922–930.
35. REHM, W. S. 1966. Ann. N.Y. Acad. **137:** 591–606.
36. SACHS, G., R. SHOEMAKER & B. E. HIRSCHOWITZ. 1965. Am. J. Physiol. **209:** 461–466.
37. SCHWARTZ, M. & T. N. MACKRELL. 1969. Proc. Soc. Exp. Biol. Med. **130:** 1048–1051.
38. DINNO, F., M. DINNO, C. LEE, M. SCHWARTZ & T. N. MACKRELL. 1971. Proc. Soc. Exp. Biol. Med. **138:** 479–481.
39. FINKELSTEIN, A. 1970. Biochim. Biophys. Acta **205:** 1–6.
40. MITCHELL, P. & J. MOYLE. 1968. Eur. J. Biochem. **4:** 530–539.
41. HEINZ, E. & R. P. DURBIN. 1958. J. Gen. Physiol. **41:** 101–117.
42. DURBIN, R. P. & E. HEINZ. 1958. J. Gen. Physiol. **41:** 1035–1047.
43. SCHULTZ, S. G. 1974. *In* Gastrointestinal Physiology. E. D. Jacobson and E. L. Shanbour, Eds. Vol. 4: 69–103 University Park Press. Baltimore, Md.
44. REHM, W. S., S. S. SANDERS, M. G. TANT, E. A. FRIDAY & J. T. TARVIN. *In* Gastric Secretion. D. Kasbekar, G. Sachs and W. S. Rehm, Eds. Marcel Dekker. New York, N.Y. In press.
45. USSING, H. H. & B. ANDERSEN. 1956. Proc. 3rd Int. Congr. Biochem.: 434–440. Academic Press. New York, N.Y.
46. SCHWARTZ, M., A. D. PACIFICO, T. N. MACKRELL, A. JACOBSON & W. S. REHM. 1968. Proc. Soc. Exp. Biol. Med. **127:** 223–225.
47. PACFICO, A. D., M. SCHWARTZ, T. N. MACKRELL, S. G. SPANGLER, S. S. SANDERS & W. S. REHM. 1969. Am. J. Physiol. **216:** 536–541.
48. HARRIS, J. B. & I. S. EDELMAN. 1964. Am. J. Physiol. **206:** 769–782.
49. SPANGLER, S. G. & W. S. REHM. 1968. Biophys. J. **8:** 1211–1227.
50. SANDERS, S. S., J. O'CALLAGHAN, C. F. BUTLER & W. S. REHM. 1972. Am. J. Physiol. **22:** 1348–1354.
51. REHM, W. S., S. S. SANDERS, R. L. SHOEMAKER, J. O'CALLAGHAN, J. T. TARVIN & E. A. FRIDAY. 1973. J. Theor. Biol. **39:** 131–153.
52. REHM, W. S. *In* Current Topics in Membranes and Transport. F. Bronner and A. Kleinzeller, Eds. Vol. 6. Academic Press. New York, N.Y. In press.
53. O'CALLAGHAN, J., S. S. SANDERS, R. L. SHOEMAKER & W. S. REHM. 1974. A. J. Physiol. **227:** 273–288.
54. HELANDER, H. F., S. S. SANDERS, L. L. SHANBOUR & W. S. REHM. Acta Physiol. Scand. In press.

Discussion

DR. HOPFER: Do you have any information about the potassium profile across the cell?

DR. REHM: The gastric cell?

DR. HOPFER: Yes, the oxyntic cell.

DR. REHM: That's like asking about religion or politics! A lot of work has been done and I'm not sure anyone agrees on the results. I do think that most of us agree on the findings of microelectrodes in the surface cells. But when it comes down to what we get when we probe deeper and try to get into an oxyntic cell, then there is not much accord.

REDOX AND ATP IN ACID SECRETION*

G. Sachs,† E. Rabon, G. Saccomani, and H. M. Sarau‡

Laboratory of Membrane Biology
University of Alabama in Birmingham
Birmingham, Alabama 35294 and
‡ Smith, Kline & French Laboratories
Philadelphia, Pennsylvania 19101

The gradient greater than six orders of magnitude produced by gastric mucosa is probably the highest one achieved in any biological system. An understanding of this proton transport system would seem to have quite general implications for proton pumps in other tissues or systems, since it appears that a proton pumping capability is a quite generalized phenomenon. In fact, one could well suggest that this pump is the original active transport system.

The history of research in gastric mucosa faithfully reflects the same problems that have been considered elsewhere, the question of whether the pump is electrogenic, or whether redox energy or ATP is utilized as the driving force. It seems, moreover, that the chemiosmotic hypothesis should find strong support or denial in this tissue, which should be the most dedicated available. Thus the pump should be electrogenic, react without formation of a chemical intermediate, and by appropriate membrane conductance modifications drive ATP synthesis or coupled transport. Since the origin of the chemiosmotic hypothesis was based on Mitchell's understanding of vectorial arrangement of proteins in membranes as applied to HCl secretion by the stomach, it would be appropriate to derive conclusive evidence for or against the hypothesis in gastric mucosa. Essentially, looking at an ATPase mechanism this requires an asymmetric membrane, one side permeable to protons, the other to hydroxyl ion. The operation of the gradient in the direction of dissipation results in ATP synthesis, or cotransport, or potential-drive transport. ATP breakdown generates a gradient, or potential. Since, as will be seen, proton gradients can be generated in subcellular membrane particles from gastric mucosa, the above considerations will become particularly relevant. An alternate means of generating a proton gradient is to utilize redox energy to separate protons and electrons (or OH^-, in the presence of O_2). The first problem to be discussed is to attempt to review some recent findings which may distinguish, or coalesce the two possible mechanisms for gradient generation, redox and ATPase.

Redox Hypothesis

The first suggestion for the energy source for proton transport in gastric mucosa or yeast was a redox mechanism similar to that illustrated in FIGURE

* This work was supported by Grant No. AM15878 from the National Institutes of Health, Grant No. GB31075 from the National Science Foundation, and a grant-in-aid from Smith, Kline & French Laboratories.

† To whom correspondence should be addressed.

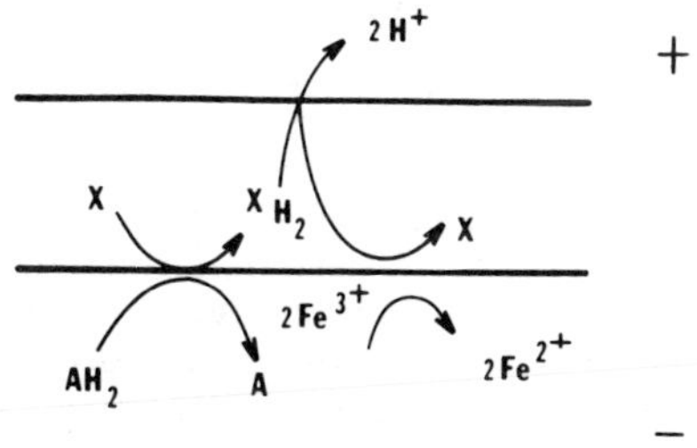

FIGURE 1. A simple redox mechanism for proton secretion by gastric mucosa, where a cytoplasmic donor reacts with a membrane component responsible for proton transport, and transfer of electrons to an acceptor such as cytochrome *c*.

1.[1] Three major points about this mechanism that still remain to be established after quarter of a century of investigation are (a) the nature of AH_2 and whether it is a cellular substrate, such as lactate, which is utilized by the bacterial system studied by Kaback and his collaborators[2] or whether it is a redox intermediate such as NADH; (b) the nature of the vectorial redox system, and whether it is a non-heme iron complexed with sulfur, a known redox component or an energizable ionophore; (c) the nature of the electron acceptor, and whether it is molecular oxygen, or part of the microsomal chain, or an acceptor that shuttles between membrane and mitochondria.

Another general property is whether the H^+ pump is electrogenic. As illustrated, the redox pump is indeed electrogenic, but this is not a necessary consequence of a redox pump. To complete the circuit containing an E_H, g_H, one may insert a g_{Cl} alone or an E_{Cl}, g_{Cl}. The latter is more usual in the mucosa, since Cl^- is also actively transported, and also in Cl^- solutions the lumen is negative. Only in the presence of SO_4^{--} substituting for Cl^- does the transmucosal potential reflect the electrogenic H^+ pump.[3] This property is important when we come to consider vesicle H^+ secretion in a later section. There is also evidence for an anion conductance in gastric membranes since a low molecular weight peptide can be extracted from such membranes and incorporated into an artificial planar bilayer, inducing voltage-dependent weakly anion-selective conductance.[4] Proton pumps studied elsewhere, such as in mitochondria[5, 9G] or bacterial vesicles,[7] are apparently electrogenic and transport of solute may be coupled to this potential[2] or cotransport with H^+ may occur.[8] These proton pumps are usually associated with redox systems, rather than being ATPase-dependent; hence the proof of the existence of a redox mechanism is very important in the stomach, particularly to see whether this portion of the Mitchell condition[9] is fulfilled.

Nature of Redox Donor

One can conceive of three types of donors: a substrate, a redox chain intermediate, or a combination of these, involving shuttling of reducing equivalents from mitochondrial to membrane system.

A glycolytic or lipolytic substrate may be considered as a candidate for the H^+ donor. Irrespective of the nature of the substrate, direct oxidation of substrate, followed by elimination of the oxidized substrate from the secretory pathway, is incompatible with the measured H^+/O_2 ratio lying somewhere between 2 and 3.[10]

For example, the oxidation of lactate to pyruvate followed by the oxidation of pyruvate to CO_2 and H_2O and the use of the initial reaction for H^+ secretion would result in an equation of the form

$$3O_2 + CH_3 \cdot CHOH \cdot CHOOH \rightarrow 2CO_2 + 2H_2O + CO_3^{--} + \underline{2H^+}$$

and give an H^+/O_2 ration of about ½ to ⅓ of the values reported in the literature. Hence, one has to invoke an additional source of H^+, such as the use of cycling through mitochondria or a multistep respiratory chain in the membrane with appropriate arrangement of the electron-transfer steps, as has been postulated in the chemiosmotic hypothesis.[9]

A shuttle system could be used, such as ketoacids or amino acids, which are thought to equilibrate cytoplasmic and mitochondrial levels of NADH or NADPH. A requirement for such a shuttle is that there be both a mitochondrial and cytoplasmic enzyme location. The dehydrogenase may be located at the cell surface or in the cytoplasm, followed by oxidation of NADH or NADPH at the cell surface. Since reduced pyridine nucleotide is more firmly bound than the oxidized form, it is more likely that the primary dehydrogenase is located at the membrane. This is illustrated schematically in FIGURE 2, using a glutamate-α-ketoglutarate shuttle. One way of deciding for or against a specific dehydrogenase is to measure metabolite levels during a large transition in H^+ rate, and to determine a metabolite crossover; another is to find the membrane-bound dehydrogenase.

Nature of Redox Transport System

Critical to a redox mechanism is the presence at some locus in the cell of a membrane-bound vectorial dehydrogenase, the emphasis on which is one of the major contributions of the chemiosmotic hypothesis. In such a scheme electrons and protons are separated across a membrane, with a consequent generation of a potential and proton gradient (proton motive force). This charge separation has to be coupled obligatorily to an equivalent but opposite charge separation. For example, the separation of K^+ from Cl^- is an endergonic reaction, which could result from the activity of the exergonic proton pump. This type of model requires coupling of proton pumping to movement of

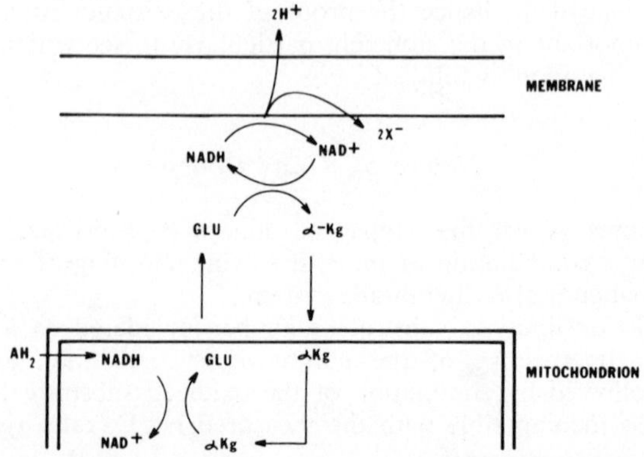

FIGURE 2. A redox scheme using a mitochondrial substrate shuttle and a membrane-bound NAD$^+$-linked dehydrogenase as a proton transport mechanism.

an ionophoric system, such as K^+ valinomycin or its natural membrane equivalent. An alternative, more commonly considered in gastric mucosa, is electrogenic extrusion of Cl^- to complete the H ion circuit. Essential to this hypothesis then is the establishment of a dehydrogenase of correct orientation in the plasma membrane and evidence for electrogenic H^+ transport and the presence of Cl^- or K^+ or some other ionic conductance.

Nature of the Electron-Acceptor

Here, too, there are quite strict requirements for the putative acceptor for the redox transport scheme. First, the redox potential of the system in the cell must be sufficiently different from the primary dehydrogenase to allow production of a $10^{6.6}$-fold gradient of H^+ ions, i.e., about 400 mv. Second, a transport system must be present in the mitochondrial membrane to allow reoxidation of the carrier by the mitochondrial redox chain. Cytochrome *c* has been suggested as such a carrier,[11] but apart from its bulk, the ensuing requirement for the presence of large quantities of extramitochondrial cytochrome *c* seems not to have been met.[12] The presence of a large concentration of —SH compounds in gastric fundus, mainly aspartyl cysteine,[13] may indeed relate to this problem. This might lead to the suggestion that an —SH component of the membrane which can participate in cyclic oxidoreduction may be involved in acid secretion. We conclude that a mobile, transportable shuttle is required.

Localization of the Redox System

Many problems could be bypassed if it were conceptually possible, as has been suggested recently,[14] to utilize the mitochondrial respiratory chain directly. In our view there are some overwhelming problems in this. Without some limiting transport system, the cytosol pH would fall to unacceptably low values. In the absence of evidence for membrane tubes extending from the mitochondrion to the cell surface, it could be that there is direct contact between the mitochondrion and tubules of the secretory cell. Another alternative is that the process is exocytotic with a cycle of formation, fusion, and resynthesis accounting for H^+ secretion. On the contrary, morphologic evidence[15] suggests that essentially all the vesicular elements disappear, forming the microtubules with onset of secretion, and hence a vesicular cycle is unlikely. In contrast to Cecropia, which secretes isotonic KCl,[16] there is no definite intimate relationship between mitochondria and the tubulovesicles. Indeed, if there were direct fusion, one would expect crosscontamination of mitochondria with membrane markers, and at least in the case of K^+ *p*-nitrophenyl phosphatase[17] this does not happen. We conclude that the redox system must be located at the secretory surface.

Evidence for a Redox Mechanism

There are various ways in which one could obtain evidence for a redox mechanism. Historically the most popular method has been to measure the

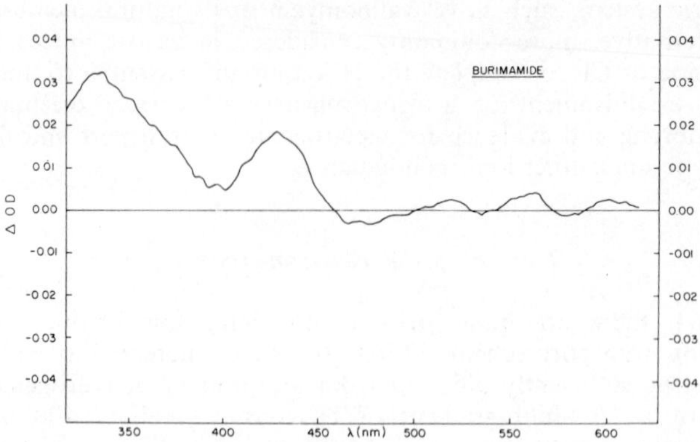

FIGURE 3. The effect of adding burimamide to a spontaneously secreting mucosa. This shows relative oxidation of all the commonly observed redox components, NAD^+, FAD, and cytochromes *b*, *c*, *a*, a_3.

H^+/O_2 ratio, which cannot exceed 4 in a simple mechanism, as illustrated in FIGURE 1. The limitations of this approach are obvious since a value less than 4 does not exclude a redox mechanism, and a value greater than 4 requires only cycling of electrons down a respiratory chain. More recently the capability of observing the redox components of the *in vitro* gastric mucosa has provided conflicting evidence as to steady-state changes in these components.[14, 18]

When a frog mucosa is mounted in a chamber there is usually spontaneous secretion. At this stage one may add an H_2 antagonist[19] which specifically blocks the hormonal component of gastric secretion.[20] The occasional spontaneously resting mucosa shows no response to H_2 antagonists. The secreting mucosa shows the changes shown in FIGURE 3. Thus there is oxidation of all the observable redox components, namely NAD^+, FAD, and cytochromes *b*, *c*, *a*, and a_3. When the tissue is restimulated by histamine there is reduction of FAD, and the cytochromes, with remarkably little effect on NAD^+ (FIGURE 4). By comparing the redox state in the presence of amytal to anaerobiosis, one can see the change from full oxidation (amytal) to full reduction and calculate the percentage of reduction induced by histamine. TABLE 1 presents those data, and in agreement with Hersey we find a considerable increase in reduction of the redox pigments subsequent to NAD^+ with onset of secre-

TABLE 1

PERCENTAGE OF REDUCTION OF REDOX COMPONENTS

Histamine (after burimamide)	
DPN^+	32*
FP	49
Cα	43
b	49
aα	49
$a_{3}\gamma$	41

* Maximal value.

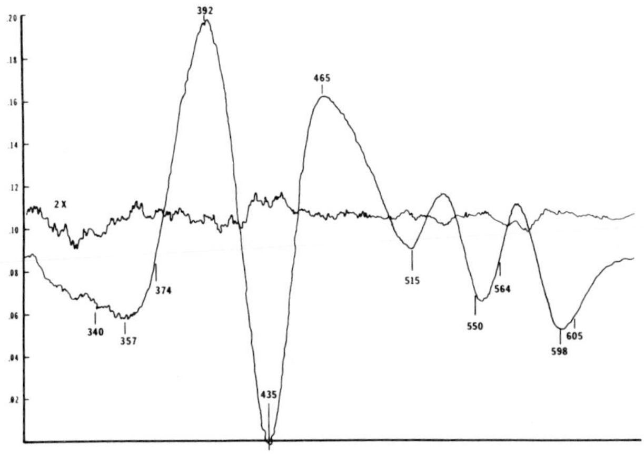

FIGURE 4. The effect of adding histamine to a nonsecreting mucosa, inhibited by burimamide, showing reduction of all the redox components. Note the unusual wavelength peak at 355 nm.

tion.[12] These findings are in contrast to what is expected from an increase in ADP, or a fall in the ATP/ADP ratio, commonly referred to as a state 4 to state 3 transition, namely, from phosphate-acceptor depleted to respiration. Hence, this result is equivocal as far as a redox hypothesis is concerned. FIGURE 5 illustrates the scheme, whereby stimulation of metabolism occurs, hence resulting in the redox changes observed. If an increase in ADP level occurs, the predicted oxidative changes would be masked by the increased access of substrate.

It should be possible to measure the change in substrate levels as transition occurs from rest to acid secretion, and thus define which metabolic pathways are stimulated, and by extending these studies to inhibition of acid secretion, eventually determine which pathway, if any, is specific to the H^+ mechanism.

This is done by studying metabolite levels in biopsy specimens from anesthetized dog gastric mucosa at rest and following challenge with histamine. The biopsy specimen is rapidly frozen and cryostat sections are cut until parietal cell areas are seen. Every fourth section is stained for succinic dehydrogenase, and the other sections are frozen-dried and trimmed to remove as much tissue as possible except the defined islands of parietal cells. This small tissue fragment

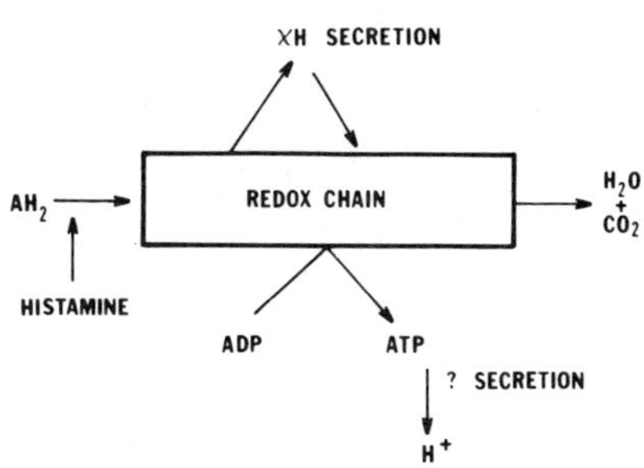

FIGURE 5. A scheme illustrating the problem involved in interpreting the spectroscopic data. If histamine stimulates substrate production, then relative reduction of the intermediates would occur, irrespective of ADP levels. A redox hypothesis should also produce oxidation of some respiratory components.

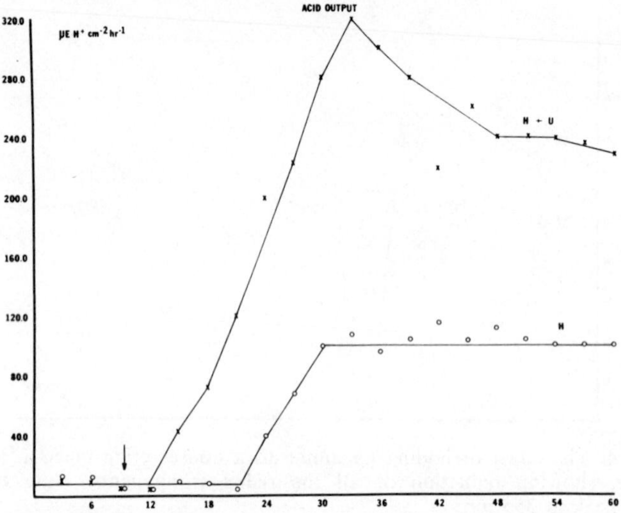

FIGURE 6. The acid transition occurring with histamine or histamine and urecholine in the anesthetized dog. Biopsy samples were taken prior to infusion of histamine or histamine and urecholine (100 μg/kg) and just prior to onset, during increase, and at steady-state acid secretion.

is placed in an oil well and the various metabolites are measured by the Lowry cycling procedures.[21] The advantage of the *in vivo* study in the dog is that in the absence of exogenous stimulus there is no measurable acid production, and as shown in FIGURE 6, in a short period of time a level as high as 300 μE H^+ cm^{-2} can be produced, a value to be compared to a 5 μE level found in the *in vivo* frog mucosa.

TABLE 2 shows results obtained for glycogen and glycolytic intermediates as well as a hexose monophosphate shunt intermediate, intermediates of the Krebs cycle, and glycerol, CoA and acetyl CoA, which are the only accessible substrates related to lipolysis.

TABLE 2A and the schematic FIGURE 7 show that there is little evidence for glycogenolysis, but an evident increase in all other glycolytic intermediates, as well as in 6-phosphogluconate. One interpretation of this is that there is stimulation of glycolysis, and the marked rise in fructose-1,6 diphosphate suggests that it is phosphofructokinase that is stimulated. There is also a notable rise in lactate, suggesting an increased reduction of cytoplasmic DPN^+, as well as some problem in flux into mitochondrial oxidation.

TABLE 2B shows that Krebs-cycle acids are significantly increased, suggesting increased cycle activity, as would be expected from the change in metabolism. FIGURE 8 illustrates this finding schematically. There is an increase in a ketoglutarate and a fall in glutamate, a change opposite to that observed for other acid pairs measured.

TABLE 2C and FIGURE 8 show the changes found in glycerol, CoA, and acetyl CoA. The constancy of glycerol level is rather compelling evidence against triglyceride breakdown being involved in a metabolic transition related to acid secretion. The rise in CoA and acetyl CoA would argue for an increase

TABLE 2A

	Resting	Blood Flow	H I	H II
Glycogen	60.18 ± .34 (n = 11)	56.75 ± 4.2 (n = 11)	51.88 ± 3.01 (n = 11)	50.35 ± 3.9 (n = 10)
Glucose	9.59 ± .82 (n = 17)	9.34 ± 6.8 (n = 13)	10.47 ± 1.60 (n = 11)	10.35 ± 1.24 (n = 10)
Glucose 6 Phosphate	0.267 ± .039 (n = 14)	.243 ± .024 (n = 12)	.335 ± .031 (n = 12)	.395 ± .021 (n = 12)
Fructose 6 Phosphate	.094 ± .036 (n = 6)	.092 ± .015 (n = 5)	.122 ± 0.13 (n = 6)	.072 ± .036 (n = 3)
Fructose 1,6 Diphosphate	.033 ± .004 (n = 9)	.044 ± .015 (n = 6)	.093 ± .011 (n = 6)	.074 ± .025 (n = 3)
Glyceraldehyde 3 phosphate	.015 ± .002 (n = 7)	.017 ± .001 (n = 6)	.024 ± .003 (n = 6)	–
Dihydroxyacetone phosphate	.045 ± .005 (n = 6)	.053 ± .007 (n = 6)	.066 ± .008 (n = 6)	–
Glycerol-phosphate	.219 ± .030 (n = 11)	.220 ± .025 (n = 7)	.475 ± .038 (n = 7)	.410 ± .072 (n = 11)
3 phosphoglycerate	.297 ± .024 (n = 3)	.271 ± .019 (n = 3)	.386 ± .077 (n = 3)	.419 ± .014 (n = 3)
Phosphoenol pyruvate	1.96 ± .12 (n = 7)	1.56 ± .20 (n = 6)	1.58 ± .19 (n = 7)	–
Pyruvate	.678 ± .032 (n = 17)	.747 ± .046 (n = 12)	.760 ± .074 (n = 11)	.817 ± .080 (n = 13)
Lactate	3.91 ± .35 (n = 17)	4.68 ± .59 (n = 12)	7.82 ± 1.19 (n = 11)	7.25 ± 1.02 (n = 13)
6 phosphogluconate	.023 ± .003 (n = 11)	.023 ± .003 (n = 4)	.050 ± .004 (n = 9)	.050 ± .005 (n = 6)

NOTE: The values are the means ±SEM of n dogs, at least four measurements having been made on each sample in each dog. Blood flow refers to the state of reddening of the mucosa, H I the sample when acid secretion was detected in that area and H II a sample at steady-state secretion. Values are given, when appropriate in mmol/kg^{-1} dry weight.

in pyruvate decarboxylation, or fatty acid oxidation. There is no evidence in any of our studies for lipase activation, *in vivo* or *in vitro,* or with the purified enzymes.[22] Thus, if fatty-acid oxidation is increased, this must be accompanied by increased uptake of fatty acid. This is also true of the increased glycolysis, in that the small change in glycogen is compatible with an increased glucose uptake by the tissue.

None of the intermediates measured shows any unusual change in level to suggest direct involvement in acid secretion.

It is also possible to use these direct-measurement techniques to confirm some of the changes observed spectroscopically and to localize these changes whether they occur in the cytoplasm, mitochondria, or both. TABLE 3 and FIGURE 9 provide information on the levels of pyridine nucleotides and 2 keto-acid couples, lactate-pyruvate and β-OH butyrate and acetoacetate, which reflect cytoplasmic and mitochondrial redox states, respectively. From this, there appears to be reduction of NAD^+ in both compartments with secretion,

TABLE 2B

	Resting	Blood Flow	H I	H II
Citrate	1.39 ± .08 (n = 8)	1.51 ± .02 (n = 8)	2.05 ± .16 (n = 9)	2.39 ± .033 (n = 8)
α-ketoglutarate	.742 ± .031 (n = 7)	.768 ± .030 (n = 7)	.880 ± .026 (n = 7)	.930 ± .086 (n = 3)
Fumarate	.075 ± .027 (n = 4)	.108 ± 0.72 (n = 4)	.212 ± .050 (n = 4)	.228 ± .054 (n = 3)
Malate	.383 ± .026 (n = 3)	.454 ± .053 (n = 3)	.912 ± .184 (n = 3)	.882 ± .166 (n = 3)
Glutamate	12.13 ± 2.04 (n = 7)	9.94 ± 2.68 (n = 7)	9.34 ± 2.38 (n = 7)	10.70 ± .79 (n = 3)

NOTE: The values are the means ±SEM of n dogs, at least four measurements having been made on each sample in each dog. Blood flow refers to the state of reddening of the mucosa, H I the sample when acid secretion was detected in that area, and H II a sample at steady-state secretion. Values are given, when appropriate in mmol/kg^{-1} dry weight.

confirming the spectroscopic data, provided β-OH butyrate dehydrogenase is sufficient.

A quite different approach to the redox problem is provided by purification of the gastric membrane and discovery then of a relevant redox system. We will deal with this later.

ATPase Hypothesis

Since the discovery of the Na + K-ATPase by Skou in 1957,[23] the history of this enzyme, superbly related by Post[24] has forced an ATPase hypothesis on the field of gastric secretion. The most popular way of establishing this has been to correlate changes in adenine nucleotide levels with acid secretion. In frog it has been shown that ATP levels correlate with acid rate[25] and measurements of ATP, ADP, and phosphocreatine among others have shown that there is a fall in ATP and a rise in ADP but an increase in phospho-

TABLE 2C

	Resting	Blood Flow	H I	H II
Glycerol	2.15 ± .32 (n = 8)	2.21 ± .27 (n = 4)	1.66 ± .20 (n = 4)	2.38 ± .50 (n = 4)
CoA	.554 ± .065 (n = 8)	.562 ± .048 (n = 4)	.711 ± .086 (n = 4)	.781 ± .090 (n = 8)
Acetyl CoA	.197 ± .034 (n = 8)	.240 ± .044 (n = 4)	.263 ± .046 (n = 4)	.280 ± .024 (n = 7)

NOTE: The values are the means ±SEM of n dogs, at least four measurements having been made on each sample in each dog. Blood flow refers to the state of reddening of the mucosa, H I the sample when acid secretion was detected in that area, and H II a sample at steady-state secretion. Values are given, when appropriate in mmol/kg^{-1} dry weight.

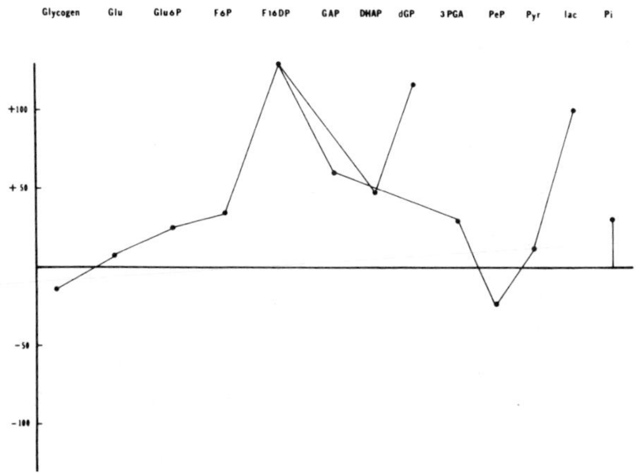

FIGURE 7. The percentage changes occurring in several glycogenolytic and glycolytic intermediates, showing changes compatible with stimulation of phosphofructokinase, and no evidence for a block in pyruvate dehydrogenase. The small change in glycogen suggests an increased uptake of glucose from the blood to account for the changes. The increase in 6-phosphogluconate is consistent with increase in hexose monophosphate shunt.

creatine.[26] These measurements were carried out *in vitro* and on whole tissue. Measurements *in vivo* on defined areas of parietal cells give somewhat different data, where, during the large transition in transport rate as shown in FIGURE 6, the data of TABLE 4 and FIGURE 10 are obtained. These data are clearly incompatible with a simple ATPase hypothesis, but can be unsatisfactorily rationalized by invoking interplay between metabolism, secretion, and ATP synthesis.

Alternatively, one can suggest a dual mechanism utilizing an ATP-redox scheme operating via an NADH- or NADPH-linked step, which also requires ATP to alter the midpoint redox potential. A possible scheme is illustrated in FIGURE 11. Here an —SH—Fe^{++} center is used, in analogy to the non-heme iron in mitochondria. This goes through a reversible O—R cycle as well as phosphorylation and dephosphorylation. The net result of this scheme is diversion of one ATP molecule for secretion, as well as loss of the coupling step at site 1, and hence, net production of one ATP molecule and an effective P/O ratio of 1 in secreting mucosa. This is only one of many possible hypotheses.

An alternative is to define an ATPase in gastric membranes and relate its function to acid secretion. The first such viable candidate was an SCN^--inhibited ATPase, stimulated by HCO_3^-,[27] which was also recognized (a) not to be confined to gastric mucosa and (b) to be similar to mitochondrial ATPase[28] but distributed differently in dog mucosa[29] but similarly in rat mucosa.[30] The presence of this enzyme suggested the possibility of a Mitchell chemiosmotic mechanism for acid secretion in the stomach, which would fit with electrogenic acid secretion, on one hand, and a more universal type proton pump,[4] on the other.

An alternative ATPase, which is K^+-stimulated and apparently contains an acetyl phosphatase or *p*-nitrophenyl phosphatase as a component reaction,[31]

TABLE 3

	Resting	Blood Flow	H I	H II
DPN^+	1.970 ± .074 (n = 17)	1.962 ± .074 (n = 13)	1.882 ± .070 (n = 11)	1.933 ± .074 (n = 13)
DPNH	.295 ± .020 (n = 17)	.315 ± .027 (n = 13)	.523 ± .074 (n = 11)	.507 ± .070 (n = 13)
DPN^+/DPNH	6.68 ± .045 (n = 17)	5.91 ± .044 (n = 13)	3.60 ± .072 (n = 11)	3.81 ± .074 (n = 13)
TPN^+	.069 ± .008 (n = 16)	.069 ± .012 (n = 7)	.064 ± .008 (n = 14)	.0461 ± .010 (n = 11)
TPNH	0.219 ± .012 (n = 20)	.223 ± .014 (n = 14)	.230 ± .013 (n = 12)	.230 ± .072 (n = 14)
TPNH/TPN^+	3.17 ± .010 (n = 16)	3.23 ± .013 (n = 7)	3.59 ± .013 (n = 12)	4.78 ± .020 (n = 17)
Lactate	3.909 ± .351 (n = 17)	4.679 ± .588 (n = 12)	7.823 ± 1.191 (n = 11)	7.249 ± 1.020 (n = 13)
Pyruvate	.678 ± .032 (n = 15)	.747 ± .046 (n = 12)	.760 ± .074 (n = 11)	.817 ± .080 (n = 13)
Lactate/pyruvate	5.77 ± .32 (n = 15)	6.26 ± .53 (n = 12)	10.29 ± 1.10 (n = 11)	8.87 ± 1.00 (n = 13)
Acetoacetate	1.379 ± .131 (n = 7)	1.309 ± .084 (n = 7)	1.377 ± .148 (n = 7)	1.25 ± .115 (n = 7)
β-hydroxybutyrate	.181 ± .013 (n = 7)	.157 ± .021 (n = 7)	.257 ± .024 (n = 7)	.333 ± .068 (n = 7)
β-OH But/AcAc	.140 ± .019 (n = 7)	.135 ± .028 (n = 6)	.211 ± .040 (n = 7)	.327 ± .113 (n = 7)

NOTE: The values are the means ±SEM of n dogs, at least four measurements having been made on each sample in each dog. Blood flow refers to the state of reddening of the mucosa, H I the sample when acid secretion was detected in that area and H II a sample at steady-state secretion. Values are given, when appropriate in $mmol/kg^{-1}$ dry weight.

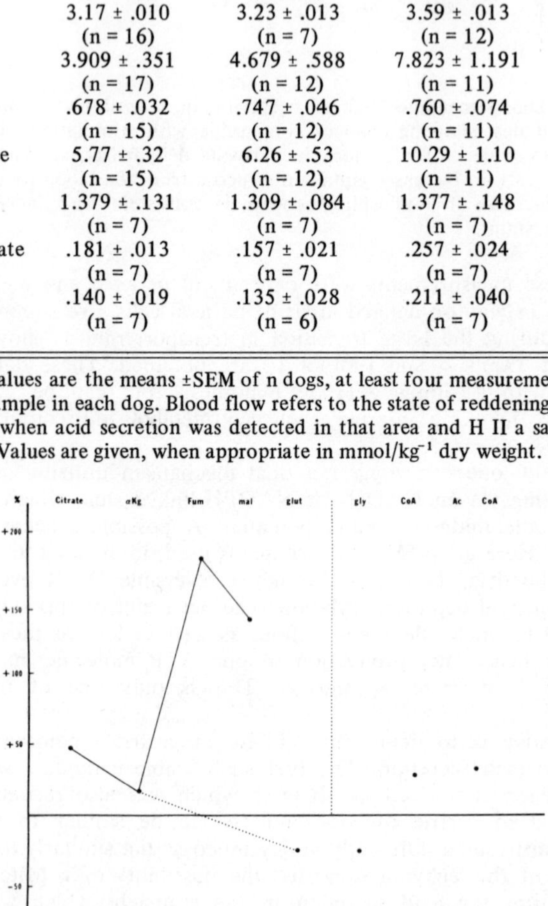

FIGURE 8. The percentage changes occurring in Krebs-cycle acids accessible to the fluorimetric cycling techniques. The increase in these is strong evidence for an increased activity of the Krebs cycle. The percentage changes in glycerol, acetyl CoA, and CoA indicate only slight alteration in lipolysis, but the key intermediates in these reactions are not amenable to fluorimetric measurement in the samples used.

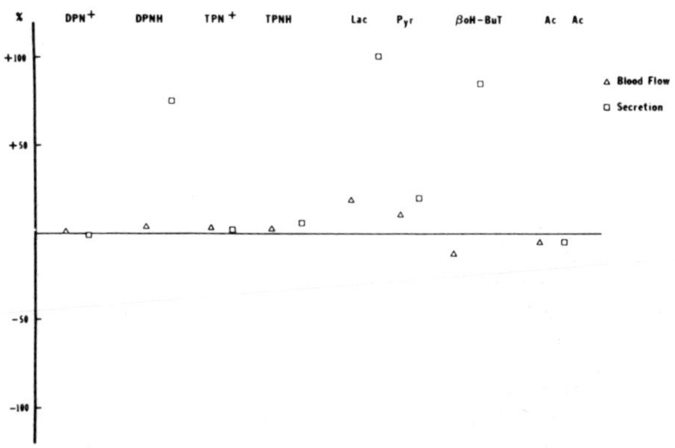

FIGURE 9. The percentage changes observed in pyridine nucleotides and lactate pyruvate, β-OH butyrate-acetoacetate, confirming the reduction of pyridine nucleotide found spectroscopically.

has been described and thus far appears specific to gastric mucosa, and the membrane fraction of that tissue. Interestingly, maximal K^+ stimulation of the ATP is obtained only in the presence of an ionophore such as valinomycin or gramicidin[32] or after hypotonic dilution, freeze-thawing, or freeze-drying. Data for a highly purified membrane fraction for hog gastric mucosa are presented in TABLE 5, showing about a six-fold stimulation by K^+ in the presence of valinomycin. This ion restriction of ATPase activity might suggest that a study of proton transport is feasible in these subcellular vesicles. Recently evidence[33] has been presented that use of the technique of Mitchell and Moyle[34] (i.e., a pH of 6.1 using glycylglycine buffer and 150 mM KCl) shows alkalinization of the medium when ATP is added to such a vesicle preparation.

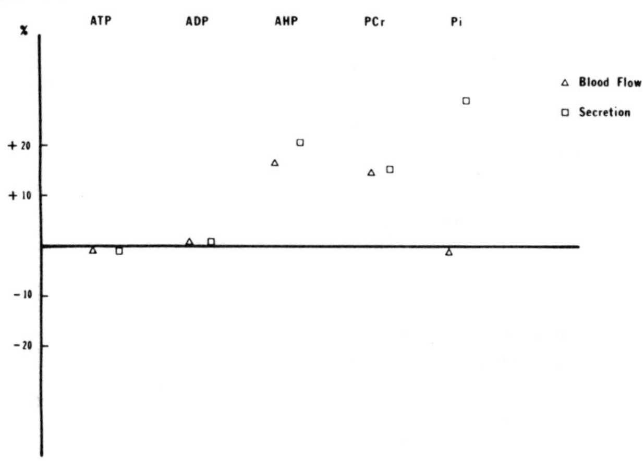

FIGURE 10. The percentage changes observed in adenine nucleotides with changes in secretion in dog gastric mucosa, as well as phosphocreatine. The remarkable constancy of ATP and phosphocreatine argues against a simple ATPase hypothesis.

TABLE 4

	Resting	Blood Flow	H I	H II
ATP	12.438 ± 0.410 (n = 17)	12.414 ± .530 (n = 13)	12.127 ± .627 (n = 11)	11.959 ± .570 (n = 13)
ADP	5.691 ± .192 (n = 17)	5.741 ± .175 (n = 13)	5.779 ± .269 (n = 11)	5.800 ± .206 (n = 13)
ATP/ADP	2.267 ± .102 (n = 17)	2.218 ± .133 (n = 13)	2.372 ± .207 (n = 11)	2.216 ± .172 (n = 13)
AMP	1.175 ± .121 (n = 15)	1.383 ± .130 (n = 10)	1.453 ± .122 (n = 11)	1.423 ± .113 (n = 11)
Total nucleotide	19.30 ± 0.18 (n = 15)	19.49 ± .23 (n = 10)	19.36 ± .31 (n = 11)	19.18 ± .27 (n = 11)
PCr	4.149 ± .318 (n = 17)	4.788 ± .434 (n = 13)	4.724 ± .720 (n = 11)	4.812 ± .690 (n = 13)
Pi	6.42 ± 0.14 (n = 5)	6.38 ± 0.37 (n = 5)	8.33 ± 1.30 (n = 5)	–

NOTE: The values are the means ±SEM of n dogs, at least four measurements having been made on each sample in each dog. Blood flow refers to the state of reddening of the mucosa, H I the sample when acid secretion was detected in that area, and H II a sample at steady-state secretion. Values are given, when appropriate, in mmol/kg^{-1} dry weight.

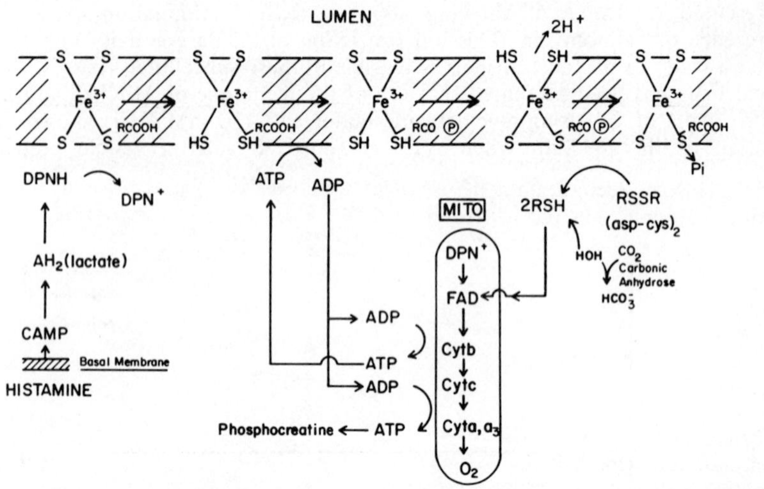

FIGURE 11. A detailed scheme to account for the evidence given in favor of both a redox and an ATPase mechanism, showing that the ATP levels in the tissue would be maintained by this type of mechanism. An Fe^{++}-SH complex is suggested as a potential proton transporter, and aspartyl cysteine as a redox shuttle between the membrane and mitochondrion.

TABLE 5

ATPASE (μMOL·MG^{-1}·HR^{-1})

	Fraction II
Mg^{++}	13.7
$Mg^{++}K^{+}$	22.7
Mg Val	15.0
Mg K^{+} Val	68.5

This apparent proton gradient is dissipated only in the simultaneous presence of valinomycin and a protonophore such as CCCP or FCCP. The rest of this paper will deal with our findings in hog gastric mucosa, using a similar preparation and deducing from protein localization studies that the vesicle preparation is inside out.[35]

Before discussing the problems inherent in this type of data, we will describe some recent results on this problem from our laboratory, working on a preparation from hog gastric mucosa.

FIGURE 12 summarizes fairly typical results on a 10^5-*g* pellet from hog mucosa. This is contaminated to at least 30% by mitochondria. The vesicles are suspended in a medium of pH 6.1, to prevent a ΔH^+ due to ATP hydrolysis, and this is done using 150 mM KCl and glycylglycine buffer. All additions are adjusted to an identical pH using a radiometer pHM 64 pH stat system with the REA 112 high-sensitivity module. The addition of ATP results in a rapid change (only in the presence of membranes) of pH so that the medium becomes alkaline. This change would be typical of inside-out mitochondria. However, one can at this stage marshal two arguments against simply mitochondrial contamination being responsible for the phenomenon. First, oligomycin does not block the reaction in contrast to the effect of this peptide on mitochondrial ATPase. Second, once the gradient is established, the addition of *m*-chlorocyano carbonyl phenylhydrazone (CCCP) (an effective uncoupler, protono-

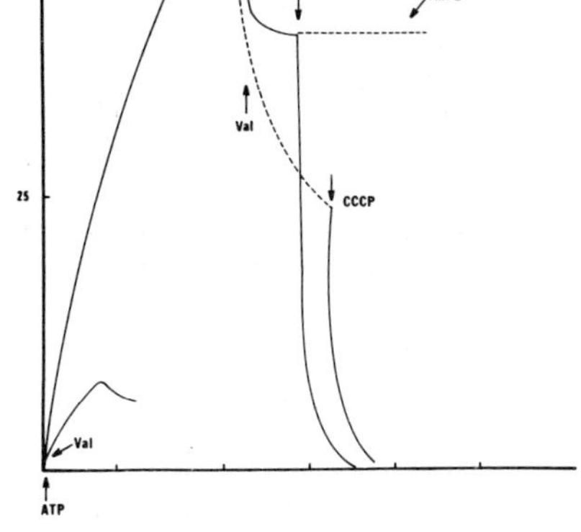

FIGURE 12. The change in H^+ ion concentration upon addition of ATP to the 10^5-*g* pellet at a concentration of 1.5 mg/ml protein, showing the effect of valinomycin, CCCP, and TMPB.

phore or —SH reagent, depending on one's point of view) does not fully dissipate the gradient. The addition of valinomycin is required. In fact valinomycin alone is considerably more effective in dissipating this gradient than is CCCP alone. The combination of both dissipates or prevents formation of the gradient. A lipid-permeable cation, such as triphenyl methyl phosphonium in (TMP^+) the presence of CCCP is without effect on the gradient. This latter finding shows that one cannot account for the effect of valinomycin and K^+ as simply due to dissipation of a potential, due to inward active Cl^-. However, the lack of activity of TMP^+ raises deeper problems than it solves. Thus, why does the val-K^+ complex dissipate the gradient in the presence of H^+ conductance, induced by CCCP, whereas TMP^+ does not. This suggests that there is something specific about val-K^+. TMP^+ does not stimulate the ATPase in any way. Moreover, gramicidin alone or in combination with CCCP does not dissipate the gradient despite the fact that gramicidin is as effective as valinomycin in enhancing the K^+ stimulation of ATPase. On the other hand, the dissipation of the gradient by these ionophores, or by triton, shows that a gradient is indeed generated by ATP addition. The requirement for both ionophores argues against a simple electrogenic mechanism for the ATP-induced gradient. If the ATP H^+ pump is electrogenic, then to complete the circuit, one can either use an electrogenic Cl^- pump and/or a K^+ conductance as illustrated in FIGURE 13. According to this circuit, the addition of valinomycin in the presence of K^+ or TMP^+ should be able to dissipate the gradient by increasing or inducing a K^+-H^+ exchange. CCCP, in the presence of a g_{Cl} should allow efflux of HCl. In the circuit it is assumed that an inward K^+ gradient exists, hence an increase of g_K will dissipate the H^+ gradient. The finding that valinomycin is more effective than CCCP may suggest a larger g_H than g_{Cl}. However, the greater efficacy of the combination of protonophore and cationophore suggests that the system is nonelectrogenic or else that g_H or g_{Cl} are very small. Nigericin alone dissipates the gradient.

Another problem is to determine whether K^+ is essential for activity. FIGURE 14 shows that removal of K^+ and substitution by Na^+ sharply reduces, but does not absolutely abolish the ATP-dependent gradient. Cl^- removal and SO_4^{--} substitution also reduce the gradient developed, which would be predicted from data for the intact mucosa (see other papers in this monograph). The

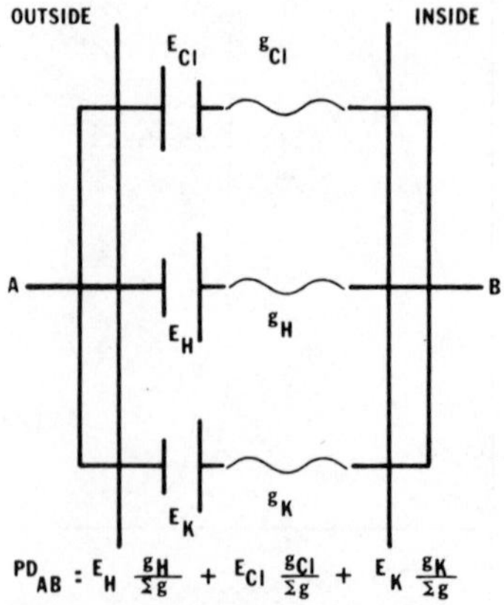

FIGURE 13. An equivalent circuit representation of the membrane vesicle, with an equation describing the potential across the membrane, containing a g_H^+, g_{Cl}, g_K^+.

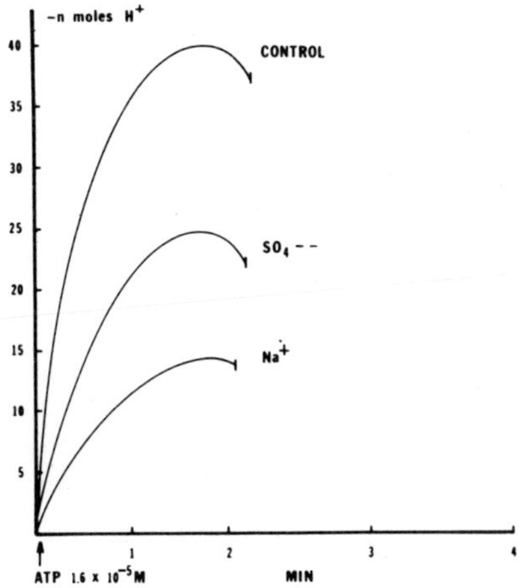

FIGURE 14. The effect of K^+ removal and Na^+ substitution and of Cl^- removal and SO_4^{--} substitution on the H^+ gradient generated.

evidence then does not exclude electrogenicity of the pump, but is not exactly in its favor. It is also possible to invert the K^+ gradient, as shown in the data in FIGURE 15. Here the vesicles are preincubated in 1 M KCl and then placed in 1 M NaCl or 1 M choline Cl medium. The high K^+-choline experiments produce a gradient on the addition of ATP, which is dissipated in part by CCCP. Addition of valinomycin produces a transient alkalinization, which is opposite to the usual effect. The addition of CCCP to K^+-preloaded vesicles, in Na^+ medium with no ATP addition, followed by valinomycin, produces a large transient alkalinization. Hence, these vesicles, with these ionophores, can undertake a K^+-H^+ exchange.

When ATP is added to these vesicles, however, acidification rather than alkalinization occurs. CCCP and valinomycin dissipate this inverse gradient. Valinomycin in fact produces an overshoot, as would be predicted from the direction of the K^+ gradient. The effect of ATP is more difficult to explain.

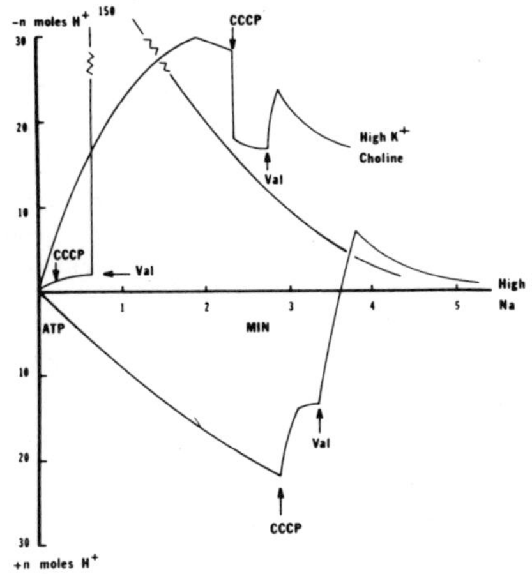

FIGURE 15. The effect of preincubating the 10^5-*g* pellet in 1 M KCl solutions, followed by dilution into 1 M choline Cl, and ATP addition, or dilution into 1 M NaCl solution followed by addition of CCCP and then valinomycin or dilution into NaCl followed by ATP addition and then CCCP and valinomycin.

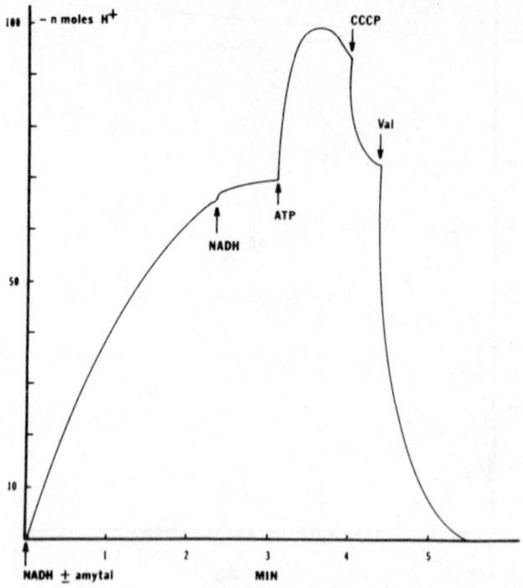

FIGURE 16. The effect of NADH and a subsequent NADH addition to the 10^5-*g* pellet, followed by the addition of ATP and CCCP and valinomycin.

If, however, there is an Na^+ conductance, and ATP allows an Na^+: H^+ exchange then an inverse gradient could be produced. Clearly, however, the situation is quite complicated.

Other nucleotides as well as γ-methylene ATP do not substitute for ATP, showing that ATP hydrolysis is required. However, in FIGURE 16, the surprising result that NADH addition can produce an amytal-insensitive gradient in the same direction as ATP is shown. Two additions of NADH do not increase the gradient and the further addition of ATP increases the gradient, which is once again dissipated by CCCP and valinomycin. However, fractionation of the 10^5-*g* pellet on a ficoll-step gradient (ficoll-sucrose gradient thus far abolishs gradient- but not valinomycin-sensitive ATPase activity) produces the data of TABLE 6, where it can be seen that the ATP effect is correlated with the presence of K^+-ATPase or K^+-PNPPase and clearly separated from NADH-dependent activity. Using this fractionation and measuring the sucrose and water space, we can estimate the vesicular volume to be about 2.0 μl mg^{-1} protein, and this gives the surprisingly low intravesicular pH, in the

TABLE 6

	10^5-*g* Pellet	Fraction I	II	III
1.6×10^{-5} M		Acid Secretion (nmol/mg^{-1})		
NADH	39	0	25	120
ATP	22.2	75	20	15
		Enzyme Activity (μmol · mg^{-1} · hr^{-1})		
K^+ ATPase	17.6	46.44	16.92	2.36
K^+ PNPPase	21.1	50.92	24.18	2.58

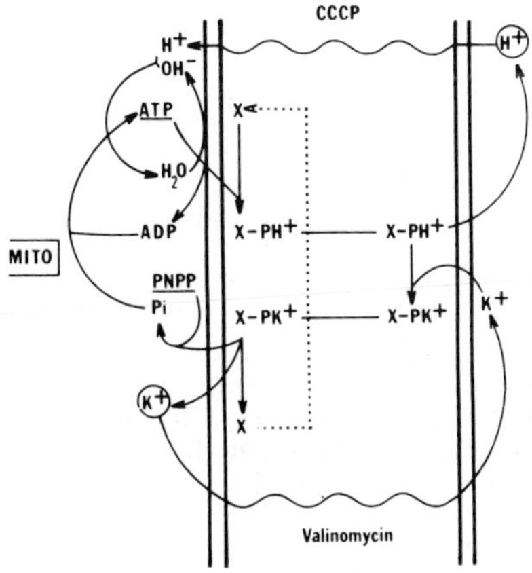

FIGURE 17. A model to account for the data presented on ATPase and gradient formation by the fraction I vesicle.

absence of buffers (but we know that the vesicles have buffering capacity), of 1.43.

Lest it be thought that the problem of proton secretion is now solved by an ATP-dependent proton pump, which is essentially nonelectrogenic, it should be noted that we do not as yet know the source of the active vesicles, nor does valinomycin affect *in vitro* acid secretion. The role of the K^+-ATPase is also not clear since high K^+ mucosal solutions do not increase the H^+ rate. ZN^{++} or F^-, which are K^+-ATPase inhibitors, prevents formation of this proton gradient, as does 10 mM SCN^-.

FIGURE 17 summarizes the data presented in this section. ATP reacts with a 100,000 catalytic subunit to form a phosphorylated intermediate, which reacts with water. There is resultant inward H^+ movement, coupled to outward K^+ movement in a 1:1 exchange. *p*-Nitrophenyl phosphate reacts only at the outer surface of the vesicle, hence the K^+ stimulation is insensitive to valinomycin, and no proton gradient results since X-P-K^+ is not formed.

The addition of valinomycin alone increased g_K, and hence outward K^+ transport that is ATP-dependent, with consequent increase of ATP hydrolysis. This increased outward K^+ movement will not result in increased H^+ movement, since valinomycin will effectively shunt the forced exchange. The addition of CCCP will does not affect the ATPase rate if the K^+-dependent step is rate-limiting. Since the purified membrane fraction is free of mitochondria and does not interact with NADH, no role of a redox system is given.

Preincubating in high K^+, then diluting into 0 K^+, with induction of a g_{K^+} with valinomycin, will drive H^+ inward in the absence of CCCP.

Thus, this vesicle preparation shows many interesting properties, but as yet, evidence linking this capacity to generate a gradient in the presence of ATP has not been linked to acid secretion in the intact stomach.

ACKNOWLEDGMENTS

We thank G. Moonsamy and J. Foley for their help in the metabolic measurements and W. S. Rehm for valuable discussions.

References

1. Conway, E. J. & T. G. Brady. 1948. Nature **162:** 456–458.
2. Kaback, H. R. 1974. Science **186:** 882–888.
3. Durbin, R. P. & E. Heinz. 1957. J. Gen. Physiol. **41:** 1035–1047.
4. Sachs, G., J. G. Spenney, G. Saccomani & M. C. Goodall. Biochem. Biophys. Acta **323:**
5. Thayer, W. S. & P. C. Hinkle. 1973. J. Biol. Chem. **248:** 5385–5402.
6. Mitchell, P. & J. Moyle. 1968. Eur. J. Biochem. **4:** 530–539.
7. Harold, F. M. 1972. Bact. Rev. **36:** 172–230.
8. Seaton, A., C. Letson & A. A. Eddy. 1973. Biochem. J. **134:** 1031–1043.
9. Mitchell, P. 1966. Biol. Rev. **41:** 445–488.
10. Moody, F. G. 1968. Am. J. Physiol. **215:** 127–131.
11. Kidder, G. W., P. F. Curran & W. S. Rehm. 1966. Am. J. Physiol. **211:** 513–519.
12. Hersey, S. J. 1971. Phil. Trans. Royal Soc. **262:** 261–277.
13. Takeuchi, S. 1972. Nipon Un. J. Med. **14:** 111–122.
14. Hersey, S. J. 1974. Biochem. Biophys. Acta **344:** 158–203.
15. Sedar, A. W. 1965. Fed. Proc. **24:** 1360–1367.
16. Hersey, W. R. & Wood, J. L. 1973. *In* Transport Mechanisms in Epithelia H. Hussing and N. Thorce, Eds.: 342–357. Munksgaard. Copenhagen, Denmark.
17. Ray, T. K. & J. G. Forte. Bioch. Biophys. Acta **363:**
18. Sachs, G., E. Rabon & W. S. Rehm. Biochm. Biophys. Acta Submitted for publication.
19. Black, J. R., W. A. M. Duncan, L. J. Durant, O. R. Ganellin & E. M. Parsons. 1972. Nature **230:** 385–387.
20. Shoemaker, R. L., E. Buchner, J. G. Spenney & G. Sachs. 1974. Am. J. Physiol. **226:** 898–902.
21. Lowry, O. H. & J. V. Passonneau. 1972. Flexible Method of Enzymatic Analysis. Academic Press. New York, N.Y.
22. Sung, C. P., L. R. Burns & G. Sachs. Unpublished observations.
23. Skou, J. C., 1957. Biochim. Biophys. Acta **23:** 394–401.
24. Post, R. L. 1974. Ann. N.Y. Acad. Sci. **242:** 6–11.
25. Forte, J. G., P. H. Adams & R. E. Davies. 1965. Biochim. Biophys. Acta **104:** 25–38.
26. Durbin, R. P., F. Michelangeli & A. Michel. 1974. Biochim. Boiphys. Acta **367:** 177–189.
27. Kasbekar, D. K. & R. P. Durbin. 1965. Biochim. Biophys. Acta **105:** 472–482.
28. Sachs, G., V. D. Wiebelhaus & B. I. Hirshcowitz. 1972. *In* Gastric Secretion. G. Sachs, E. Heinz, and K. J. Ullrich, Eds.: 321–345. Academic Press. New York, N.Y.
29. Spenney, J. G., A. Strych, A. H. Price, H. F. Helander & G. Sachs. 1973. Biochim. Biophys. Acta **311:** 545–564.
30. Soumarmon, A., M. Lewin, A. M. Cherct & S. Bonfils. 1974. Biochim. Biophys. Acta **339:** 403–414.
31. Tanisawa, A. J. & J. G. Forte. 1971. Arch. Biochem. Biophys. **147:** 166–176.
32. Ganser, A. L. & J. G. Forte. 1973. Biochem. Biophys. Res. Commun. **54:** 690–696.
33. Lee, J., G. Simpson & P. Scholes. 1974. Biochem. Biophys. Res. Commun. **60:** 825–830.
34. Mitchell, P. & J. Moyle. 1968. Eur. J. Biochem. **4:** 530–539.
35. Saccomani, G., G. Shah, J. G. Spenney & G. Sachs. J. Biol. Chem. In press.

DISCUSSION

DR. HEINZ: I'm not so well informed on the present stage of your subject, but doesn't Hersey find that during stimulation of acid secretion all his redox parts are reduced? Can you confirm that?

DR. SACHS: Yes. In the controversy between Kidder and Hersey, we confirm Hersey's findings.

DR. REHM: I might note that on the basis of new work Kidder is now leaning towards Hersey's conclusions.

Part X. The Lipid Bilayer Membrane as an Assay for Ion-Conductance-Inducing Materials Isolated from Biological Membranes

THE INTERACTION OF HEMOCYANIN AND LYMPHOCYTES WITH LIPID BILAYER MEMBRANES

Robert Blumenthal

Laboratory of Theoretical Biology
National Cancer Institute
National Institutes of Health
Bethesda, Maryland 20014

Introduction

Two different examples of the use of the lipid bilayer as an assay for ion-conductance-inducing materials are shown. In the case of hemocyanin, a giant respiratory protein dissolved in the hemolymph of Mollusca and Arthropoda, Pant and Conrad[1] had shown that the aggregated form produces electrical conductance increases of as much as six orders of magnitude when added to the aqueous solution after forming the lipid bilayer membrane. Alvarez, Diaz, and Latorre[2] carried out discrete current fluctuation analysis of the hemocyanin-induced conductance and showed that the ion transport is mediated by a channel mechanism. The dissociated form of hemocyanin also induces conductance changes in lipid bilayer membranes, which is more likely to be mediated by a "defect" mechanism.[3] In the case of lymphocytes, it had been postulated that the killing of target cells is caused by an initial permeability change to cations[4] and subsequent colloid osmotic lysis. The bilayer experiments indicate that the permeability change can be induced without any contribution of target cell cytoplasm or membrane proteins.[5]

Hemocyanin

Molluscan hemocyanins are cylindrical molecules with a diameter of about 340 Å and a height ranging from 140 Å to 680 Å, built up from 3 to 12 rows of subunits.[6] Hemocyanin readily dissociates and aggregates upon changing the pH, Ca^{2+} concentration, and hemocyanin concentration.[7] Aggregated hemocyanin, which has a sedimentation coefficient of about 100 S, dissociates into half-molecules at pH above 7 or below 5, and next into smaller units, which are tenth molecules. At still higher pH values the tenths dissociate into units, which are found to be twentieth molecules. Three-dimensional image reconstruction[8] of hemocyanin indicates that the subunits are arranged in a helical order giving rise to a structure with an internal polar core or "channel."

The differences between the dose response curves of associated (A) and dissociated (D) hemocyanins are shown in FIGURE 1. The associated form is a dispersion of commercial (Schwarz-Mann) key-hole limpet hemocyanin in deionized water. The dispersion contains a mixture of twentieth,

FIGURE 1. Conductance-concentration curve for associated (A) and dissociated (D) hemocyanin in 100 mM KCl, pH = 7. The vertical bars (*left*) represent conductance increases within four minutes after increasing the associated hemocyanin concentration; the dots (*right*) represent dissociated hemocyanin-induced conductance.

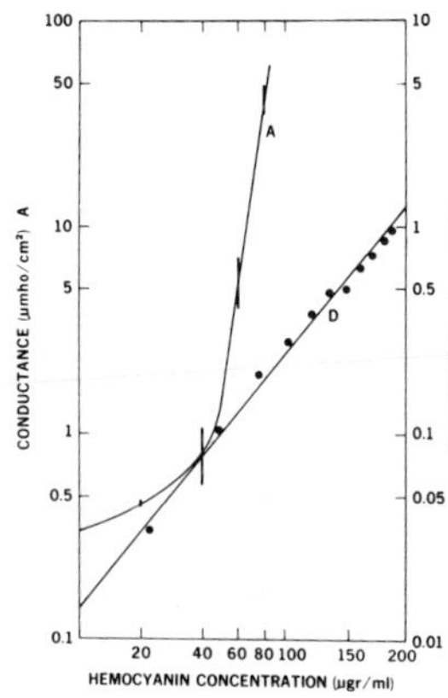

tenth, half, and whole aggregates.[7] The higher-order concentration-dependence of the conductance indicates that a number of units aggregate to form a channel. The straight portion of the conductance-concentration curve yields a slope of six on a logarithmic scale which is an average value for the insertion of one, two, ten and twenty units into the bilayer to form a channel. The higher order concentration-dependence has been observed in the case of the other well-known channel-formers, nystatin[9] and alamethicin.[10] Dissociated hemocyanin, prepared according to the method of Weigle[11] consists of twentieth units, which have irreversibly lost their capacity to reaggregate.

The dose-response curves in FIGURE 1 show three marked differences between associated and dissociated hemocyanin. For the same hemocyanin concentration at 80 μg/ml there is about two orders of magnitude difference in conductance increase with associated hemocyanin. This effect is even more dramatic when we take into account that the dissociated hemocyanin is about tenfold more concentrated than the associated hemocyanin on a molar basis, because the molecular weight of associated hemocyanin is about tenfold that of dissociated hemocyanin. The other difference is that upon addition of the dissociated hemocyanin to the solution, a steady-state conductance level is reached immediately, whereas with associated hemocyanin concentrations greater than 40 μg/ml the conductance keeps increasing at a slower rate after a large stepwise increase after addition of the hemocyanin in the external solution. The later conductance increase is actually indicated in FIGURE 1 by the vertical bars, which represent a conductance change within four minutes. The third difference is that the straight portion of the conductance-concentration curve yields a slope of six on a logarithmic scale in the case of associated hemocyanin, whereas the dissociated hemocyanin yields a slope of one.

Upon addition of associated hemocyanin to the bathing solution, the current increases in discrete steps ("formation bumps") at a given voltage, with conductances of about 0.1 nmho in 100 mM KCl (FIGURE 2). Discrete conductance steps induced by associated hemocyanin were originally observed by Alvarez, Diaz, and Latorre,[2] who showed that the voltage-dependence of the conductance-step fluctuation correlates very well with the voltage-dependence of the conductance of a many-channel membrane. Discrete conductance steps were not observed with dissociated hemocyanin.

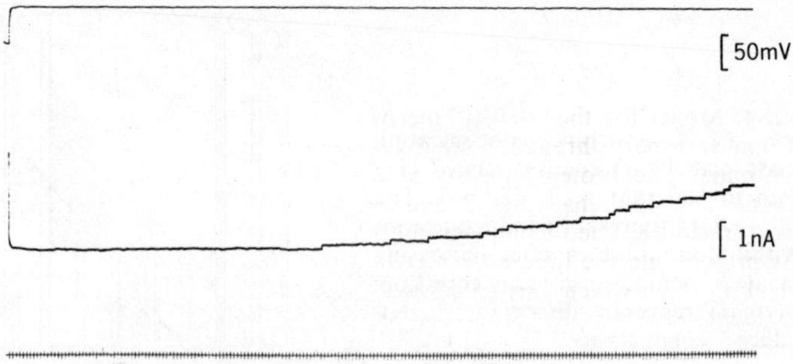

FIGURE 2. Stepwise increase in current (*middle trace*), driven by a constant voltage (*upper trace*), after adding 20 μg/ml associated hemocyanin in 100 mM KCl, pH = 7. The lower trace marks one-second intervals.

The rectification in the current-voltage characteristic for associated hemocyanin, which had been observed by Pant and Conrad[1] and by Alvarez *et al.*,[2] is shown in FIGURE 3. Dissociated hemocyanin, however, exhibits a linear current-voltage characteristic as shown in FIGURE 3.

The nature of the conductance induced into the lipid bilayer membrane by associated hemocyanin (discrete conductance fluctuations, higher-order dose-response curve and rectification) and its three-dimensional structure[8] indicate that ion transport is mediated by a channel mechanism. The single subunit of dissociated hemocyanin, on the other hand, which has no possibility of forming a polar core, together with the nature of its conductance induced into the lipid bilayer (absence of discrete conductance steps, first-order dose-response curve and linear current-voltage curve) points to an alternative mechanism mediating the ion transport. As shown in FIGURE 4 insertion of a single subunit into the lipid matrix could give rise to a "crack" and conductance could be induced by a "defect" mechanism. In that case the lipids do not completely surround the inserted protein and resulting leaks could cause ion permeation. The protein could possess some hydrophobic moiety that attaches to the bilayer. The possibility of a "defect" mechanism for ion transport

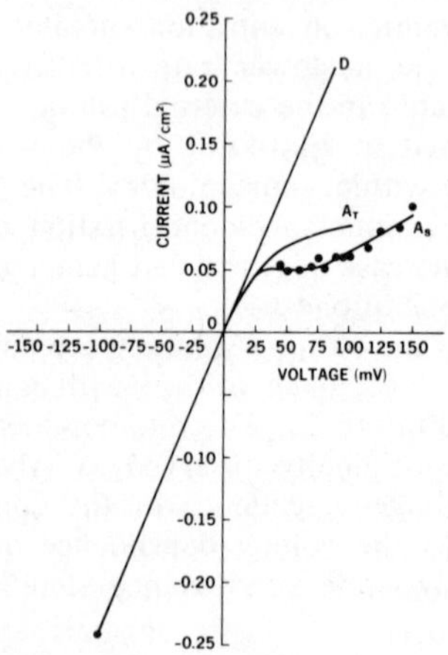

FIGURE 3. Current-voltage curves for dissociated (D) and associated (A) hemocyanin-doped membranes. The curve marked A_T represents the current and voltage taken from a triangular voltage sweep at a rate of 15.2 mV/sec and the curve marked A_S is obtained from the steady-state current, driven by voltage pulses lasting 20 seconds. The experiments were in 100 mM KCl, pH = 7; associated hemocyanin was at 40 μg/ml, dissociated hemocyanin at 100 μg/ml. The current of the dissociated hemocyanin-doped membrane was normalized to that of associated hemocyanin-doped membrane.

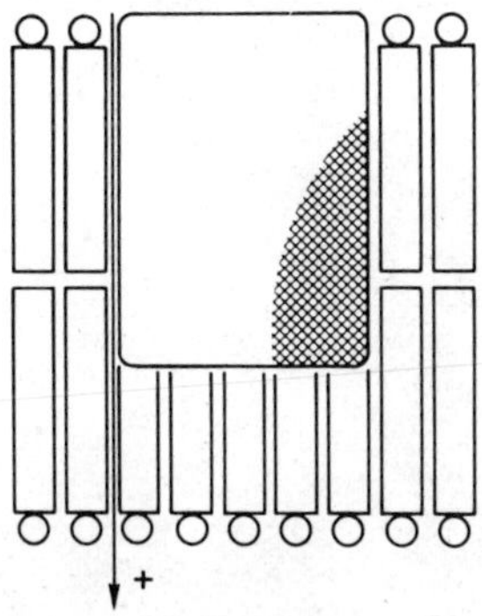

FIGURE 4. Model for the "defect" mechanism of ion transport through a lipid bilayer membrane. The protein consists of a single subunit and contains a small hydrophobic area (*shaded*). The incomplete insertion of the protein into the lipid matrix gives rise to a leak through which ions could flow.

of the "closed" channel of excitability-inducing material (EIM) has been suggested by the experiments of Latorre *et al.*[12] on the temperature-dependence of the conductance of the "open" and "closed" channels of EIM. Latorre *et al.*[12] found that the conductance of the open channel increased as expected from the temperature-dependence of the mobility of ions in water, whereas the conductance of the closed channel *decreased* with increasing temperature. The experiments indicate that the closed channel conductance is mediated by a "defect" mechanism and that at higher temperature the increased lipid fluidity "seals" the crack around the protein.

Del Castillo *et al.*[13] had reported previously on the conductance induced into thin lipid films by enzymes or antibodies and the enhancement of the conductance caused by substrate-enzyme or antigen-antibody interaction. I believe that those conductances are caused by a defective mechanism and that the interaction with substrate or antigen enhances the size of the defect.

LYMPHOCYTES

In order to study more closely the first stages of lymphocyte-mediated cytolysis of target cells, Henkart and Blumenthal[5] prepared black lipid membranes consisting of oxidized cholesterol and DNP-phosphatidylethanolamine, synthesized from bacterial phosphatidylethanolamine by the method of Kinsky.[14] The lipid bilayers were formed across a 330-μ hole in a thin bottom Teflon cup that is immersed in a small lucite chamber. The antigenic membrane was coated by injecting 5-$\mu\lambda$ volume of 1–2 mg/ml purified anti-TNP *antibody* directly above the membrane followed by 20-minute incubation. Lymphocytes were then injected in a 5-μl volume (of a suspension of 10^7 cells/ml) down onto the bilayer membrane, became visible in the microscope immediately, and finished settling within five minutes as shown in FIGURE 5. The lipid bilayer could support dozens of lymphocytes for periods of more than an hour before membrane-breakage occurred. The conductance of the lipid bilayer membrane was monitored at all stages of antibody and lymphocyte incubation. As shown in FIGURE 6, the "antigenic" oxidized cholesterol-DNP phosphatidylethanolamine bilayer had a conductance between 10^{-6} and 10^{-8} mho/cm^2. Exposure to the antibody had no effect on this conductance. (The antigen-antibody-induced conductance observed by Del Castillo *et al.*[13] is very dependent upon

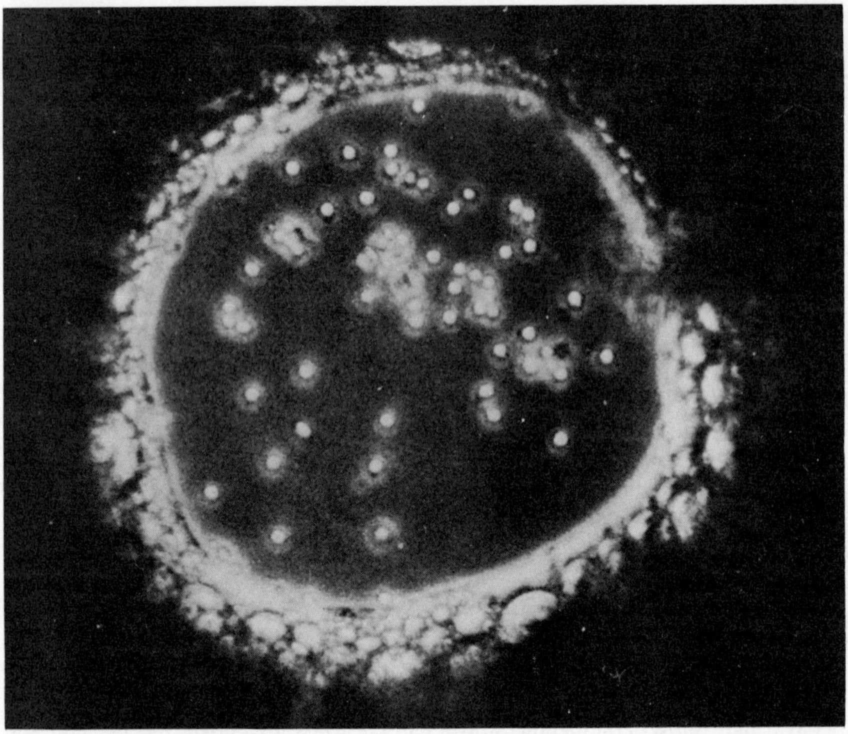

FIGURE 5. Lymphocytes resting on the antibody-coated lipid bilayer. This photograph was taken through the phase microscope during the course of an experiment. A torus containing a thick layer of lipid, including bubbles, surrounds the bilayer, which appears dark under phase optics. Approximately 50 lymphocytes can be seen on the bilayer.

the lipid forming the bilayer. With the lipids we used we never observed a significant antibody-antigen-induced conductance change). A marked conductance change was observed within 2–20 minutes after lymphocytes settled on the membrane. Two types of patterns of conductance that change with time, when the bilayer membrane was clamped at a fixed voltage, are shown in FIGURE 6. The conductance increase was voltage-dependent: the increases only occurred when the membrane potential was positive on the lymphocyte side of the membrane. Since the electrical potential difference across cellular membranes is generally positive on the outside with respect to the inside of the cell, the polarity conditions for lymphocyte-mediated killing are the same in the lipid bilayer as in the target cell. Observation of discrete conductance steps and the absence of a lymphocyte-mediated conductance change in thick membranes indicated that the response is mediated by a channel mechanism. The system has all its controls built in, as shown in TABLE 1. Lymphocytes resting on the bilayer in the absence of antibody did not induce significant conductance changes, which is analogous to the cytotoxicity of target cells, which is also antibody-dependent. The other controls shown in TABLE 1 ($F(ab)_2$ antibodies, which do not contain the Fc fragment recognized by the lymphocyte

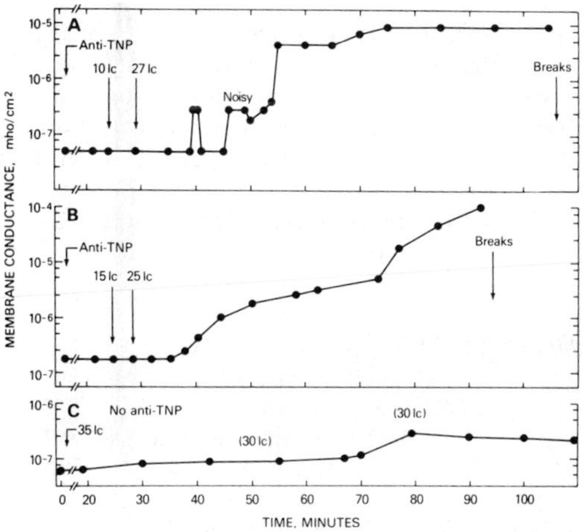

FIGURE 6. Conductance changes caused by lymphocytes acting on antibody-coated lipid bilayers. A and B are typical experiments in which anti-TNP is added to the medium over the lipid bilayer shortly after formation of the bilayer. Twenty minutes later 5×10^4 lymphocytes are injected into the medium over the membrane, and the number of cells landing on the bilayer (FIGURE 5) is recorded (Lc here stands for lymphocyte). At 30 minutes the cells have all settled. The experiment shown in C is a control experiment with no antibody; the cells were added shortly after membrane formation. Thirty lymphocytes remained on the bilayer throughout the experiment. The membrane voltages were held constant at +20 to +60 mV until the conductance had increased 100×.

and lymphocytes lacking the Fc receptor) also correlated well with the target-cell-killing experiments. The experiments show that lymphocyte- and antibody-dependence of membrane conductance increase is identical to that of target cell lysis and the results support the hypothesis that the membrane-permeability increase is the initial event in lymphocyte-mediated cytolysis. The experiments do not distinguish yet whether the channel-forming material is released from

TABLE 1
ANTIBODY- AND LYMPHOCYTE-DEPENDENCE OF MEMBRANE CONDUCTANCE INCREASE

Antibody	Lymphocytes	Percentage of ^{51}Cr Release by Tnp-RBC (No. of Exp.*)	Membrane Conductance Increase (No. of Exp.)
Rabbit anti-Tnp	Human PBL	100	250×–20,000× (12)
None	Human PBL	0–2 (4)	0–10× (6)
Anti-Tnp-F(ab)$_2$	Human PBL	4–5 (2)	0 (1)
Anti-Tnp	None	–1–8 (4)	0–5× (2†)
Anti-Tnp	Fc receptor negative	3–15 (4)	0 (1)

* Relative to complete system (line 1).

† In each experiment the membrane is incubated with antibody alone for 20 minutes; conductance increases greater than 5× were never seen. In two experiments the membranes were incubated with antibody alone for one hour.

the lymphocyte and subsequently inserted into the bilayer or whether the conductance change is induced while the channel-forming material is still attached to the lymphocyte surface. The experiments on the interaction of lymphocytes with the antigen-antibody-coated lipid bilayer membrane clearly show that the permeability change and consequently cytolysis can be induced without any involvement of target-cell cytoplasm or membrane proteins. The system thus provides us with a new approach to study the molecular nature of lymphocyte-mediated cytolysis and the molecular basis of the specificity of the response.

References

1. Pant, H. C. & P. Conrad. 1972. J. Membr. Biol. **8:** 357–362.
2. Alvarez, O., E. Diaz & R. Latorre. 1975. Biochim. Biophys. Acta **389:** 444–448.
3. Blumenthal, R. 1975. Submitted for publication.
4. Cerottini, J. C. & K. T. Brunner. 1974. Adv. Immunol. **18:** 67–132.
5. Henkart, P. & R. Blumenthal. 1975. Proc. Nat. Acad. Sci. U.S.A. **72:** 2789–2793.
6. Fernandez-Moran, H., E. F. J. Van Bruggen & M. Ohtsuki. 1966. J. Mol. Biol. **16:** 191–207.
7. Konings, W. N., R. J. Siezen & M. Graber. 1969. Biochim. Biophys. Acta **194:** 376–385.
8. Mellema, J. E. & A. Klug. 1972. Nature (London) **239:** 146–150.
9. Finkelstein, A. & A. Cass 1968. J. Gen. Physiol. **52:** 145S–172S.
10. Mueller, P. & D. O. Rudin. 1968. Nature (London) **217:** 713–719.
11. Weigle, W. O. 1964. Immunochemistry **1:** 295–302.
12. Latorre, R., O. Alvarez & P. Verdugo. 1974. Biochim. Biophys. Acta **376:** 361–365.
13. Del Castillo, J. A. Rodriquez, C. A. Romero & V. Sanchez. 1966. Science **153:** 185–188.
14. Uemura, K. & S. C. Kinsky. 1972. Biochemistry **11:** 4085–4094.

THE LIPID BILAYER MEMBRANE AS AN ASSAY FOR ION-CONDUCTANCE-INDUCING MATERIALS ISOLATED FROM BIOLOGICAL MEMBRANES*†

Robert Blumenthal

Laboratory of Theoretical Biology
National Cancer Institute
National Institutes of Health
Bethesda, Maryland 20014

Adil E. Shamoo

Department of Radiation Biology and Biophysics
University of Rochester
School of Medicine and Dentistry
Rochester, New York 14642

Any investigator who has had experience with black lipid membranes (BLM) knows that nonspecific conductance changes can easily be achieved with proteins extracted from biomembranes. Those changes are frequently followed by rupture of the membrane, and one can readily observe that such nonspecific changes in BLMs do not yield significant information about the protein action in bilayers.

What approaches are therefore to be taken to achieve meaningful reconstitutions of ion conductance from biological membranes? One of the prerequisites is to start with purified material from a system with known physiology and pharmacology, such as the acetylcholine receptor or sodium channels in nerve, or purified enzyme known to be associated with ion transport. There one has very strong a priori notions of what one should get. For example, Dr. Finkelstein isolated from excitable tissue of garfish a detergent extract of the membrane which he hoped would contain the sodium channels of nerve. When the impure extract was tested on bilayer, the only things they were lacking were sodium selectivity, inhibition by tetrodotoxin, and voltage-dependence.

Should the bilayer response then mimic the known physiological condition? It is known that with the antibiotic channel-formers such as alamethicin and EIM the kinetics depend very much on the lipids. By the same token the unnatural lipid environment for such systems as sodium channels or acetylcholine receptor might change its properties considerably. On the other hand, some of the reported reconstitutions have been so far from the physiological response that it is hard to deduce any information from them. For instance,

* Summary of panel discussion held February 27, 1975. The panel members were Drs. R. Blumenthal (Chairman), A. Finkelstein, V. T. Ivanov, P. Mueller, A. E. Shamoo, and D. C. Tosteson.

† This paper is based on work performed under contract with the United States Energy Research and Development Administration at the University of Rochester Biomedical and Environmental Research Project and has been assigned Report No. UR-3490-770. The work was also supported in part by National Science Foundation Grant No. NIH 1 R01 AM 17571.

in the case of rodopsin one can get a slowly increasing conductance change upon addition of the protein to the bilayer and that rate is speeded up in the presence of light. A similar phenomenon has been reported with the acetylcholine receptor: in the presence of acetylcholine receptor the rate of conductance increase is speeded up when the acetylcholine is added. The experiments of Eldefrawi and Shamoo with the acetylcholine receptor reported at this meeting come closer to the physiological response in the sense that they are cation-selective, activated by acetylcholine, and inhibited by curare.

Shamoo and MacLennan used a single purified protein—the sarcoplasmic reticulum $Ca^{++} + Mg^{++}$-ATPase—to incorporate into lipid bilayer. They found a Ca^{++}-dependent and -selective ionophore associated with both the intact protein and with one of the peptidic fragments of the enzyme. It was pointed out that ligand-protein interaction such as antigen-antibody can cause a nonspecific increase in bilayer conductance. However, it was also pointed out that a variety of proteins' interaction with their ligand caused no increase in conductance. Some of these proteins are calcium-binding proteins (see Dr. Shamoo's paper) and amino-acid-binding proteins (see Dr. Oxender's paper).

Apart from using the bilayer to obtain additional information such as carrier versus channel mechanism on ion-conducting systems about which much is known *in vivo,* could the bilayer be used to answer questions about ion-conductance generating systems, such as rhodopsin, of which we do not know the *in vivo* characteristics? This was certainly an aim of Mueller and Rudin, who originally developed the black lipid membrane. In view of the difficulty in reconstituting known systems the general consensus for this kind of approach did not seem too hopeful.

What then have been the difficulties in the bilayer reconstitution experiments? The insertion problem is an inherent difficulty. Originally the extracted membrane material has been lipid-soluble and only water-soluble in the presence of detergents. How do we get it back into the membrane since the hydrophobic material faces first the hydrophilic groups of the bilayer? If there is too much detergent, the bilayer ruptures. How does the cell get around this problem? For instance, in the axoplasm of squid axons there is no protein synthesis. How are channels then inserted into nerve after being transported down the microtubules? How is cytochrome oxidase incorporated in the inner membrane of mitochondria? Some mitochondrial membrane proteins are synthesized outside the membrane and consequently transported. Rhodopsin might be another example of a membrane protein synthesized at the inner segment and subsequently inserted into the disc membrane.

The reconstitution methods of Dr. Racker as applied to vesicles are promising. Both the lipids and proteins are dispersed in cholate and then sonicated; then the cholate is removed by dialysis. Mueller hopes to apply a similar technique to large vesicular membranes. Then, the opinions concerning the lipid composition necessary for reconstitution were divided. On the one hand, the *in vivo* environment, such as the asymmetrical distribution of membrane lipids, was considered to be very important; on the other hand, the lipid forming the black lipid membrane was considered just a matrix.

Why are we not satisfied with the reconstitution experiments in sonicated vesicles, such as those presented by Racker and Goldin at this meeting; or, as Pressman puts it, why not use nature's own lipid bilayer—mitochondria or erythrocytes—for reconstitutions? Essentially almost all of the ionophores were discovered using mitochondria.

Reconstitution is not an aim in itself. The rationale of black lipid membrane reconstitutions is to study the transport system in a very controlled way, which one cannot achieve in biological systems. In the black lipid membrane set-up, the compositions of both sides as well as the membrane potential can be controlled. It is impossible otherwise to monitor time- and voltage-dependent conductance changes. Moreover, as has been indicated in a number of presentations in this meeting, clear distinctions between carrier and channel permeation mechanisms can only be made in a black lipid membrane. The tryptic digest of the microsomal fraction of the electric organ of the eel preparation of Dr. Shamoo did not have any effect on mitochondria. We must also keep in mind, as Dr. Ivanov pointed out at this meeting, that there are intermediate forms of transport between carriers and channels, namely, sandwich complexes.

During this meeting a point of view was presented by Green and by Blondin that certain compounds, such as hexokinase and the Beechey protein, could partition ions into an organic phase. How far are these related to physiological function? It has been shown that phospholipid, tryptic digest fragments of bovine serum albumin, and many other compounds can partition ions into an organic phase.

The point that emerged clearly in the conference is that bulk phase extractions are good measurements of equilibrium binding or complexation, but they do not indicate whether they will mediate ion transport through bilayers or natural membranes. Even the electrically silent carriers, such as X537A and A23817, have been shown by tracer methods to transport ions through lipid bilayer membranes.

In summary, meaningful experiments on bilayer reconstitutions should start with purified material with a well-defined biochemical basis. The bilayer experiments should closely mimic the physiological response. Bulk phase experiments should complement bilayer measurements. Reproducibility and effect of the extracted protein on bilayer rupture should be reported.